The neurobiology of dopamine systems

Studies in neuroscience

Series editor: William Winlow, *Lecturer in Physiology, The University, Leeds LS2 9NQ, UK*

Titles in this series:

The neurobiology of pain
Editors: A. V. Holden and W. Winlow

The neurobiology of dopamine systems
Editors: W. Winlow and R. Markstein

Working methods in neuropsychopharmacology
Editors: M. Joseph and J. Waddington

The neurobiology of the cardiorespiratory system
Editor: E. W. Taylor

Aims and methods in neuroethology
Editor: D. Guthrie

The neurobiology of dopamine systems

edited by W. Winlow *and* R. Markstein

Second Symposium of the
Northern Neurobiology Group,
held at Leeds from 16 to 19 July 1984

Manchester University Press

Published by
Manchester University Press
Oxford Road, Manchester M13 9PL, UK
and
51 Washington Street, Dover, NH 03820, USA

British Library cataloguing in publication data
Northern Neurobiology Group, *Symposium (2nd: 1984: Leeds)*
   The neurobiology of dopamine systems.—(Studies in neuroscience)
   1. Dopamine
   I. Title   II. Winlow, W.   III. Markstein, R.
   IV. Series
   547.7'5   QP563.D66

Library of Congress cataloging in publication data
Northern Neurobiology Group (Great Britain).
   Symposium (2nd : 1984 : Leeds, West Yorkshire)
   The neurobiology of dopamine systems.
   1. Dopamine—Physiological effect—Congresses.
   2. Dopamine—Therapeutic use—Congresses.
   3. Dopamine—Receptors—Congresses.
   4. Schizophrenia—Physiological aspects—Congresses.
   5. Tardive dyskinesia—Congresses.
   I. Winlow, W. (William)   II. Markstein, R. (Rudolf)   III. Title.
   [DNLM: 1. Dopamine—physiology—Congresses.
   2. Dopamine—therapeutic use—Congresses.
   W3 N078 2nd 1984n/WK 725 N874 1984n]
   QP563.D66N67   1984   599'.0188   85 13754

ISBN 0-7190-1799-8 *cased only*

Printed in Great Britain by
A. Wheaton & Co. Ltd., Exeter, England.

Contents

Preface   ix

Acknowledgements   xi

Part I   **Basic biology of dopamine systems**   **1**

1   Biosynthesis, storage and release of dopamine   3
R. J. Walker

2   Brain dopamine receptors   25
S. B. Freedman and G. N. Woodruff

3   Pharmacological characterisation of central dopamine receptors using functional criteria   40
R. Markstein

4   Dopamine systems in invertebrates   53
G. A. Cottrell

5   Localisation and developmental aspects of dopaminergic neurones in the vertebrate retina   68
N. N. Osborne

6   Dopamine neurones of the vertebrate brain: some aspects of anatomy and pathology   87
D. M. A. Mann

7   The D-1 dopamine receptor and ageing: behavioural and neurochemical studies   104
A. G. Molloy, K. M. O'Boyle and J. L. Waddington

8   Properties of dopamine receptors of identified neurones in the brain of the gastropod mollusc, *Helix aspersa*   108
A. J. Bokisch and R. J. Walker

9   Differential postsynaptic potential larencies on cells monosynaptically connected to the giant dopamine-containing neurone of *Lymnaea stagnalis*   112
W. Winlow

10   Modulation by dopamine and 5-hydroxytryptamine of rhythmic activity in buccal motoneurones of the pond snail, *Lymnaea stagnalis*   116
M. D. Tuersley and C. R. McCrohan

Part II Neuroendocrinology of dopamine   **121**

11   Dopaminergic control of anterior pituitary function   123
         C. Dieguez, S. M. Foord, J. R. Peters, B. M. Lewis,
                          P. Harris and M. F. Scanlon
12   Dopaminergic control of growth hormone secretion   137
                                              Daniela Cocchi
13   Dopamine and disorders of growth hormone secretion   146
                                                  S. M. Shalet
14   Dopaminergic regulation of the prolactin-secreting cells: the
     role of calcium and cyclic AMP   156
                   G. Schettini, E. Hewlett, M. J. Cronin,
                           T. Yasumoto and R. M. MacLeod
15   Post-receptor mechanisms mediating dopaminergic inhibi-
     tion of prolactin secretion   178
           B. L. Brown, J. G. Baird, J. E. Merritt and P. R. M. Dobson
16   Recent dopaminergic agents   185
                             E. Flückiger and R. Markstein
17   Dopaminergic control of somatostatin and thyrotropin-
     releasing   hormone   release   from   perfused   adult   rat
     hypothalami   194
               B. M. Lewis, C. Dieguez, S. M. Foord, R. Hall
                                     and M. F. Scanlon
18   Lack of influence of endogenous cortisol on the dopaminergic
     inhibition of thyrotropin and prolactin   198
               J. Salvador, I. Cano, A. Rodriguez, J. J. Barberia
                                      and E. Moncada
19   Oestrogens   increase   the   sensitivity   of   prolactin   to   the
     inhibitory   actions   of   dopamine   in   hyperprolactinaemic
     patients   203
                 R. Valcavi, R. Elli, P. E. Harris, C. Diegues, S. M. Foord,
                             J. R. Peters, R. Hall and M. F. Scanlon
20   CU 32085 is a dopamine agonist *in vivo* and a dopamine
     antagonist *in vitro*   207
                 C. A. Edwards, G. Shewring, C. Dieguez, S. M. Foord,
                               J. R. Peters and M. F. Scanlon
21   The influence of dopamine neurones in the zona incerta on
     sexual behaviour in the rat   211
                         J. P. Grierson and C. A. Wilson

Part III Schizophrenia   **217**

22   Clinical aspects of schizophrenia   219
                                         R. H. S. Mindham

23 Some philosophical reservations on the neurobiology of
schizophrenia 230
F. A. Jenner and S. S. Johl

24 Site of action of anti-schizophrenics 240
A. El-Sobky

25 Acute and chronic consequences of persistently disturbing
cerebral dopamine function 253
B. M. Costall, A. M. Domeney and R. J. Naylor

26 A reappraisal of abnormal, involuntary movements (tardive
dyskinesia) in schizophrenia and other disorders: animal
models and alternative hypotheses 266
*J. L. Waddington, H. A. Youssef, K. M. O'Boyle
and A. G. Molloy*

27 The nature of drug delivery can alter the short- and long-term
consequences of administering a dopamine agonist 287
B. M. Costall, A. M. Domeney and R. J. Naylor

28 A functional role for dopamine in the habenula nucleus:
effects of intracerebral injection of a dopamine agonist 290
E. W. Thornton and J. A. C. Evans

29 Lack of tolerance of striatal dopamine metabolism to chronic
haloperidol in the rat brain 294
E. F. Marshall, W. N. Kennedy, C. Iniguez and D. Eccleston

30 Anomalous relationship between serum chlorpromazine and
prolactin in tardive dyskinesia 299
*G. Hall, F. J. Rowell, C. G. Rich, F. Hassanyeh,
A. F. Fairburn and D. Eccleston*

31 Topography of action of dihomo-γ-linolenic acid in the
guinea-pig striatum to antagonise dyskinesias 303
B. Costall, M. E. Kelly and R. J. Naylor

Part IV Basal ganglia disturbances and extra-pyramidal syndrome **307**

32 Huntington's chorea 309
E. G. S. Spokes

33 Dopamine deficiency and dopamine substitution in
Parkinson's disease 319
O. Hornykiewicz

34 Dopaminergic trasplantation in an animal model of
Parkinson's disease 331
S. B. Dunnett and A. Björklund

35 Long-term complications of levodopa therapy 341
D. B. Calne

36 Parkinsonism induced by 1-methyl-4-phenyl-1,2,3,6-tetra-
hydropyridine (MPTP) in man and animals 350
P. Jenner and C. D. Marsden

37 Delineation of a particular subregion within the neostriatum
involved in the expression of muscular rigidity   370
B. Ellenbroek, M. Schwarz, K.-H. Sontag,
R. Jaspers and A. Cools

38 Use of the 2-deoxyglucose technique to study the neural
mechanisms that mediate basal ganglia dyskinesias   374
I. J. Mitchell, A. Jackson, M. A. Sambrook
and A. R. Crossman

39 Unilateral infusion of 1-methyl-4-phenyl-1,2,3,6-tetrahydro-
pyridine (MPTP) into the rat substantia nigra: behavioural
and biochemical consequences   378
A. J. Bradbury, B. Costall, A. M. Domeney, P. Jenner,
C. D. Marsden and R. J. Naylor

Part V   Further roles for dopamine   **383**

40 Dopamine in the kidney   385
B. C. Williams

41 Dopamine super-reactivity in the 'quasi-abstinence' syndrome   402
F. Sicuteri, M. Fanciullacci, E. Baldi and M. G. Spillantini

42 Dopamine as a deuteragonist in migraine: implications for
clinical pharmalogy of dopamine agonists   415
R. Horowski, H. Wachtel and R. Dorow

43 Use of dopaminergic agents in the treatment of senile
dementia   427
R. J. McDonald

44 Can dopaminergic substances be useful in patients with
symptoms of senile dementia?   439
I. H. Suchy

45 Dopamine D-1 and D-2 antagonism: changes in fluid intake
and saline preference in the water-deprived rat   459
D. B. Gilbert and S. J. Cooper

46 Involvement of the striatum in dopamine agonist-induced
yawning   463
C. T. Dourish, S. J. Cooper and P. H. Hutson

Part VI Epilogue   **467**

47 The end of the beginning   469
W. Winlow and R. Markstein

Index   471

Preface

This book is derived from an international symposium held by the *Northern Neurobiology Group* at the University of Leeds from 16 to 19 July 1984. Many books about dopamine have been written in the past, but mostly they only cover one of the many aspects of the transmitter. This book is intended to fill a gap in that it provides a series of short surveys of the various biochemical, pharmacological, physiological and clinical aspects of dopamine. It should therefore be of value to scientists working in the neurosciences, to clinicians and to postgraduate students.

As a result of the development of new and more refined techniques and methodologies, our knowledge about the various pharmacological, physiological and clinical aspects of dopamine is growing rapidly. It is therefore quite natural that even experts working in these fields have difficulties in keeping their knowledge up to date. These difficulties are even greater for newcomers, students or clinicians, who want to have a short but comprehensive overview about the present state of dopamine research. We have therefore divided the book into five sections, each of which contains review articles and shorter research papers originally presented as posters at the meeting:

    I Basic biology of dopamine systems
    II Neuroendocrinology of dopamine
    III Schizophrenia
    IV Basal ganglia disturbances and extra-pyramidal syndrome
    V Further roles for dopamine

We hope that *The Neurobiology of Dopamine Systems* will appeal to a wide audience of both clinical and basic neuroscientists, and will serve as a useful and necessary framework and background to their future research efforts.

W. Winlow
R. Markstein

Acknowledgements

In particular we would like to thank Dr David Harris of Sandoz Products Ltd, Horsforth, Leeds, for his help, advice and expertise prior to and during the conference on which this book is based. We are also grateful to Miss Debbie Girdlestone and Mr David Harrison for their organisational assistance. We are deeply indebted to the financial support provided by the following companies and institutions, without which neither the conference nor this book would have been possible: Sandoz Products Ltd; Harvard Apparatus; Stag Instruments; Britannia Pharmaceuticals; Ormed; Gould Instruments; Digitimer; Bioscience; Manchester University Press; Tektronix UK; Goodfellow Metals; Biodata; and Leeds University School of Medicine.

For permission to reproduce text figures and tables from a number of publications we would like to thank the authors who are mentioned in figure captions and the following: American Academy of Neurology (*Neurology*) (figs. 35.2 and 44.2); American Association of Neuropathologists, Inc. (*J. Neuropathol. exp. Neurol.*) (fig. 6.4); American Medical Association (*Arch. Neurol.*) (fig. 44.1); American Physiological Society (*Am. J. Physiol.*) (figs. 40.1, 40.2 and 40.3); Ankho Internation, Inc. (*Neurobiology of Ageing*) (fig. 44.4); Associated Book Publishers (UK) Ltd (*J. Neurocytol.*) (fig. 4.3); Company of Biologists (*J. exp. Biol.*) (fig. 9.2 and table 9.1; Editiones Roche (fig. 35.3); Elsevier Science Publishers BV (Amsterdam) (figs. 4.2 and 44.3) (*Brain Res.*), (table 12.1 and fig. 16.3) (*Eur. J. Pharmacol*); Elsevier Scientific Publishers (Ireland) Ltd (*Mech. Ageing Dev.*) (fig. 6.3); National Research Council of Canada (*Can. J. Neurol. Sci.*) (figs. 35.1 and 35.4; Pergamon Press (*Neuropharmacology*) (fig. 4.4), (*Neurochemistry International*) (figs. 5.2 to 5.12 and 5.17 to 5.19) and (*Comp. Biochem. Physiol. A*) (fig. 9.1); The Physiological Society (*J. Physiol.*) (fig. 4.5); Piccin Medical Books, Padua (*Biochem. Exp. Biol.*) (fig. 6.2); Raven Press (fig. 2.3 and table 2.4) (*J. Neurochem.*) (figs 24.2 and 24.3); The Royal Society (*Proc. R. Soc. Lond. B*) (figs. 4.6, 4.7 and 4.8); Sandoz Products Ltd (fig. 24.4); and Williams and Wilkins (*Endocrinology*) (fig. 12.1).

Part I

Basic biology of dopamine systems

1

Biosynthesis, storage and release of dopamine

R. J. Walker

Department of Neurophysiology, School of Biochemical and Physiological Sciences, University of Southampton, Southampton SO9 3TU, UK

1.1. Introduction

There is excellent evidence for dopamine as a transmitter in the vertebrate central nervous system and the dopamine systems are probably more complex than those for noradrenaline. In contrast to the situation with noradrenaline, there are a number of distinct brain areas which contain dopamine cell bodies. Some dopamine neurones have long axon projections from their cell bodies to their terminals whereas others are very localised. In terms of the topics to be covered in this chapter, it has in general been assumed that these anatomically divergent dopamine neurones are a homogeneous population. However, it must be remembered that many of the studies have been carried out primarily on neurones in the substantia nigra–corpus striatum system, and should be verified on the other dopamine systems.

1.2. Biosynthesis of dopamine

One of the problems in establishing a specific role for dopamine as a transmitter as distinct from the other major endogenous catecholamines was the finding that dopamine is on the major synthetic pathway for noradrenaline and adrenaline. This pathway is illustrated in Fig. 1.1 and uses tyrosine as the precursor. The sequence of enzyme reactions was first postulated by Blaschko (1939) and has subsequently been confirmed (Nagatsu *et al.*, 1964). Tyrosine may be absorbed from the diet or it may be synthesised from dietary phenylalanine, mainly in the liver. Both tyrosine and phenylalanine are present in the blood and the central nervous system at a concentration of around 5×10^{-5} M. Tyrosine is taken up by an active transport process into the catecholamine-containing neurones (Guroff *et al.*, 1961). The rate of synthesis of catecholamine is dependent on the concentration of extracellular tyrosine, particularly at low concen-

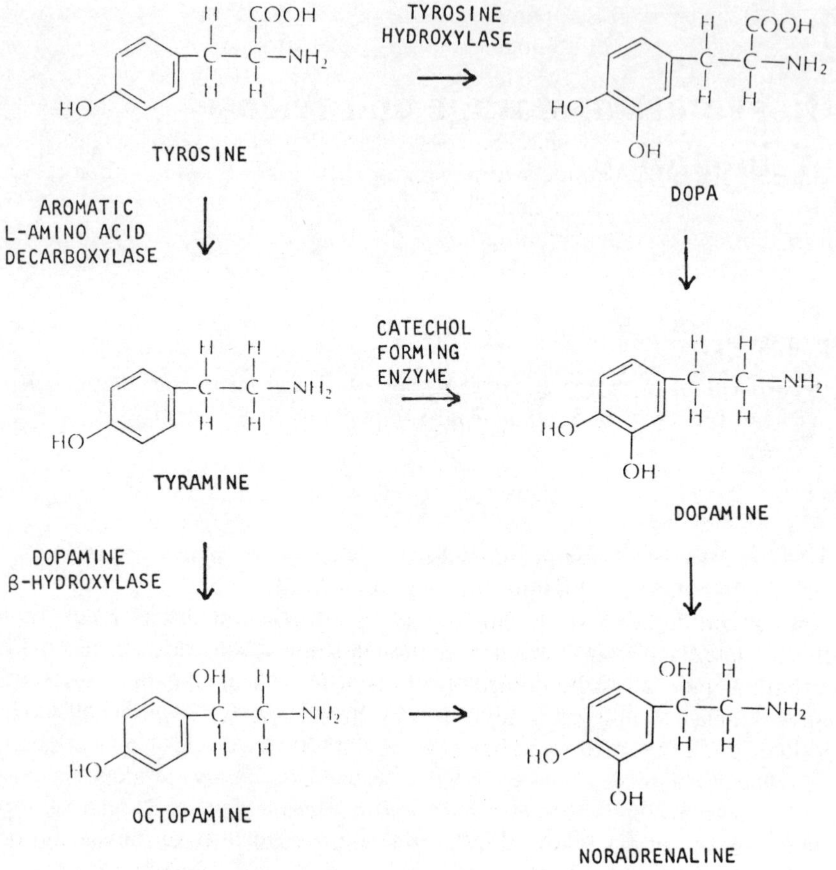

Fig. 1.1. Summary of the pathways for the synthesis of dopamine, octopamine and noradrenaline from tyrosine. (After Cooper *et al.*, 1982.)

trations of tyrosine, although this is probably not of physiological importance at normal levels of extracellular tyrosine.

The rate-limiting stage in the synthesis of dopamine occurs at the level of the conversion of tyrosine to 3,4-dihydroxyphenylalanine (DOPA) by tyrosine hydroxylase. This enzyme is present in all neurones which are capable of synthesising catecholamines and its presence is an excellent marker for the identification of catecholamine-containing neurones. In the absence of this enzyme, tyrosine may be converted to tyramine and then to octopamine (Fig. 1.1). Tyrosine hydroxylase requires a tetrahydropteridine cofactor and exhibits a high degree of substrate specificity (Cooper *et al.*, 1982). It would appear that tyrosine hydroxylase and phenylalanine hydroxylase are distinct enzymes since the latter does not hydroxylate tyrosine and is unaffected by inhibitors of the former. Tyrosine hydroxylase is not located in the dopamine vesicles but appears to be present in the particulate fraction of homogenates and is present in synaptosomes. The K_m for the conversion of tyrosine to DOPA is around 10 μM and tyrosine hydroxylase activity is about 100 to 1000 times lower than that of aromatic L-amino acid decarboxylase. Since tyrosine hydroxylase represents the rate-limiting step in dopamine synthesis, manipulation of brain dopamine levels can best be achieved by activation or inhibition of this enzyme. A wide range of compounds including amino acid analogues, catechol derivatives, tropolones and selective ion chelators have been used as inhibitors. For *in vivo* studies, α-methyl-*p*-tyrosine is one of the best inhibitors of dopamine synthesis. The enzyme dihydropteridine reductase is closely associated with tyrosine hydroxylase in the formation of DOPA. Reduced pteridines are required for tyrosine hydroxylase activity and the reductase causes reduction of the quinonoid dihydropteridine which has been oxidised in the formation of DOPA.

The next enzyme involved in the synthesis of dopamine is DOPA decarboxylase (Fig. 1.1). However, this enzyme will act on a wide range of substrates including tryptophan, 5-hydroxytryptophan, phenylalanine, DOPA, tyrosine and histidine, and so is more correctly called aromatic L-amino acid decarboxylase. This enzyme is not associated with any specific particulate fraction in the neurones though some of its activity is located in synaptosomes. Thus the decarboxylation of DOPA to dopamine occurs in the cytoplasm. This enzyme possesses high activity, requires pyridoxal phosphate, as a cofactor and has a K_m of 400 μM.

Once synthesised in the cytoplasm, the dopamine is taken up into granules or vesicles located in the nerve terminals where it is stored until required for release. It is probable that these granules are synthesised in the cell body of the neurone and transported to the nerve terminals.

The regulation of the synthesis of catecholamines occurs at the level of tyrosine hydroxylase activation. The activity of this enzyme is dependent on the concentration of dopamine in the nerve terminals. This is an example of feedback inhibition. However, the activity of tyrosine hydroxylase is also dependent on the extent of cell activity. Thus electrical activity of neurones leads to the synthesis of additional amounts of catecholamine. It has been suggested that in response to increased cell activity, more dopamine is released and a possible regulatory pool of dopamine which can influence tyrosine hydroxylase activity is removed, so removing end product inhibition (Roth *et al.*, 1974). This causes an increase in tyrosine hydroxylase activity. This activation of the enzyme is associated partly with an increased affinity for its substrate and for the pteridine cofactor, and partly with a decrease in affinity for the end-product inhibitor, dopamine. The changes in affinity for the cofactor and dopamine may both be important physiological effects since both occur with an increase in cell activity. The actual way in which cell excitation can induce these changes is not clear though both calcium and cyclic AMP may be involved. Tyrosine hydroxylase is sensitive to phosphorylation and so this process may be involved in the regulation of the enzyme activity. There is also evidence that under prolonged neuronal activation there is induction of tyrosine hydroxylase. It is also likely that circulating hormones may regulate the activity of the enzyme. The regulation of tyrosine hydroxylase has been reviewed by Mandell (1978, 1984).

The overall rate at which the total dopamine store is replaced is termed the turnover, and the turnover rate may not be the same as the rate of biosynthesis. For example, the turnover times in hours for dopamine in the rat brain and rabbit caudate are 3·6 and 3·0 respectively, but the corresponding rates of synthesis for the two systems are very different, that is, 0·21 and 2·85 μg g^{-1} h^{-1}, respectively. In general the rates of synthesis and turnover times for dopamine and 5-hydroxytryptamine are faster than for noradrenaline. If the tyrosine hydroxylase activity is measured, it is found to be higher in areas associated with dopamine than in noradrenaline areas. For example, in the dopamine cell bodies of the substantia nigra, enzyme activity is 17·5 nmol DOPA per milligram of protein per hour, whereas for noradrenaline cell bodies in the locus coeruleus the corresponding value is 4·6. The steady-state catecholamine content for the two areas is 1·7 μg g^{-1} and 0·99 μg g^{-1}, respectively, and the catecholamine turnover rates are 1·13 μg g^{-1} h^{-1} and 0·22 μg g^{-1} h^{-1}, respectively. Tyrosine hydroxylase activity is higher in cell bodies than in dopamine nerve terminals but the steady-state dopamine content of the latter is normally greater (Bacopoulos and Bhatnagar, 1977) (see Table 1.1). Thus the activity of tyrosine hydroxylase adapts to the functional demands of catecholamine neurones for transmitter synthesis. Brain areas with high intrinsic rates of catecholamine turnover also possess high levels of tyrosine hydroxylase activity, that is, tyrosine hydroxylase activity is coupled to rates

Table 1.1.   Tyrosine hydroxylase activity and dopamine turnover in specific areas of the rat brain. (After Bacopoulos and Bhatnagar, 1977.)

Brain Area	Tyrosine hydroxylase activity (nmol DOPA mg^{-1} protein h^{-1})	Steady-state dopamine content (μg g^{-1})	Dopamine turnover rate (μg g^{-1} h^{-1})
Dopamine cell bodies in substantia nigra	17·5	1·7	1·1
Dopamine terminals in			
(a) caudate nucleus	12·0	7·4	2·1
(b) olfactory tubercle	11·1	4·3	1·2
(c) nucleus accumbens	7·7	2·6	0·6
(d) amydala	4·3	0·9	0·3
(e) hypothalamus	3·1	0·6	0·2

of transmitter utilisation. It is not possible to say whether differences in turnover between areas is due to differences in cell firing or to differences in innervation density. Bacopoulos and Bhatnagar (1977) conclude that in brain areas that contain mainly one type of catecholamine, tyrosine hydroxylase is a reliable index of transmitter turnover.The high tyrosine hydroxylase activity in cell bodies is associated with enzyme synthesis since it is not associated with the highest turnover values. However, the study of Bacopoulos and Bhatnagar (1977) does not consider slow and fast pools of dopamine, which may be associated with very different rates of tyrosine hydroxylase activity. For example, it has been suggested that the turnover rate for the functional pool of dopamine in the rat striatum is $10 \cdot 6$ μg g^{-1} h^{-1} whereas that for the storage pool is $2 \cdot 6$ μg g^{-1} h^{-1} (Javoy and Glowinski, 1971). Release studies on the isolated striatum of the rat show that newly synthesised [^{3}H]dopamine from [^{3}H]tyrosine is released spontaneously in preference to [^{3}H]dopamine stored for a longer period (Besson *et al.*, 1969). The evidence for this was that the amount of [^{3}H] dopamine collected from the rat striatum in the first 15 min was about 30 times greater than the [^{3}H]dopamine content collected after 105 min whereas the tissue [^{3}H]dopamine level at the beginning of the experiment and after 2 h was about the same. A critical assessment of the methods employed for determining monoamine synthesis and turnover rates *in vivo* has been made by Weiner (1974). Neuroleptics can increase the turnover rate of dopamine in the striatum. This increase is associated with a change in the affinity constants of soluble tyrosine hydroxylase (Costa *et al.*, 1974). The affinity constant for the particle-bound tyrosine hydroxylase remains constant. The change in affinity constant for the enzyme may be related to enhanced neuronal activity.

1.3. Storage of dopamine

The storage of dopamine has many features in common with the storage mechanisms associated with noradrenaline. Dopamine is stored in granules in a complex with chromogranins, divalent ions (for example, magnesium and calcium) and ATP. A small fraction of the dopamine pool is not complexed in storage granules. There is evidence for more than one pool of dopamine. There may be a soluble or free storage pool in the cytoplasm and this is important in controlling dopamine synthesis through changing the activity of tyrosine hydroxylase. The stability of the dopamine–ATP–protein–metal ion complex can be disrupted by compounds that act as chelators of magnesium ions, for example reserpine and tetrabenazine. This means that dopamine is no longer stored in the granules and leaks out into the cytoplasm where it is inactivated by monoamine oxidase. The action of reserpine is irreversible and so the dopamine levels are impaired until new granules appear through axoplasmic transport from the cell body.

Dopamine release is by exocytosis, is calcium dependent and occurs in response to nerve stimulation or drugs such as amphetamine.

In terms of different stores of dopamine, amphetamine appears to release newly synthesised dopamine preferentially (Shore and Dorris, 1975) but it may also do so in such a way that some of the released dopamine enters the neuronal storage system (Miller and Shore, 1982). There is evidence that newly synthesised dopamine has a greater role in striatal function than does the dopamine of the main striatal storage pool. Amphetamine, in the presence of haloperidol, decreased release of [^{3}H]DOPAC but increased [^{3}H]dopamine even in the presence of monoamine oxidase inhibitors (Miller and Shore, 1982).

1.4. Metabolism of dopamine

The two main enzymes involved in the metabolism of dopamine are monoamine oxidase (MAO) and catechol-O-methyltransferase (COMT). MAO is localised on the outer membrane of the mitochondria and converts dopamine to 3,4-dihydroxyphenylacetaldehyde which in turn is converted to 3,4-dihydroxyphenylacetic acid (DOPAC). DOPAC is converted to 3-methoxy-4-hydroxyphenylacetic acid (homovanillic acid, HVA) by COMT. There is also reported to be some MAO located outside the dopamine neurone. Brain MAO occurs in at least two forms, that is, isoenzymes, designated type A and type B. This division is based on substrate specificity and sensitivity to inhibitors. For example, type A has a preference for serotonin and noradrenaline, and clorgyline is a specific inhibitor. In contrast, type B has a preference for β-phenylethylamine and benzylamine, and deprenyl is a selective inhibitor (Cooper *et al.*, 1982). Apparently, neither form of the enzyme appears to have a preference for dopamine as a substrate, though there is evidence linking type B with dopamine. COMT is a soluble enzyme present in the cytoplasm, and may also be present extracellularly. It catalyses the transfer of methyl groups from S-adenosyl methionine to the *m*-hydroxy group of dopamine. It is relatively non-specific. Under the influence of COMT, dopamine is converted to 3-methoxytyramine, which in turn is converted to 3-methoxy-4-hydroxyphenyl-acetaldehyde by MAO. This compound can then either be converted to 3-methoxy-4-hydroxyphenylethanol by aldehyde reductase or to homovanillic acid by aldehyde dehydrogenase. These methods for the enzymatic inactivation of dopamine are summarised in Fig. 1.2.

The main dopamine metabolites found in the brain are HVA and DOPAC, and the accumulation of HVA in either the brain or the cerebrospinal fluid (CSF) can be used as a monitor of the activity of dopamine neurones. Administration of compounds that increase the activity of dopamine neurones causes an increase in the turnover of dopamine and levels of HVA. Electrical stimulation of the pathway from the substantia

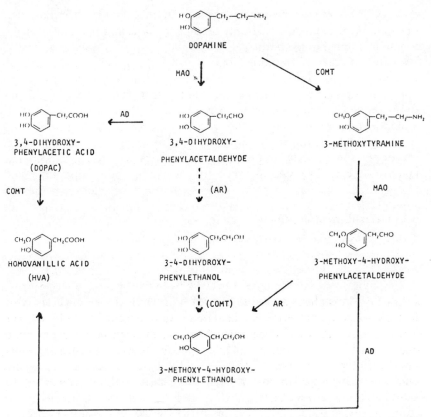

Fig. 1.2. Summary of the pathways for the metabolism of dopamine. (After Cooper *et al.*, 1982.) MAO, monoamine oxidase; COMT, catechol-O-methyltransferase; AR, aldehyde reductase; AD, aldehyde dehydrogenase.

nigra to the striatum also results in enhanced release of HVA (Bacopoulos *et al.*, 1979). Levels of DOPAC have also been used as an indicator of dopamine neurone activity. For example, acute lesions in the nigrostriatal pathway lead to a rapid decline in the level of DOPAC in the striatum, and electrical stimulation of the pathway leads to an increase in striatal DOPAC. Drugs that enhance or block activity in the nigrostriatal pathway also induce changes in DOPAC levels in the striatum. However, these changes in DOPAC levels are particularly relevant to the rat brain since in primates DOPAC is only a minor metabolite. DOPAC is a major metabolite of dopamine in the rat brain but there is evidence that the metabolism of dopamine differs throughout the brain (Karoum *et al.*, 1977). This evidence suggests that the metabolism of dopamine at nerve terminals may be different from that in the cell bodies. Thus changes in HVA and DOPAC in different brain areas should be evaluated with caution. Karoum *et al.* (1977) found that the accumulation of DOPAC and HVA in the brain after treatment with probenid only accounted for a minor fraction of the dopamine formed in the brain.

As in the case of noradrenaline, dopamine is partly inactivated by presynaptic uptake into the nerve terminals. This uptake system is a high-affinity, energy-dependent, active transport system. The preferred conformations for block of dopamine uptake have been investigated by Koe (1976). A number of compounds that act to release dopamine also act as inhibitors of this neuronal dopamine uptake. For example, (+)-amphetamine, benztropine and nomifensine are all blockers of dopamine uptake. The tricyclic antidepressants, such as imipramine, desimipramine and amitryptiline, are less potent as dopamine uptake blockers than as noradrenaline and serotonin uptake blockers. It has been suggested that most of the dopamine taken up into the nerve terminals is oxidatively deaminated to DOPAC (Demarest *et al.*, 1983).

1.5. Role of electrical stimulation in dopamine synthesis

The rate of dopamine synthesis and release is dependent on the activity in the nigrostriatal pathway. Either an increase or a decrease in activity can lead to an increase in dopamine synthesis. Activation of presynaptic autoreceptors on the nerve terminals by dopamine and dopamine agonists leads to a decrease in dopamine synthesis through suppression of tyrosine hydroxylase. These observations have been elucidated by Roth and his colleagues (Nowycky and Roth, 1978), who have been able to modify nerve activity either by mechanical lesions of the dopamine axons or through administration of γ-butyrolactone, both events abolishing activity; they have also brought about enhanced activity through electrical stimulation. Once the rate of activity in the axons has been stabilised, any effects of drugs on synthesis must be related to activation of the presynaptic receptors. Dopamine agonists were found to depress synthesis whereas dopamine antagonists reversed this effect. The activation of these receptors by dopamine in the synaptic cleft leads to modulation of the synthesis of dopamine which occurs normally in response to the amount of activity in the dopamine nigrostriatal pathway. It is not clear whether the receptors that modify release of dopamine are the same as those that modify synthesis. It has been suggested that changes in the release and synthesis of dopamine do not follow the same time course (Westfall *et al.*, 1976).

Earlier evidence suggested that the rate of activity in the nigrostriatal pathway can modify the physical properties of tyrosine hydroxylase. Thus enzyme obtained from the striatum of rats after electrical stimulation of their nigrostriatal pathway had an increased affinity for the pteridine cofactor and substrate and a decreased affinity for dopamine (Murrin *et al.*, 1976). This activation required 10 min of continuous stimulation to become maximal and lasted for 10–15 min following termination of the stimulus. The extent of tyrosine hydroxylase activation can be measured *in vivo* by following the short-term accumulation of DOPA after administration of Ro4-4602, an aromatic L-amino acid decarboxylase inhibitor. A clear quantitative relationship exists between activity in the nigrostriatal pathway and striatal tyrosine hydroxylase activity, measured as short-term accumulation of DOPA. A maximum post-stimulation activation is obtained following continuous stimulation of the tract for 10 min at 15 Hz. When there is no activity in the pathway, this relationship no longer exists since DOPA accumulation is greater than at maximum stimulation parameters. This fact was first observed by Carlsson and his co-workers in 1972 (Kehr *et al.*, 1972). They observed a rise in brain dopamine and an increase in the rate of tyrosine hydroxylation following cerebral hemisection. A direct link between cessation of activity in the nigrostriatal tract and activation of tyrosine hydroxylase is shown by the finding that, if impulse activity is restored in the dopamine cells by low-frequency stimulation, the activation of tyrosine hydroxylase is prevented (Murrin and Roth, 1976). It is hard to reconcile this finding with the observation that enhanced activity in the nigrostriatal system enhances tyrosine activity. It was for this reason that Carlsson postulated the presence of presynaptic autoreceptors for the modulation of dopamine synthesis (Carlsson, 1975). Cessation of activity in the dopamine axons would lead to a fall in dopamine release; this in turn means less activation of the presynaptic receptors which, when activated by dopamine, reduce dopamine synthesis. This removes the normal influence of the presynaptic receptors on tyrosine hydroxylase activity and so dopamine synthesis increases. Thus hemisection and lesion experiments increase dopamine levels. This effect can be prevented by administration of dopamine agonists which can be countered by then administering dopamine antagonists. Using synaptosomal preparations, a similar relationship can be established. For example, apomorphine depresses tyrosine hydroxylase activity in synaptosomes to a greater extent than it does the purified soluble enzyme. Haloperidol will partially reverse this effect of apomorphine on synaptosomal tyrosine hydroxylase (Christiansen and Squires, 1974).

The question to be answered is: what is the mechanism whereby lack of nerve activity can lead to a change in tyrosine hydroxylase activity? The pharmacological evidence would all suggest that the mechanism is associated with the presynaptic terminal rather than any postsynaptic mediated system. Cessation of activity in the dopamine axons causes changes in the kinetic properties of tyrosine hydroxylase. These changes differ from those produced following stimulation of dopamine neurones. Striatal tyrosine hydroxylase from rats subjected to lesion of the nigrostriatal tract has an increased affinity for substrate and cofactor and a decreased affinity for dopamine, but the decreased affinity for dopamine is much greater than the decrease observed following electrical stimulation (Morgenroth *et al.*, 1976). This loss in susceptibility to inhibition by dopamine can explain at least partly how enhanced synthesis can proceed in the presence of raised levels of dopamine. The kinetic changes that occur in the activity of tyrosine can be reversed *in vitro* by the addition of calcium (Nowycky and Roth, 1978). In addition, the calcium chelator EGTA, when added to control striatum, causes changes in the properties of tyrosine hydroxylase such that the enzyme now behaves as though it had been obtained from a rat in which activity in the nigrostriatal tract had been blocked. This effect of EGTA can be blocked by the addition of excess calcium. Thus it can be postulated that during the absence of activity in the dopamine axons, calcium entry into the nerve terminal and dopamine release are blocked, and changes in intracellular free calcium may cause changes in the properties of the enzyme such that it is less sensitive to dopamine. *In vivo* administration of apomorphine to rats treated with γ-butyrolactone reverses the effects of γ-butyrolactone on the tyrosine hydroxylase activity such that it now resembles that isolated from control rats. Thus *in vivo* experiments confirm *in vitro* studies that dopamine agonists restore the ability of endogenous dopamine to inhibit tyrosine hydroxylase activity. These experiments suggest that a role for presynaptic receptors is to gate the entry of calcium to dopamine nerve terminals and so regulate tyrosine hydroxylase activity. An extension of this is that depolorisation of the nerve terminals by activity in the axons activates a voltage-dependent calcium ionophore which allows calcium to enter the terminal. Calcium levels in the nerve terminal are raised, which increases release of dopamine into the synaptic cleft. This dopamine activates the presynaptic autoreceptors which then reduce further calcium entry and so dopamine release is reduced.

Let us consider the possible sites at which feedback mechanisms could operate to reduce the acceleration of synthesis of dopamine that follows administration of tyrosine. The possibilities are summarised in Fig. 1.3. The first possibility is that this could occur within the dopamine presynaptic terminal itself, that is, direct inhibition by dopamine of the rate-limiting enzyme tyrosine hydroxylase, indicated as site (1) in Fig. 1.3. Secondly, feedback could include an extracellular link, site (2). In this case, dopamine released into the synaptic cleft activates the presynaptic autoreceptors located on the nerve terminal of the dopamine neurone. Activation of these receptors informs the neurone that dopamine is being released and activates intracellular mechanisms, for example allosteric changes in tyrosine hydroxylase, which in turn slow down dopamine synthesis. Alternatively, a multisynaptic reflex arc may be involved through one or more interneurones which reduces the firing rate of the dopamine cell, site (3). This site could be located either on the nerve terminal within the caudate nucleus or on the dendrites of the dopamine cell body in the substantia nigra. Another

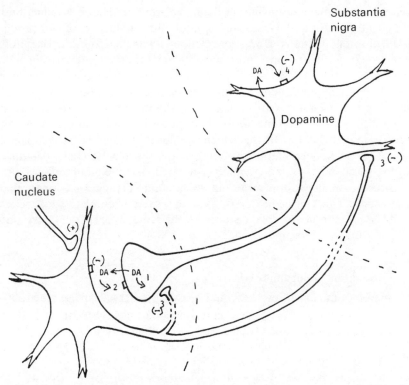

Fig. 1.3. Sites at which feedback mechanism might operate to reduce the acceleration of dopamine synthesis that follows administration of tyrosine. (1) Intraneuronal inhibition by dopamine of tyrosine hydroxylase activity. (2) Release of dopamine and activation of presynaptic autoreceptors. (3) Multisynaptic reflex to reduce activity of dopamine cell. (4) Release of dopamine from dendrites and activation of somatic autoreceptors.

possibility is that dopamine could be released from the dopamine cell dendrites within the nigra thus activating autoreceptors also located on the dendrites of the dopamine neurones, site (4).

The effect of amphetamine on the rate of dopamine synthesis in the terminals of nigrostriatal and mesolimbic pathways has been investigated *in vivo* by Demarest *et al.* (1983) by measuring the accumulation of DOPA. Low to moderate doses of amphetamine, while activating dopamine synthesis in the striatum, did not alter synthesis in the nucleus accumbens and olfactory tubercle. However, amphetamine did reverse the reduction in dopamine synthesis by apomorphine in all three areas. Dopamine synthesis was increased in all three areas by γ-butyrolactone. These authors discuss the possible mode of action of amphetamine. They suggest that amphetamine releases dopamine and reduces uptake, thus there is more extracellular dopamine and less intracellular. This means there is a reduction in the 'strategic' intracellular dopamine pool and a reduction of end product inhibition on synthesis. Dopamine synthesis will therefore increase. But a rise in extracellular dopamine increases activation of presynaptic auto-receptors, which leads to a decrease in dopamine synthesis. The auto-receptors on mesolimbic dopamine neurones may be more sensitive than those in the striatum. Activation of postsynaptic receptors would activate the inhibitory long neurone feedback loops to the nigra and reduce activity in the nigrostriatal pathway which would lead to a fall in dopamine synthesis. The concentration of amphetamine would be vital in terms of which mechanism is preferentially activated. Amphetamine affects both release and uptake of dopamine and it is useful to distinguish between these two events. This can be achieved by using $[^{14}C]$dopamine to study release and $[^{3}H]$dopamine for uptake studies in the same preparation, for example rat striatum synaptosomes (Bonnett *et al.*, 1984).

1.6. Regulation of dopamine release

A possible summary of events would be: when a nerve impulse reaches the nerve terminals it induces a depolarisation of the terminal membrane potential. This depolarisation activates voltage-dependent ionophores, particularly calcium ionophores, and calcium ions enter the nerve terminal. This increase in the entry rate of calcium activates the dopamine vesicles and enhances the rate of dopamine release, primarily by exocytosis. This process can be regulated at the level of presynaptic autoreceptors. The evidence suggests that normally activation of these autoreceptors by dopamine and dopamine agonists reduces the release of dopamine, whereas blockade of these receptors by dopamine antagonists, for example haloperidol, induces an enhanced release of dopamine. If this sequence is correct, then it must be assumed that autoreceptor activation reduces calcium influx through the voltage-dependent calcium ionophores.

Dopamine release and synthesis are modulated by extracellular feedback mechanisms involving both presynaptic autoreceptors and changes in neuronal firing rates. For example, it was shown that incubation of rat neostriatal slices with $[^{3}H]$dopamine followed by electrical stimulation results in the overflow of radioactivity which can be measured (Farnebo and Hamberger, 1971). If the presynaptic uptake of dopamine is inhibited with cocaine, the amount of stimulation-induced overflow is increased by about 35 %. If the dopamine receptors are activated with apomorphine, there is a decrease in dopamine release. In contrast, if the dopamine receptors are blocked with a neuroleptic, for example pimozide, there is an increase in stimulation-induced overflow associated with increased release of dopamine. This was the first evidence for a local effect of dopamine at the level of the synapse through a negative feedback mechanism. However, the initial experiments of Farnebo and Hamberger (1971) were not confirmed by subsequent investigators, and the role of presynaptic autoreceptors in the control of dopamine release was questioned (Raiteri *et al.*, 1978). It was suggested that presynaptic dopamine autoreceptors inhibited synthesis but not release. However, good evidence for the presence of presynaptic inhibitory dopamine receptors that regulate dopamine release through a negative feedback system was obtained by Starke *et al.* (1978). They found that although tetrodotoxin (TTX), apomorphine, bromocriptine, chlorpromazine, haloperidol or omission of calcium had little effect on spontaneous release of $[^3H]$dopamine from rabbit caudate, these compounds modified electrically evoked released. The electrically evoked release was abolished by tetrodotoxin and calcium-free Ringer; apomorphine and bromocriptine reduced release, and chlorpromazine and haloperidol enhanced release. Chlorpromazine and haloperidol antagonised the inhibitory effect of apomorphine and bromocriptine. The calcium-dependent release of dopamine elicited by electrical stimulation of the rabbit caudate nucleus is modulated by presynaptic dopamine receptors, and the calcium-independent release elicited by amphetamine or tyramine is not (Kamal *et al.*, 1981; Aribilla *et al.*, 1982). The amount of $[^3H]$dopamine released by electrical stimulation depends on the frequency of stimulation (Aribilla *et al.*, 1982). The inhibition of dopamine release by apomorphine in caudate slices is greater when the slice is stimulated at 3 Hz compared with 10 Hz. The enhancement of $[^3H]$dopamine release by haloperidol was only observed at 3 Hz. Thus the modulation of dopamine release by presynaptic autoreceptors is less at higher frequencies of stimulation. Another possibility is that an interneurone is involved in the regulation of dopamine release but this is unlikely. For example, the inhibition of $[^3H]$dopamine release induced by apomorphine and the enhancement due to haloperidol occurred from rabbit caudate nucleus in presence of TTX (Jackisch, *et al.*, 1980). The effect of tyrosine on the release of endogenous dopamine from the rat striatum was investigated by Milner and Wurtman (1984). They found that dopamine release could be maintained as long as 50 μM tyrosine was present in the superfusate. Thus tyrosine levels are able to influence the release of dopamine.

There is evidence that a large number of compounds can modify the release of dopamine and these are listed by Starke in his review on presynaptic receptors (Starke, 1981). It would appear that the release of most if not all transmitters can be modified by presynaptic receptor activation. Starke considers that it is easier to evaluate the role of these receptors using *in vitro* as opposed to *in vivo* techniques. It is also apparent that there are tissue and species variations, and it cannot be assumed that all dopamine neurones contain the same range of receptors. For example, it is clear that not all dopamine neurones possess autoreceptors. There would appear to be two populations of mesocortical cells, a medial group that projects exclusively to cortical regions and is fast-firing, and a lateral group that innervates cortical, limbic and striatal regions and is slow-firing (Shepard and German, 1984). For example, mesocortical dopamine cells which project to the prefrontal or cingulate cortices are devoid of synthesis-modulating nerve terminals and impulse-regulating somatodendritic auto-receptors (Chiodo *et al.*, 1984). The firing rate of these cells is not altered by intravenous apomorphine or iontophoretic dopamine. The firing rate of these cells was also faster than those possessing autoreceptors. γ-Butyrolactone increased levels of dopamine in prefrontal, cingulate and piriform cortices and in the caudate nucleus. Pretreatment with apomorphine reversed this effect in the piriform cortex and caudate but not in the prefrontal and cingulate cortices. These observations raise the question as to how these dopamine systems are regulated.

There is evidence that a number of peptides may alter the release of dopamine. For example, *in vivo* endogenous release of dopamine from the rat nucleus accumbens is suppressed by the sulphated form of chole-cystokinin octapeptide (Voight and Wang, 1984).

Although it is not really part of the present topic, it should be noted that a number of workers have studied the release of acetylcholine from caudate and hippocampal slices and the influence of dopamine on acetylcholine release (Strittmather *et al.*, 1982; Starke *et al.*, 1983). For example, apomorphine reduced the release of labelled acetylcholine whereas halo-

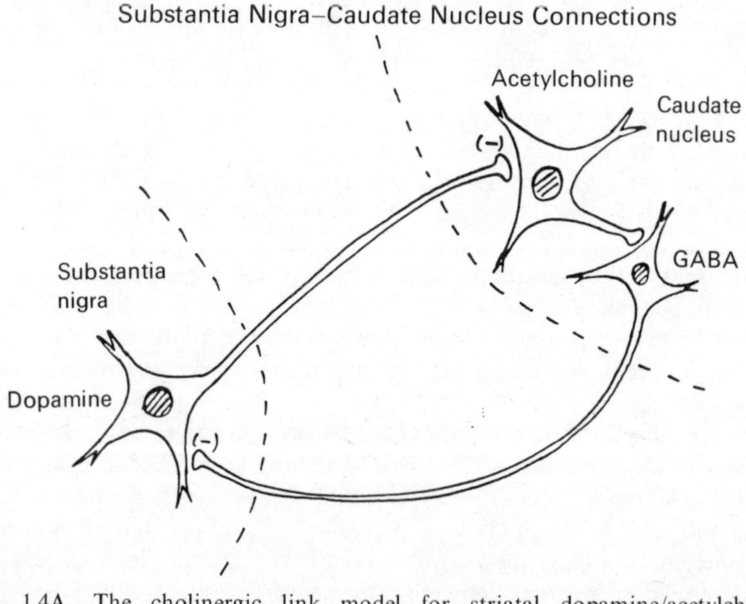

Fig. 1.4A. The cholinergic link model for striatal dopamine/acetylcholine interactions.

peridol antagonised this effect. It is normally assumed that dopamine exerts its action postsynaptically on to the dendrites of the acetylcholine neurones in the caudate nucleus (Fig. 1.4A). However, recently Langer and his co-workers have collected evidence that challenges this assumption (Lehmann and Langer, 1983). They conclude that the dopamine released from nerve terminals acts presynaptically on to the acetylcholine nerve terminals, so modifying the release of acetylcholine (Fig. 1.4B). They term these presynaptic receptors Type II and the postsynaptic receptors Type I. The former type act at the level of the voltage-dependent calcium channels and Type I act on voltage-independent ion channels to alter the membrane potential. Lehmann and Langer (1983) find that both the dopamine receptor that regulates acetylcholine release and that regulating dopamine release are very similar pharmacologically, whereas dopamine autoreceptors are generally about ten times more sensitive to apomorphine than are the postsynaptic receptors. For example, apomorphine has a similar if not identical potency and efficacy in inhibiting the release of labelled dopamine and acetylcholine from cat caudate (Lehmann and Langer, 1983). Starke *et al.* (1983) have also found that dopamine presynaptic autoreceptors and the receptors that mediate reduction in acetylcholine release in the rabbit striatum are very similar in properties and are both D-2 type receptors. The pharmacological properties of the receptors that release dopamine and acetylcholine in the striatum have also been found to be similar by Hoffman

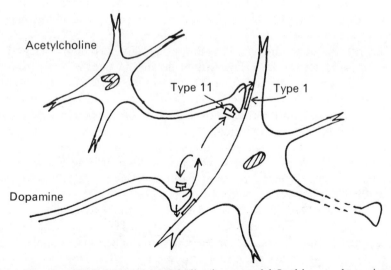

Fig. 1.4B.   Parallel dopamine/acetylcholine input model. In this case, dopamine can modulate the release of acetylcholine through autoreceptors on the acetylcholine nerve terminals. (After Lehmann and Langer, 1983.)

and Cubeddu (1984) and are of the D-2 type. The effect of dopamine on the release of endogenous acetylcholine from striatal slices and synaptosomes also suggests that presynaptic dopamine receptors may act on acetylcholine terminals to regulate acetylcholine release (Belleroche et al., 1982).

1.7. Dopamine receptors located on dopamine dendrites and soma in nigra

The possible occurrence of dopamine autoreceptors on the dendrites of dopamine neurones in the zona compacta of the substantia nigra was postulated by Groves et al. (1975) to explain a possible site of action of amphetamine. Amphetamine acts to release dopamine from vesicles and also blocks re-uptake of dopamine into nerve terminals. Applied systemically, amphetamine inhibits cell firing in the striatum and cell firing of dopamine cells in the substantia nigra whereas haloperidol enhances cell firing in both the striatum and the substantia nigra. Haloperidol blocks the action of amphetamine at these sites. A 'neuronal feedback loop' hypothesis was put forward by Carlsson and Lindqvist (1963) to explain this. This loop involves dopamine cells in the substantia nigra which inhibit cholinergic neurones in the striatum which in turn activate GABA in the striatum which projects back to the substantia nigra (Fig. 1.4A). An alternative hypothesis would involve autoreceptors on the dendrites of dopamine neurones in the substantia nigra which would release dopamine that would inhibit dopamine cell activity directly. The sensitivity of these nigral autoreceptors differs from that of the postsynaptic receptors located in the caudate nucleus (Skirboli et al., 1979). The nigral autoreceptors are more sensitive to iontophoretically applied dopamine and intravenous apomorphine than are those in the caudate.

The role of dopamine in the substantia nigra has been reviewed by Cheramy et al. (1981). They suggest that dopamine, once released from the dendrites of the nigrostriatal dopamine cells in the nigra, can be involved at a number of sites. For example, it may be involved in the self-regulation of the dopamine cells, in the control of the release of transmitters from nigral afferent fibres, and in influencing the activity of nigral non-dopamine cells. Through their role in both the striatum and nigra, dopamine cells can influence information entering the striatum by the corticostriatal and thalamostriatal pathways, and pathways leaving the nigra through projections to the thalamus and pontine nuclei.

Geffen et al. (1976), using slices of the rat nigra, have demonstrated the release of labelled dopamine from the dendrites. They induced this release by raising external potassium concentration in the Ringer to depolarise the dendrites and this release was calcium dependent. They found that nigral slices released a higher proportion of their labelled dopamine than did striatal slices. Geffen et al. (1976) demonstrated that there was release from dendrites by performing experiments where they used only the pars reticulata, the cell bodies being in the pars compacta. They found that a substantial portion of the dopamine released from whole nigral slices came from the dendrites in the pars reticulata. The authors suggest that this dendritic release could be involved in controlling the excitability of dopamine cells in the nigra and that these receptors could be on the dopamine cells themselves and/or located presynaptically on nerve terminals innervating the dopamine cells. The release of transmitter close to the cell bodies could be involved in their metabolism. Local release could also be involved in the control of plastic and regenerative sprouting of new synapses. Since the dopamine cells are sensitive to dopamine, they clearly possess dopamine receptors (Bunney and Aghajanian, 1973; Grace and Bunney, 1983). Dopamine receptors are also located on nerve terminals that innervate the dopamine cells (Reubi *et al.*, 1977). These authors investigated the release of labelled GABA and substances P from slices of nigra. They found that whereas dopamine enhanced the release of GABA from nerve terminals, it had no effect on the release of substance P. There is a dopamine-sensitive adenylate cyclase in the rat nigra, and Reubi *et al.* (1977) suggest it may be localised at least partly on GABA-containing axon terminals. There is also evidence for dopamine receptors on GABA-releasing nerve terminals in the nucleus accumbens (Beart *et al.*, 1980). The veratridine-evoked release of labelled dopamine has been demonstrated and this is sensitive to tetrodotoxin, indicating that the dopamine dendrites have fast sodium channels (Tagerud and Cuello, 1979; but see below).

Homogenisation of nigral dopamine dendrites can produce particles that behave like synaptosomes. These paricles can be termed dendrosomes and they take up labelled dopamine which can be released by potassium depolarisation (Silbergeld and Walters, 1979). Amphetamine induces release of dopamine from these dendrosomes. GABA potentiates the potassium-induced release of labelled material from nigral slices, and substance P and glycine enhance spontaneous efflux of labelled dopamine (Cheramy *et al.*, 1981).

Dopamine release has also been demonstrated *in vivo* through the use of push-pull cannulae in the nigra (Cheramy *et al.*, 1981). The nigra was continually superfused with labelled tyrosine. Dopamine release was calcium dependent, and raised potassium greatly enhanced release. Tetrodotoxin enhanced the spontaneous release of $[^3H]$dopamine from the nigra, which contrasts with the observation that it reduces or has little effect on the spontaneous release from striatal nerve endings. This would suggest that dopamine is not released from nigral nerve terminals and that an afferent input inhibits the spontaneous release of dendritic dopamine. Since tetrodotoxin is unable to prevent the release of dendritic dopamine, it would appear that the dopamine is not released linked with fast sodium channels and that another mechanism, possibly calcium, is involved. Evidence in support of this comes from recent experiments where calcium currents have been found on nigral dendritic processes (Llinas *et al.*, 1984). Glycine, acetylcholine and 5-hydroxytryptamine all enhanced dopamine release *in vivo*, whereas substance P reduced dopamine release from the nigra. Muscimol but not GABA potentiated the release of labelled dopamine. Electrical stimulation of the motor or visual cortex also changed the amount of dopamine released. Thus input into the nigra can alter dopamine release from the dendrites. Nigral application of dopamine reduces the spontaneous release of dopamine in the ipsilateral striatum, whereas the application of dopamine antagonists to the nigra leads to a rise in the release of striatal dopamine. Application of amphetamine to the striatum increased dopamine release in the striatum and also stimulated dendritic release in the nigra. Thus changes in dopamine transmission in the striatum will alter activity in the nigra dopamine cells through the nigrostriatal pathway on to the dopamine dendrites. It is suggested that substance P cells may be responsible for activation and GABA neurones for inhibition of dopamine dendrites. The possible importance of the nigrostriatal pathway in regulating nigral cell activity was investigated by Doudet *et al.* (1984). These authors destroyed intrinsic striatal neurones with kainic acid and recorded the activity of nigral cells after the lesion. They found that the pattern of activity became greatly disorganised and could not simply be explained by the removal of inhibitory GABA input on to the nigral cells.

1.8. **Electrochemical detection of dopamine release**

The technique of voltammetry is potentially a very powerful tool in the study of dopamine release since it can be used in both *in vitro* and *in vivo* studies. Electrodes that can detect changing levels of amines can be implanted chronically (Conti *et al.*, 1978). This method for measuring release from brain tissue is an extension of the method employed for measuring changes in amine levels in cerebrospinal fluid. The type of electrode and its testing and calibration are fully described by Conti *et al.* (1978). These authors used the method to monitor the release of dopamine induced by amphetamine in the rat caudate.

One of the problems with this technique is that it may measure changes in levels of compounds in addition to amines, for example ascorbic acid. To overcome this problem, differential pulse voltammetry has been developed. This means that separate peak currents can be obtained for ascorbic acid and dopamine. It has been suggested that the peak obtained for the amine may be measuring the metabolite 3,4-dihydroxyphenylacetic acid (DOPAC) (Gonnon *et al.*, 1980). A further modification of the technique is the use of stearate-modified graphite paste electrodes which do not respond to ascorbic acid and DOPAC (Blaha and Lane, 1984). These authors studied the release of dopamine from rat striatum in anaesthetised animals following unilateral depletion of dopamine from the nigra with 6-

hydroxydopamine. Both haloperidol and chlorpromazine enhanced release of dopamine in normal animals but this release was abolished in animals pretreated with 6-hydroxydopamine. Using this technique the release of dopamine in the striatum of freely moving hamsters has also been detected (Forni and Nieoullon, 1984). The release of dopamine was reduced following 6-hydroxydopamine pretreatment whereas amphetamine enhanced dopamine release. The ability to record changes in amine levels from non-anaesthetised animals means that the changes in the release of amines can be monitored during different behaviour patterns and during normal physiological states.

1.9. Summary

Dopamine is an important and widespread transmitter in the central nervous system. The rate-limiting enzyme for the synthesis of dopamine is tyrosine hydroxylase. Tyrosine hydroxylase activity is influenced by the degree of activation of dopamine neurones. Activation of presynaptic autoreceptors on dopamine nerve terminals can influence both release and synthesis of dopamine. However, these autoreceptors are not present on all dopamine neurones. Autoreceptors located on dopamine cell dendrites and possibly on cell bodies also influence dopamine cell activity and so in turn influence synthesis. The level of free intracellular calcium is important in the regulation of dopamine levels. It is possible that presynaptic dopamine receptors on nerve terminals act at the level of voltage-dependent calcium channels. Presynaptic autoreceptors would appear to play an important role in the regulation of transmitter function in the central nervous system.

Acknowledgements

I am grateful to Noel Harris and Steve Long for helpful discussions during the preparation of this chapter.

References

Aribilla, S., Nowak, J. Z. and Langer, S. Z. (1982). Presynaptic autoregulation of dopamine release. In *Presynaptic Receptors: Mechanism and Function.* J. de Belleroche (ed.), pp. 30–45. Ellis Horwood, Chichester.

Bacopoulos, N. G. and Bhatnagar, R. K. (1977). Correlation between tyrosine hydroxylase activity and catecholamine concentration or turnover in brain regions. *J. Neurochem.* **29,** 639–43.

Bacopoulos, N. G., Hattox, S. E. and Roth, R. H. (1979). 3,4-Dihydroxyphenylacetic acid and homovanillic acid in rat plasma: possible indicators of central dopaminergic activity. *Eur. J. Pharmacol.* **56,** 225–36.

Beart, P. M., Kuppers, D. and Louis, W. J. (1980). Evidence for dopamine receptors on GABA-releasing nerve terminals in rat nucleus accumbens. *Brit. J. Pharmacol.* **68,** 160–1P.

Belleroche, J. de, Coutinho-Netto, J. and Bradford, H. F. (1982). Dopamine inhibition of the release of endogenous acetylcholine from corpus striatum and cerebral cortex in tissue slices and synaptosomes: a presynaptic response? *J. Neurochem.* **39**, 217–22.

Besson, M. J., Cheramy, A., Feltz, P. and Glowinski, J. (1969). Release of newly synthesized dopamine from dopamine-containing terminals in the striatum of the rat. *Proc. nat. Acad. Sci. USA* **62**, 741–8.

Blaha, C. D. and Lane, R. F. (1984). Direct *in vivo* electrochemical monitoring of dopamine release in response to neuroleptics. *Eur. J. Pharmacol.* **98**, 113–17.

Blascho, H. (1939). The specific action of L-DOPA decarboxylase. *J. Physiol.* **96**, 50–1P.

Bonnet, J. J., Lemasson, M. H. and Costentin, J. (1984). Simultaneous evaluation by a double labelling method of drug-induced uptake inhibition and release of dopamine in synaptosomal preparation of rat striatum. *Biochem. Pharmacol.* **33**, 2129–35.

Bunney, B. S. and Aghajanian, G. K. (1973). Comparison of effects of L-DOPA, amphetamine and apomorphine on firing rate of rat dopaminergic neurones. *Nature New Biol. (Lond.)* **245**, 123–5.

Carlsson, A. (1975). Receptor-mediated control of dopamine metabolism. In *Pre- and Postsynaptic Receptors.* E. Usdin and W. E. Bunney (eds), pp. 49–63. Marcel Dekker, New York.

Carlsson, A. and Lindqvist, M. (1963). Effect of chlorpromazine or haloperidol on the formation of 3-methoxytryramine and normetanephrine in mouse brain. *Acta Pharm. Toxicol.* **20**, 140–4.

Cheramy, A., Levial, V. and Glowsinski, J. (1981). Dendritic release of dopamine in the substantia nigra. *Nature (Lond.)* **289**, 537–42.

Chiodo, L. A., Bannon, M. J., Grace, A. A., Roth, R. H. and Bunney, B. S. (1984). Evidence for the absence of impulse-regulating somatodendritic and synthesis-modulating nerve terminal autoreceptors on subpopulations of mesocortical dopamine neurons. *Neurosci.* **12**, 1–16.

Christiansen, J. and Squires, R. F. (1974). Antagonistic effects of apomorphine and haloperidol on rat striatal synaptosomal tyrosine hydroxylase. *J. Pharm. Pharmacol.* **26**, 367–9.

Conti, J. C., Strope, E., Adams, R. N. and Marsden, C. A. (1978). Voltammetry in brain tissue: chronic recording of stimulated dopamine and 5-hydroxytryptamine release. *Life Sci.* **23**, 2705–16.

Cooper, J. R., Bloom, F. E. and Roth, R. H. (1982). *The Biochemical Basis of Neuropharmacology,* pp. 367. Oxford University Press, Oxford.

Costa, E., Guidotti, A. and Zivkovic, B. (1974). Short- and long-term regulation of tyrosine hydroxylase. In *Neuropsychopharmacology of Monomines and their Regulatory Enzymes.* E. Usdin (ed.), pp. 161–75. Raven Press, New York.

Demarest, K. T., Lawson-Wendling, K. L. and Moore, K. E. (1983). *d*-Amphetamine and γ-butyrolactone alteration of dopamine synthesis in the terminals of nigrostriatal and mesolimbic neurons. *Biochem. Pharmacol.* **32**, 691–7.

Doudet, D., Gross, C., Seal, J. and Bioulac, B. (1984). Activity of nigral dopaminergic neurons after lesion of the neostriatum in rats. *Brain Res.* **302**, 45–55.

Farnebo, L.-O. and Hamberger, B. (1971). Drug-induced changes in the release of ^{3}H-monoamines from field stimulated rat brain slices. *Acta physiol. Scand. Suppl.* *371,* 35–44.

Forni, C. and Nieoullon, A. (1984). Electrochemical detection of dopamine release in the striatum of freely moving hamsters. *Brain Res.* **297**, 11–20.

Geffen, L. B., Jessel, T. M., Cuello, A. C. and Iversen, L. L. (1976). Release of dopamine from dendrites in rat substantia nigra. *Nature (Lond.)* **260**, 258–60.

Gonnon, F., Buda, M., Cespuglio, R., Jouvet, M. and Pujol, J. E. (1980). *In vivo* electrochemical detection of catechols in the neostriatum of anaesthetised rats: dopamine or DOPAC? *Nature (Lond.)* **286**, 902–4.

Grace, A. A. and Bunney, B. S. (1983). Intracellular and extracellular electrophysiology of nigral dopaminergic neurons—I. Identification and characterization. *Neurosci.* **10**, 301–15.

Groves, P. M., Wilson, C. J., Young, S. J. and Rebec, G. V. (1975). Self-inhibition by dopamingergic neurons. *Science* **190**, 522–9.

Guroff, C., King, W. and Udenfriend, S. (1961). The uptake of tyrosine by rat brain *in vitro. J. biol. Chem.* **236**, 1773–7.

Hoffman, I. S. and Cubeddu, L. X. (1984). Differential effects of bromocriptine on dopamine and acetycholine release modulatory receptors. *J. Neurochem.* **42**, 278–82.

Jackisch, R., Zumstein, A., Hertting, G. and Starke, K. (1980). Interneurons are probably not involved in the presynaptic dopaminergic control of dopamine release in the rabbit caudate nucleus. *Naunyn-Schmiedeberg's Arch. Pharmacol.* **314**, 129–33.

Javoy, F. and Glowinski, J. (1971). Dynamic characteristics of the 'Functional Compartment' of dopamine in dopaminergic terminals of the rat striatum. *J. Neurochem.* **18**, 1305–11.

Kamal, L., Aribilla, S. and Langer, S. Z. (1981). Presynaptic modulation of the release of dopamine from the rabbit caudate nucleus: differences between electrical stimulation, amphetamine and tyramine. *J. Pharm. exp. Ther.* **216**, 592–8.

Karoum, F., Neff, N. H. and Wyatt, R. J. (1977). The dynamics of dopamine metabolism in various regions of rat brain. *Eur. J. Pharmacol.* **44**, 311–18.

Kehr, W., Carlsson, A., Lindqvist, M., Magnusson, T. and Atack, C. V. (1972). Evidence for a receptor-mediated feedback control of striatal tyrosine hydroxylase activity. *J. Pharm. Pharmacol.* **24**, 744–7.

Koe, B. K. (1976). Molecular geometry of inhibition of the uptake of catecholamines and serotonin in synaptosomal preparations of rat brain. *J. Pharm. exp. Ther.* **199**, 649–61.

Lehmann, J. and Langer, S. Z. (1983). The striatal cholinergic interneuron: synaptic target of dopaminergic terminals? *Neurosci.* **10**, 1105–20.

Llinas, R., Greenfield, S. A. and Jahnsen, H. (1984). Electrophysiology of pars compacta cells in the *in vitro* substantia nigra—a possible mechanism of dendritic release. *Brain Res.* **294**, 127–32.

Mandell, A. J. (1978). Redundant mechanisms regulating brain tyrosine and tryptophan hydroxylases. *Ann. Rev. Pharm. Toxicol.* **18**, 461–93.

Mandell, A. J. (1984). Non-equilibrium behaviour of some brain enzyme and receptor systems. *Ann. Rev. Pharm. Toxicol.* **24**, 237–74.

Miller, H. H. and Shore, P. A. (1982). Effect of amphetamine and amfonelic acid on the disposition of striatal newly synthesised dopamine. *Eur. J. Pharmacol.* **78**, 33–44.

Milner, J. D. and Wurtman, R. J. (1984). Release of endogenous dopamine from electrically stimulated slices of rat striatum. *Brain Res.* **301**, 139–42.

Morgenroth, V. H., Walters, J. R. and Roth, R. H. (1976). Dopaminergic neurons: alteration in the kinetic properties of tyrosine hydroxylase after cessation of impulse flow. *Biochem. Pharmacol.* **25**, 655–61.

Murrin, L. C. and Roth, R. H. (1976). Dopaminergic neurons: Reversal of effects by gamma-butyrolactone by stimulation of the nigro-striatal pathway. *Naunyn-Schmiedeberg's Arch. Pharmacol.* **295**, 15–20.

Murrin, L. C., Morgenroth, V. H. and Roth, R. H. (1976). Dopaminergic neurons: effects of electrical stimulation on tyrosine hydroxylase. *Mol. Pharmacol.* **12**, 1070–81.

Nagatsu, T., Levitt, M. and Udenfriend, S. (1964). Tyrosine hydroxylase: the initial step in norepinephrine biosynthesis. *J. biol. Chem.* **239**, 2910–17.

Nowycky, M. C. and Roth, R. H. (1978). Dopaminergic neurons: role of presynaptic receptors in the regulation of transmitter biosynthesis. *Progr. Neuro-psychopharm.* **2**, 139–58.

Raiteri, M., Cervoni, A. M., Del Carmine, R. and Levi, G. (1978). Do presynaptic auto-receptors control dopamine release? *Nature (Lond.)*, **274**, 706–8.

Reubi, J. C., Iversen, L. L. and Jessel, T. M. (1977). Dopamine selectively increases ^{3}H-GABA release from slices of rat substantia nigra *in vitro*. *Nature (Lond.)* **268**, 652–4.

Roth, R. H., Walters, J. R. and Morgenroth, V. H. (1974). Effects of alterations in impulse flow on transmitter metabolism in central dopaminergic neurons. In *Neuropsychopharmacology of Monoamines and their Regulatory Enzymes*. E. Usdin (ed.), pp. 369–84. Raven Press, New York.

Shepard, P. D. and German, D. C. (1984). A subpopulation of mesolimbic dopamine neurons possesses autoreceptors. *Eur. J. Pharmacol.* **98**, 455–6.

Shore, P. A. and Dorris, R. L. (1975). On a prime role for newly synthesised dopamine in striatal function. *Eur. J. Pharmacol.* **30**, 315–18.

Silbergeld, E. K. and Walters, J. R. (1979). Synaptosomal uptake and release of dopamine in substantia nigra. Effects of γ-aminobutyric acid and Substance P. *Neurosci. Lett.* **12**, 119–26.

Skirboli, L. R., Grace, A. A. and Bunney, B. S. (1979). Dopamine auto- and postsynaptic receptors. Electrophysiological evidence for differential sensitivities to dopamine agonists. *Science* **206**, 80–(2.

Starke, K. (1981). Presynaptic receptors. *Ann. Rev. Pharm. Toxicol.* **21**, 7–30.

Starke, K., Reimann, W., Zumstein, A. and Hertting, G. (1978). Effect of dopamine receptor agonists and antagonists on the release of dopamine in the rabbit caudate nucleus *in vitro*. *Naunyn-Schmiedeberg's Arch. Pharmacol.* **305**, 27–36.

Starke, K., Spath, L., Lang, J. D. and Adelung, C. (1983). Further functional *in vitro* comparison of pre- and postsynaptic dopamine receptors in the rabbit caudate nucleus. *Naunyn-Schmiedeberg's Arch. Pharmacol.***323**, 298–306.

Strittmather, H., Jackisch, R. and Hertting, G. (1982). Role of dopamine receptors in the modulation of acetylcholine release in the rabbit hippocampus. *Naunyn-Schmiedeberg's Arch. Pharmacol.* **321**, 195–200.

Tagerud, S. E. O. and Cuello, A. C. (1979). Dopamine release from the rat substantia nigra *in vitro*. Effect of raphe lesions and veratridine stimulation. *Neurosci.* **4**, 2021–9.

Voigt, M. M. and Wang, R. Y. (1984). *In vivo* release of dopamine in the nucleus accumbens of the rat: modulation by cholecystokinin. *Brain Res.* **296**, 189–93.

Weiner, N. (1974). A critical assessment of methods for the determination of monoamine synthesis turnover rates *in vivo*. In *Neuropsychopharmacology of Monoamines and their Regulatory Enzymes*. E. Usdin (ed.), pp. 143–59. Raven Press, New York.

Westfall, T. C., Besson, M. J., Giorguieff, M. F. and Glowinski, J. (1976). The role of presynaptic receptors in the release and synthesis of ^{3}H-dopamine by slices of rat striatum. *Naunyn-Schmiedeberg's Arch. Pharmacol.* **292**, 279–87.

2

Brain dopamine receptors

S. B. Freedman* and G. N. Woodruff

Merck Sharp & Dohme Research Laboratories, Neuroscience Research Centre, Terlings Park, Eastwick Road, Harlow, Essex CM20 2QR, UK

2.1. Introduction

Attempts to characterise dopamine receptors have been in progress for more than twenty years. Yet it is only very recently that unambiguous evidence has emerged for the existence of dopamine receptor subtypes. In the mammalian peripheral system the dopamine receptors present on pre-synaptic sympathetic nerve terminals can be clearly distinguished from those at postjunctional vessels (Goldberg and Kohli, 1983). Unfortunately, the situation in the central nervous system is still far from clear, with workers variously suggesting one (Laduron, 1980) or as many as four dopamine receptors (Seeman, 1980; Sokoloff et al. 1980). Progress in this areas has been hampered by the lack of specific agonists and antagonists for the proposed receptor subtypes and a lack of specific functional models.

The finding that dopamine stimulated the formation of cyclic AMP in striatal homogenates (Kebabian et al., 1972) and that this response could be antagonised by neuroleptic drugs such as chlorpromazine, haloperidol and *cis*-flupenthixol led to suggestions that dopamine receptors in the brain were coupled to adenylate cyclase and that the increase in cyclic AMP production was physiologically relevant in mediating the actions of dopamine. However, doubts soon began to arise. For example, the concentrations of butyrophenone neuroleptics required to block the stimulatory effects of dopamine on adenylate cyclase were much higher than those present at therapeutic doses in man or active in receptor binding assays. Furthermore, substituted benzamides such as sulpiride, which are potent dopamine receptor antagonists in a number of pharmacological tests, were completely devoid of activity in the adenylate cyclase model (Trabucchi et al., 1975). In addition, ergot alkaloids, which had previously been shown to act as antagonists of invertebrate dopamine receptors, were

---

*To whom all correspondence should be addressed.

found to have dopamine-agonist-like activity in the mammalian central nervous system yet to have weak or negligible activity as agonists on the dopamine-stimulated adenylate cyclase (Woodruff, 1978). One explanation for these findings was that there existed at least two types of dopamine receptors, and that at least one of these was not linked to adenylate cyclase (Spano *et al.*, 1978). This suggestion was developed by Kebabian and Calne (1979), who defined D-1 dopamine receptors as being linked to adenylate cyclase and D-2 dopamine receptors as being cyclase independent.

Since this original suggestion, the absence of specific research tools has led to confusion in the literature and the introduction of a number of different methods for classifying dopamine receptors (review by Seeman, 1980). However, the past few years have seen the development of several drugs which are of great potential value in probing central dopamine receptors.

Sulpiride and zetidoline are compounds of diverse chemical structure which have in common the property that they are virtually inactive as dopamine antagonists on the dopamine-sensitive adenylate cyclase, yet are similar to typical dopamine receptor antagonists in terms of their effects on prolactin levels and in behavioural tests (Jenner and Marsden, 1979; Barone *et al.*, 1982). In contrast, the recently introduced benzazepine, (R)-(+)-8-chloro-2,3,4,5-tetrahydro-3-methyl-5-phenyl-1H-3-benzazepine-7-ol (SCH 23390), which is related to the dopamine agonist SKF 38393 (Pendleton *et al.*, 1978), is a very potent antagonist of the stimulatory adenylate cyclase, but is only a very weak displacer of [^{3}H]butyrophenone binding (Hyttel, 1983).

We have used these compounds to study the relevance of putative multiple dopamine receptors using behavioural, electrophysiological and biochemical studies in mammalian brain and also biochemical studies in neuronal clonal cell lines.

2.2. Methods

2.2.1. *Electrophysiological studies*

Detailed methodology has been described previously (Harris *et al.*, 1985). In brief, male Wistar rats (150–180 g) were anaesthetised with chloral hydrate (350 mg kg^{-1}) and secured in a stereotaxic frame. Zona compacta cells were identified by their sensitivity to iontophoretically applied dopamine and their responses to antidromic stimulation. Extracellular recordings were made using eight barrelled protruding-tip electrodes. At the end of each experiment, Pontamine sky blue was ejected from the recording barrel. Electrode position was subsequently verified histologically.

2.2.2. *Biochemical studies*

[^{3}H]*Sulpiride binding studies.* The binding of [^{3}H]sulpiride (26 Ci mmol^{-1}) to partially purified rat striatal synaptic membranes was measured as previously described (Freedman *et al.*, 1981, Woodruff and Freedman, 1981). Membranes and ligand were incubated in 50 mM Tris–Krebs buffer pH 7·4 in the presence of test compounds at 37°C for 10 min. Bound radioactivity was separated by rapid filtration and washing through Millipore HAWP 0024 filters. Under these conditions specific binding (as defined by 1 μM sulpiride) was 60–70% of the total binding, with filter blanks (counts bound on the filter and displaced by S-(−)-sulpiride in the absence of tissue) representing 1–3% of this.

Clonal cell line. Neuroblastoma × Chinese hamster brains explant hybrid (NCB-20) cells were grown as monolayers in Costar 35 mm plastic tissue culture wells in modified Eagle's medium supplemented with 10% newborn (bobby) calf serum as described previously (Dawson *et al.*, 1972).

Measurement of cyclic AMP levels. Cyclic AMP was measured in monolayer cultures as previously described (Berry-Kravis *et al.*, 1984). Determination of cyclic AMP was by radioimmunoassay via competition with cyclic AMP derivative [^{125}I] for binding to cyclic AMP antisera according to the method of Harper and Brooker (1975).

2.3. Results

2.3.1. *Electrophysiological experiments*

The dopamine-containing cell bodies in the zona compacta of the substantia nigra can be identified electrophysiologically following electrical stimulation of the striatum (Bunney, 1979). Apomorphine, injected intravenously, caused a dose-dependent depression of the firing rate of the cells. The effect of apomorphine was mimicked by the intravenous injection of Ru 24926 (*N-n*-propyl-di-β(3-hydroxyphenyl)ethylamine hydrochloride), Ru 24213 (*N-n*-propyl-N-phenylethyl-*p*(3-hydroxyphenyl ethylamine hydrochloride) and pergolide. The dopamine agonist, ADTN, which is known not to penetrate the blood-brain barrier, and the dibenzoyl ester of ADTN were both inactive following intravenous administration. The intravenous injection of the dopamine antagonists and *cis*-flupenthixol reversibly blocked the actions of the agonists on the zona compacta neurones. The newly introduced zetidoline was similarly an antagonist of dopamine receptors in this system, causing a reversal of the depressant effect of dopamine and apomorphine on these cells and blocking the effect of subsequent doses of the agonist.

When applied iontophoretically on to the zona compacta neurones, dopamine causes dose-related depressions of the firing rate of the cells. This effect of dopamine is mimicked by the application of dopamine agonists. In particular, the dopamine agonist, ADTN, applied iontophoretically, is extremely potent in depressing the firing of zona compacta neurones (Woodruff and Pinnock, 1981). The iontophoretic applications of sulpiride

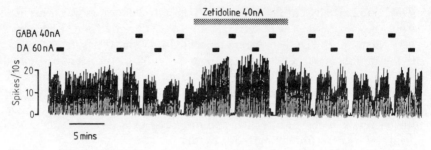

Fig. 2.1.   Effect of zetidoline upon substantia nigra neurones. Continuous ratemeter recording showing the effect of zetidoline on *in vivo* responses of zona compacta neurones to dopamine and γ-aminobutyric acid. All drugs were applied micro-iontophoretically using the ejecting currents indicated.

onto zona compacta neurones causes a selective and reversible antagonism of the inhibitory effect of dopamine. Zetidoline is also a selective and reversible antagonist of dopamine receptors following iontophoretic application in this study (Fig. 2.1). Conversely, intravenous injections of SCH 23390 were ineffective in blocking the actions of dopamine agonists on these cells (N. C. Harris, personal communication).

2.3.2. *Behavioural experiments*

We have measured functional dopamine receptors using the behavioural model of locomotor activity. The bilateral injection of the dopamine receptor agonist ADTN (2-amino-6, 7-dihydroxy-1,2,3,4-tetrahydro-naphthalene) into the rat nucleus accumbens produces a long-lasting stimulation of locomotor activity for up to 20 h. This response is potently blocked by typical neuroleptics and by the substituted benzamide sulpiride (Andrews and Woodruff, 1978). The benzazepine SKF 38393 (2,3,4,5-tetrahydro-7,8-dihydroxy-1-phenyl-1H-3-benzazepine), a suggested D-1 selective agonist, will also produce locomotor activity following bilateral injection into the rat nucleus accumbens. Interestingly, this response is blocked both by typical neuroleptics such as fluphenazine and also by the substituted benzamides sulpiride and tiapride (Freedman *et al.*, 1979). Recently, Christensen and colleagues (1984) have reported that the selective D-1 antagonist SCH 23390 will also antagonise the response to ADTN stimulation.

2.3.3. *Biochemical studies*

[³H]*Sulpiride binding.* [³H]Sulpiride bound specifically to partially purified rat striatal synaptic membranes, showing saturable binding in the ligand concentration range of 1–40 nM. When analysed by Scatchard analysis, the data identify a single binding site with an affinity constant (K_d) of 8.8 ± 1.0 nM and a maximal binding capacity (B_{max}) of 433 ± 27 fmol μg^{-1} protein. The largest binding capacity was found in regions of known dopaminergic innervation, namely the striatum and nucleus accumbens. In contrast, only very low levels of specific binding were found in the hippocampus, pons medulla and cerebellum.

A large number of dopamine receptor agonists and antagonists were found to potently and stereospecifically displace [³H]sulpiride binding (see Table 2.1). The most potent displacers of binding were the neuroleptic drugs from all major classes, with the phenothiazine fluphenazine, the butyro-phenone, spiroperidol and the thioxanthene *cis*-flupenthixol among the most active compounds. The substituted benzamides were only slightly weaker, with clebopride and S-(−)-sulpiride being relatively potent. The novel butyrophenone derivative, domperidone, a peripheral dopamine receptor antagonist which does not cross the blood–brain barrier (Laduron and Leysen, 1979), was also a potent displacer of binding. Domperidone, like sulpiride and zetidoline, does not antagonise the dopamine-senstivive adenylate cyclase. In contrast, SCH 23390, the recent benzazepine derivative which is a potent antagonist of the dopamine-sensitive adenylate cyclase (Hyttel, 1983; Iorio *et al.*, 1983), was extremely weak as a displacer of [³H]sulpiride binding, with an IC_{50} value of 5 μM. (+)-Bulbocapnine is an effective albeit weak antagonist of the dopamine-sensitive adenylate cyclase

Table 2.1. Displacement of [³H]sulpiride binding to synaptic membranes from rat striatum. (Assays were performed using 15 nM [³H]sulpiride and with 1 μM S-(−)-sulpiride to define specific binding.)

Compound	IC_{50} (nM)
Spiroperidol	0.2
Lisuride	0.7
Domperidone	0.8
Fluphenazine	0.9
cis-Flupenthixol	1.6
Clebopride	2.0
(−)-Sulpiride	11.0
ADTN	28.0
Ru 24926	32.0
Apomorphine	32.0
trans-Flupenthixol	40.0
Ru 24213	110.0
Ergometrine	220.0
Dopamine	580.0
SCH 23390	5000.0
SKF 38393	⩾10000.0
Bulbocapnine	⩾10000.0

(Miller *et al.*, 1976). However, this compound did not displace [^{3}H]sulpiride binding.

The dopamine receptor agonists examined were at least an order of magnitude less active than the neuroleptics, with ADTN more than 20 times more potent than dopamine itself. The ergot alkaloids which behave as potent agonists on central dopamine receptors, but which act as weak partial agonists or antagonists in the cyclase assay, were potent displacers of binding. Two *N*-diphenethylamine derivatives, Ru 24213 and Ru 24926, have potent dopamine receptor activity at the striatal and anterior pituitary level (Euvrard *et al.*, 1980) but are completely devoid of activity on the adenylate cyclase model. Both compounds were potent inhibitors of binding. In contrast, SKF 38393 the benzazepine derivative which is a partial agonist on the adenylate cyclase (Setler *et al.*, 1978) was an extremely weak displacer of binding. This result is particularly interesting since SKF 38393 induced locomotor activity (Freedman *et al.*, 1979) and renal vasodilation (Pendleton *et al.*, 1978; Marcou *et al.*, 1982) are both potently blocked by sulpiride.

It is well documented that guanine nucleotides can alter the affinity of ligands whose binding sites are linked to an adenylate cyclase system (Rodbell, 1980). Although it was initially suggested that D-2 receptors were not non-cyclase linked and were not regulated by guanine nucleotides (Snyder and Goodman, 1980), it is now clear that this view is incorrect.

Studies with [^{3}H]sulpiride show that GTP, GDP and the stable GTP analogue Gpp(NH)p can significantly reduce the affinity of dopamine receptor agonists for the [^{3}H]sulpiride binding site (Table 2.2). In contrast, antagonist affinity for the [^{3}H]sulpiride binding site is unaffected (Fig. 2.2). As other investigators have found with [^{3}H]spiroperidol binding, agonist competition curves do not obey mass action kinetics and have Hill coefficients significantly less than 1·0. In the presence of guanine nucleotides, the Hill coefficients significantly increase towards unity (Table

Table 2.2. The effect of guanine nucleotides on the affinity of dopaminergic drugs for [^{3}H]sulpiride binding to rat striatal membranes.

| | IC_{50} (*nM*) | | |
| | | Presence of 100 μM | |
Drug	*Control*	*Gpp(NH)p*	*Shift*
Dopamine	580 ± 80	3800 ± 450	6·6*
ADTN	28 ± 5	84 ± 10	3·0*
Apomorphine	32 ± 5	140 ± 30	4·4*
Fluphenazine	0·88 ± 0·16	1·0 ± 0·15	1·1
cis-Flupenthixol	1·6 ± 0·15	1·9 ± 0·6	1·2
S-(−)-Sulpiride	11·0 ± 2	14·5 ± 2·5	1·3

* Represents a statistically significant shift.

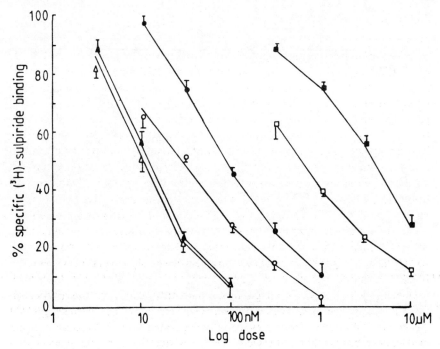

Fig. 2.2.   Effect of guanine nucleotides upon agonist and antagonist displacement of [³H]sulpiride binding to rat striatal membranes. Increasing concentrations of compounds are shown in the presence and absence of 100 μM GppNHp. The results are expressed as mean±SEM ($n = 3–9$). △, S-(−)-Sulpiride; ▲, S-(−)-Sulpiride + 100 μM GppNHp; ○, ADTN; ●, ADTN + 100 μM GppNHp; □ dopamine; ■ dopamine + 100 μM GppNHp.

2.3). These results are similar to those reported with other hormone and neurotransmitter receptors and suggest that the dopamine receptor exists in multiple conformational states which are differentiated by high and low agonist affinity and are sensitive to guanine nucleotides. Creese and colleagues (Sibley *et al.*, 1982) have recently proposed an elegant two-step three-component ternary complex model to explain these properties.

Table 2.3.   The effect of guanine nucleotides on the co-operativity of agonist inhibition of [³H]sulpiride binding.

	Hill coefficient	
Agonist	*Control*	*Presence of 100 μM GTP*
Dopamine	0·76	0·91
ADTN	0·70	0·90
Apomorphine	0·66	0·90

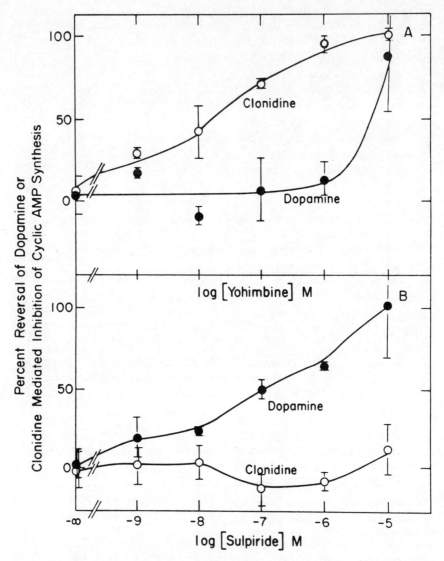

Fig. 2.3. Selective antagonism of inhibitory adenylate cyclase in NCB-20 cells. (A) Antagonism of dopamine (●) and clonidine (○) induced inhibition of serotonin-stimulated cyclic AMP synthesis by yohimbine. (B) Antagonism of dopamine (●) and clonidine (○) induced inhibition of serotonin-stimulated cyclic AMP synthesis by sulpiride. Serotonin-stimulated cyclic AMP synthesis was measured in the presence of 10 μM serotonin. Maximal inhibition of serotonin-stimulated cyclic AMP synthesis was obtained with 10 μM dopamine and 1 μM clonidine. Data reproduced from Berry-Kravis *et al.*, 1984.

These findings clearly suggest that the [^{3}H]sulpiride-labelled receptor is associated with adenylate cyclase and is subject to guanine nucleotide regulation. It is of interest that, subsequent to these findings, studies with pituitary tissue have shown that stimulation of the sulpiride-labelled receptor is associated with an inhibitory adenylate cyclase (Onali *et al.*, 1981; Cote *et al.*, 1982).

Clonal cell lines. The presence of a dopamine-sensitive inhibitory adenylate cyclase has recently been demonstrated in a clonal hybrid cell line, the neuroblastoma × Chinese hamster brain explant cell line NCB-20 (Berry-Kravis *et al.*, 1984). Both clonidine (IC$_{50}$, 70–100 nM) and dopamine (IC$_{50}$, 1–2 μM) inhibited basal, serotonin-stimulated and prostaglandin E$_1$-stimulated levels of cyclic AMP in NCB-20 cells. Dopamine at a concentration of 10 μM or more completely reversed the serotonin-stimulated cyclic AMP synthesis while inhibiting basal and prostaglandin E$_1$-stimulated levels by 40%.

Sulpiride (EC$_{50}$, 30 nM) dose-dependently (Fig. 2.3.) reversed the inhibitory effect of dopamine on both basal and serotonin-stimulated cyclic AMP levels but had no effect on the clonidine inhibition of either basal or serotonin-stimulated levels. In contrast, yohimbine (EC$_{50}$, 12 nM), a relatively specific α-adrenoceptor antagonist, reversed the clonidine inhibition of both basal and serotonin-stimulated cyclic AMP but had no effect on the dopamine inhibition at concentrations below 10 μM. In the presence of 1 μM yohimbine, which blocks any effect through α-adrenoceptors, a number of dopamine antagonists were found to reverse the dopamine-induced inhibition of cyclic AMP (Table 2.4). Stereospecific reversal was shown by sulpiride, butaclamol and flupenthixol.

Table 2.4. Potency of dopamine receptor antagonists in reversing dopamine-mediated inhibition of cyclic AMP synthesis in NCB-20 cells.*

Compound	IC$_{50}$ (μM)
(−)-Sulpiride	0·005
(±)-Sulpiride	0·03
Spiroperidol	0·06
(+)-Sulpiride	0·5
cis-Flupenthixol	0·6
(+)-Butaclamol	1·0
trans-Flupenthixol	10·0
(−)-Butaclamol	>10·0

* Methods are as described in the text. Each IC$_{50}$ value represents the concentration of antagonist required to reverse by 50% the maximal inhibition of 5-HT or PGE$_1$ stimulated cyclic AMP synthesis achieved by 10 μM dopamine. Experiments were performed in the presence of 1 μM yohimbine. (For further details see Berry-Kravis *et al.*, 1984.)

2.4. Discussion

Early conclusions on the central pharmacological actions of dopamine receptor antagonists and agonists came mainly from behavioural studies. From these it was believed that apomorphine was direct dopamine receptor agonist whereas the neuroleptic drugs were dopamine receptor antagonists. Evidence in support of these proposals began to accumulate with the introduction of direct methods to measure dopamine receptor activity.

During the last few years the advent of radioreceptor binding studies has led to a profusion of studies suggesting the multiplicity of dopamine receptors. The keystone of the idea will remain the proposal by Kebabian and Calne (1979) that two dopamine receptors exist, but it has been suggested that as few as one (Laduron, 1980) or as many as four, D-1, D-2, D-3, D-4, exist (Seeman, 1980). This has led to the view that this area is one of high controversy.

The problems may have arisen because of the criteria that have been used for defining potential dopamine receptors. These have included the differing affinity of displacing drugs in binding studies, biphasic and shallow displacement curves, and lesion studies using a variety of neurotoxic agents. The problem of using such criteria as a sole definition has been illustrated in recent studies which have questioned the relevance of at least two of these suggested receptors.

Creese and colleagues (Sibley *et al.*, 1982) have suggested that the D-2 and D-4 binding sites represent interconvertible states of the same molecule (the D-2 receptor), since the distinction between them is solely based on the use of agonist affinity. Examples of such interconversions have been clearly shown in other receptor systems. In rather elegant studies, Bacopoulos (1983) and others have suggested that 'the evidence supporting the existence of presynaptic "D-3" sites is artefactual'. Thus lesioning of dopamine terminals by 6-hydroxydopamine which produces a loss of 'D-3' sites can be mimicked by dopamine-depleting agents such as reserpine and can be reversed by preincubation of membranes with added dopamine. Thus, current debate concentrates on the suggested existence of two dopamine receptors (D-1, D-2) as first suggested by Kebabian and Calne.

The most reliable demonstration of multiple receptors must come from two areas; specific functional model systems and the development of specific agonists and antagonists. Of the latter, several compounds have been available for a number of years. Sulpiride, domperidone and more recently zetidoline are compounds which do not block the dopamine-stimulated adenylate cyclase model yet are very potent and specific dopamine receptor antagonists in a number of behavioural, electrophysiological and bio-chemical models. In our laboratories, using these models we have demonstrated that these compounds are potent antagonists in the striatum, nucleus accumbens and substantia nigra. The functional receptor can be clearly labelled in our binding studies with [^{3}H]sulpiride and has all of the properties one would expect of a functional receptor. In particular, we find potent blocking activity with virtually all clinically useful dopamine agonists and antagonists. Of particular interest is the very potent activity seen by Ru 24926 and Ru 24213, two compounds that do not interact with the stimulatory adenylate cyclase model.

The receptor labelled by sulpiride corresponds with the D-2 receptor in the Kebabian and Calne model, though clearly the original suggestion that the D-2 receptors are not linked to adenylate cyclase must be modified to a negative coupling model. One piece of evidence in support of this suggestion is our finding that guanine nucleotides can modify the displacement by agonists of [^{3}H]sulpiridine binding. Our results on the inhibitory adenylate cyclase present in the NCB-20 cell line are in accord with findings from other tissues, including the anterior and intermediate lobes of the pituitary gland (Onali *et al.*, 1981; Cote *et al.*, 1982). Further study of cell lines such as NCB-20 may be very rewarding since cell culture offers a homogenous cell population and an ability to study cells under a strictly controlled environment over a long period of time.

Many other functional models are available to measure responses to this receptor. The secretion of prolactin from the anterior pituitary and α-MSH from the intermediate lobe are both responsive to compounds such as sulpiride, domperidone, Ru 24926 and Ru 24213 (Hofmann *et al.*, 1979; Euvrard *et al.*, 1980). Similarly, the activity of these compounds can be correlated with their ability to interact with an inhibitory adenylate cyclase in isolated tissue from these organs. Other examples include the release of ACh from striatal tissue, release of dopamine from the nigrostriatal pathway and the inhibition of chemosensory discharge in the rabbit carotid gland (Mir *et al.*, 1984).

In contrast the D-1 receptor has been termed 'the receptor in search of a function' (Laduron, 1980). The demonstration that homgenates of dopamine-rich regions of the brain contained an adenylate cyclase that was activated by dopamine led to suggestions that dopamine effects were mediated by increased cyclic AMP. However, it is now evident that there are complications in such a model. The anomalies of sulpiride, domperidone, ergot alkaloids and others in the adenylate cyclase assay have suggested that not all dopamine receptors are linked to adenylate cyclase. The question is now whether the putative dopamine receptor linked to stimulation of adenylate cyclase (D-1) really exists as a functional entity. One view says not. Because apomorphine or ergot alkaloids act as partial agonists, agonists or antagonists in different systems, it does not necessarily follow that dopamine receptors in these tissues are different. Kenakin (1984) has recently produced an exmple of this in the α-receptor field. 'Thus, an agonist with a high affinity but low efficacy may be a full agonist in a tissue with a highly efficient stimulus response mechanism, but since efficacy is the prime determinant of agonist activity, the low efficacy of the same drug may be insufficient to produce any agonist response in a tissue with an inefficient stimulus–response mechanism'. In this context the adenylate cyclase *in vitro* assay may be a tissue of the latter category.

However, the significance of the D-1 receptor must ultimately depend on functional models and the production of selective compounds. In this regard the last two years have seen some important steps forward. By far the best model system for the D-1 receptor remains the bovine parathyroid gland (Brown *et al.*, 1980) where receptor activation leads to enhanced adenylate cyclase activity, activation of protein kinase and consequent secretion of parathyroid hormone. A similar model system has been claimed to exist in the retina of several mammalian species (Watling and Dowling, 1981). However, perhaps the most exciting recent development has been the discovery in Greengard's laboratory of DARPP-32 (Walaas and Greengard, 1984), a dopamine and adenosine 3'5'-monophosphate regulated phospho-protein, which is believed to be a protein occuring in cells that possess a D-1 receptor.

Selective compounds for the D-1 receptor have been a major obstacle for a number of years. Although originally suggested as such (Kebabian and Calne, 1979), it is now clear that thioxanthenes such as *cis*-flupenthixol and *cis*-clopenthixol can recognise the D-2 receptor both in pharmacological models and in binding assays with [^{3}H]sulpiride. SKF 38393 is perhaps the compound with a pharmacological profile closest to being called a D-1 selective agonist. This compound is a partial agonist, although it has high affinity, on the stimulatory adenylate cyclase model (Setler *et al.*, 1978) yet is inactive in the pituitary. Interestingly, this compound is virtually inactive on the [^{3}H]sulpiride binding assay. Recently, SCH 23390 has been introduced as a selective antagonist of the D-1 receptor (Hyttel, 1983). This compound is the 3-methyl-7-chloro analogue of SKF 38393, whose pharmacological profile includes a potent inhibition of dopamine-stimulated adenylate cyclase activity, but does not increase serum prolactin levels or antagonise emesis (Iorio *et al.*, 1983). This compound, in a manner analogous to SKF 38393, only interacts with [^{3}H]sulpiride binding at very high concentrations (IC$_{50}$, 5000 nM). This would suggest that at last selective compounds do now exist for the D-1 receptor.

The agonist SKF 38393 and the antagonist SCH 23390 clearly distinguish between dopamine receptor subtypes in the periphery, being active at postsynaptic receptors mediating renal vasodilation but inactive on nerve terminal receptors which modify transmitter release (Goldberg and Kohli, 1983). These same compounds show a similar selective profile in radio-receptor binding assays.

The dopamine autoreceptors on zona compacta neurones and striatal dopaminergic terminals are blocked by sulpiride and zetidoline but are unaffected by SKF 38393 or SCH 23390. Thus, according to the Kebabian and Calne classification, these receptors would be of the D-2 subtype. Our receptor binding studies clearly indicate that a similar receptor, negatively linked to adenylate cyclase, is present in postsynaptic neurones in the striatum and nucleus accumbens.

Functional studies, however, reveal a more complex situation. Thus locomotor activity can be stimulated by 'non-selective' agonists such as ADTN, 'selective D-1 agonists', such as SKF 38393, and 'selective D-2 agonists' such as ergot alkaloids. The locomotor activity induced by intra-accumbens injection of SKF 38393 can be potently blocked by both sulpiride and tiapride (D-2 antagonists; Freedman *et al.*, 1979). Similarly, locomotor activity induced by D-2 agonists can be blocked by SCH 23390 (Christensen *et al.*, 1984; O'Molloy and Waddington, 1984). These results suggest that even if D-1 and D-2 receptors are separate entities, they are functionally connected.

Thus the last few years have seen a tremendous advance with the development of new selective compounds for dopamine receptors. It is hoped that these compounds will assist in our understanding of dopaminergic neurotransmission in the central nervous system.

Acknowledgements

We acknowledge Dr Glyn Dawson and Elizabeth Berry-Kravis (University of Chicago) for their studies on the inhibitory adenylate cyclase in NCB-20 cells; Drs Noel Harris (University of Southampton) and Robert Pinnock (MRC, Cambridge) for the electrophysiological studies.

References

Andrews, C. D. and Woodruff, G. N. (1978). Effect of the (+)- and (−)-enantiomers of sulpiride on ADTN induced hyperactivity in the rat. *Brit. J. Pharmacol.* **64**, 434p.

Bacopoulos, N. G. (1983). Dopaminergic ^{3}H-agonist receptors in rat brain. New evidence on localization and pharmacology. *Life Sci.* **34**, 307–15.

Barone, D., Corsico, N., Dienna, A., Restelli, A., Glaser, A. and Rodenghi, F. (1982). Biochemical and pharmacological activities of zetidoline (DL 308-IT): a new antidopaminergic agent. *J. Pharm. Pharmacol.* **34**, 129–32.

Berry-Kravis, E., Freedman, S. B. and Dawson, G. (1984). Specific receptor-mediated inhibition of cyclic AMP synthesis by dopamine in a neuroblastoma × brain hybrid cell line NCB-20. *J. Neurochem.* **43**, 413–20.

Brown, E. M., Attie, M. F., Reen, S., Gardner, D. G., Kebabian J. W. and Aurbach, G. D. (1980). Characterisation of dopaminergic receptors in dispersed bovine parathyroid cells. *Mol. Pharmacol.* **18**, 335–40.

Bunney, B. S. (1979). The electrophysiological pharmacology of midbrain dopaminergic systems. In *The Neurobiology of Dopamine*. A. S. Horn, J. Korf and B. H. C. Westerink (eds), pp. 417–52. Academic Press, London.

Christensen, A. V., Arnt, J., Hyttel, J., Larsen, J-J. and Svendsen, O. (1984). Pharmacological effects of a specific dopamine D-1 antagonist SCH 23390 in comparison with neuroleptics. *Life Sci.* **34**, 1529–40.

Cote, T. E., Eskay, R. L., Frey, E. A., Grewe, C. W., Munemura, M., Stoof, J. C., Tsurata, K. and Kebabian, J. W. (1982). Biochemical and physiological studies of the β-adrenoceptor and the D-2 dopamine receptor in the intermediate lobe of rat pituitary glands; a review. Neuroendocrinology 35, 217–24.

Dawson, G., Matalon, R. and Dorfman, A. (1972). Glycosphingolipids in cultured human skin fibroblasts. J. Biol. Chem. 247, 5944–51.

Euvrard, C., Farland, L., DiPaolo, T., Beaulieu, M., Labrie, F., Oberlander, C., Raynaud, J. P. and Boissier, J. R. (1980). Activity of two new potent dopaminergic agonists at the striatal and anterior pituitary levels. Neuropharmacology 19, 379–86.

Freedman, S. B., Wait, C. P. and Woodruff, G. N. (1979). Effects of dopamine receptor agonists and antagonists in the rat nucleus accumbens. Brit. J. Pharmacol. 67, 430–1.

Freedman, S. B., Poat, J. A. and Woodruff, G. N. (1981). ^{3}H-Sulpiride, a ligand for neuroleptic receptors. Neuropharmacology 20, 1323–6.

Goldberg, L. I. and Kohli, J. D. (1983). Differentiation of dopamine receptors in the periphery. In Dopamine Receptors. C. Kaiser and J. W. Kebabian (eds), pp. 101–13. American Chemical Society, Washington, D.C.

Harper, J. and Brooker, G. (1975). Femtomole-sensitive RIA for cAMP and cAMP after 2'0-acetylation by acetic anhydride in aqueous solution. Cyclic Nucl. Res. 1, 207–18.

Hartis, N. C., Pinnock, R. D. and Woodruff, G. N. (1985). Zetidoline blocks dopamine autoreceptors in the rat substantia nigra. Neuropharmacology. 24, 33–6.

Hofmann, M., Jommi, G. C., Montefusco, O., Tonon, G. C., Spano, P. F. and Trabucchi, M. (1979). Stereospecific effects of (−)-sulpiride on brain dopamine metabolism and prolactin release. J. Neurochem. 32, 1547–50.

Hyttel, J. (1983). SCH 23390–the first selective dopamine D-1 antagonist. Eur. J. Pharmacol. 91, 153–4.

Iorio, L. C., Burnett, A., Leitz, F. H., Houser, V. P. and Korduba, C. A. (1983). SCH 23390, a potential benzazepine antipsychotic with unique interactions on dopaminergic systems. J. Pharmacol. Exp. Ther. 226, 462–8.

Jenner, P. and Marsden, C. D. (1979). The substituted benzamides, a novel class of dopamine antagonists. Life Sci. 25, 479–86.

Kebabian, J. W. and Calne, D. B. (1979). Multiple receptors for dopamine. Nature (Lond.) 277, 93–6.

Kebabian, J. W., Petzold, G. L. and Greengard, P. (1972). Dopamine-sensitive adenylate cyclase in the caudate nucleus of rat brain and its similarity to the 'dopamine receptor'. Proc. nat. Acad. Sci. USA 69, 2145–9.

Kenakin, T. P. (1984). The relative contribution of affinity and efficacy to agonist activity: organ selectivity of noradrenaline and oxymetazoline with reference to the classification of drug receptors. Brit. J. Pharmacol. 81, 131–42.

Laduron, P. M. (1980). Dopamine receptor: from an in vivo concept towards molecular characterisation. Trends Pharmacol. Sci. 1, 471–4.

Laduron, P. M. and Leysen, J. E. (1979). Domperidone, a specific in vitro dopamine antagonist, devoid of in vivo central dopaminergic activity. Biochem. Pharmacol. 28, 2161–5.

Marcou, M., Munday, A., Horn, N. and Woodruff, G. N. (1982). A study of dopamine receptors in guinea-pig renal tissue. Arch. Int. Pharmacodyn. Ther., 259, 59–71.

Miller, R. J., Kelly, P. H. and Neumeyer, J. L. (1976). Aporphines 15. Action of aporphine alkaloids on dopaminergic mechanisms in rat brain. Eur. J. Pharmacol. 35, 77–83.

Mir, A. K., McQueen, D. S., Pallot, D. J. and Nahorski, S. R. (1984). Direct biochemical and neuropharmacological identification of dopamine D-2 receptors in the rabbit carotid body. *Brain Res.* **291**, 273–83.

O'Molloy, A. G. and Waddington, J. L. (1984). Dopaminergic behaviour stereospecifically promoted by the D-1 agonist R-SKF 38393 and selectively blocked by the D-1 agonist SCH 23390. *Psychopharmacologia* **82**, 409–10.

Onali, P., Schwartz, J. P. and Costa, E. (1981). Dopaminergic modulation of adenylate cyclase stimulation by vasoactive intestinal peptide in anterior pituitary. *Proc. nat. Acad. Sci. USA* **78**, 6531–4.

Pendleton, R. G., Samler, L., Kaiser, C. and Ridley, P. T. (1978). Studies on renal dopamine receptors with a new agonist. *Eur. J. Pharmacol.* **51**, 19–28.

Rodbell, M. (1980). The role of hormone receptors and GTP-regulatory proteins in membrane transduction. *Nature (Lond.)* **284**, 17–22.

Seeman, P. (1980). Brain dopamine receptors. *Pharmacol. Rev.* **32**, 229–313.

Setler, P. E., Sarau, H. M., Zirkle, C. L. and Saunders, H. L. (1978). The central effects of a novel dopamine agonist. *Eur. J. Pharmacol.* **50**, 419–30.

Sibley, D. R., DeLean, A. and Creese, I. (1982). Anterior pituitary dopamine receptors; a demonstration of the interconvertible high and low affinity state of the D-2 dopamine receptor. *J. Biol. Chem.* **257**, 6351–61.

Snyder, S. H. and Goodman, R. R. (1980). Multiple neurotransmitter receptors. *J. Neurochem.* **35**, 5–15.

Sokoloff, P., Martres, M. P. and Schwartz, J. C. (1980). Three classes of dopamine receptor (D-2, D-3, D-4) identified by binding studies with ^{3}H-apomorphine and ^{3}H-doperidone. *Naunyn-Schmiedeberg's Arch. Pharmacol.* **315**, 89–102.

Spano, P. F., Govoni, S. and Trabucchi, M. (1978). Studies on the pharmacological properties of dopamine receptors in various areas of the central nervous system. In *Advances in Biochemistry and Psychopharmacology*, Vol. 19. P. J. Roberts, G. N. Woodruff and L. L. Iversen (eds), pp. 155–65. Raven Press, New York.

Trabucchi, M., Longoni, R., Fresia, R. and Spano, P. F. (1975). Sulpiride: a study of the effects on dopamine receptors in rat neostriatum and limbic forebrain. *Life Sci.* **17**, 1551–6.

Walaas, S. I. and Greengard, P. (1984). DARPP-32, a dopamine and adenosine 3'5'-monophosphate-regulated phosphoprotein enriched in dopamine-innervated brain regions. I. Regional and cellular distribution. *J. Neurosci.* **4**, 84–8.

Watling, K. J. and Dowling, J. E. (1981). I. Dopamine-sensitive adenylate cyclase in homogenates of carp retina. Effects of agonists, antagonist and ergots. *J. Neurochem.* **36**, 559–68.

Woodruff, G. N. (1978). Biochemical and pharmacological studies on dopamine receptors. In *Advances in Biochemistry and Psychopharmacology*, Vol. 19. P. J. Roberts, G. N. Woodruff and L. L. Iversen (eds), pp. 89–119. Raven Press, New York.

Woodruff, G. N. and Freedman, S. B. (1981). Binding of ^{3}H-sulpiride to purified rat striatal synaptic membranes. *Neurosci.* **6**, 407–10.

Woodruff, G. N. and Pinnock, R. D. (1981). Some central actions of ADTN, a potent dopamine receptor agonist. In *Apomorphine and Other Dopaminomimetics*, vol. 1. G. L. Gessa and G. U. Corsini (eds), pp. 241–52. Raven Press, New York.

3

Pharmacological characterisation of central dopamine receptors using functional criteria

R. Markstein

Preclinical Research, Sandoz Ltd, CH-4002 Basle, Switzerland

3.1. Introduction

Dopamine acts as a neurotransmitter in the central and probably also the peripheral nervous system of vertebrates. In the central nervous system, dopamine has been shown to be involved in the regulation of movements, emotional processes, endocrine systems and other physiological events. In the periphery, dopamine inhibits sympathetic neurotransmission, induces relaxation of renal, mesenteric and coronary arteries, and decreases intragastric pressure an oesophageal tone.

Presently, it is widely accepted that the various physiological effects of dopamine are mediated by more than one dopamine receptor type. However, there is much less agreement about their real number, pharmacological properties, physiological roles or terminology (for review see: Seeman, 1980; Beart, 1982; Creese, 1982). This problem is not only of scientific interest but is also of clinical importance. For instance, dopamine receptor agonists have been used to treat Parkinson's disease, and dopamine receptor antagonists to treat schizophrenia.

However, dopamine receptor agonists may not only counteract akinesia in Parkinson's disease, but also produce unwanted side effects such as hypotension, nausea, hallucinations and inhibition of prolactin secretion. Dopamine receptor antagonists in turn may not only exert antipsychotic effects, but often may also produce movement disorders. The development of new therapeutically active dopamine receptor agonists and antagonists without unwanted side-effects is therefore highly desirable. A prerequisite for such an attempt is a better knowledge of the pharmacological properties of the dopamine receptors involved in mediating the various physiological effects of dopamine in the central nervous system and the periphery.

3.2. Classification of dopamine receptors in the central nervous system

In the central nervous system, dopamine receptors have been subdivided on

Table 3.1.   Proposed classification of dopamine receptors in the central nervous system.

Criterion	*Authors*
Behavioural	
DA-1; DA-2	Costall and Naylor (1981)
DA_e; DA_i	Cools and van Rossum (1980)
Biochemical	
Linkage and non-linkage to adenylate cyclase; D-1; D-2	Kebabian and Calne (1979)
Ligand affinities: D-1, D-2, D-3, D-4	Sokoloff *et al.* (1980)
Anatomical	
Presynaptic: Postsynaptic	Carlsson (1975)

the basis of behavioural observations, biochemical criteria, affinity differences to radioligands, or anatomical localisation (Table 3.1).

However, many pitfalls are inherent in attempts to characterise receptors *in vivo* or by ligand binding studies. For instance, drug effects *in vivo* may be produced indirectly or by metabolites. Differences of drug actions may also be due to differences in absorption, distribution or metabolic conversion.

Binding studies usually have the disadvantage of not providing clear information about whether a drug is an agonist or an antagonist. Furthermore, the membrane constituents to which radioligands bind in a specific manner may not be true receptors. Moreover, differences in dopamine receptor-like binding sites may be simulated by differences in the physicochemical properties of the ligand used.

Dopamine receptors are not only present on elements postsynaptic to the dopamine neurones, but also occur on the dopaminergic neurones themselves. For these receptors which are localised on the soma, dendrites and axons of the dopamine neurone, the term dopamine autoreceptor has been suggested (Carlsson, 1975). These dopamine autoreceptors form part of a negative feedback system by which the release and synthesis of dopamine and the firing rate of the neurone are regulated (Aghajanian and Bunney, 1977; Starke *et al.*, 1978; Roth, 1979).

The recent classification of dopamine receptors into D-1 and D-2 types according to Kebabian and Calne (1979) has gained the greatest acceptance. By definition, D-1 receptors mediate stimulation of adenylate cyclase. They have a micromolar potency for dopamine and a low affinity for butyophenone derivates. D-2 receptors in turn are either unassociated with adenylate cyclase or mediate its inhibition. They have a nanomolar potency for dopamine and a high affinity for butyrophenone derivates.

The classical way to characterise a receptor is to quantify in an appropriate tissue preparation a response elicited by receptor activation, thereby allowing the establishment of rank orders of agonists and antagonists. In the central nervous system, dopamine mediates several responses suitable for quantifying drug effects. For instance, at the postsynaptic level, it has been shown to stimulate adenylate cyclase in brain regions having a dopaminergic innervation such as the striatum (Kebabian *et al.*, 1972), the nucleus accumbens, the tuberculum olfactorium, the median eminence (Clement-Cormier and Robinson, 1977), the hippocampus (Barbaccia *et al.*, 1983) and also the retina (Brown and Makman, 1972). In contrast, in the anterior pituitary of the rat, dopamine has been shown to inhibit basal or vasoactive intestinal peptide (VIP) stimulated adenylate cyclase activity (Onali *et al.*, 1981; Enjalbert and Bockaert, 1983), and in the intermediate lobe of the pituitary, dopamine-attenuated β-adrenergic receptor induced stimulation of adenylate cyclase (Cote *et al.*, 1981).

Furthermore, dopamine has been reported to modulate the release of various transmitters. For instance, in the striatum it has been shown to inhibit acetylcholine release (Stoof *et al.*, 1979) and in the hypothalamus to inhibit noradrenaline release (Galzin *et al.*, 1983). Despite the fact that these release-modulating receptors have a postsynaptic localisation with regard to the dopamine neurone, they are frequently designated as presynaptic dopamine receptors. To distinguish these dopamine receptors occurring on non-dopaminergic nerve terminals from those occurring on dopaminergic nerve terminals, the term presynaptic autoreceptor should be used for the latter.

At the presynaptic level, dopamine inhibits its own release (Starke *et al.*, 1978) and the activity of tyrosine hydroxylase, the rate-limiting enzyme of dopamine biosynthesis (Kehr *et al.*, 1972).

Any attempt to characterise receptor subtypes depends not only on the availability of a suitable test system, but also on the availability of drugs that distinguish between receptor subtypes. Recently, with SKF 38–393, a selective agonist, and with SCH 23–390, a selective antagonist, for dopamine receptors involved in stimulating adenylate cyclase in the rat striatum became available (Setler *et al.*, 1978; Christensen *et al.*, 1984). In contrast, bromocriptine and LY 141865 are considered to stimulate, and (—)-sulpiride to antagonise selectively dopamine receptors which, according to the nomenclature proposed by Kebabian and Calne (1979), are D-2 types. The following section describes an attempt to characterise dopamine receptors in the central nervous system by comparing drug effects on adenylate cyclase in homogenates of rat striatum, rat hippocampus, rat pituitary and bovine retina and on evoked dopamine and acetylcholine release from slices of rat striatum.

3.3. Characterisation of the dopamine receptor involved in stimulating adenylate cyclases

To study drug effects on adenylate cyclase activity a method was used which is based on that described by Kebabian *et al.* (1972). Tissue homogenates were preincubated with test substances and the enzyme reaction was initiated by adding the substrate ATP together with GTP, a necessary cofactor for the signal transfer from the receptor to the catalytic unit. The reaction was terminated by heating the mixture. Subsequently, cAMP content was measured by a radioimmunoassay.

Figure 3.1 illustrates the effects of various agonists and antagonists in the adenylate cyclase test in homogenates of rat striatum. Agonist effects were quantified by determining the maximal response and the concentration where half maximal stimulation occurred (EC_{50} value). The maximal stimulatory effect of a drug is expressed as a percentage of that of dopamine.

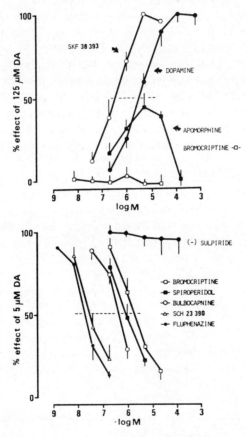

Fig. 3.1. Effects of various agonists on basal, and various antagonists on dopamine stimulated adenylate, cyclase activity in homogenates of rat striatum. Basal cAMP production was 117 ± 0.62 pmol per milligram of protein per minute, rising to 222.5 ± 0.98 pmol per milligram of protein per minute in the presence of 125 μM. (Means $\pm$ S.E. of 27 determinations.)

The apparent affinity of antagonists was calculated either from the shift of the dose–response curve of dopamine induced by a constant concentration of a competitive antagonist (Van Rossum, 1963) or from the IC_{50} value according to Cheng and Prusoff (1973). SKF 38-393 stimulated adenylate cyclase in homogenates of rat striatum with the same maximal effect as dopamine but had a lower EC_{50} value. Apomorphine induced only about 40 % of the maximal stimulatory effect of dopamine. Bromocriptine was devoid of stimulating effects but in turn showed measurable antagonistic properties. SCH 23-390 and fluphenazine were potent antagonists, whereas (−)-sulpiride was inactive (Fig. 3.1, Table 3.2). Similar results were obtained using homogenates of bovine retina. Figure 3.2 illustrates that there exists a significant correlation between affinities of various drugs estimated in the adenylate cyclase test of rat striatum and bovine retina.

Another significant correlation was found between drug effects on adenylate cyclase in homogenates of rat striatum and data reported from Watling and Dowling (1981), who used homogenates of carp retina (Fig. 3.2). Furthermore, similar affinities for dopamine, fluphenazine, chlorpromazine and clozapine were reported from other investigations using the adenylate cyclase test in homogenates of other rat brain areas receiving a dopaminergic innervation such as the nucleus accumbens, the tuberculum olfactorium or the median eminence (Clement-Cormier and Robinson, 1977). This suggests that there exists a class of dopamine receptors mediating stimulation of adenylate cyclase in different areas of the rat brain and the retina of different species which are pharmacologically very similar.

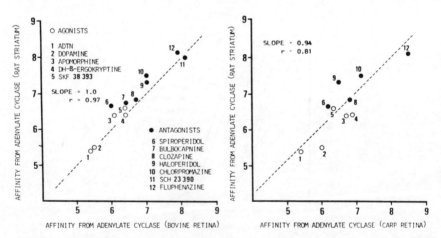

Fig. 3.2. Correlation between the affinities of various agonists (−log EC_{50}) and antagonists (−log K_i) to dopamine receptors mediating stimulation of adenylate cyclase in homogenates of rat striatum and bovine retina or carp retina. Data on carp retina taken from Watling and Dowling (1981). The correlation is statistically significant ($P < 0.001$).

According to Kebabian and Calne (1979), these dopamine receptors are D-1 receptors.

Rat hippocampus is another brain area receiving a dopaminergic innervation from the midbrain and possessing also a dopamine-sensitive adenylate cyclase (Scatton *et al.*, 1982; Barbaccia *et al.*, 1983). However, no significant correlation was found between the affinities of various drugs for dopamine receptors mediating stimulation of adenylate cyclase in hippocampus and striatum (see Table 3.2). For instance, dopamine had about a 10 times higher EC_{50} value in homogenates of rat hippocampus than in

Table 3.2. Comparison of drug effects in different functional dopamine receptor tests using preparations of rat brain.

Agonists	$Striatum^a$		Adenylate cyclase $Hippocampus^b$		$Anterior^c$ pituitary		Acetylcholine release $Striatum^d$	
	E_r	$-log\ EC_{50}$	E_r	$-log\ EC_{50}$	E_r	$-log\ EC_{50}$	$E_r{}^*$	$-log\ EC_{50}$
Dopamine	100	5·5	100	4·6	100	6·3	—	—
ADTN	82	5·5	74	5·1	—		−83	7·3
Apomorphine	40	6·4	66	6·1	100	7·0	−64	7·6
SKF 38-393	94	6·6	43	5·1	0	0	0	0
Isoprenaline	20	5·9	25	5·6	0	0	0	0
DH-Ergocornine	23	6·1	—		40	8·7	−70	8·5
DH-α-Ergokryptine	0		—		40	9·1	−33	8·7
DH-β-Ergokryptine	52	6·1	—		30	8·7	−51	8·4
Bromocriptine	0	$(pK_i=6·2)$	0		100	7·9	−62	8·3
CH 29-717	0		0		43	8·9	−76	8·3
LY 141865	0		0		100	6·8	−85	7·3
Lergotrile	0		—		100	7·0	−80	6·7
(−)3-PPP	0		—		—		—	
Antagonists			$-log(K_i)$					
SCH 23-390	8·0		6·9		—			5·8
Bulbocapnine	6·7		7·0		—			<5
Fluphenazine	8·1		7·0		7·8			7·6
Chlorpromazine	7·5		—		7·2			7·4
Spiroperidol	6·6		6·7		9·1			10·0
(−)Sulpiride	4·0		4·0		7·4			8·2
Clozapine	6·8		—		—			—
Haloperidol	7·3		—		8·0			9·1
Methiothepine	8·3		7·9		—			—

E_r, Maximal stimulation as percentage of that of dopamine.
$E_r{}^*$, Maximal inhibition as percentage of control (S_2/S_1) ratio.
[a,b] Adenylate cyclase assay was performed in the presence of 10 μM GTP.
[c] Data from Enjalbert and Bockaert (1983).
[d] Rats were pretreated twice with reserpine (2.5 mg kg^{-1} s.c.) 12 and 18 h before sacrifice.

striatum. Therefore, it cannot be excluded that dopamine receptors stimulating adenylate cyclase in rat hippocampus are pharmacologically different from those stimulating adenylate cyclase activity in homogenates of rat striatum, and bovine or carp retina. However, in order to prove the existence of subtypes of D-1 receptors, further investigations would be required with drugs which will more readily discriminate between potential subtypes than those used in the present investigation.

3.4. Characterisation of dopamine receptors mediating inhbition of neurotransmitter release

In the striatum, dopamine has been shown to inhibit its own release and that of acetylcholine (see Starke *et al.*, 1983). The receptors mediating these effects appear to be independent of an adenylate cyclase (see Eitan and Herschkowitz, 1977; Euvrard *et al.*, 1979).

To measure drug effects on dopamine receptors mediating inhibition of evoked dopamine or acetylcholine release in the striatum, rat striatal slices were preincubated either with [^{3}H]dopamine or [^{3}H]choline and superfused with oxygenated Krebs–bicarbonate buffer. Under normal conditions, only small amounts of tritium activity are released from such prelabelled slices. Electrical stimulation caused a marked increase in tritium efflux which was Ca^{2+} dependent and tetrodotoxin sensitive, and was, therefore, considered to reflect action potential-induced release of newly taken up [^{3}H]dopamine or [^{3}H]acetylcholine, newly synthesised from [^{3}H]choline. In the experiments, neurotransmitter release was elicited by two periods of electrical stimulation (S_1, S_2) at 2 Hz for 2 min. EC_{50} values of agonists correspond to the drug concentrations that produce 50 % of the maximal inhibition observed with the respective drug. K_i values of antagonists were calculated either from IC_{50} values against apomorphine or from the shift of the dose–response curve of apomorphine induced by a constant concentration of the antagonist.

Figure 3.3 illustrates the effects of various drugs on electrically evoked acetylcholine release from rat striatal slices. To exclude the possiblity that endogeneous dopamine interferes with the drug effects, slices of reserpinised animals were used.

Apomorphine, bromocriptine, LY 141865 and ADTN decreased electrically evoked acetylcholine release in a dose-dependent manner, which is consistent with agonistic effects. In contrast, SKF 38-393, which was very potent in stimulating adenylate cyclase of rat striatum, had no effect on evoked acetylcholine release from rat striatal slices. Spiroperidol, (−)-sulpiride, haloperidol and fluphenazine antagonised the effect of apomorphine at concentrations around 10^{-9} M. In contrast, SCH 23-390 and bulbocapnine had detectable antagonistic effects only at concentrations as high as 10^{-6} M. Similar results were obtained when the effects of the same

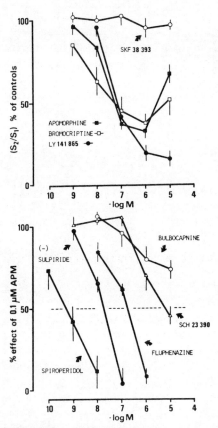

Fig. 3.3. Effect of various agonists alone and various antagonists in combination with 0·1 μM apomorphine on electrically evoked acetylcholine release from striatal slices of reserpine pretreated rats. (S_2/S_1) represents the ratio of electrically evoked release of [³H]acetylcholine; S_1 represents the percentage of total tissue tritium activity released by a first period of electrical stimulation (2H₂, 2 ms; 2 min); and S_2 that released by a second stimulation period in the presence of test drugs. (S_2/S_1) ratio in control experiments was 0·86 ± 0·02. (Means ± S.E. of 27 experiments.) For experimental details, see Markstein and Lahaye (1983).

drugs on dopamine release from rat striatal slices were measured. Figure 3.4 shows that there also exists a significant correlation between the affinities of various dopamine receptor agonists and antagonists estimated in the dopamine and acetylcholine release test. Several other dopamine receptor agonists and antagonists have also been reported to be unable to distinguish between presynaptic release-inhibiting dopamine autoreceptors and post-synaptic dopamine receptors inhibiting acetylcholine release in the caudate nucleus of the rabbit (Starke *et al.*, 1983). Interestingly, the (−)1-enantiomer of 3-PPP, a new compound claimed to be a selective dopamine autoreceptor

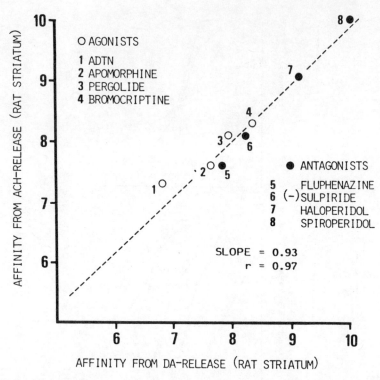

Fig. 3.4. Correlation between the affinities of various agonists ($-\log EC_{50}$) and antagonists ($-\log K_i$) to dopamine receptors mediating inhibition of electrically evoked dopamine and acetylcholine release from rat striatal slices. Identical experimental conditions were used to measure drug effects on electrically stimulated dopamine and acetylcholine release, except that reserpine-pretreated animals were used in the acetylcholine release experiments. The correlation is statistically significant ($2P < 0.01$). For experimental details, see Markstein and Lahaye, 1983.

agonist (Hjorth *et al.*, 1981, 1982), had no effect either on dopamine or on acetylcholine release from rat striatal slices.

Altogether these findings suggest that presynaptic release-inhibiting dopamine autoreceptors and postsynaptic dopamine receptors inhibiting acetylcholine release in the striatum are closely similar. It therefore remains doubtful whether compounds can be developed which, at least in the striatum, have a selective effect on presynaptic, release-inhibiting dopamine autoreceptors. However, it cannot be excluded that there exist presynaptic dopamine autoreceptors pharmacologically distinct from those inhibiting dopamine release. For instance, there is considerable evidence that the activity of tyrosine hydroxylase in dopaminergic terminals in the striatum is also regulated by dopamine autoreceptors (Walters and Roth, 1976) which are possibly pharmacologically distinct from those regulating dopamine

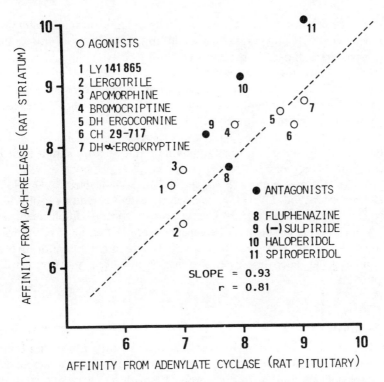

Fig. 3.5. Correlation between drug affinities for the dopamine receptor mediating inhibition of acetylcholine release in the striatum and that mediating inhibition of adenylate cyclase in the anterior pituitary of the rat. Data for adenylate cyclase inhibition are taken from Enjalbert and Bockaert (1983). The correlation is statistically significant ($2P < 0.01$).

release (Westfall *et al.*, 1976). For instance, 3-PPP, which had no effect on dopamine release, has been reported to inhibit tyrosine hydroxylase activity in rat striatal synaptosomes (Haubrich and Pflueger, 1982). In homogenates of the anterior pituitary of the rat, there exists a dopamine receptor mediating inhibition of adenylate cyclase activity (Enjalbert and Bockaert, 1983).

Figure 3.5 illustrates that a significant correlation also exists between data on drug affinities to dopamine receptors mediating inhibition of acetylcholine release in the striatum and the data reported by Enjalbert and Bockaert (1983) who used the adenylate cyclase test on rat pituitary.

3.5. Conclusion

The present findings suggest that the pharmacological properties of dopamine receptors are not dependent on the anatomical localisation or linkage to an effector system. Based on results obtained in funtional *in vitro* tests, it appears that presynaptic, release-inhibiting dopamine autoreceptors and postsynaptic dopamine receptors inhibiting acetylcholine release in the rat striatum or adenylate cyclase activity in the rat pituitary are pharmacologically very similar. The properties of these receptors agree well with those of the D-2 receptors as defined by Kebabian and Calne (1979). The dopamine receptors involved in stimulating adenylate cyclase in homogenates of rat striatum are pharmacologically very similar to those stimulating adenylate cyclase in the bovine retina or carp retina fulfil the criteria of D-1 receptors.

SKF 38-393 is a selective agonist and SCH 23-390 a selective antagonist for D-1 receptors, and LY 141865 is a selective agonist and (−)-sulpiride a selective antagonist for the D-2-like receptors. In contrast, fluphenazine and chlorpromazine are unable to distinguish between D-1 and D-2 receptors. In homogenates of rat hippocampus, there exists a dopamine receptor involved in stimulating an adenylate cyclase which is possibly a subtype of a D-1 receptor.

3.6. Summary

The classical way to characterise receptors is to quantify a response elicited by receptor activation, allowing rank orders of agonists and antagonists to be established. Using such an approach, evidence was obtained that dopamine receptors involved in regulating evoked dopamine and acetylcholine release in the rat striatum and those inhibiting adenylate cyclase activity in the anterior pituitary are pharmacologically very similar and fulfil the criteria of D-2 receptors. The dopamine receptors stimulating adenylate cyclase in rat striatum, and bovine and carp retina are pharmacologically similar and fulfil the criteria for D-1 receptors. In contrast, in homogenates of rat hippocampus there exist dopamine receptors involved in stimulating an adenylate cyclase which are probably subtypes of D-1 receptors.

References

Aghajanian, G. K. and Bunney, B. S. (1977). Dopamine autroreceptors: pharmacological characterization by micro-iontophoretic single cell recording studies. *Naunyn-Schmiedeberg's Arch. Pharmacol.* **291**, 1–7.

Barbaccia, M. L., Brunello, N., Chuan, G. M. and Costa, E. (1983). Serotonin-elicited amplification of adenylate cyclase activity in hippocampal membranes from adult rats. *J. Neurochem.* **40**, 1671–9.

Beart, P. M. (1982). Multiple dopamine receptors—new vistas. *TIPS* **3**, 100–2.

Brown, J. H. and Makman, M. H. (1972). Stimulation by dopamine of adenylate cyclase in retinal homogenates and of adenosine 3′5′ cyclic monophosphate formation in intact retina. *Proc. nat. Acad. Sci. USA* **69**, 535–43.

Carlsson, A. (1975). Receptor-mediated control of dopamine metabolism. In *Pre- and Postsynaptic Receptors*. E. Usdin and W. E. Bunney (eds), pp. 49–63, Marcel Dekker, New York.

Cheng, Y. C. and Prusoff, W. H. (1973). Relationship between the inhibition constant (K_i) and the concentration of inhibitor which causes 50 per cent inhibition (I_{50}) of an enzymatic reaction. *Biochem. Pharmacol.* **22**, 3099–108.

Christensen, A. V., Arnt, J., Hyttel, J., Lars, J.-J. and Svendsen, O. (1984). Pharmacological effects of a specific D-1 antagonist SCH 23-390 in comparison with neuroleptics. *Life Sci.* **34**, 1529–40.

Clement-Cormier, Y. C. and Robinson, G. A. (1977). Adenylate cyclase from various dopaminergic areas of the brain and the action of antipsychotic drugs. *Biochem. Pharmacol.* **26**, 1719–22.

Cools, A. R. and van Rossum, J. M. (1980). Multiple receptors from brain dopamine in behavior regulation: Concept of dopamine-E and dopamine-I receptors. *Life Sci.* **27**, 1237–53.

Costall, B. and Naylor, R. J. (1981). The hypothesis of different dopamine receptor mechanisms. *Life Sci.* **28**, 215–29.

Cote, T. E., Crewe, C. W. and Kebabian J. W. (1981). Stimulation of D_2 dopamine receptor in the intermediate lobe of the rat pituitary decreases the responsiveness of the β-adrenoceptor: biochemical mechanism. *Endocrinology* **108**, 420–6.

Creese, J. (1982). Dopamine receptors explained. *TINS* **5**, 40–3.

Eitan, A. and Hershkowitz, M. (1977). The effects of dibutyryl cyclic AMP, theophylline and papaverine on the release of ^{3}H-catecholamines from rat brain striatal and cortical synaptosomes.

Enjalbert, A. and Bockaert, J. (1983). Pharmacological characterization of the D_2 dopamine receptor negatively coupled with adenylate cyclase in rat anterior pituitary. *Mol. Pharmacol.* **23**, 576–584.

Euvrard, C., Premont, J., Oberlander, C., Boissier, J. R. and Bockaert, J. (1979). Is dopamine-sensitive adenylate cyclase involved in regulating the activity of striatal cholinergic neurons? *Naunyn-Schmiedeberg's Arch. Pharmacol.* **309**, 241–5.

Galzin, A. M., Dubocovich, M. L. and Langer, S. L. (1983). Presynaptic inhibition by dopamine receptor agonists of noradrenergic neurotransmission in the rabbit hypothalamus. *J. Pharmacol. exp. Ther.* **221**, 461–71.

Haubrich, D. R. and Pflueger, A. B. (1982). The autoreceptor control of dopamine synthesis. An *in vitro* and *in vivo* comparison of dopamine agonist. *Mol. Pharmacol.* **21**, 114–20.

Hjorth, S., Carlsson, A., Wikström, H., Lindberg, P., Sanchez, D., Hacksell, U., Arvidsson, L. E., Svenson, U. and Nilsson, J. L. G. (1981). 3-PPP, a new centrally acting DA receptor agonist with selectivity for autoreceptors. *Life Sci.* **28**, 1225–38.

Hjorth, S., Carlsson, A., Clark, D., Svensson, K., Wikström, H., Sanchez, D., Lindberg, P., Hacksell, U., Arvidsson, L. E., Johansson, A. and Nilsson, J. L. G. (1982). Central dopamine receptor agonist and antagonist actions of the enantiomers of 3-(3-hydroxyphenyl)-N-n-propylpiperidine (3-PPP). In *Basic and Clinical Aspects of Molecular Neurobiology. Proceedings of the Fourth Meeting of the European Society for Neurochemistry*. A. M. Gruffrida Stella, G. Combos, G. Benzi and H. S. Bacheland (eds), pp. 327. Fondazione Internazionale Menarini, Milano.

Kebabian, J. W. and Calne, D. B. (1979). Multiple receptors for dopamine. *Nature* (*Lond.*) **277**, 93–6.

Kebabian, J. W., Petzold, G. L. and Greengard, P. (1972). Dopamine-sensitive adenylate cyclase in caudate nucleus of rat brain, and its similarity to the 'dopamine' receptor. *Proc. nat. Acad. Sci. USA* **69**, 2145–9.

Kehr, W., Carlsson, A., Lindquist, M., Magnusson, T. and Atack, C. (1972). Evidence for a receptor-mediated feedback control of striatal tyrosine hydroxylase activity. *J. Pharm. Pharmacol.* **24**, 744–7.

Markstein, R. and Lahaye, D. (1983). In vitro effect of the racemic mixture and the (−)enantiomer of N-n-propyl-3-(3-hydroxyphenyl)-piperidin (3-PPP) on postsynaptic dopamine receptors and on a presynaptic dopamine autoreceptor. *J. Neural Transm.* **58**, 43–53.

Onali, P., Schwartz, J. P. and Costa, E. (1981). Dopaminergic modulation of adenylate cyclase stimulation by vasoactive intestinal peptide in anterior pituitary. *Proc. nat. Acad. Sci. USA* **78**, 6531–4.

Roth, R. H. (1979). Dopamine autoreceptors: pharmacology and comparison with postsynaptic dopamine receptors. *Commun. Psychopharmacol.* **3**, 429–45.

Scatton, B., D'Ambrosio, A., Javoy-Agid, F., Agid, Y., Bischoff, S., Simon, H. and Le Moal, M. (1982). Evidence for the existence of a dopaminergic innervation of the rat and human hippocampal formation. In *Advances in Dopamine Research*. M. Kohsaka, T. Shonmori, Y. Tsukada and G. M. Woodruff (eds), pp. 377–82. Pergamon Press, Oxford.

Seeman, P. (1980). Brain dopamine receptors. *Pharmacol. Rev.* **32**, 229–313.

Setler, P. E., Sarau, H. M., Zirkle, C. L. and Saunders, H. L. (1978). The central effects of a novel dopamine agonist. *Eur. J. Pharmacol.* **50**, 419–30.

Sokoloff, P., Martres, M. P. and Schwartz, J. C. (1980). Three classes of dopamine receptors (D-2, D-3, D-2) identified by binding studies with ^{3}H-apomorphine and ^{3}H-doperidone. *Naunyn-Schmiedeberg's Arch. Pharmacol.* **315**, 89–102.

Starke, K., Reimann, W., Zumstein, A. and Hertting, G. (1978). Effect of dopamine receptor agonists and antagonists on release of dopamine in the rabbit caudate nucleus in vitro. *Naunyn-Schmiedeberg's Arch. Pharmacol.* **305**, 27–36.

Starke, K., Späth, L., Lang, J. D. and Adelung, C. (1983). Further functional in vitro comparison of pre- and postsynaptic dopamine receptors in the rabbit caudate nucleus. *Naunyn-Schmiedeberg's Arch. Pharmacol.* **323**, 298–306.

Stoof, J. C., Thieme, R. E., Vrijmoed-de Vriesand and Mulder, A. (1979). *In vitro* acetylcholine release from rat caudate nucleus as a new model for testing drugs with dopamine-receptor activity. *Naunyn-Schmiedeberg's Arch. Pharmacol.* **309**, 119–24.

Van Rossum, J. M. (1963). Cumulative dose–response curves. II. Technique for the making of dose-response curves in isolated organs and the evaluation of drug parameters. *Arch. int. Pharmacodyn.* **143**, 299–330.

Walters, J. R. and R. H. Roth, (1976). Dopaminergic neurones: An in vivo system for measuring drug interactions with presynaptic receptors. *Naunyn-Schmiedeberg's Arch. Pharmacol.* **296**, 5–14.

Watling, K. J. and Dowling, J. E. (1981). Dopaminergic mechanisms in the teleost retina. I. Dopamine-sensitve adenylate cyclase in homogenates of carp retina; effects of agonists, antagonists and ergots. *J. Neurochem.* **36**, 559–68.

Westfall, T., Besson, M., Giorguieff, M. F. and Glowinski, J. (1976). The role of presynaptic receptors in the release and synthesis of ^{3}H-dopamine by slices of rat striatum. *Naunyn-Schmiedeberg's Arch. Pharmacol.* **299**, 279–87.

4

Dopamine systems in invertebrates

G. A. Cottrell

Department of Physiology and Pharmacology, University of St. Andrews, Fife KY16 9TS, UK

4.1. Introduction

Dopamine occurs commonly in invertebrates, often in high concentrations. In many of the species of the different phyla investigated, dopamine is the predominant catecholamine within the nervous system (Cottrell, 1967; Cottrell and Laverack, 1968).

Studies on the cellular and membrane actions of neuronal chemical messengers are greatly facilitated in some invertebrate species by the presence of large and readily identifiable neurones (see, for example Cottrell, 1980). The most detailed studies to date have been conducted on identified neurones of gastropod molluscs. Many central neurones of these animals are very sensitive to applied dopamine (Walker *et al.*, 1968; Ascher, 1972). Some neurones are excited, others inhibited, and some show a biphasic response to iontophoresed dopamine (Ascher, 1972).

An amine–histofluorescence study by Marsden and Kerkut (1970) of the central ganglia of the water snail *Planorbis corneus* revealed a large primary catecholamine-containing neurone in the left pedal ganglion. Subsequent chemical analysis of the contents of this neurone established the presence of dopamine but no noradrenaline (Powell and Cottrell, 1974). Because of the lack of a simple dopaminergic system in the vertebrates comparable to the peripheral cholinergic or noradrenergic systems, a detailed study has been made of the anatomy and synaptic connexions made by this neurone, labelled for convenience the giant dopamine neurone (GDN). This work forms the major subject of this chapter.

4.2. Giant dopamine neurone system

4.2.1. *Location and amine content of the GDN*

In *Planorbis corneus*, the GDN lies on the posterio-ventral surface of the left pedal ganglion (Fig. 4.1). In most specimens, it is the largest neurone

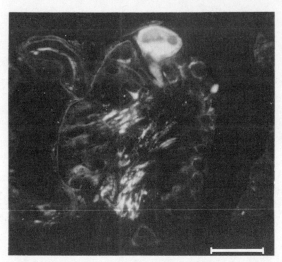

Fig. 4.1. Section through the left pedal ganglion of *Planorbis corneus* showing formaldehyde-induced fluorescence in the GDN, seen at the top of the section. Dopamine fluorescence is also seen in the central part of the ganglion, i.e. the neuropile. Scale bar 200 μm.

perikaryon in this region and therefore may be readily identified in living preparations.

Analysis of individually dissected GDN perikarya revealed a level of dopamine of $5 \cdot 4 \pm 0 \cdot 6$ pmol per perikaryon, but no detectable noradrenaline—i.e. less than about 1 pmol per perikaryon (Powell and Cottrell, 1974). The concentration of dopamine in the GDN perikaryon is therefore in the millimolar range, since the perikaryon approximates to a sphere with a diameter about 180 μm, although, of course, a large proportion of the dopamine is contained in vesicles (see below).

4.2.2. *Processes leading from the GDN perikaryon*

Dye and isotope injection experiments of the perikaryon have revealed a complex system of processes leading from the left pedal ganglion through the left and right pleural and parietal ganglia and the visceral ganglion (Fig. 4.2). Both of the main processes were revealed autoradiographically following injection of the perikaryon with [3H]-dopamine, but only the pathway passing through the left ganglia and into the visceral ganglion was detected with injection of dyes such as cobalt and Procion Yellow (Berry *et*

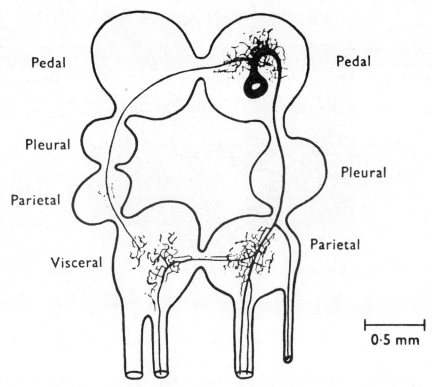

Fig. 4.2. Processes of the GDN observed from the ventral surface of the ganglionic ring. Note the two major processes and the large number of fine terminals to the visceral and left parietal ganglia. (Taken from Berry *et al.*, 1974*a*.)

al., 1974a). A dense network of fine processes was observed in both the visceral and right parietal ganglia, suggesting the presence of numerous presynaptic terminals of the GDN in these ganglia.

Probable dopaminergic terminals have been elegantly studied using electron microscopic autoradiographic techniques by Pentreath and Berry (1975). After intracellular injection of the GDN perikaryon with small amounts of tritiated dopamine, radioactivity was observed over small granulated vesicles in fine axonal processes. Such processes have the appearance of *en passant* terminals (Fig. 4.3).

Uptake experiments, again made using autoradiographic techniques, with a number of tritiated transmitters and their precursors, have shown that the GDN (but also some 5HT-containing neurones) takes up L-DOPA, whereas tyramine is more selectively accumulated by the GDN. Tyrosine appeared to be accumulated non-selectively by many neurones (Turner and Cottrell, 1978).

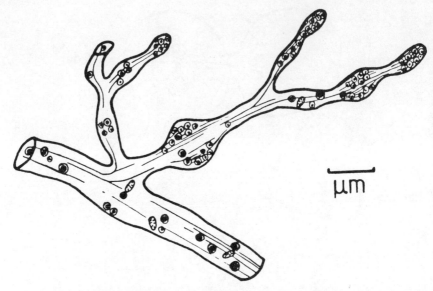

Fig. 4.3. Main anatomical features of the terminal process of the GDN in the visceral and parietal ganglia. The terminal branches are often varicose in appearance and contain numerous dense-cored vesicles, ranging from 50 to 250 nm. (Scale bar represents 1 μm.) (Taken from Pentreath and Berry, 1975.)

4.2.3. *Electrical properties and chemical sensitivity of the GDN*

The resting membrane potential of the GDN is usually close to -50 mV, recorded with KCl, K_2SO_4 and K-acetate electrodes. Frequently, overshooting action potentials are observed as well as ongoing 'spontaneous' synaptic activity. In some cells, damaged in some way, only small axon spikes are recorded. In many preparations, it is necessary to hold the membrane potential more negative than the resting level to block impulse activity.

The GDN is depolarised by acetylcholine (ACh) and 5-hydroxytryptamine (51-HT), and hyperpolarised by γ-amino butyric acid (GABA) (Logan and Cottrell, 1975).

4.3. Synaptic responses mediated by the GDN and influence of dopamine on the follower neurones

Small synaptic responses are recorded from several neurones in the visceral and left parietal ganglia in response to activation of the GDN (Fig. 4.4). Four different types of response are seen: (1) slow hyperpolarising potentials, (2) slow depolarising potentials, (3) fast depolarising potentials, (4) biphasic depolarising–hyperpolarising potentials. Each of these is

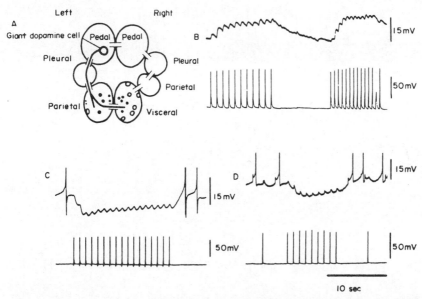

Fig. 4.4. (A) View of the ganglionic ring from the *dorsal* surface showing only one of the main branches, the one passing through the left ganglia. Recordings B, C and D show GDN evoked e.p.s.p.s, i.p.s.p.s and biphasic potentials, respectively. Recording B was made with the parietal follower cell hyperpolarised to −90 mV to accentuate the e.p.s.p.s. The follower cells of C and D were located in the visceral ganglion. (Taken from Cottrell *et al.*, 1974.)

described in detail below and compared with the responses evoked by dopamine iontophoresis on to these same follower neurones.

Evidence has been obtained using a number of criteria that each type of synaptic potential is mediated directly by a monosynaptic connection from the GDN (see Berry and Cottrell, 1973a; Berry and Pentreath, 1976).

4.3.1. *Hyperpolarising synaptic potentials*

Many cells on the dorsal surface of the visceral ganglion show an i.p.s.p. following each spike in the GDN (Fig. 4.5). Recordings have been made from as many as fifteen different postsynaptic neurones responding in this way in a single preparation. GDN-evoked i.p.s.p.s range up to 7 mV in amplitude and have a duration of 0·3 to 5 s. Sometimes individual i.p.s.p.s cannot be recorded but a smooth hyperpolarisation is seen when the GDN is stimulated repetitively at frequencies of about 3 Hz or above. Dopamine applied to such a follower cell also causes hyperpolarisation and inhibition (see Fig. 4.5).

Many of the neurones receiving the inhibitory input are connected to one another by non-rectifying electrotonic synapses, but the strength of coupling

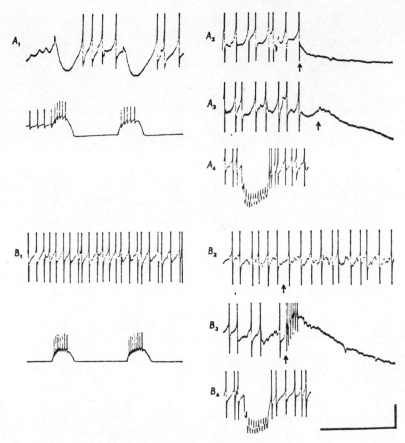

Fig. 4.5.   This complex figure, taken from Berry and Cottrell (1975), shows GDN-induced hyperpolarisation on a neurone of the visceral ganglion, A_1, and the response to applied dopamine, A_2. Also shown are responses to applied glutamate, A_3, and repetitive stimulation of a parietal nerve, A_4, both of which serve as controls for testing the effect of ergometrine. Recordings B were made during exposure to 2·5 μM ergometrine. Note that the response to GDN stimulation and dopamine are blocked, whereas the responses to glutamate, B_3, and nerve trunk stimulation, B_4, (both non-dopaminergic) are not blocked. Calibrations 10 s, 110 mV (presynaptic) and 10 mV (postsynaptic).

is very variable. Often, artificially increasing the membrane potential reduces the amplitude of the i.p.s.p.s and the response to dopamine, but rarely can these responses be inverted. Difficulty in inverting the responses is probably due to the distant location of the synaptic connections and the coupling occurring among the follower cells (see Fig. 4.6). Certainly the reduction in i.p.s.p.s and responses to dopamine with hyperpolarisation is consistent with mediation by an increase in membrane conductance. When conductance changes were estimated by injecting brief hyperpolarisation

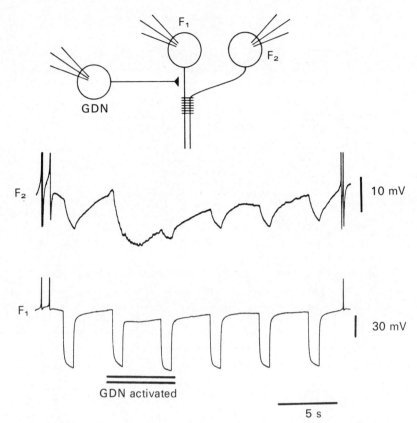

Fig. 4.6. The diagram at the top shows the arrangement of recording electrodes. The hyperpolarising pulse applied to the follower neurone F_1 was transmitted to F_2 by an electrotonic junction. During activation of the GDN (shown by the double lines) there was only a small reduction in the voltage deflections in F_1 but a much larger reduction in the pulses electronically transmitted to F_2, suggesting that the connection made by the GDN on to these two cells is close to the site of the electronic junction. (Modified from Berry and Cottrell, 1979.)

currents, a small fall in input resistance did in fact occur. In order to gain more information on conductance charges, the transmission of applied current between the perikarya of neighbouring postsynaptic neurones was observed during stimulation of the GDN and with dopamine application. With both responses the electrotonically transmitted pulses were greatly attenuated compared with the reduction in these in the neurone in which the current was applied—see Figs 4.6 and 4.7 (Berry and Cottrell, 1979). This indicates a large increase in conductance in the region close to the electrotonic connections. Ion substitution experiments, postsynaptic injection of tetraethylammonium and cooling suggest that the hyperpolarising responses are mediated by potassium ions.

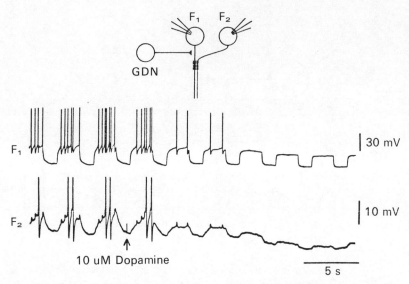

Fig. 4.7.   Brief hyperpolarising current pulses were applied to follower neurone F_1. These pulses resulted in large voltage deflections in F_1 and small ones in F_2, which was electronically coupled to F_1. Application of dopamine hyperpolarised both follower neurones but produced a greater reduction in pulse size in F_2 than in F_1, indicating that dopamine produced large increase in membrane conductance at a region distant from the perikaryon. (From Berry and Cottrell, 1979.)

In a significant proportion of the follower neurones, artificial hyper-polarisation produced an anomalous *potentiation* of the i.p.s.p.s (Fig. 4.8). At potential as high as -100 mV or more, the i.p.s.p.s were very much larger than at the resting membrane level. This observatiohn suggests a conductance decrease and it is possible that these i.p.s.p.s are brought about by a different ionic mechanism from those above. However, direct hyperpolarisation of these neurones produced a sharp increase in membrane resistance and it was often possible to detect a small conductance increase during input from the GDN in spite of the fact that the resulting hyperpolarisation tended to increase the membrane resistance. It is possible that the response results from a combination of ionic mechanisms. The electronic coupling among the follower cells complicates the analysis.

4.3.2.   *Slow excitatory postsynaptic potentials*
Stimulation of the GDN produces e.p.s.p.s in several neurones of the parietal ganglia. The e.p.s.p.s are about 1 s in duration. The response to iontophoresed dopamine is slow, reaching a peak at about 10 s and lasting up to 1 min (Fig. 9). Bath application of dopamine leads to prolonged desensitisation. The amplitude of these e.p.s.p.s is increased by artificial

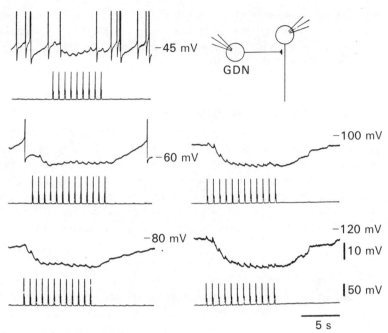

Fig. 4.8. An example of the anomalous increase in the amplitude of the hyperpolarising synaptic response with *increasing* membrane potential of the follower neurone. The arrangement of electrodes is shown in the top right. In this particular example the response was biphasic. (From Berry and Cottrell, 1979.)

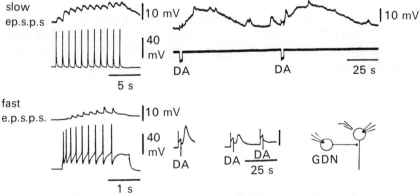

Fig. 4.9. Example of slow and fast responses to iontophoresed dopamine on neurones which show the slow and fast e.p.s.p.s respectively (shown on left). The slow responses are shown above and the fast ones below. In the records showing the fast iontophoretic response, there is a large current artifact. Here, the second application of dopamine was made after the first and produced a response which was reduced compared with the first. The third application produced a negligible response due to rapid desensitisation. Note the different time scales on the record of the GDN mediated e.p.s.p.s.

hyperpolarisation, suggesting that they are mediated by an increase in conductance, probably to more than one ion. A conductance increase underlying the response was also indicated from pulse injection experiments (Berry and Cottrell, 1975, 1979).

4.3.3. *Fast excitatory postsynaptic potentials*

In some postsynaptic neurones e.p.s.p.s of a shorter duration, about 150 ms, have been observed. These often have considerable variation in amplitude in any one neurone. Iontophoretic application of dopamine on such neurones produces a depolarising response which peaks at 3 s and lasts for 10 s or less. This response was much more rapidly desensitised than that observed on cells showing the slower e.p.s.p. As with the slow e.p.s.p., increasing the membrane potential increases the amplitude of the fast e.p.s.p.s.

4.3.4. *Biphasic synaptic response*

In some cells of the visceral ganglion, a biphasic synaptic potential, consisting of a depolarising then a hyperpolarising part, is observed following each action potential elicited in the GDN. The size of the depolarising phase is increased as the potential is increased, as with the e.p.s.p.s, whereas the hyperpolarising phase is generally decreased. The amplitude of the depolarising phase rarely exceeds 2 mV. Although the first few depolarisations often summated during a burst of GDN activity, they never appear to shift the membrane potential to the threshold for firing. During a prolonged burst of GDN spikes, the overall effect is hyperpolarising and inhibitory (Berry and Cottrell, 1975). With careful positioning of the iontophoretic electrode (cf. Ascher, 1972), it is often possible to mimic the biphasic synaptic potential with local application of dopamine (Berry and Cottrell, 1979).

The depolarising phase of each biphasic potential appears to be identical to the fast e.p.s.p.; both are rapidly desensitised and blocked by curare (see below). The hyperpolarising phase appears to be due to an increased conductance to potassium ions, as is the i.p.s.p. described above.

4.3.5. *GDN of Lymnaea stagnalis*

A large homologous dopamine-containing neurone has been detected in another water snail, *Lymnaea stagnalis* (Cottrell *et al.*, 1979). The amine content of the perikaryon of this neurone is similar to that of the GDN of *Planorbis*. The morphology and synaptic connections of the *Lymnaea* GDN, which are very similar to those of *Planorbis*, have been studied by Haydon and Winlow (1981) and Winlow *et al.* (1981) (see ch. 9 for further details).

4.4. **Pharmacology of the different responses observed in** *Planorbis*

4.4.1. *Ergometrine*

Ergometrine potently antagonises dopamine on snail neurones (Walker *et al.*, 1971) but may also produce a conductance change (Ascher, 1972) and thus be non-specific in higher doses. In low doses (1 to 10 μM), ergometrine reduced or completely abolished the i.p.s.p.s in a selective manner (Fig. 4.5) (Berry and Cottrell, 1975). In slightly higher doses (10 μM and above), ergometrine also had an antagonistic effect on the e.p.s.p.s, whereas ACh-induced depolarisation is not affected, suggesting a selective action of ergometrine on dopamine-induced depolarisations too.

4.4.2. *Curare*

Depolarising responses to a number of transmitter agents are known to be antagonised in molluscan neurones by curare, probably by a blocking action on ion channels (Carpenter *et al.*, 1977). In *Planorbis* the e.p.s.p.s produced by stimulating the GDN were antagonised with 100 μM curare. The depolarising phase of the biphasic response was also blocked with 100 μM curare but the hyperpolarising phase was unaffected even at 1 mM curare (Berry and Cottrell, 1975).

4.4.3. *6-Hydroxydopamine*

During the course of a series of experiments on the effects of 6-hydroxydopamine on the amine content of the GDN, it was observed that this drug also had an effect on the postsynaptic receptors. Exposure of preparations to a concentration of 250 μM to 1 mM 6-hydroxydopamine quickly abolished the GDN-mediated hyperpolarising synaptic potential and the hyperpolarising response to dopamine; both recovered after washing for 10–60 min. A similar but weaker effect was observed on the dopaminergic e.p.s.p.s. The effects appeared to be specific because non-dopaminergic responses were not antagonised in the same way (Berry and Cottrell, 1973*b*; Berry *et al.*, 1974).

4.4.4. *Neuroleptic drugs*

The effects of haloperidol, fluphenazine and metoclopramide have been tested on both excitatory and inhibitory responses (McDonald and Berry, 1971). Some blocking effect of these drugs was seen at concentrations ranging from 50 to 100 μM but the effects were often variable.

4.4.5. *Antagonist 'D-1' and 'D-2' receptors*

Stoof *et al.* (1984) have recently investigated the possibility that snail dopamine receptors may be separated using the classification made in

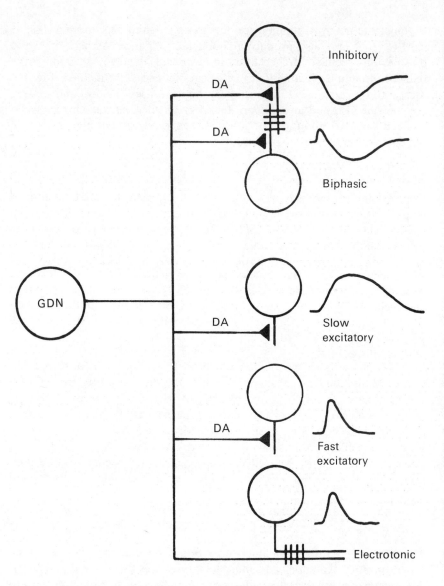

Fig. 4.10. Summary of the different types of postsynaptic potentials mediated by the dopamine neurone. *Note*: A few examples of electronic connections have been detected and are indicated at the bottom of the figure; they have not been discussed here. (See Berry and Cottrell, 1979.) DA = dopamine.

Table 4.1. Evidence for D-1 and D-2 receptors in *Lymnaea*.

Receptor	D-1	D-2
Agonist	SKF 38393 Increased the excitability of the GHCs	LY-14865 Hyperpolarised the GHCs
Antagonist	SCH-23390 Blocked increase in excitability produced by SKF-38393	(−)-Sulpiride Blocked hyperpolarising effect of LY-14865 and dopamine

vertebrates of D-1 (linked to adenylate cyclase in a stimulatory way) and D-2 (either unlinked or inhibitory on adenylate cyclase) (Stoof and Kebabian, 1981). Evidence was obtained for both D-1 and D-2 receptors on the growth hormone producing neurones (GHCs) of *Lymnaea stagnalis* which are hyperpolarised by iontophoresed dopamine. Their data are summarised in Table 4.1. Therefore, the hyperpolarising action of dopamine on these cells appears to be mediated by D-2 type receptors.

Stoof *et al.* (1984) also showed that IBMX (phosphodiesterase inhibitor) and cyclic AMP increase excitability of the GHCs in line with the view that the D-1 agonists are working via the adenylate cyclase system in these neurones (see also ch. 8).

4.5. Conclusions

The experimental results described provide clear evidence for a transmitter role for dopamine in the identified GDN of *Planorbis corneus* and most probably in *Lymnaea stagnalis* too. A summary of different types of GDN-mediated synaptic potentials is shown in Fig. 4.10. These systems offer major advantages for studying the cellular and membrane activities of dopamine and also perhaps certain drugs thought to influence the actions of dopamine.

References

Ascher, P. (1972). Inhibitory and excitatory effects of dopamine on *Aplysia* neurones. *J. Physiol.* **225,** 173–209.

Berry, M. S. and Cottrell, G. A. (1973*a*). Dopamine: excitatory and inhibitory transmission from a giant dopamine neurone. *Nature (Lond.)* **242,** 250–3.

Berry, M. S. and Cottrell, G. A. (1973*b*). Post-synaptic blockade of dopaminergic transmission by 6-hydroxydopamine. *J. Pharm. Pharmacol.* **25,** 1010.

Berry, M. S. and Cottrell, G. A. (1975). Excitatory, inhibitory and biphasic synaptic potentials mediated by an identified dopamine-containing neurone. *J. Physiol.* **244,** 589–612.

Berry, M. S. and Cottrell, G. A. (1979). Ionic bases of different synaptic potentials mediated by an identified dopamine-containing neurone in *Planorbis*. *Proc. R. Soc. Lond. B* **203**, 427–44.

Berry, M. S. and Pentreath, V. W. (1976). Criteria for distinguishing between monosynaptic and polysynaptic transmission. *Brain Res.* **105**, 1–20.

Berry, M. S., Cottrell, G. A. and Pentreath, V. W. (1974*a*). Intracellular injection of radioactive transmitter for marking the processes and terminals of identified neurones for light and electron microscopy. *J. Physiol.* **246**, 30–1P.

Berry, M. S., Pentreath, V. W., Turner, J. D. and Cottrell, G. A. (1974*b*). Effects of 6-hydroxydopamine on an identified neurone in the central nervous system of *Planorbis corneus*. *Brain Res.* **76**, 309–24.

Carpenter, D. O., Swann, J. W. and Yarowsky, P. J. (1977). Effect of curare on responses to different putative neurotransmitters in *Aplysia* neurones. *J. Neurobiol.* **8**, 119–32.

Cottrell, G. A. (1967). Occurrence of dopamine and nonadrenaline in the nervous system of some invertebrate species. *Brit. J. Pharmacol.* **29**, 63–9.

Cottrell, G. A. (1980). Identified amine-containing neurones and their synaptic connexions. In *Commentaries in the Neurosciences*. A. D. Smith, R. Llinas and P. G. Kostyuk (eds). Published for IBRO by Pergamon Press, Oxford.

Cottrell, G. A. and Laverack, M. S. (1968). Invertebrate pharmacology. *Ann. Rev. Pharmacol.* **8**, 273–96.

Cottrell, G. A., Berry, M. S. and Macon, J. B. (1974). Synapses of a giant serotonium neurone and a giant dopamine neurone: studies using antagonists. *Neuropharmacol.* **13**, 431–9.

Cottrell, G. A., Abernethy, K. B. and Barrand. M. A. (1979). Large amine-containing neurones in the central ganglia of *Lymnaea stagnalis*. *Neurosci.* **4**, 685–9.

Haydon, P. G. and Winlow, W. (1981). Morphology of the giant dopamine-containing neurone R.PE D.1, in *Lymnaea stagnalis* revealed by Lucifer Yellow CH. *J. exp. Biol.* **94**, 149–57.

Logan, S. D. and Cottrell, G. A. (1975). Response of an identified dopamine-containing neurone to iontophoretically applied drugs. *Neuropharmacol.* **14**, 453–5.

McDonald, J. F. and Berry, M. S. (1977). Further identification of multiple responses mediated by DA in the CNS of *Planorbis corneus*. *Canad. J. Physiol. Pharm.* **56**, 7–18.

Marsden, C. A. and Kerkut, G. A. (1970). The occurrence of monoamines in *Planorbis corneus*: a fluorescence microscopic and microspectrometric study. *Comp. gen. Pharmacol.* **1**, 101–16.

Pentreath, V. W. and Berry, M. S. (1975). Ultrastructure of the terminals of an identified dopamine-containing neurone marked by intracellular injection of radioactive dopamine. *J. Neurocytol.* **4**, 249–60.

Powell, B. and Cottrell, G. A. (1974). Dopamine in an identified neurone of *Planorbis corneus*. *J. Neurochem.* **22**, 605–6.

Stoof, J. C. and Kebabian, J. W. (1981). Opposing roles for D-1 and D-2 dopamine receptors in effecting cyclic AMP from rat neostriatum. *Nature (Lond.)* **294**, 366–8.

Stoof, J. C., de Vlieger, Th.A. and Lodder, J. C. (1984). D-1 receptor stimulation decreases the excitability of growth hormone producing cells in the snail *Lymnaea stagnalis*. *Eur. J. Pharmacol.* **106**, 431–435.

Turner, J. D. and Cottrell, G. A. (1978). Cellular accumulation of amines and amino acids in the central ganglia of a gastropod mollusc *Planorbis corneus*: an autoradiographic study. *J. Neurocytol.* **7**, 759–76.

Walker, R. J., Woodruff, G. N., Glaizner, B., Sedden, C. B. and Kerkut, G. A. (1968). The pharmacology of *Helix* dopamine receptors on specific neurones in the snail *Helix aspersa. Comp. Biochem. Physiol.* **24,** 455–69.

Walker, R. J., Ralph, K. L., Woodruff, G. M. and Kerkut, G. A. (1971). Evidence of a dopamine inhibitory postsynaptic potential in the brain of *Helix aspersa. Comp. Biochem. Physiol.* **24,** 455–70.

Winlow, W., Haydon, P. G. and Benjamin, P. R. (1981). Multiple actions of the giant dopamine neurone R. Pe. D.1 of *Lymnaea stagnalis. J. exp. biol.* **94,** 137–48.

5

Localisation and developmental aspects of dopaminergic neurones in the vertebrate retina

N. N. Osborne

Nuffield Laboratory of Ophthalmology, University of Oxford, Walton Street, Oxford OX2 6AW, UK

5.1. Introduction

The retina is not a peripheral organ, but a direct extension of the central nervous system, being embryologically derived from the neuroectoderm, the part of the ectoderm that gives rise to the brain. The structural organisation and physiological diversity of the cellular components of the retina are complex and vary in retinas from different species. Nevertheless, the retina can be isolated rapidly without injury and kept for hours in a viable condition (Ames and Gurian, 1960; Ames and Nesbett, 1981). The response of the tissue can be monitored by recording intracellularly from specific cells or the electroretinogram. Different parts of the retina corresponding to specific cell types or parts of cell types are well defined (Fig. 5.1), which facilitates the necessary correlation of neurochemical, morphological, histological, autoradiographical and neurophysiological, findings. Furthermore, the retina is very thin, with short diffusion pathways; the extracellular fluid therefore equilibrates rapidly with the bathing fluid. These are obvious advantages in the use of the retina for pharmacological studies and release mechanisms of putative transmitter substances. In addition, specific populations of nerve endings and cell types can be isolated from the retina (Redburn, 1977; Lam, 1978; Osborne and Neuhoff, 1983), which is enormously helpful for studies designed to establish the type of chemicals involved in neurotransmission.

Many of the putative transmitters of the brain, including acetylcholine, γ-aminobutyric acid (GABA), dopamine, glycine and substance P, are also found in a subpopulation of neurones, particularly amacrine cells, in the retina. However, despite the advantages offered by the retinal preparation over other nervous tissue (brain, spinal cord, various ganglia), the identification of the various transmitters employed at the different synapses has proved a difficult task. Gerschenfeld and Piccolino (1979) have even suggested that only for dopamine is the evidence sufficient for it to be

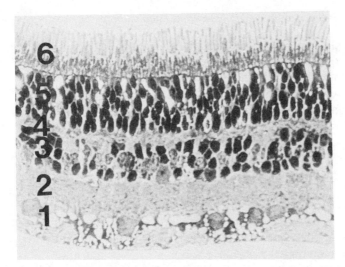

Fig. 5.1. Section through a guinea pig retina showing its layered structure. 1 = Ganglion cell layer; 2 = inner plexiform layer; 3 = inner nuclear layer; 4 = outer plexiform layer; 5 = outer nuclear layer; 6 = photoreceptor layer.

considered a retinal transmitter. The evidence for dopamine having a neurotransmitter function in the vertebrate retina is persuasive, as it is indeed for acetylcholine (Neal, 1983) and GABA (Voaden, 1976).

This chapter is primarily concerned with developmental aspects of dopaminergic neurones in the retina, and their occurrence in relation to other amacrine cells in different species. As a means of providing a background, some of the reasons for the generally accepted view that dopamine is a retinal transmitter are briefly discussed.

5.2. Reasons for believing dopamine to be a retinal transmitter

It has been suggested by Werman (1966) that, for a substance to be a neurotransmitter, it should fulfil the following criteria:

the compound must be present in the neurone's endings;
corresponding synthesising enzymes should be present;
a suitable method for the termination of the action of the compound at the postsynaptic site should operate;
stimulation of the neurone should lead to a release of the compound;
the action of the suspected transmitter should be identical in every way to the natural transmitter.

There is evidence that dopamine fulfils all these criteria in the retina, and some of the relevant data are presented in brief.

5.2.1. *Presence of dopamine in neurones and endings*

Catecholamine-containing amacrine neurones were first observed in the rabbit retina by Häggendahl and Malmfors (1965) using the process of Falck and Hillarp. With the use of spectrofluorimetric procedures it was subsequently shown by Nichols *et al.* (1967) that dopamine was the dominant, if not exclusive, catecholamine in these cells. Dopaminergic amacrine neurones have since been described in the retina of many animal species, including humans (see Ehinger, 1982; Frederick *et al.*, 1982). In addition, it has been found that in some animals, such as New World monkeys, rats and teleost fish, the dopaminergic neurones form a separate type of retinal neurone, called interplexiform cells (see Ehinger, 1973; Nguyen-Legros *et al.*, 1982). These cells have their cell bodies in the innermost part of the inner nuclear layer, like the dopamine amacrine cells of most species, but their processes ramify widely in both inner and outer plexiform layers (see Fig. 5.7).

The dopaminergic retinal neurones can also be visualised in the electron microscope using a labelling technique with 5,6-dihydroxytryptamine. After intravitreal injection, this substance is taken up by dopaminergic neurones following specific treatments (see Ehinger and Florén, 1978) and causes characteristic ultrastructural changes in them. Thus the ultrastructural changes make it possible to identify the dopamine neurones.

5.2.2. *Synthesis of dopamine*

Dopamine is synthesised in the retina, and exposure to light produces more than a two-fold enhancement of tyrosine hydroxylase activity in previously dark-adapted retinas. This effect is obvious whether tyrosine hydroxylase is measured *in vitro* (Iuvone *et al.*, 1978) or *in vivo* by measuring the accumulation of DOPA following inhibition of L-aromatic amino acid decarboxylase (de Prada, 1977). The turnover of dopamine as measured by the decline of amine following the inhibition of dopamine synthesis with α-methyl-p-tyrosine is, for example, increased three- to four-fold, following the stimulation of dark-adapted retinas in rats *in vivo* with light (Iuvone *et al.*, 1978; Morgan & Kamp, 1980). The increase in dopamine turnover in the retina in response to light is not a nerve or cell phenomenon but is peripheral to the intensity of light reaching the retina (see Morgan, 1982).

It has also been shown that the activity of the rate-limiting enzyme tyrosine hydroxylase is influenced by GABA and its antagonists or agonists (Iuvone *et al.*, 1978; Kamp and Morgan, 1981; Morgan, 1982). This is of interest because there is evidence that retinal dopaminergic neurones often receive input from GABA ergic neurones (Marshburn and Iuvone, 1981; Ehinger, 1983).

5.2.3. *Dopamine uptake (inactivation mechanism)*

The ability of specific neurones in the retina to take up exogenous dopamine is well documented. The uptake is of the active, saturable, high-affinity type with K_ms in the order of 2·6 to 5·6 × 10^{-7} M (Ehinger and Floren, 1978; Sarthy and Lam, 1979; Kato *et al.*, 1981). High-affinity uptake mechanisms are now generally thought to be the major method of inactivating released transmitter substances (Iversen, 1971; Bennett *et al.*, 1973). However, high-affinity uptake mechanisms need not necessarily be associated with neurones, as there have been reports of glial cells possessing high-affinity uptake systems (Wilkins *et al.*, 1974; Iversen *et al.*, 1975). In the case of the retina it has been demonstrated, by both fluorescence histochemistry and autoradiography (see Figs 5·8–5·12), that exogenous dopamine is taken up by neuronal elements (see Ehinger, 1983).

5.2.4. *Stimulated release of dopamine*

A variety of experiments have shown that stimulation of the dopamine cells, either by potassium depolarisation (Sarthy and Lam, 1979), or by means of light flashes (Kramer, 1971, 1976), causes a release of the amine which is dependent on calcium. The photic stimulation experiments are of particular physiological relevance, and are therefore important. In the experiments by Kramer (1971, 1976) it was observed that dopamine amacrine neurones of the intact, dark-adapted retina of cats, release levels of radioactive dopamine proportional to the stimulus frequency of a flashing light. A flashing-light stimulus also augmented dopamine release from the light-adapted retina, but the rate was no longer proportional to that of the stimulus. Since bilateral section of the optic nerve did not affect the light-induced release of radioactive dopamine from the retina, it was concluded that the release of dopamine was mediated at the level of the retina and was not secondary to efferent activation from the brain.

5.2.5. *Receptors for dopamine*

Some electrophysiological evidence suggests that retinal dopamine neurones may supply an inhibitory input to retinal ganglion cells (Straschill and Perwein, 1969, 1975). The transjunctional effects of dopamine are mediated, at least partly, via an adenylate cyclase-coupled mechanism (Brown and Makman, 1972; Watling *et al.*, 1979). Dopamine-sensitive adenylate cyclase activity is localised in the inner layers of the retina (Lolley *et al.*, 1974) and haloperidol, a dopamine antagonist, blocks the effect of dopamine on adenylate cyclase (Spano *et al.*, 1977). Dopamine receptors, as identified by high-affinity spiroperidol or ADTN binding, have also been characterised in retinas of various species (Magistretti and Schorderet, 1979; Makman *et al.*, 1980; Redburn *et al.*, 1980; Osborne, 1981). The evidence, therefore, is quite convincing for the existence of dopamine receptors in the retina, some linked to adenylate cyclase.

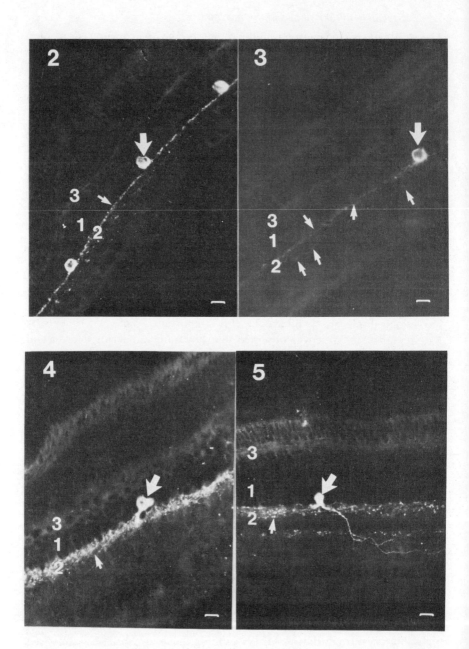

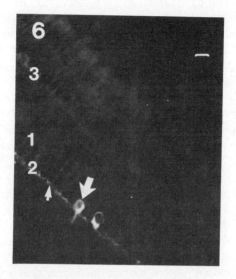

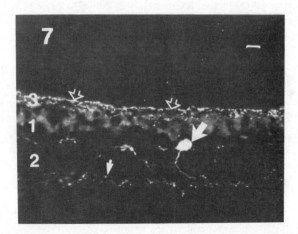

Figs. 5.2.–5.7. Immunohistochemical localisation of tyrosine hydroxylase immunoreactivity in retinas of guinea pig (2), monkey (*Macaca mulatta*) (3), cow (4), pigeon (5), frog (6) and goldfish (7). In each instance there are some cell bodies (large arrows) situated in the inner nuclear layer (1) and some processes (small arrows) situated in the inner plexiform layer (2) which contain immunoreactivity. In the case of the goldfish retina, immunoreactive processes (open arrows) are also present in the outer plexiform layer (3). The scale bars are 20 μm. (From Osborne and Patel, 1984.)

5.3. Localisation of dopaminergic neurones in different retinas

Two types of dopamine neurones exist in the vertebrate retina. One type is an amacrine cell and the other an interplexiform cell (see Ehinger, 1983). Interplexiform cells were first described in the goldfish and *Cebus* monkey by Dowling and Ehinger (1975), and have only recently been found in the rat (Nguyen-Legros *et al.*, 1982). Interplexiform cells containing other transmitter-like substances such as substance P (Brecha *et al.*, 1982; Osborne, 1984a) also exist in certain retinas. The main morphological difference between dopamine amacrine and interplexiform cells is based on their dendrite ramification: the former send out processes exclusively in the inner plexiform layer, and the latter have abundant processes and terminals in both the inner and outer plexiform layers of the retina.

The presence of dopamine-containing neurones in the retina was first indicated by Häggendahl and Malmfors (1965) using pharmacological and histochemical approaches. Following this report, a number of studies based on the use of the fluorescence histochemical procedure by Falck and Hillarp (Falck *et al.*, 1962) have described the occurrence of dopaminergic neurones in a variety of retinas (see Ehinger, 1983). However, the Falck and Hillarp method lacks specificity, and in view of the finding that noradrenaline exists in the retina (Wyse and Lorscheider, 1981; Nesselhut and Osborne, 1982; Hadjiconstantinou *et al.*, 1983), alternative methods have been used. The first is the autoradiographical localisation of the uptake of [^{3}H]-dopamine and the other employs immunohistochemical procedures to localise the major synthesising enzyme for dopamine, tyrosine hydroxylase. A combination of these two procedures has recently been employed by Osborne *et al.* (1984a), to show that a correlation exists between retinal structures that take up [^{3}H]-dopamine and 'stain' for tyrosine hydroxylase. These findings are summarised in Figs 5.2–5.12.

In all the retinas examined for tyrosine hydroxylase immunoreactivity (goldfish, monkey, frog, rabbit, guinea pig and pigeon), amacrine cell bodies situated in the inner nuclear layer and terminal processes in the inner plexiform layer were found to contain immunoreactivity (Figs 5.2–5.7 and 5.19). No immunoreactive material was observed in the outer nuclear layer or in the ganglion cells, nor was immunoreactivity associated with Müller cells. Only in the goldfish retina (Fig. 5.7) was immunoreactivity associated with some fibres in the outer plexiform layer. All other retinas examined did not contain any tyrosine hydroxylase activity in the outer plexiform layers. Immunoreactivity for tyrosine hydroxylase was observed in both central and peripheral retinal regions, and a limited analysis would not distinguish obvious differences in the distribution of immunoreactive-positive amacrine cells in the various areas of a specific retina.

The concentrations of tyrosine hydroxylase positive amacrine perikarya in the retinas of all species studied were similar, although their processes observed in the inner plexiform layers, and therefore the morphology of the

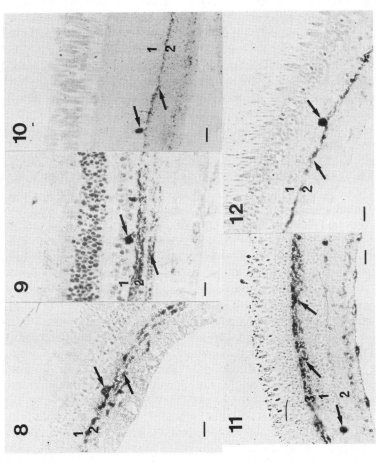

Figs 5.8–5.12. Light microscopy autoradiography of [³H]dopamine uptake by rabbit retina (8), bovine retina (9), frog retina (10), goldfish retina (11), and monkey retina (12). Perikarya (arrows) in the inner nuclear layer (1) and nerve endings in the inner plexiform layer (2) are the sites that are labelled for radioactivity in rabbit, bovine, frog, and monkey retinas. In the goldfish retina, perikarya in the inner nuclear layer and processes in the outer plexiform layer (3) are the sites that are labelled for radioactivity. Scale bar 20 μm. (From Osborne and Patel, 1984.)

amacrine cells, varied. The immunoreactive pattern of processes in the inner plexiform layer was very similar in rabbit (Fig. 5.19), guinea pig (Fig. 5.2) and monkey (Fig. 5.3), with two bands (laminae 1 and 3) in about the same position. Both bovine (Fig. 5.4) and frog (Fig. 5.6) retinas contain a single band of immunoreactivity situated in the outermost part of the inner plexiform layer (laminae 1), although the degree of immunoreactivity in the bovine retina is greater.

In the pigeon retina (Fig. 5.5), immunoreactive processes occur in a variety of layers, with the most intense band situated nearest to the inner nuclear layer. The goldfish retina (Fig. 5.7) contains some immunoreactivity, mainly in two bands in the inner plexiform layer, but most of the immunoreactivity processed is in the outer plexiform layer. A striking generalised feature of the distribution of the immunoreactive perikarya in all the retinas analysed is their position, which is close to the border between the inner nuclear and plexiform layers.

A correlation between tyrosine hydroxylase immunoreactivity and [^{3}H]dopamine uptake was revealed by comparing the described results with the autoradiographical data for the uptake of [^{3}H]dopamine in rabbit, frog, cow, goldfish and monkey (Figs 5.8–5.12). The uptake sites for [^{3}H]dopamine are very similar to, but not identical with, the distribution of tyrosine hydroxylase activity. All amacrine cell bodies containing radioactive grains were situated very close to the border of the inner plexiform and nuclear layers. In the rabbit retina, two intense bands (corresponding to laminae 1 and 3) accumulated exogenous radioactivity. Only a single band of radioactivity in the inner plexiform layer was discernible in the retinas of monkey, frog and cow. In contrast, the goldfish retina showed a single band in the outer plexiform layer taking up radioactive dopamine.

The position and number of either tyrosine hydroxylase positive neurones or neurones which take up [^{3}H]dopamine in the different retinas are almost identical. A similar distribution of immunoreactive-positive processes and terminals which have accumulated [^{3}H]dopamine in the inner plexiform layers of the various retinas also exists, although some minor differences may be found. For example, only a single layer of fibres was observed in the monkey retina which took up exogenous [^{3}H]dopamine (Fig. 5.12), and two layers containing tyrosine hydroxylase immunoreactivity (Fig. 5.3) were revealed. In contrast, the distributions of tyrosine hydroxylase immunoreactivity and sites containing radioactive grains in the inner plexiform layer of the rabbit were identical. The greatest difference between the correlation of tyrosine hydroxylase immunoreactivity and [^{3}H]dopamine uptake sites occurs in the goldfish retina (see Figs 5.7 and 5.11), with radioactive grains visible only in the outer plexiform layer. It is possible that the known dopaminergic inner plexiform cells of fish retinas (see Ehinger, 1983) have most of the presynaptic elements in the outer plexiform layer, which would then be the areas of the neurones involved with the specific uptake process of exogenous amine. In contrast, tyrosine hydroxylase would be fairly evenly distributed throughout the inner plexiform cells. This would seem a plausible explanation of the findings reported here, and also supports the opinion that the morphology of catecholaminergic neurones is studied best by localising tyrosine hydroxylase rather than using autoradiography to localise the uptake of [^{3}H]dopamine.

Immunocytochemical studies have also provided persuasive evidence that the tyrosine hydroxylase postive cells in the retinas described are dopaminergic and not noradrenergic (Ballesta *et al.*, 1984; Osborne and Patel, 1985). In the two studies reported, antisera to dopamine-β-hydroxylase never gave a positive immune reaction in the retinal tyrosine hydroxylase positive neurones. Osborne and Patel (1985) did show, however, that some dopamine-β-hydroxylase positive cells occur in the bovine retina, but these were not associated with the tyrosine hydroxylase positive amacrine cells of the retina.

5.4. Relationship between dopaminergic and other amacrine cells

Retinal cells for which the greatest number of putative transmitters have been identified are the amacrine cells whose functional contribution to the physiological operation of the retina is least understood. The following is a list of compounds generally associated with amacrine cells of retinas from different species: substance P, enkephalin, neurotensin, somatostatin, cholecystokinin, serotonin, glucagon, dopamine and acetylcholine. These substances are absent, except in one or two instances, from other retinal cells, viz. photoreceptors, horizontal cells, bipolar cells and ganglion cells in different species. By comparing the distinctive morphology of amacrine cells containing the various putative transmitters, all of which can be localised

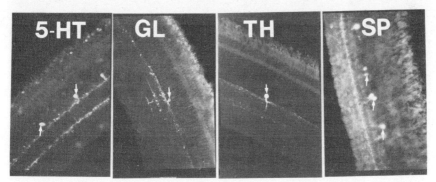

Fig. 5.13. Localisation of serotonin (5-HT), glucagon (GL), tyrosine hydroxylase (TH) and substance P (SP) immunoreactivities in pigeon retina. It can be seen that all of these substances are restricted to amacrine cells. From the position of immunoreactive perikarya (arrows) in the inner nuclear layer and processes in the inner plexiform layer, it seems certain that none of the substances co-exists with another.

with suitable antisera, it has been observed that none of the putative transmitters co exists in the same amacrine cells. This is exemplified in Fig. 5.13 with a comparison of the localisation of glucagon, substance P, serotonin and tyrosine hydroxylase immunoreactivities in the pigeon retina. Although all the substances are restricted to amacrine cells, their distribution profile shows that they are different. The different morphology of tyrosine hydroxylase-positive amacrine cells can be particularly well illustrated by using double-labelling procedures with different fluorophores, such as fluorescence isothiocyanate (FITC) and rhodamine. This is illustrated in Figs 5.14–5.16, where sections of frog, goldfish and chick retinas were double stained to localise either serotonin and tyrosine hydroxylase, or substance P and tyrosine hydroxylase. It can be seen clearly that there is no co existence of tyrosine hydroxylase with either serotonin or substance P.

The apparent findings that putative transmitter substances do not coexist in amacrine cells of different retinas is surprising in view of what is known about other nervous systems where the coexistence of neurotransmitter-like substances in certain neurones seems to be a general phenomenon (Osborne, 1984*b*).

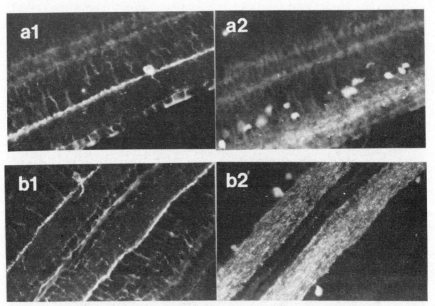

Fig. 5.14. Immunohistochemical localisation of tyrosine hydroxylase (a1) with substance P (a2) and tyrosine hydroxylase (b1) with serotonin (b2) on the same section in frog retina. Tyrosine hydroxylase, in a section, was viewed with a fluorescein filter, and changing over to a rhodamine filter either substance P or serotonin could be observed. From the nature of the fluorescence in the sections, it is clear that tyrosine hydroxylase immunoreactive neurones are different from either serotonin or substance P immunoreactive neurones.

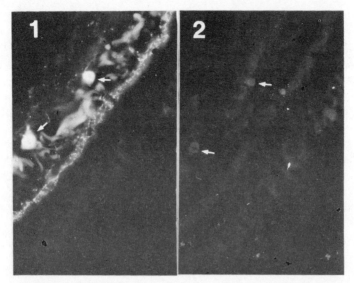

Fig. 5.15. Immunohistochemical localisation of tyrosine hydroxylase (1) with serotonin (2) in the same section of goldfish retina. The serotonin immunoreactive cell bodies (arrows) are weak compared with the tyrosine hydroxylase ones (arrows) and have a different distribution. It is clear that tyrosine hydroxylase does not co-exist with serotonin in the same cells in the goldfish retina. Section 1 was viewed with a fluorescein filter and section 2 with a rhodamine filter.

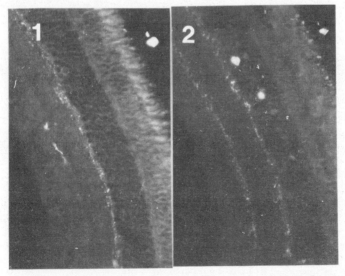

Fig. 5.16. Immunohistochemical localisation of tyrosine hydroxylase (1) with serotonin (2) in the same section of chick retina. Section 1 was viewed with a fluorescein filter and section 2 with a rhodamine filter. No evidence could be found for the co-existence of serotonin with tyrosine hydroxylase.

5.5 Developmental aspects of dopaminergic retinal neurones

Identification of developmental events for specific types of neurone is a prerequisite for the understanding of tissue maturation. A number of reports have therefore been conducted on retinal dopaminergic neurones and I shall concentrate on studies on rabbit tissues.

Studies by Fung *et al.* (1982) on the prenatal development of dopamine neurones in the rabbit retina showed that specific neurones at embryonic day 27, have the capacity to take up exogenous dopamine, to synthesise it and to release [^{3}H]dopamine following a potassium stimulation. These positive dopamine neurones are probably immature because even at birth newborn retinas contain very low levels of tyrosine hydroxylase activity and endogenous dopamine (Lam *et al.*, 1981). The activity of tyrosine hydroxylase is extremely low between days 0 and 6 after birth, increasing slowly to 30 % of the adult level at day 25. The concentration of dopamine in the developing retinas is closely linked to the tyrosine hydroxylase activity. The conclusion drawn from the studies by Lam and collaborators (Lam *et al.*, 1981; Fung *et al.*, 1982) is that neurones in the rabbit retina are committed to be dopaminergic prenatally, but that these cells can only be considered mature at about 25 days after birth.

Studies by Osborne *et al.* (1984*a*) and Redburn *et al.* (1982) are consistent with the conclusions drawn by Lam and collaborators. Osborne *et al.* (1984*a*) reported that the first tyrosine hydroxylase positive immunoreactive neurones observed in the rabbit retina were at postnatal day 3, although they were too weak to be photographed. At postnatal day 5, immuno-reactive perikarya were more discernible, as shown in Fig. 5.17. At this stage there were no fluorescent processes to be observed in the clearly distinct inner plexiform layer. At 10 days after birth, the tyrosine hydroxylase immunoreactive perikarya appeared to be more intense (Fig. 5.18), and more clearly defined processes were associated with the perikarya. The two layers of immunoreactivity in the adult inner plexiform layer (Fig. 5.19) only began to appear distinctly in retinas from animals 22 to 28 days postnatal. Before this period, i.e. 10 to 22 days postnatal, varying degrees of immunoreactive processes could be observed in the inner plexiform layer, although the ordered two-band structure was not apparent.

The studies by Redburn *et al.* (1982) on [^{3}H]spiroperidol binding to the dopamine receptor and dopamine-stimulated adenylate cyclase activity displayed a developmental profile. At birth, no appreciable amount of specific [^{3}H]spiroperidol binding was observed and dopamine had no effect on cAMP production. Dramatic increases in these two parameters were noted between days 5 and 9, followed by a levelling-off by day 11, which was about the adult level.

All the studies discussed so far are consistent with the idea that specific neurones in the retina of the rabbit are coded to be dopaminergic at birth, but the various aspects associated with functional dopaminergic neurones

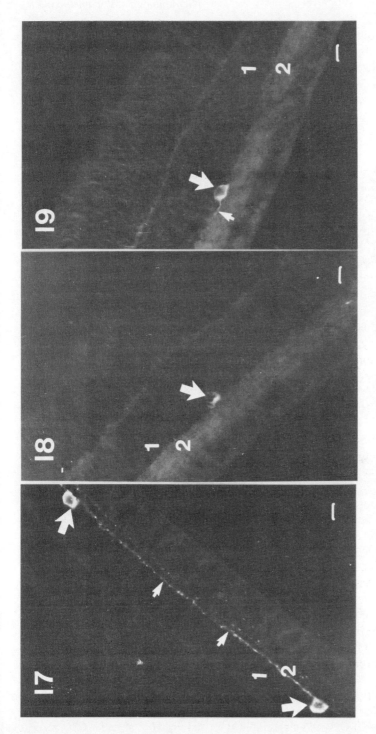

Figs 5.17–5.19. Localisation of tyrosine hydroxylase immunoreactivity in rabbit retinas from adult (17), 5-day-old (18) and 10-day-old (19) animals. Immunoreactive perikarya (large arrows) are observed in the inner nuclear layer (1) of all retinas whereas terminal processes are only clearly observed in the inner plexiform layer (2) of the adult. Scale bar 20 μm. (From Osborne and Patel, 1984.)

develop at different stages and seem to be generally unrelated to the eye-opening at about 12 days postnatal. It would appear that the dopaminergic neurones are fully developed between the 22nd and 28th postnatal day. Whether the neurones could function normally at an earlier stage is not known.

Recent studies on rabbit retinal cultures (Osborne *et al.*, 1984*b*) are in concordance with some of the data already described. Retinal cultures from

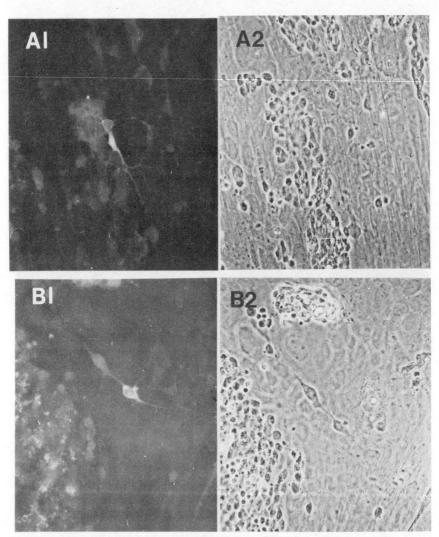

Fig. 5.20. Tyrosine hydroxylase immunoreactivity in rabbit retinal culture. Matching fluorescence and phase-contrast photomicrographs (A1 and A2, B1 and B2) show that the majority of cell bodies and neuronal processes do not stain.

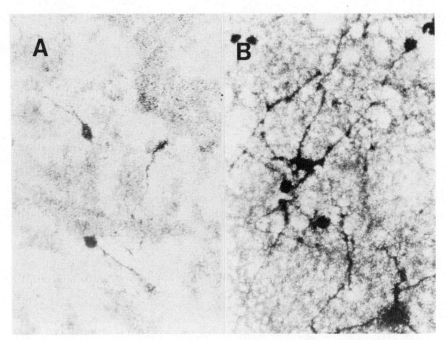

Fig. 5.21.   Autoradiography of [³H]dopamine (A) and [³H]GABA (B) by 10-day-old rabbit retinal cultures. Cultures were produced from two-day-old rabbits. It can be seen that only a few cells take up [³H]dopamine. Many more cells take up exogenous [³H]GABA.

2-day-old rabbits were prepared and tyrosine hydroxylase immunoreactive positive neurones were subsequently demonstrated (see Fig. 5.20). Staining was only weak, was restricted to a few neurones and could not be observed in cultures of less than 10 days. Autoradiographical analyses of [³H]-dopamine and [³H]GABA uptake (Fig. 5.21) by the same retinal cultures strongly suggest that neurones containing tyrosine hydroxylase immunoreactivity have the capacity to take up exogenous dopamine, and that these cells are different from the GABA-accumulating neurones.

5.6. Summary

All the vertebrate retinas so far analysed contain dopamine neurones. Evidence that the amine functions as a neurotransmitter in this tissue is most persuasive. The suitability of the retina as a preparation in neurobiological research has allowed various aspects of the dopaminergic neurones to undergo detailed study. For example, there is clear evidence that dopaminergic neurones have the capacity to take up exogenous dopamine. The morphology and development of these neurones in different retinas have also been subjected to detailed analysis.

Acknowledgements

The work reported in this article has been supported by the Stiftung
Volkswagenwerk and Wellcome Trust.

References

Ames, A. III and Gurian, B. S. (1960). Measurement of function in an *in vitro*
preparation of mammalian central nervous tissues. *J. Neurophysiol.* **23**, 676–91.
Ames, A. III and Nesbett, F. B. (1981). *In vitro* retina as an experimental model of the
central nervous system. *J. Neurochem.* **37**, 867–77.
Ballesta, J., Terenghi, G., Thibault, J. and Polak, J. (1984). Putative dopamine-
containing cells in the retina of seven species demonstrated by tyrosine
hydroxylase immunohistochemistry. *Neuroscience* **12**, 1147–56.
Bennett, J. P. Jr, Logan, W. J. and Snyder, S. H. (1973). Amino acids as central
nervous transmitters: the influence of ions, amino acid analogues, and ontogeny
on transport systems for glutamic acid, L-aspartic acid and glycine in central
nervous synaptosomes of the rat. *J. Neurochem.* **21**, 1533–50.
Brecha, N., Hendrickson, A., Floren, I. and Karten, H. J. (1982). Localisation of
substance P-like immunoreactivity within monkey retina. *Invest. Ophthal. and Vis.
Sci.* **23**, 147–53.
Brown, J. H. and Makman, M. H. (1972). Stimulation by dopamine of adenylate
cyclase in retinal homogenates and of adenosine 3,5′-cyclic monophosphate
formation in intact retina. *Proc. nat. Acad. Sci. USA* **69**, 539–43.
de Prada, M. (1977). Dopamine content and synthesis in retina and n. acumbens
septi: pharmacological and light-induced modifications. *Adv. Biochem. Psycho-
pharmacol.* **16**, 311–19.
Dowling, J. E. and Ehinger, B. (1975). Synaptic organization of the amine-containing
interplexiform cells of the goldfish and *Cebus* monkey retinas. *Science* **188**, 270–3.
Ehinger, B. (1982). Neurotransmitter systems in the retina. *Retina* **2**, 305–21.
Ehinger, B. (1983). Functional role of dopamine in the retina. In *Progress in Retinal
Research*, vol. 2. N. N. Osborne and G. J. Chader (eds), pp. 213–32. Pergamon,
Oxford.
Ehinger, B. and Florén, I. (1978). Quantitation of the uptake of indoleamine and
dopamine in the rabbit retina. *Exp. Eye Res.* **26**, 1–11.
Falck, B., Hillarp, N. A., Thieme, G. and Thorp, A. (1962). Fluorescence of
catecholamines and related compounds with formaldehyde. *J. Histochem.
Cytochem.* **10**, 348–54.
Frederick, J. M., Rayborn, M. E., Laties, A. M., Lam, D. M. K. and Hollyfield, J. G.
(1982). Dopaminergic neurones in the human retina. *J. Comp. Neurol.* **210**, 65–79.
Fung, S. C., Kong, Y-C. and Lam, D. M. K. (1982). Prenatal development of
GABAergic, glycinergic, and dopaminergic neurones in the rabbit retina. *J.
Neuroscience* **2**, 1623–32.
Gerschenfeld, H. M. and Piccolino, M. (1979). Pharmacology of the connections of
cones and L-horizontal cells in the vertebrate retina. In *The Neurosciences; 4th
Study Program*. F. O. Schmitt and F. G. Worden (eds), pp. 213–26. MIT Press,
Cambridge, Mass.
Hadjiconstantinou, M., Cohen, J. and Neff, N. H. (1983). Epinephrine: a potential
neurotransmitter of the retina. *J. Neurochem.* **41**, 1440–4.
Häggendahl, J. and Malmfors, T. (1965). Identification and cellular localization of
the catecholamines in the retina and choroid of the rabbit. *Acta. Physiol. Scand.*
64, 58–66.

Iuvone, P. M., Galli, C. L., Garrison-Gund, C. K. and Neff, N. H. (1978). Light stimulates tyrosine hydroxylase and dopamine synthesis in retinal amacrine neurones. *Science* **200**, 901–2.

Iversen, L. L. (1971). Role of transmitter uptake mechanisms in synaptic neurotransmission. *Brit. J. Pharmacol.* **41**, 571–91.

Iversen, L. L., Dick, F., Kelly, J. S. and Shon, F. (1975). Uptake and localization of transmitter amino acids in the nervous system. In *Metabolic Compartmentation and Neurotransmission*. S. Bert, D. D. Clarke and D. Schneider (eds), pp. 65–78. Plenum, New York.

Kamp, C. W. and Morgan, W. W. (1981). GABA antagonists enhance dopamine turnover in the rat retina *in vivo*. *Eur. J. Pharmacol.* **69**, 273–9.

Kato, S., Kunyama, K., Ito, Y., Teranishi, T. and Negishi, K. (1981). ^{3}H-dopamine uptake in dark-adapted fish retina. *Vision Res.* **21**, 1189–92.

Kramer, S. G. (1971). Dopamine: a retinal neurotransmitter, I. Retinal uptake, storage and light-stimulated release of ^{3}H-dopamine *in vivo*. *Invest. Ophthalmol.* **10**, 438–52.

Kramer, S. G. (1976). Dopamine in retinal neurotransmission. In *Transmitters in the Visual Process*. S. P. Bonting (ed.), pp. 165–98. Pergamon, Oxford.

Lam, D. M. K. (1978). Physiological and biochemical studies of identified cells in the vertebrate retina. In *Biochemistry of Characterised Neurones*. N. N. Osborne (ed.), pp. 239–60. Pergamon, Oxford.

Lam, D. M. K., Fung, S.-C. and Kong, Y-C. (1981). Postnatal development of dopaminergic neurones in the rabbit retina. *J. Neurosci.* **1**, 1117–32.

Lolley, R. N., Schmidt, S. Y. and Farber, D. B. (1974). Alterations in cyclic AMP metabolism associated with photoreceptor cell degeneration in the C3H mouse. *J. Neurochem.* **22**, 701–7.

Magistretti, P. J. and Schorderet, M. (1979). Dopamine receptors in bovine retina: characterization of the ^{3}H-spiroperidol binding and its use for screening dopamine receptor affinity of drugs. *Life Sci.* **25**, 1675–86.

Makman, M. H., Dvorkin, B., Horowitz, S. G. and Thal, L. J. (1980). Properties of dopamine agonist and antagonist binding sites in mammalian retina. *Brain Res.* **194**, 403–18.

Marshburn, P. B. and Iuvone, P. M. (1981). The role of GABA in the regulation of the dopamine/tyrosine hydroxylase-containing neurones of the rat retina. *Brain Res.* **214**, 335–47.

Morgan, W. W. (1982). Dopamine neurones in the retina: a new pharmacological model. In *Cell Biology of the Eye*, pp. 533–54. Academic Press, New York.

Morgan, W. W. and Kamp, C. W. (1980). A GABAergic influence on the light-induced increase in dopamine turnover in the dark-adapted rat retina *in vivo*. *J. Neurochem.* **34**, 1082–6.

Neal, M. J. (1983). Cholinergic mechanisms in the vertebrate retina. In *Progress in Retinal Research*, vol. 2. N. N. Osborne and G. J. Chader (eds), pp. 191–212. Pergamon, Oxford.

Nesselhut, T. and Osborne, N. N. (1982). Is noradrenaline a major catecholamine in the bovine retina? *Neurosci. Lett.* **28**. 41–5.

Nguyen-Legros, J., Berger, B., Vigny, A. and Alvarez, C. (1982). Presence of interplexiform dopamine neurones in the rat retina. *Brain Res. Bull.* **9**, 379–81.

Nichols, C. W., Jacobowitz, D. and Hottenstein, M. (1967). The influence of light and dark in the catecholamine content of the retina and choroid. *Invest. Ophthalm.* **6**, 642–6.

Osborne, N. N. (1981). Binding of ^{3}H-ADTN, a dopamine agonist, to membranes of bovine retina. *Cell. Molec. Neurobiol.* **1**, 167–74.

Osborne, N. N. (1984a). Substance P in the bovine retina: localization, identification, release, uptake and receptor anaylsis. J. Physiol. (Lond.) 349, 83–93.

Osborne, N. N. (1984b). Transmitter specificity in neurones. In Handbook of Neurochemistry, vol. 6. A. Lajtha (ed.), pp. 511–26. Plenum, New York.

Osborne, N. N. and Neuhoff, V. (1983). Characterization and isolation of mammalian retinal neurones. Biochem. Soc. Trans. 11, 682–4.

Osborne, N. N. and Patel, S. (1985). The presence of dopamine-β-hydroxylase like enzyme in the vertebrate retina. Neurochem. Int. 7, 51–6.

Osborne, N. N., Patel, S. and Vigny, A. (1984a). Dopaminergic neurones in various retinas and the postnatal development of tyrosine-hydroxylase immunoreactivity in the rabbit retina. Histochemistry 80, 389–93.

Osborne, N. N., Beaton, D. W., Vigny, A. and Neuhoff, V. (1984b). Localization of tyrosine-hydroxylase immunoreactive cells in rabbit retinal cultures. Neurosci. Lett. 50, 117–20.

Redburn, D. (1977). Analysis of ^{14}C-GABA uptake and release from rabbit retina synaptosomes. Exp. Eye Res. 25, 265–75.

Redburn, D., Clement-Cormier, Y. and Lam, D. M. K. (1980). GABA and dopamine receptor binding in retinal synaptosomal fractions. Neurochem. Int. 1, 167–82.

Redburn, D. A., Mitchell, C. K. and Hampton, C. K. (1982). Developmental analysis of neurotransmitter systems in the rabbit retina. In Proceedings of the NATO Advance Study Institute on Compartmentation and Neurotransmitter Interactions. H. F. Bradford (ed.), pp. 79–97. Plenum, New York.

Sarthy, P. V. and Lam, D. M. K. (1979). The uptake and release of ^{3}H-dopamine in the goldfish retina. J. Neurochem. 32, 1269–77.

Spano, P. F., Govoni, S., Hofmann, M., Kumakura, K. and Trabucchi, M. (1977). Physiological and pharmacological influences on dopaminergic receptors in the retina. Adv. Biochem. Psychopharmacol. 16, 307–10.

Straschill, M. and Perwein, J. (1969). The inhibition of retinal ganglion cells by catecholamines and γ-amino butyric acid. Pflügers Arch. Eur. J. Physiol. 312, 45–54.

Straschill, M. and Perwein, J. (1975). Effects of biogenic amines and amino acids on cat's retinal ganglion cells. In Proceedings of the Golgi Centennial Symposium. M. Santini (ed.), pp. 583–91. Raven Press, New York.

Voaden, M. J. (1976). Gamma-aminobutyric acid and glycine as retinal neuro-transmitters. in Transmitters in the Visual Process. S. L. Bonting (ed.), pp. 107–25. Pergamon, Oxford.

Watling, K. J., Dowling, J. E. and Iversen, L. L. (1979). Dopamine receptors in the retina may all be linked to adenylate cyclase. Nature (Lond.) 281, 578–80.

Werman, R. (1966). Criteria for the identification of central nervous transmitter systems. Comp. Biochem. Physiol. 18, 745–66.

Wilkins, G., Wilson, J. E., Balasz, R., Shon, F. and Kelly, J. S. (1974). How selective is the high affinity uptake of GABA into inhibitory nerve terminals? Nature (Lond.) 252, 397–9.

Wyse, J. P. H. and Lorscheider, F. L. (1981). Low retinal dopamine and serum prolactin levels indicate an inherited dopaminergic abnormality in BW rats. Exp. Eye Res. 32, 541–51.

6

Dopamine neurones of the vertebrate brain: some aspects of anatomy and pathology

D. M. A. Mann

Department of Pathology, University of Manchester, Oxford Road, Manchester M13 9PT, UK

6.1. Anatomy and histology

The dopamine-containing nerve cells of the human central nervous system, together with their noradrenaline-containing counterparts, form islands within an archipelago of pigmented nerve cell groups that extends bilaterally along the midbrain and brainstem from a position central to the motor nuclei of the oculomotor nerve and terminating at the caudal medulla in direct continuity with the interomediolateral grey column of the spinal cord (Bazelon *et al.*, 1967; Bogerts, 1981; Saper and Petito, 1982). The mass of information that has been gathered over the past decade from fluorescent histochemical, microbiochemical and more recently immunohistochemical and protein tracer studies has led to the mapping of dopamine pathways within the brain, both in humans and in several animal species. These data have been well reviewed by others (Dahlstrom and Fuxe, 1964; Ungerstedt, 1971; Lindvall and Bjorklund, 1978; Moore and Bloom, 1978; Moore, 1982). However, as an introduction to this section, the main findings are summarised.

Although most of the primary data that describe the organisation and cytology of the dopaminergic nerve cells are derived from studies of animal brain, the systems they comprise appear to change little in mammalian phylogeny and most statements are probably equally applicable to humans.

In humans the main group of dopaminergic cells is the substantia nigra (A9), whose 300 000 to 400 000 cells (McGeer *et al.*, 1977; Bogerts *et al.*, 1983) are organised into the densely populated pars (or zona) compacta and the less populated pars reticulata and pars lateralis, and whose fibres project mainly to the corpus striatum (caudate nucleus and putamen) and globus pallidum. A smaller group of about 5000 cells (A10) (Bogerts *et al.*, 1983) is located within the ventral tegmental area of the mesencephalon, adjacent to the substantia nigra, and its fibres project upon a series of phylogenetically older cortical regions which include the nucleus accumbens, olfactory tubercle, amygdaloid nucleus, septum, stria terminalis and piriform cortex as well as with neocortical areas such as medial frontal and anterior cingulate cortex. Such pathways form the mesolimbic and mesocortical tracts, respectively. The A10 region seems much less developed in humans than in rat, where it is the largest dopamine cell group (Dahlstrom and Fuxe, 1964). Other cells in the hypothalamus and thalamus (A11) project to the spinal cord (Lindvall *et al.*, 1983), and a proportion of the cells in the dorsal motor vagus (A1, A2) may be dopaminergic and form efferent fibres within the vagus nerve (Armstrong *et al.*, 1982). Dopamine-containing cells are also found in the hypothalamic/hypophyseal axis, the retina and the olfactory bulb.

Since most of the histological and cytological studies made on dopaminergic cells have been made on those of the substantia nigra, the following descriptions are strictly only relevant to cells of that area, although it is implied that, unless otherwise specified, such descriptions are also applicable to those cells of the other regions. Nissl and Golgi studies (Juraska *et al.*, 1977; Domesick *et al.*, 1983) show the substantia nigra and ventral tegmental area to contain medium-sized (15–50 μm) triangular or fusiform cells which contain many block-like masses of Nissl substance and a large (12 μm × 10 μm) smoothly outlined nucleus in which the single nucleolus (3–4 μm in diameter) is prominently basophilic (Fig. 6.1a). The cell body gives rise to two to six, rather long, thin dendrites, which divide once or twice shortly after emerging from the cell body and then pursue a relatively unbranched course; such dendrites are poorly furnished with spines. Both cell bodies and dendrites are orientated along a rostrocaudal axis. Each cell gives rise to an extensive, highly collateralised plexus of axon terminals. For example, in the striatum, dopamine cells provide about 20 % of all terminals and it has been estimated that the total length of axon may exceed 70 cm, with some 500 000 terminals (Anden *et al.*, 1966).

With such extensive cell processes to support, it is not surprising that the most prominent structures seen in the nigral cell under the electron microscope are the masses of rough endoplasmic reticulum (which form the well organised Nissl bodies seen in light microscopy) and the many free ribosomes, which extend into the proximal parts of the dendrites, and the well organised Golgi apparatus. The highly indented nuclear envelope is often continuous with the rough ER and ribosomes, and ER is seen in these deep, narrow indentations. The processes of astrocytes cover most of the cell body and the dendritic trunks. Boutons are scarce on the cell body itself, but cover an increasing proportion of the dendrite surface as one proceeds from the cell body; these are separated from the neighbouring neuropil by astrocytic processes (for complete descriptions see Bak, 1967; Rinvik and Grofova, 1970; Domesick *et al.*, 1983).

Nerve cells of the human and other primate substantia nigra seem different from most other species in as much as they contain lipoprotein

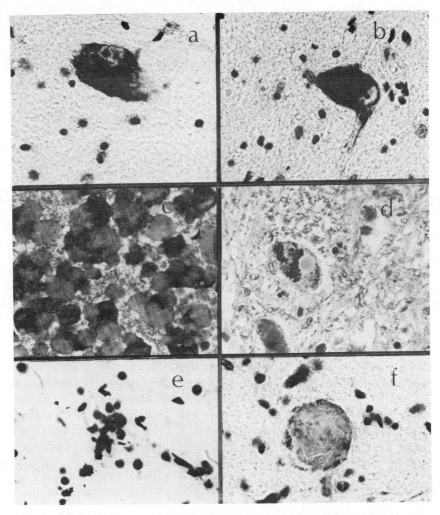

Fig. 6.1.   Nerve cells of the human substantia nigra in a young adult (a), a 60-year-old person (b, c) and an 83-year-old person (d, e, f), showing those cytological changes associated with ageing which are described in the text.

pigment granules known as neuromelanin (Fig. 6.1b) (hereafter referred to as melanin). These are especially numerous in adult and elderly people (see later) (Foley and Baxter, 1958; Bazelon *et al.*, 1967; Fenichel and Bazelon, 1968; Mann and Yates, 1974). Other primate species show less melanin (Scherer, 1939; Adler, 1942); cats and dogs show variable degrees of pigment (Brown, 1943; Marsden, 1961; Brayda-Bruno and Levi, 1979) but melanin is apparently absent, even in aged animals, in nigral cells of the laboratory rat and mouse (Marsden, 1961; Bak, 1967; Domesick *et al.*, 1983). Melanin granules are also present in normal amounts in the nigral cells of human albinos (Foley and Baxter, 1958). These persons lack the enzyme tyrosinase which is responsible for the enzymatic oxidation of tryosine to melanin in the skin and retina, and the presence of melanin in their nerve cells clearly indicates a different mode of formation. In fact, noradrenaline, adrenaline and serotonin, along with dopamine, can all be incorporated into the melanin polymer, through auto-oxidative (non-enzymatic) pathways (Graham *et al.*, 1978; Graham, 1979). However, these pathways also form, as by-products, peroxides, superoxide anions, free radicals and sulphydryl reagents, which, without the protecting influence of enzymes such as superoxide dismutase, catalase and glutathione, have the potential to damage cell membranes and organelles; consequently, levels of these latter substances are normally high in substantia nigra. Why these melanin-forming pathways should seemingly predominate in humans and other primates is not known.

Under the electron microscope, the melanin granules lie not within melanosomes, as in the melanocytes of the skin and retina, but free within the cytoplasm. Each (Fig. 6.1c) is composed of a globular lipid fragment with electron-dense material and other coarse granules (derived from dopamine breakdown), all being bounded by a limiting membrane (Duffy and Tennyson, 1965). The similarity of this structure to the more familiar nerve cell pigment, lipofuscin, implies that melanin is a melanised form of lipofuscin (Barden, 1969), both pigments having similar origins in the lysosomal compartment of the cell. Essentially, melanin seems to be accumulated as a waste product of metabolism; no useful purpose has been demonstrated and indeed its presence in such large quantities in middle age and later life may actually be detrimental to the well-being of the cell (see later).

Another feature apparently restricted to humans is the presence of hyaline (Lewy) inclusion bodies within the cytoplasm of nerve cells of the substantia nigra (Fig. 6.1d) and other monoamine-containing nerve cell types (Ohama and Ikuta, 1976). These are seen occasionally in apparently normal elderly people (Lipkin, 1959; Woodward, 1962; Forno, 1969) but their incidence is greatly increased in Parkinson's disease (Greenfield and Bosanquet, 1953; Lipkin, 1959; Jager and Bethlem, 1960). Electron microscopy shows them to be composed of filamentous protein material radiating from a dense central core and ending without a membrane separating the body from the cytoplasmic organelles and the melanin granules (Duffy and Tennyson, 1965; Forno, 1969). In humans, small protein-rich bodies (intracytoplasmic acidophilic granules, IAGs) are also normally abundant in nigral neurones (Issidorides *et al.*, 1978; Sekiya *et al.*, 1983) as well as in other cell types, including the other monoaminergic cell groups (Sekiya *et al.*, 1983). These IAGs display similar staining reactions to Lewy bodies (Issidorides *et al.*, 1978; Sekiya *et al.*, 1982) and are thought to be derived from mitochondria

(Sekiya *et al.*, 1982). The decrease or disappearance of IAGs in cells containing Lewy bodies (Issidorides *et al.*, 1978) may be due to their coalescence into Lewy bodies. Other rod-shaped inclusions made up of thin filaments orientated in parallel paracrystalline arrays (Schochet *et al.*, 1970) and staining immuno–histochemically for actin (Goldman and Horoupian, 1982) have also been described in humans, though their relevance is unknown.

One final feature of the structure of the substantia nigra (again apparently only in humans and other primates) is the proximity of blood vessels to nerve cell perikarya and dendrites (Issidorides, 1971; Felten and Crutcher, 1979; Scheibel and Tomiyasu, 1980). The basement membrane of capillaries, small arterioles and venules are often seen in direct apposition to the perikaryon and dendrite, without any intervention of glial processes (Felten and Crutcher, 1979; Scheibel and Tomiyasu, 1980); such close association may allow the function of these cells to be closely influenced by blood-borne factors.

Whether this close vascular relationship, or indeed the presence of features such as melanin granules, Lewy bodies, IAGs and other inclusions, are indeed all peculiar to humans and other primates is not known: studies on other animal species, and especially on aged animals, are needed to establish this.

6.2. Changes in the human substantia nigra with ageing and in Parkinson's disease

In humans at birth the nerve cells of the substantia nigra are small with little cytoplasmic RNA (Nissl substance) and few melanin granules. As they age, the cell body increases in size such that in young adults the cells are plump and contain much cytoplasmic RNA with a strongly basophilic nucleolus within a large open nucleus in which the chromatin is finely dispersed (Fig. 6.1a). At this age, only small amounts of melanin are present, though this increases such that by 50 to 60 years most cells are densely pigmented (Fig. 6.1b). With age, atrophic cells containing much pigment and showing shrinkage of the cell body, loss of RNA and decreases in the size of the nucleus and nucleolus begin to appear (Fig. 6.1b), and these become increasingly common towards old age. Eventual heterolysis leaves aggregations of melanin pigment being freely deposited within the neuropil or lying within clusters of macrophages (Fig. 6.1e). Sometimes, and especially in the very old person, nerve cells containing Lewy bodies (Fig. 6.1d) or neurofibrillary tangles (Fig. 6.1f) are present. No obvious changes in glial cells or blood vessels in the vicinity of the substantia nigra are usually seen. Whether similar changes occur in the dopamine-containing cells in the ventral tegmental area and dorsal motor vagus is not known.

Quantitative studies bear out these histological observations. For example, the number of nerve cells declines with age (McGeer *et al.*, 1977; Mann *et al.*, 1984), the decrease becoming disproportionately greater towards old age such that by 80 years of age some 30 % (Table 6.1; Mann *et al.*, 1984) to 60 % (McGeer *et al.*, 1977) of the original complement has been lost. Nucleolar volume also decreases with age in this disproportionate way, reductions of 15 % being found at 80 years of age (Table 6.1) (Mann *et al.*, 1984). Presynaptically acting cholinergic cells are decreased by 20 % in striatum of aged humans (Bugiani *et al.*, 1978).

Loss of nerve cells from substantia nigra and reductions in activity of remaining cells combine to give age-correlated decreases in dopamine in striatum and other brain regions (Carlsson and Winblad, 1976; Riederer and Wuketich, 1976; Winblad *et al.*, 1978; Adolfsson *et al.*, 1979*a,b*; Spokes, 1979) and the number of dopaminergic binding sites in caudate nucleus is also diminished (Severson *et al.*, 1982). By contrast, levels of homovanillic acid (HVA) in basal ganglia are unaltered with age (Winblad *et al.*, 1978) but increase in cerebrospinal fluid (Bowers and Gerbode, 1968; Gottfries *et al.*, 1971), suggesting that out-transport of HVA from cerebral tissue is impaired in old people.

This progressive cell loss from the substantia nigra, together with the decline in activity of remaining cells, which combine to cause the level of dopamine in the striatum to fall with age, are probably partly reflected in the difficulties in control and co-ordination of muscular activity that lead to an uncertainty of movement and the adoption of a shuffling gait in many old people. Any exaggeration of these clinical signs would indicate a diagnosis of Parkinson's disease, implying that the changes in the nervous system that characterise this latter condition may be qualitatively similar, although much more pronounced in severity, to those we have described in ageing alone. Exacerbation of other age-related variables in Parkinson's disease, such as a greater slowing of reaction time (Evarts *et al.*, 1981) and an enhanced late reflex response of muscles to stretch (Tatton and Lee, 1975) re-emphasise this point.

As might be expected, therefore, in Parkinson's disease cell numbers in substantia nigra and ventral tegmental area are grossly depleted (Pakkenberg and Brody, 1965; Bogerts *et al.*, 1983; Mann and Yates, 1983*a*) (Table 6.1), and further reductions (for age) in nucleolar volume (Table 6.1) (Mann and Yates, 1983*a*) and tyrosine hydroxylase activity (Javoy-Agid *et al.*, 1981) are seen, together with extreme loss of striatal dopamine (Hornykiewicz, 1966). Lewy bodies in remaining nigral cells are common. Such changes account for the characteristic tremor, rigidity and akinesia of Parkinson's disease. Loss of tyrosine hydroxylase activity in the ventral tegmental area (Javoy-Agid *et al.*, 1981) and reductions in dopamine in limbic structures (Price *et al.*, 1978) parallel the degeneration of A10 dopaminergic neurones (Bogerts *et al.*, 1983); changes here may play an important role in certain aspects of akinesia (defect in initiation and

Table 6.1 Mean values of nerve cell number, nucleolar volume and melanin content in substantia nigra of three control groups of patients and a group of patients with Parkinson's disease. (Data from Mann and Yates, 1983a, and Mann et al., 1984.)

	Control group 1 (n = 4)	Control group 2 (n = 13)		Control group 3 (n = 13)		Parkinson's disease (n = 8)	
Age (years)	15·0 ± 0·2	65·1 ± 0·6		83·9 ± 0·6		68·2 ± 6·8	
Nerve cell number	610·5 ± 20·9	567·4 ± 14·8	(7·1*)	440·0 ± 36·3	(27·9***)	90·8 ± 16·0	(84·0†††)
Nucleolar volume (μm^3)	52·9 ± 0·4	51·7 ± 0·5	(2·3*)	44·8 ± 0·4	(15·4***)	42·5 ± 3·0	(17·8††)
Melanin content (arbitrary units)	22·9 ± 0·3	48·6 ± 1·8		40·2 ± 1·4		38·4 ± 2·4	(21·0††)

*, *** denote significantly different from control group 1 value $P < 0.05$, < 0.001.
††, ††† denote significantly different from control group 2 value $P < 0.01$, < 0.001.
(Values in parentheses indicate percentage loss compared with either group 1 or group 2 mean value (where appropriate).)

strategy of movement) and in the cognitive impairment commonly seen later in the illness.

6.3. Other conditions

In contrast to the extensive damage that occurs in Alzheimer's disease to the noradrenergic system (see Mann *et al.*, 1984, for review) and serotonin system, (Mann and Yates, 1983*b*) dopamine cells of the substantia nigra, at least, are only mildly damaged and then usually in younger patients (i.e. under 75 years of age), who also show a greater level of change in the above other two systems (Mann *et al.*, 1984). Measured losses of dopamine and HVA (Adolfsson *et al.*, 1979*b*; Mann *et al.*, 1980; Winblad *et al.*, 1981) and dopamine receptors (Resine *et al.*, 1978) in caudate nucleus probably stem from a decreased cholinergic activity in this region, which is also more pronounced in younger patients (Yates *et al.*, 1979). Such changes in caudate nucleus may partly explain why a high proportion of patients, especially those in the later stages of the illness, show extrapyramidal signs such as rigidity and hypokinesia (Sulkava, 1982). Changes in the dopamine system similar to those in Alzheimer's disease are seen in patients with Down's syndrome who survive past 35 years and whose brains also contain numerous senile plaques and neurofibrillary tangles (Mann *et al.*, 1985). In Huntington's chorea there is no apparent loss of cells from substantia nigra (McGeer *et al.*, 1977; Mann, unpublished observations), though nucleolar volume is decreased (Mann, unpublished observations), as are dopamine and HVA levels in striatum (Bernheimer, 1973); changes which again are probably secondary to damaged cholingergic neurones in caudate nucleus.

In these three conditions it seems, therefore, that alterations of the dopamine system are probably secondary to damaged basal ganglia and are not involved in the primary degenerative events of the diseases.

In rarer neurological illnesses, such as progressive supranuclear palsy, nigrostriatal degeneration, Shy–Drager syndrome and in the traumatic encephalopathy of boxers, extrapyramidal signs similar to those of Parkinson's disease are often present; changes which are also due to severe cell loss from substantia nigra. It is not clear, however, whether one or all of these conditions share a common causal relationship with Parkinson's disease, forming part of a spectrum of disease with this, or whether they simply reflect the common outcome of basically unrelated pathological processes.

6.4. Possible relationship between cell damage and melanin pigmentation

In contrast to humans, losses of brain dopamine and the activities of associated enzymes in laboratory rats are of a lesser magnitude and begin to decline at a point much later in the average life expectancy (see McNeill *et al.*, 1984). One possible reason for this apparently greater decline in function with age of the human dopaminergic system may lie with the accumulation of melanin; a change that is much less marked in other primates, variable in carnivores, and absent in the common laboratory animals (see previously). This may also be part of the reason for the failure to produce effective models of Parkinson's disease in these latter species. We have investigated the effects of pigment accumulation on the functional capacity of nigral neurones in ageing and in Parkinson's disease (Mann and Yates, 1974, 1982, 1983*a*; Mann *et al.*, 1977*a*, 1977*b*) and these findings are now summarised.

The melanin content of cells of the substantia nigra increases linearly in amount up to 60 years of age (Mann and Yates, 1974; Graham, 1979; Kaiya, 1980), but after this time, mean pigment levels progressively decrease (Fig. 6.2) (Mann and Yates, 1974). Comparisons of distribution of pigment in remaining cells of elderly persons (Mann *et al.*, 1977*a*; Mann and Yates, 1983*a*) (Fig. 6.3) favour a process whereby these average reductions in later life stem from a preferential and progressive loss of those nerve cells that contain the most pigment. Increases in melanin are first accompanied by a slow loss of cytoplasmic RNA (Mann *et al.*, 1977*b*) (Fig. 6.4a) but later, when pigment content surpasses a certain critical value, by reductions in nucleolar volume also (Fig. 6.4b) together with an increased rate of RNA loss (Fig. 6.4a) (Mann *et al.*, 1977*b*). Such relationships suggest an atrophic process

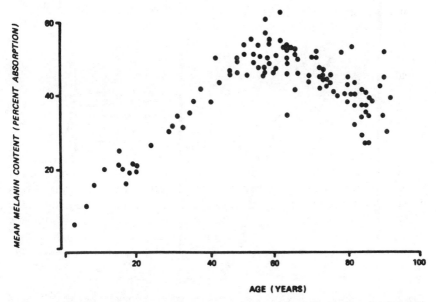

Fig. 6.2. Mean melanin content plotted against age for nerve cells of substantia nigra in 110 patients of age range 18 months to 91 years. (From Mann *et al.*, 1977*a*.)

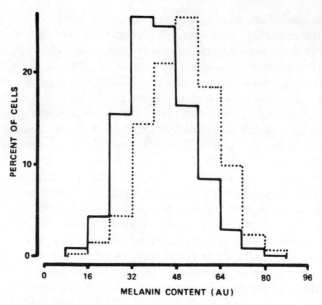

Fig. 6.3. Composite distribution of melanin content in nerve cells of substantia nigra from 20 patients of mean age 68 years (· · · · ·) and 20 patients of mean age 85 years (———). (From Mann and Yates, 1983*a*.)

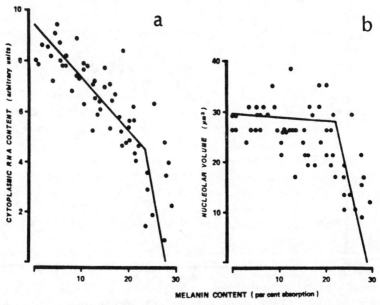

Fig. 6.4. Cytoplasmic RNA content (a) and nucleolar volume (b) plotted against melanin content for 60 nerve cells of human substantia nigra in a 56-year-old patient. (From Mann *et al.*, 1977*b*.)

occurring with age that eventually culminates in later life in a continuing loss of the more heavily pigmented cells. Although mechanical disorganisation of subcellular components by this increasing pigment bulk may contribute to this atrophy, a more serious cause is likely to be the damage that the breakdown products of dopamine (peroxides, superoxide anions, free radicals) do, as times passes, in the relative absence of protective catalase, glutathione and superoxide dismutase, to biologically active membranes and organelles (Graham *et al.*, 1978). With their metabolism weakened in this way, the heavily pigmented cells of the substantia nigra would, by late middle age, not only be less able to respond quickly in times of activity and stress, but would also be those cells least capable of withstanding the effects of exposure to pathogenic agents. Were this to happen on a large scale, striatal dopamine levels would quickly fall and extrapyramidal signs (indicative of Parkinson's disease) would ensue. Neither the age changes nor the effects of secondary factors alone would usually produce a full-blown picture of parkinsonism, though in conjunction such an outcome would be highly likely. Supporting this argument are findings that the melanin content in surviving cells of substantia nigra in Parkinson's disease is on average 20 % lower than expected at that age (Table 6.1) due to further loss of the most heavily pigmented cells (Mann and Yates, 1982, 1983*a*), and that the levels of protecting glutathione and catalase in substantia nigra are very low (Ambani *et al.*, 1975). Moreover, cell loss is always less severe in the locus coeruleus in Parkinson's disease (Greenfield and Bosanquent, 1953) where melanin concentration is lower (Mann and Yates, 1979) and the products of noradrenaline metabolism are less toxic (Graham *et al.*, 1978).

The facts that only a relatively small fraction of middle-aged persons develop Parkinson's disease, that the incidence declines past 80 years of age (Hoehn and Yahr, 1967) and that identical twins, who apparently age at the same rate, have a low concordance rate for Parkinson's disease (Ward *et al.*, 1983) all point to the conclusion that although age changes may form the basis for Parkinson's disease, not everyone would acquire the illness were they to live long enough; exposure to a secondary factor is also needed.

What this secondary agent might be is still unkown (if indeed it is the same factor in all cases). Although Parkinsonian signs can be occasioned experimentally by ischaemia (Lavyne *et al.*, 1975) or virus infection (Lycke and Roos, 1969), and patients present in clinical practice when such signs are related to previous infection by virus (encephalitis lethargica) or to episodic or chronic exposure to heavy metals such as manganese (Cook *et al.*, 1974), none of these factors appears to present a generally applicable aetiology. Findings that antibody levels in Herpes simplex are commonly elevated in patients with Parkinson's disease (Marttila *et al.*, 1981) implicate this virus as a possible candidate, though the failures (Wetmur *et al.*, 1979; Mann *et al.*, 1981) to directly demonstrate this, or the presence of other viral material within brain tissues, weaken such an argument. An important clue to the possible nature of this factor has recently emerged from the case histories of young persons taking illicit drugs and developing a severe form of the disease (Langston *et al.*, 1983; Wright *et al.*, 1983). The toxic agent in question appears to be 1-methyl-4-phenyl 1,2,3,6-tetrahydropyridine (MPTP) (Langston and Ballard, 1983). Its injection into primates causes a severe and selective destruction of cells of substantia nigra (Burns *et al.*, 1983; Langston *et al.*, 1984) accompanied by a clinical syndrome of akinesia and rigidity that responds to L-dopa and dopamine agonists. Interestingly enough, MPTP has no apparent effect in rodents (Burns *et al.*, 1983; Langston *et al.*, 1984) whose nigral cells are pigment free, thereby strengthening a possible link between this substance and those cellular reactions involved in the degradation of dopamine via melanin pathways.

Furthermore, the noradrenergic cells of the locus coeruleus are undamaged by MPTP (Burns *et al.*, 1983; Langston *et al.*, 1984), possibly because the breakdown of any noradrenaline that may be accumulated through damage to terminals produces less toxic molecules (Graham *et al.*, 1978; Graham, 1979). These latter cells are always less affected in Parkinson's disease than cells of the substantia nigra, again suggesting a lessened vulnerability.

MPTP is a simple molecule and it is possible that it or a similar substance is widely present in the environment, or is a factor in some people's diets, or might even be produced in the body. Such possibilities clearly merit examination, as does the long-term follow-up, with post-mortem analysis, of those persons (and experimental animals) exposed to MPTP, since such work would at least provide a new model of the illness and might even give a clue as to its cause and pathogenesis.

6.5. Summary

The major groups of dopaminergic neurones in humans are located in the substantia nigra (A9) and ventral tegmental area (A10), and form the nigroneostriatal, mesolimbic and mesocortical pathways, respectively. In young persons these neurones contain much RNA and have a strongly basophilic nucleolus within a large open nucleus. With age the amounts or size of these features progressively decline, marking an atrophic process that causes dopamine levels in striatum and cortex to fall in parallel and one that results in the original cell complement being reduced by 30–60 % at 80 years of age. Such alterations are probably reflected in the difficulties in control and co-ordination of movement seen in many old people. In Parkinson's disease this pattern of change is exacerbated; cell loss from substantia nigra and ventral tegmental area exceeds 75 % and nucleolar volume is further reduced. These changes cause gross depletions of striatal and cortical dopamine and result in symptoms of rigidity, akinesia and tremor on the one hand, and on the other may partly invoke the cognitive impairment present late in the illness. It is suggested that those processes that lead in humans to the formation and accumulation of melanin in dopamine cells not only bring about the cell loss of old age, but also, by middle age, leave many of the remaining cells in such a weakened condition that they are 'easy prey' for pathogenic agents whose destructive action leads to gross cell loss and onset of Parkinsonian symptomatology.

Acknowledgements

The author wishes to thank Professor P. O. Yates for a critical reading of the manuscript, and Mrs P. Bellinger for its careful preparation.

References

Adler, A. (1942). Melanin pigment in the brain of a gorilla. *J. Comp. Neurol.* **76**, 501–7.

Adolfsson, R., Gottfries, C. G., Roos, B. E. and Winbald, B. (1979*a*). Postmortem distribution of dopamine and homovanillic acid in human brain, variations related to age and a review of the literature. *J. Neural Transm.* **45**, 81–105.

Adolfsson, R., Gottfries, C. G., Roos, B. E. and Winblad, B. (1979*b*). Changes in the brain catecholamines in patients with dementia of Alzheimer type. *Brit. J. Psychiat.* **135**, 216–23.

Ambani, L. M., van Woert, H. M. and Murphy, S. (1975). Brain peroxidase and catalase in Parkinson's disease. *Arch. Neurol.* **32**, 114–18.

Anden, N. E., Fuxe, K., Hamberger, B. and Hokfelt, T. (1966). A quantitative study of the nigro-neostriatal dopamine neurone system in the rat. *Acta Physiol. Scand.* **67**, 306–12.

Armstrong, D. M., Ross, C. A., Pickel, V. M., Joh, T. H. and Reis, D. J. (1982). Distribution of dopamine, noradrenaline and adrenaline-containing cell bodies in rat medulla oblongata. *J. Comp. Neurol.* **212**, 173–87.

Bak, I. J. (1967). The ultrastructure of the substantia nigra and caudate nucleus of the mouse and the cellular localization of catecholamines. *Exp. Brain Res.* **3**, 40–57.

Barden, H. (1969). The histochemical relationship of neuromelanin and lipofuscin. *J. Neuropathol. Exp. Neurol.* **28**, 419–41.

Bazelon, M., Fenichel, G. M. and Randall, J. (1967). Studies on neuromelanin. I. A melanin system in the human adult brain stem. *Neurology* **17**, 512–19.

Bernheimer, H., Birkmayer, W., Hornykiewicz, O., Jellinger, K. and Seitelberger, F. (1973). Brain dopamine and the syndromes of Parkinson and Huntington. *J. Neurol. Sci.* **20**, 415–55.

Bogerts, B. (1981). A brain stem atlas of catecholaminergic neurones in man, using melanin as a natural marker. *J. Comp. Neurol.* **197**, 63–80.

Bogerts, B., Hantsch, J. and Herzer, M. (1983). A morphometric study of the dopamine-containing cell groups in the mesencephala of normals, Parkinson patients and schizophrenics. *Biol. Psychiat.* **18**, 951–69.

Bowers, M. B. and Gerbode, F. A. (1968). Relationship of monamine metabolites in human cerebrospinal fluid to age. *Nature (Lond,)* **219**, 1256–7.

Brayda-Bruno, M. and Levi, A. C. (1979). Ultrastructure of neuromelanin granules in dog substantia nigra at different ages. *Boll. Soc. It. Biol. Sper.* **55**, 1902–8.

Brown, J. O. (1943). Pigmentation of substantia nigra and locus coeruleus in certain carnivores. *J. Comp. Neurol.* **79**, 393–405.

Bugiani, O., Salvarini, S., Perdelli, F., Mancardi, G. L. and Leonardi, A. (1978). Nerve cell loss with ageing in the putamen. *Eur. Neurol.* **17**, 286–91.

Burns, R. S., Chiueh, G. C., Markey, S. P., Ebert, M. H., Jacobowitz, D. M. and Kopin, I. J. (1983). A primate model of parkinsonism: selective destruction of dopaminergic neurones in pars compacta of substantia nigra by N-methyl-4-phenyl-1,2,3,6-tetrahydropyridine. *Proc. Nat. Acad. Sci. USA* **80**, 4546–50.

Carlsson, A. and Winblad, B. (1976). Influence of age and time interval between death and autopsy on dopamine and 3-methoxytyramine levels in human basal ganglia. *J. Neural Transm.* **38**, 271–6.

Cook, D. G., Fahn, S. and Brait, K. A. (1974). Chronic manganese intoxication. *Arch. Neurol.* **30**, 59–64.

Dahlstrom, A. and Fuxe, K. (1964). Evidence for the existence of monoamine-containing neurones in the central nervous system. *Acta Physiol. Scand.* (Suppl. 232), 1–55.

Domesick, V. B., Stinus, L. and Paskevich, P. A. (1983). The cytology of dopaminergic and non-dopaminergic neurones in the substantia nigra and ventral tegmental area of the rat: a light and electron microscopic study. *Neuroscience* **8**, 743–65.

Duffy, P. E. and Tennyson, V. M. (1965). Phase and electron microscope observations of Lewy bodies and melanin granules in the substantia nigra and locus coeruleus in Parkinson's disease. *J. Neuropathol. Exp. Neurol.* **24**, 398–414.

Evarts, E. V., Teravainen, H. T. and Calne, D. B. (1981). Reaction time in Parkinson's disease. *Brain* **104**, 167–86.

Felten, D. L. and Crutcher, K. A. (1979). Neuronal–vascular relationships in the raphe nuclei, locus coeruleus and substantia nigra in primates. *Amer. J. Anat.* **155**, 467–82.

Fenichel, G. M. and Bazelon, M. (1968). Studies on neuromelanin. II. Melanin in the brains of infants and children. *Neurology* **18**, 817–20.

Foley, J. M. and Baxter, D. (1958). On the nature of the pigment granules in the cells of the locus coeruleus and substantia nigra. *J. Neuropathol. Exp. Neurol.* **17**, 586–98.

Forno, L. S. (1969). Concentric hyaline intraneuronal inclusions of Lewy type in brains of elderly persons (50 additional cases and their relationship to Parkinsonism). *J. Amer. Geriat. Soc.* **17**, 557–75.

Goldman, J. E. and Horoupian, D. S. (1982). An immuno-cytochemical study of intraneuronal inclusions of the caudate nucleus and substantia nigra. Reaction with an anti-actin antiserum. *Acta Neuropathol.* **58**, 300–302.

Gottfries, C. G., Gottfries, I., Johansson, B., Olsson, R., Persson, T. and Roos, B. E. (1971). Acid monoamine metabolites in human cerebrospinal fluid and their relations to age and sex. *Neuropharmacology* **10**, 665–72.

Graham, D. G. (1978). Oxidative pathways for catecholamines in the genesis of neuromelanin and cytotoxic quinones. *Mol. Pharmacol.* **14**, 633–43.

Graham, D. G. (1979). On the origin and significance of neuromelanin. *Arch. Pathol. Lab. Med.* **103**, 359–62.

Graham, D. G., Tiffany, S. M. and Bell, W. R. (1978). Auto-oxidation versus covalent binding of quinones as the mechanism of toxicity of dopamine, 6-hydroxy-dopamine and related compounds towards C1300 neuroblastoma cells in vitro. *Mol. Pharmacol.* **14**, 644–53.

Greenfield, J. G. and Bosanquet, F. D. (1953). The brain stem lesions in Parkinsonism. *J. Neurol. Neurosurg. Psychiat.* **16**, 213–26.

Hoehn, M. M. and Yahr, M. D. (1967). Parkinsonism: onset, progression and mortality. *Neurology* **17**, 427–42.

Hornykiewicz, O. (1966). Dopamine and brain function. *Pharmacol. Rev.* **8**, 925–64.

Issidorides, M. R. (1971). Neuronal and vascular relationships in the zona compacta of normal and Parkinsonian substantia nigra. *Brain Res.* **25**, 289–99.

Issidorides, M. R., Mytilineou, C., Whetsell, W. O. and Yahr, M. D. (1978). Protein-rich cytoplasmic bodies of substantia nigra and locus coeruleus. *Arch. Neurol.* **35**, 633–7.

Jager, D. H. and Bethlem, J. (1960). The distribution of Lewy bodies in cerebral autonomic nervous system in idiopathic paralysis agitans (Parkinson's disease). *J. Neurol. Neurosurg. Psychiat.* **23**, 283–90.

Javoy-Agid, F., Plotska, A. and Agid, Y. (1981). Microtopography of tyrosine hydroxylase, glutamic acid decarboxylase, choline acetyl transferase in the substantia nigra and ventral tegmental area of control and Parkinsonian brains. *J. Neurochem.* **37**, 1218–27.

Juraska, J. M., Wilson, C. J. and Groves, P. M. (1977). The substantia nigra of the rat: a Golgi study. *J. Comp. Neurol.* **172**, 585–600.

Kaiya, H. (1980). Neuromelanin, neuroleptics and schizophrenia. Neuropsychobiology **6**, 241–8.

Langston, J. W. and Ballard, P. A. (1983). Parkinson's disease in a chemist working with 1-methyl 4-phenyl 1,2,3,6 tetrahydropyridine. *New Engl. J. Med.* **309**, 310.

Langston, J. W., Ballard, P. A., Tetrud, J. W. and Irwin, I. (1983). Chronic parkinsonism in humans due to a product of meperidine analogue synthesis. *Science* **219**, 979–80.

Langston, J. W., Forno, L. S., Rebert, C. S. and Irwin, I. (1984). Selective nigral toxicity after systemic administration of 1-methyl 4-phenyl 1,2,5,6-tetrahydropyridine (MPTP). *Brain Res.* **292**, 390–4.

Lavyne, M. H., Moskowitz, M. A. and Larin, F. (1975). Brain ^{3}H-catecholamine metabolism in experimental cerebral ischaemia. *Neurology* **25**, 483–5.

Lindvall, O. and Bjorklund, A. (1978). Organization of catecholamine neurons in rat central nervous system. In *Handbook of Psychopharmacology*, vol. 9. L. Iversen, S. Iversen and S. N. Snyder (eds), pp. 139–231. Plenum, New York.

Lindvall, O., Bjorklund, A. and Skagerberg, G. (1983). Dopamine-containing neurones in the spinal cord: anatomy and some functional aspects. *Ann. Neurol.* **14**, 255–60.

Lipkin, L. E. (1959). Cytoplasmic inclusions in ganglion cells associated with Parkinsonian states; a neurocellular change studied in 53 cases and 206 controls. *Amer. J. Pathol.* **35**, 1117–33.

Lycke, E. and Roos, B. E. (1969). Some virological and biochemical aspects of the pathogenesis of Parkinson's disease. In *Third Symposium on Parkinson's Disease.* F. J. Gillingham and P. M. Donaldson (eds), pp. 16–21. Livingstone, Edinburgh.

McGeer, P. L., McGeer, E. G. and Suzuki, J. S. (1977). Aging and extrapyramidal function. *Arch. Neurol.* **34**, 33–5.

McNeill, T. H., Koek, L. L. and Haycock, J. W. (1984). Age-correlated changes in dopaminergic nigrostriatal perikarya of the C57BL/GNNia mouse. *Mech. Ageing. Dev.* **24**, 293–308.

Mann, D. M. A. and Yates, P. O. (1974). Lipoprotein pigments and their relationship to ageing in the human nervous system. II. The melanin content of pigmented nerve cells. *Brain* **97**, 489–98.

Mann, D. M. A. and Yates, P. O. (1979). The effects of ageing on the pigmented nerve cells of the human locus coeruleus and substantia nigra. *Acta Neuropathol.* **47**, 93–7.

Mann, D. M. A. and Yates, P. O. (1982). The pathogenesis of Parkinson's disease. *Arch. Neurol.* **39**, 545–9.

Mann, D. M. A. and Yates, P. O. (1983*a*). Possible role of neuromelanin in the pathogenesis of Parkinson's disease. *Mech. Ageing Dev.* **21**, 193–201.

Mann, D. M. A. and Yates, P. O. (1983b). Serotonin nerve cells in Alzheimer's disease. *J. Neurol. Neurosurg. Psychiat.* **46**, 96.

Mann, D. M. A., Yates, P. O. and Barton, C. M. (1977a). Variations in melanin content with age in the human substantia nigra. *Biochem. Exp. Biol.* **13**, 137–9.

Mann, D. M. A., Yates, P. O. and Barton, C. M. (1977b). Melanin and RNA in cells of the human substantia nigra. *J. Neuropathol. Exp. Neurol.* **36**, 379–83.

Mann, D. M. A., Yates, P. O. and Marcyniuk, B. (1984). Monoaminergic neurotransmitter systems in presenile Alzheimer's disease and in senile dementia of Alzheimer type. *Clin. Neuropathol.* **3**, 199–205.

Mann, D. M. A., Yates, P. O., Marcyniuk, B. and Ravindra, C. R. (1985). Pathological evidence for neurotransmitter deficits in Downs Syndrome of middle age. *J. Ment. Defic. Res.* **29**, 125–35.

Mann, D. M. A., Yates, P. O., Davies, J. S. and Hawkes, J. (1981). Viruses, Parkinsonism and Alzheimer's disease. *J. Neurol. Neurosurg. Psychiat.* **44**, 651.

Mann, D. M. A., Lincoln, J., Yates, P. O., Stamp, J. E. and Toper, S. (1980). Changes in monoamine-containing neurones of the human CNS in senile dementia. *Brit. J. Psychiat.* **136**, 533–41.

Marsden, C. D. (1961). Pigmentation in the substantiae nigrae of mammals. *J. Anat.* **95**, 256–61.

Marttila, R. J., Rinne, U. K. and Halonen, P. E. (1981). Herpes viruses and Parkinsonism. *Arch. Neurol.* **38**, 19–21.

Moore, R. Y. (1982). Catecholamine neurone systems in brain. *Ann. Neurol.* **12**, 321–7.

Moore, R. Y. and Bloom, F. E. (1978). Cerebral catecholamine neurone systems: anatomy and physiology of the dopamine systems. *Ann. Rev. Neurosci.* **1**, 129–69.

Ohama, E. and Ikuta, F. (1976). Parkinson's disease: distribution of Lewy bodies and monoamine system. *Acta Neuropathol.* **34**, 311–19.

Pakkenberg, H. and Brody, H. (1965). The number of nerve cells in the substantia nigra in paralysis agitans. *Acta Neuropathol.* **5**, 320–4.

Price, K. S., Farley, I. J. and Hornykiewicz, O. (1978). Neurochemistry of Parkinson's disease. In *Advances in Biochemical Psychopharmacology*, Vol. 19. P. J. Roberts (ed.), pp. 293—300. Raven Press, New York.

Resine, T. D., Yamamura, H. I., Bird, E. D. Spokes, E. and Enna, S. J. (1978). Pre- and post-synaptic neurochemical alterations in Alzheimer's disease. *Brain Res.* **159**, 477–81.

Riederer, P. and Wuketich, S. T. (1976). Time course of nigrostriatal degeneration in Parkinson's disease. *J. Neural Transm.* **38**, 277–301.

Rinvik, E. and Grofova, I. (1970). Observations on the fine structure of the substantia nigra in the cat. *Exp. Brain Res.* **11**, 229–48.

Saper, C. B. and Petito, C. L. (1982). Correspondence of melanin-pigmented neurones in human brain with A1–A14 catecholamine cell groups. *Brain* **105**, 87–101.

Scheibel, A. B. and Tomiyasu, U. (1980). A dendritic–vascular relationship in the substantia nigra. *Exp. Neurol.* **70**, 717–20.

Scherer, H. J. (1939). Melanin pigmentation of the substantia nigra in primates. *J. Comp. Neurol.* **71**, 91–8.

Schochet, S. S., Wyatt, R. D. and McCormick, W. F. (1970). Intracytoplasmic acidophilic granules in the substantia nigra. *Arch. Neurol.* **22**, 550–5.

Sekiya, S., Tanaka. M., Hayashi, S. and Oyanagi, S. (1982). Light and electron microscopic studies of intracytoplasmic acidophilic granules in the human locus coeruleus and substantia nigra. *Acta Neuropathol.* **56**, 78–80.

Sekiya, S., Tanaka, M., Hayashi, S. and Oyanagi, S. (1983). Distribution of intracytoplasmic acidophilic granule-containing neurones in the human brain. *Acta Neuropathol.* **60,** 145–8.

Severson, J. A., Marcusson, J., Winblad, B. and Finch, C. E. (1982). Age-correlated loss of dopaminergic binding sites in human basal ganglia. *J. Neurochem.* **39,** 1623–31.

Spokes, E. G. S. (1979). An analysis of factors influencing measurements of dopamine, noradrenaline, glutamate decarboxylase and choline acetylase in human post mortem brain tissue. *Brain* **102,** 333–46.

Sulkava, R. (1982). Alzheimer's disease and senile dementia of Alzheimer type: a comparative study. *Acta Neurol. Scand.* **65,** 636–50.

Tatton, W. G. and Lee, R. G. (1975). Evidence for abnormal long loop-reflexes in rigid Parkinsonian patients. *Brain Res.* **100,** 671–6.

Ungerstedt, U. (1971). Stereotaxic mapping of the monoamine pathways in the rat brain. *Acta Physiol. Scand.* **82** (Suppl. 367), 1–48.

Ward, C. D., Duvoisin, R. C., Ince, S. E., Nutt, J. D., Eldridge, R. and Calne, D. B. (1983). Parkinson's disease in 65 pairs of twins and in a set of quadruplets. *Neurology* **33,** 815–24.

Wetmur, J. B., Schwartz, J. and Elizan, T. S. (1979). Nucleic acid homology studies of viral nuclei acids in idiopathic Parkinson's disease. *Arch. Neurol.* **36,** 462–4.

Winblad, B., Adolfsson, R., Gottfries, C. G., Oreland, L. and Roos, B. E. (1978). Brain monoamines, monoamine metabolites and enzymes in physiological ageing and senile dementia. In *Recent Advances in Mass Spectrometry in Biochemistry and Medicine.* A. Frigerio (ed.), pp. 253–7. Academic Press, New York.

Winblad, B., Adolfsson, R., Carlsson, A., Gottfries, C. G. and Oreland, L. (1981). Brain biogenic amines in dementia of Alzheimer type. In *Biological Psychiatry 1981.* C. Perris, G. Struwe and B. Jansson (eds), pp. 965–9. Elsevier North-Holland, Amsterdam.

Woodward, J. S. (1962). Concentric hyaline inclusion body formation in mental disease. An analysis of 27 cases. *J. Neuropathol. Exp. Neurol.* **21,** 442–50.

Wright, J. M., Wall, R. A., Perry, T. L. and Paty, D. W. (1983). Chronic parkinsonism secondary to intranasal administration of a product of meperidine analogue synthesis. *New Engl. J. Med.* **310,** 325.

Yates, C. M., Allison, Y., Simpson, J., Maloney, A. F. J. and Gordon, A. (1979). Dopamine in Alzheimer's disease and senile dementia. *Lancet* **ii,** 851–2.

7

The D-1 dopamine receptor and ageing: behavioural and neurochemical studies

A. G. Molloy, K. M. O'Boyle and J. L. Waddington†

Department of Clinical Pharmacology, Royal College of Surgeons in Ireland, St. Stephen's Green, Dublin 2, Ireland

7.1. Introduction

There is considerable current interest in changes in dopaminergic systems occurring during ageing and senescence, especially in relation to motor function. Much of this work has involved the study of brain dopamine receptors, via behavioural responses such as stereotypy induced by the typical agonist apomorphine, and using radioligands such as [³H]spiperone (Finch *et al.*, 1981). These indices are thought to reflect primarily changes in D-2 dopaminergic function (Creese *et al.*, 1983). D-1 dopaminergic function, apart from its associated neurochemical response of stimulation of adenylate cyclase activity, has been far less systematically investigated, primarily because of the lack of agents that selectively influence this system. We have recently identified the *R*-enantiomer of the racemic compound SK&F 38393 as a stereoselective D-1 agonist (O'Boyle and Waddington, 1984*a*) and have described behavioural responses to this compound which appear to be mediated through D-1 receptors (Molloy and Waddington, 1984; Molloy *et al.*, 1984). The present studies extend these findings to comparisons between young and aged animals.

7.2. Methods

Young (3-month) and aged (22-month) male Sprague-Dawley rats were habituated to the test cage for 2·5 hr before subcutaneous injection of *R*-SK &F 38393 (1·25–20·0 mg kg^{-1}) or vehicle. Thereafter they were observed for 5 s periods at 1 min intervals over five consecutive minutes, and this cycle was repeated at 10 min intervals; on each occasion the individual behaviours observed were recorded using a behavioural checklist and group prevalences of these behaviours were determined (Molloy and Waddington, 1984).

†To whom correspondence should be addressed.

Behavioural responses were evaluated in comparison with radioligand binding studies of D-1 and D-2 receptors, using [³H]piflutixol and [³H]spiperone, respectively, in young and aged animals over a similar age range (O'Boyle and Waddington, 1984*b*).

7.3. Results

R-SK&F 38393 induced episodes of sniffing, locomotion, rearing and particularly prominent grooming behaviour, without inducing stereotypy. The responses of sniffing, grooming and intense grooming were reliably induced by 1·25–20·0 mg kg^{-1} *R*-SK&F 38393, and were indistinguishable between young and aged animals. Locomotion in response to 20·0 mg kg^{-1} *R*-SK&F 38393 was significantly more prevalent ($P < 0·05$) in aged animals, and a similar trend was noted for rearing. Chewing was also significantly more prevalent ($P < 0·05$) in aged animals at this dose (Fig. 7.1).

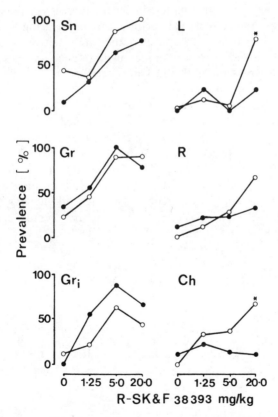

Fig. 7.1.   Prevalences (% of $n = 8$–9 per group) of sniffing (Sn), grooming (Gr), grooming intensely (Gr$_i$), locomotion (L), rearing (R) and chewing (Ch) 40 min after challenge with vehicle or 1·25–20·0 mg kg^{-1} *R*-SK & F 38393 in young (●) and aged (○) animals. *, $P < 0·05$.

We have recently reported (O'Boyle and Waddington, 1984*b*) that whereas striatal D-2 receptors labelled by [^{3}H]spiperone are lost from the brain with ageing (see Finch *et al.*, 1981), there are no changes in the characteristics of the [^{3}H]piflutixol-labelled D-1 receptor over this same age range.

7.4. Discussion

Episodes of non-stereotyped sniffing and grooming induced by *R*-SK&F 38393 appear to be principally mediated through D-1 receptors; they are sensitive to blockade by the D-1 antagonist SCH 23390 but are resistant to attenuation by the D-2 antagonist metoclopramide (Molloy and Waddington, 1984; Molloy *et al.*, 1984). The ability of *R*-SK&F 38393 to induce these behaviours to an indistinguishable extent in young and aged animals is consistent with our binding studies, which indicate that D-1 receptor characteristics are unaltered over the present age range.

Episodes of non-stereotyped locomotion and rearing induced by *R*-SK&F 38393 appear more complicated; they are sensitive to antagonism by both SCH 23390 and metoclopramide, suggesting an associated role of D-2 systems in the expression of this D-1-initiated response (Molloy *et al.*, 1984). It is interesting that these responses were more prevalent in aged animals in which the density of D-2 receptors but not D-1 receptors is reduced and hence the ratio of D-1 to D-2 receptors is perturbed.

Rosengarten *et al.* (1983) have described the induction of repetitious opening and closing of the mouth and high-frequency clonic jaw movements, by racemic SK&F 38393. We have found *R*-SK&F 38393 to induce perioral movements, and have noted chewing movements distinct from apomorphine-induced stereotyped biting and gnawing (Waddington and Gamble, 1980). This response was also more common in aged than in young animals, where the ratio of D-1 to D-2 receptors is elevated. It appears that the characteristics of the D-1 receptor may not change over the present age range; however, loss of D-2 receptors may influence D-1- initiated behaviours if D-2 systems modulate their expression.

Acknowledgements

We thank the following for financial support: the Medical Research Council of Ireland; the Royal College of Surgeons in Ireland; the Royal College of Physicians of Ireland; Sanity; and Smith Kline & French for gifts of drugs.

References

Creese, I., Sibley, D. R., Hamblin, M. R. and Leff, S. E. (1983). *Ann. Rev. Neurosci.* **6**, 43–71.
Finch, C. E., Marshall, J. F. and Randall, P. K. (1981). *Ann. Rev. Gerontol. Geriatr.* **2**, 49–87.

Molloy, A. G. and Waddington, J. L. (1984). *Psychopharmacology* **82**, 409–10.
Molloy, A. G., O'Boyle, K. M. and Waddington, J. L. (1984). *Brit. J. Pharmacol.* **81**, 21P.
O'Boyle, K. M. and Waddington, J. L. (1984*a*). *Eur. J. Pharmacol.* **98**, 433–6.
O'Boyle, K. M. and Waddington, J. L. (1984*b*). *Irish J. Med. Sci.* **153**, 225.
Rosengarten, H., Schwitzer, J. W. and Friedhoff, A. J. (1983). *Life Sci.* **33**, 2479–82.
Waddington, J. L. and Gamble, S. J. (1980). *Eur. J. Pharmacol.* **68**, 387–8.

8

Properties of dopamine receptors of identified neurones in the brain of the gastropod mollusc, *Helix aspersa*

A. J. Bokisch and R. J. Walker

School of Biochemical and Physiological Sciences, University of Southampton, Southampton SO9 3TU, UK

8.1. Introduction

On *Helix* central neurones, dopamine can have either an inhibitory or an excitatory effect, depending on the type of neurone impaled. Qualitative work on a *Helix* inhibitory dopamine response and inhibitory postsynaptic potential (Kerkut *et al.*, 1969) showed that the response was due to an increase in potassium permeability and that ergometrine would antagonise the response. Batta (1979) reported that neuroleptics, including the racemate of sulpiride, would antagonise dopamine inhibition in *Helix*. 2-Amino-6,7-dihydroxy-1,2,3,4-tetrahydro-naphthalene (ADTN), which conforms to the *b*-rotameric configuration of dopamine, is a potent agonist of the dopamine inhibitory response in *Helix* (Batta, 1979). For optimal agonist activity on the dopamine inhibitory receptor, the requirements are two hydroxyl groups on the 3- and 4-positions of the benzene ring and no larger substitution on the terminal nitrogen than one methyl group (Woodruff and Walker, 1969). The aim of this study was to investigate quantitatively the pharmacology of the dopamine response of certain identified neurones in *Helix* and to investigate the ions involved in the response using voltage clamp techniques.

8.2. Methods

Intracellular microelectrode recordings were made from identified neurones in the isolated central nervous system of the common garden snail *Helix aspersa*. The ganglia were prepared as described by Walker (1968) and placed in an experimental bath of 5 ml volume. Neuronal activity was amplified and displayed by conventional neurophysiological techniques, and for some experiments a Dagan 8100 single-electrode voltage clamp system was used. The cells used were F3, F4, F5, F6, E1 and E2, as described in the mapping work of Kerkut *et al.* (1975).

Antagonists were added to the bath in a volume of 1 ml and were allowed to equilibrate for 5 to 10 min prior to testing. Agonists were either bath-applied in a volume of 0·4 ml, or dopamine was iontophoresed from a second micropipette containing a 0·5 M solution, at pH 4·0. Quantitative determinations were made of antagonist potencies by constructing log dose–response curves to iontophoretic dopamine in the absence and presence of the antagonist and calculating pA2 values from these. In experiments where the ionic composition of the extracellular medium was altered, the preparation was allowed to equilibrate for approximately 3 min before the effect of the alteration was tested. The normal potassium concentration of *Helix* Ringer is 4 mM and in these experiments 0 mM and 24 mM extracellular potassium were also used.

The drugs used in these experiments were: apomorphine HCl, Sigma; bromocriptine mesylate, Sigma; dopamine HCl, Sigma; ergonovine maleate, Sigma; (+)-sulpiride, Ravizza; (−)-sulpiride, Ravizza; (±)-sulpiride, Ravizza; zetidoline, Ravizza.

8.3. Results

The ions involved in mediating the dopamine inhibitory response on the identified neurones used in this study were investigated using the single-electrode voltage clamp. The dopamine response was found to be associated with an outward current and increase in membrane conductance at the normal resting potential of the cell. The null point of the response occurred between −75 and −80 mV, but at potentials more negative than this, no reversal of the response was obtained, and the conductance increase was no longer apparent (cf. ch. 4). In high-potassium Ringer (24 mM) the dopamine inhibitory response was abolished, and in potassium-free Ringer the response was potentiated.

The pharmacology of this inhibitory dopamine receptor was investigated quantitatively using a range of dopamine antagonists. Table 8.1 summarises the pA2 values obtained in these experiments. Each value is expressed as the mean ± the standard error of the mean for at least five experiments.

Table 8.1. pA2 values for dopamine antagonists on *Helix* neurones.

Antagonist	pA2	n
(±)-Sulpiride	5·8 ± 0·1	5
(+)-Sulpiride	4·7 ± 0·1	5
(−)-Sulpiride	6·0 ± 0·1	5
Zetidoline	5·3 ± 0·1	6
Apomorphine	8·4 ± 0·1	5
Bromocriptine	6·2 ± 0·2	5
Ergometrine	12·5 ± 0·1	5

As can be seen from Table 8.1, the *Helix* inhibitory dopamine receptor exhibits stereoselectivity for sulpiride, with the (−)-enantiomer being approximately 20 times more potent than the (+)-enantiomer and twice as potent as the racemate. The antagonism produced by sulpiride was readily reversible following washing, and the response quickly returned to control values. Zetidoline was also an easily reversible antagonist of dopamine inhibition with a pA2 of 5·3, making it five times less potent than the racemate of sulpiride.

Ergometrine was the most potent antagonist tested so far, giving a pA2 of 12·5, and was completely reversible at low concentrations, that is, lower than 5 pM. Bromocriptine and apomorphine were, respectively, six and four orders of magnitude less potent than ergometrine and even a partial block of the dopamine response by these compounds proved only partly reversible.

8.4. Discussion

The dopamine inhibitory response on the identified neurones used in this study is mediated via an increase in the membrane conductance to potassium ions. The ion channels involved in this response are voltage sensitive and no clear reversal of the response can be obtained (*cf.* ch. 4).

The stereoselectivity of sulpiride antagonism found in *Helix* mirrors that found in most mammalian dopamine test systems except the canine renal vascular bed where the (+)-enantiomer is the more potent (Goldberg *et al.*, 1978). The degree of selectivity also compares well with data from mammalian experiments; for example, Andrews and Woodruff (1978) found that (−)-sulpiride was 25 times more potent than (+)-sulpiride in behavioural studies. Zetidoline has been reported to be more potent than sulpiride in displacing [³H]-sulpiride binding (Holden-Dye *et al.*, 1983) and also to be a potent antagonist of the dopamine autoreceptors in the substantia nigra zona compacta (Harris and Woodruff, 1983). In this study, however, zetidoline was found to be about five times less potent than the racemate of sulpiride in antagonising dopamine inhibition.

The extreme potency of ergometrine in antagonising the dopamine inhibitory response is somewhat surprising and requires further investigation. Marek and Roth (1980) reported that bromocriptine may interact non-competitively or irreversibly with presynaptic dopamine receptors in the neostriatum and olfactory tubercles, which could be the cause of the apparent lack of reversibility of bromocriptine antagonism of the dopamine hyperpolarising response in *Helix*.

The results from these experiments indicate that the phamacological profile of the dopamine inhibitory receptor in *Helix* does not fit into that of the mammalian D-1, D-2 receptor classification of Kebabian and Calne (1979). Under their scheme, apomorphine and the dopamine-like ergots would be antagonists on the D-1 receptor where sulpiride would be inactive, and on the D-2 receptor type the ergots would be agonists and sulpiride an antagonist. In these experiments bromocriptine and sulpiride proved to be approximately equipotent in antagonising dopamine inhibition. The pharmacology of this dopamine receptor does comply with that of the DAi receptor of Cools and van Rossum (1976), which was originally defined in *Helix* and has also been proposed to occur in the caudate nucleus and nucleus accumbens.

Acknowledgement

We are grateful to the University of Southampton for financial support for this project.

References

Andrews, C. D. and Woodruff, G. N. (1978). *Brit. J. Pharmacol.* **64**, 434P.
Batta, S. (1979). Ph.D. Thesis, University of Southampton.
Cools, A. R. and van Rossum, J. M. (1976). *Psychopharmacologia* **45**, 243–54.
Goldberg, L. I., Kohli, J. D., Kotake, A. N. and Volkman, P. H. (1978). *Fed. Proc.* **37**, 2396–402.
Harris, N. C. and Woodruff, G. N. (1983). *Brit. J. Pharmacol.* **80**, 426P.
Holden-Dye, L., Poat, J. A., Senior, K. A. and Woodruff, G. N. (1983). *Brit. J. Pharmacol.* **80**, 427P.
Horn, N. and Walker, R. J. (1969). *Comp. Biochem. Physiol.* **30**, 1061–74.
Kebabian, J. W. and Calne, D. B. (1979). *Nature (Lond.)* **277**, 93–6.
Kerkut, G. A., Horn, N. and Walker, R. J. (1969). *Comp. Biochem. Physiol.* **30**, 1061–74.
Kerkut, G. A., Marek, K. L. and Roth, R. H. (1980). *Eur. J. Pharmacol.* **62**, 137–46.
Walker, R. J. (1968). *Exp. Phsyiol. Biochem.* **1**, 331–45.
Woodruff, G. N. and Walker, R. J. (1969). *Int. J. Neuropharm.* **8**, 279–89.

9

Differential postsynaptic potential latencies on cells monosynaptically connected to the giant dopamine-containing neurone of *Lymnaea stagnalis*

W. Winlow

Department of Physiology, School of Medicine, University of Leeds, Leeds LS2 9JT, UK

9.1. Introduction

The soma of the giant dopamine-containing neurone, R. Pe.D.1, of *Lymnaea stagnalis* lies on the dorsal surface of the right pedal ganglion and is homologous to the giant dopamine-containing neurone of *Planorbis corneus* (see Ch. 4). It is monosynaptically coupled to electrophysiologically identified neurones of the right parietal and median visceral ganglia (Benjamin and Winlow, 1981) and has different postsynaptic actions on each cell type (Winlow *et al.*, 1981). In this chapter, evidence for considerable differences in the latency to onset of the different postsynaptic potential (p.s.p.) types is presented.

9.2. Materials and methods

Specimens of *Lymnaea stagnalis* (L.), in the 2–6 g range, were obtained from animal suppliers, kept in tap water at room temperature and fed on lettuce. Brains were removed and maintained in Hepes-buffered saline as described by Benjamin and Winlow (1981). Individual cells were penetrated with glass microelectrodes filled with the supernatant from a saturated solution of K_2SO_4. Signals were conventionally amplified, displayed and recorded according to the methods of Benjamin and Winlow (1981).

9.3. Results and discussion

R.Pe.D.1 makes three types of connection with its follower cells whose locations are shown in Fig. 9.1. It makes excitatory connections to the giant cells VD2 and VD3, most of the cells of A group (A gp), the H cells and the K cells. It has biphasic actions (excitation followed by inhibition) on all the cells of G gp, the I cells, and I-type cells scattered within A gp. In addition it has inhibitory actions on the J cells (Winlow *et al.*, 1981). As can be seen

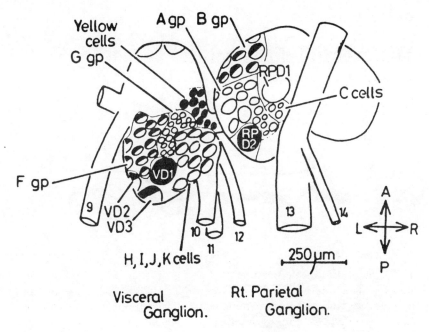

Fig. 9.1. Dorsal view of the median visceral and right parietal ganglia of *Lymnaea* showing the positions of identified neurones and cell clusters, many of which are follower cells of R.Pe.D.1. Shaded cells are white and unshaded cells are orange, with part-shaded cells representing varying degrees of paleness of colour. B and F gp extend on to the ventral surface of the visceral and right parietal ganglia, respectively. 9, Cutaneous pallial nerve; 10, intestinal nerve; 11, anal nerve; 12, genital nerve; 13, right internal parietal nerve; 14, right external parietal nerve. (After Benjamin and Winlow, 1981.)

from Table 9·1, the latency to onset of p.s.p.s induced by discharge of R.Pe.D.1 varies greatly from one cell type to the next. The e.p.s.p.s on A gp, H and K cells, and the A gp b.p.s.p.s appear to have similar latencies to onset (*c*. 17–19·5 ms) whereas those found on VD2/3 are considerably greater and compare with the excitatory phases of the biphasic p.s.p.s on G and I cells (*c*. 38–44 ms). The i.p.s.p.s on the J cells have a mean latency to onset of 182 ms and this appears to be comparable with the inhibitory phase of the G and I cell b.p.s.p.s. Thus at least two types of excitatory receptor and one type of inhibitory receptor to dopamine appear to occur on follower cells of R.Pe.D.1.

The large differences in p.s.p. latencies to onset shown in Table 9.1 cannot be accounted for by differences in path length or conduction velocity since extracellularly recorded axon spikes of R.Pe.D.1 in the nerve trunks of the visceral and right parietal ganglion occur 11–14 ms after the intracellular spike recorded in the soma (Fig. 9.2). A conservative estimate of the transit

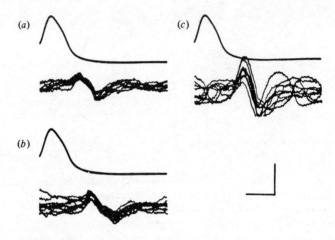

Fig. 9.2. Latency of peripheral nerve action potential to soma spike action potentials of R.Pe.D.1. (a–c) Ten superimposed sweeps. Oscilloscope sweep triggered by the ascending phase of the intracellularly recorded spike, showing 1:1 intra-extracellularly recorded spikes with constant latency. Upper, intracellular spike. Lower, extracellular spike. (a) External parietal nerve. (b) Internal parietal nerve. (c) Anal nerve. In each pair of traces, ten superimposed sweeps are shown. The oscilloscope sweep was triggered by the extracellular stimulus pulse. Voltage calibration: upper traces 50 mV; lower traces 500 μV. Time calibration: 10 ms. (After Haydon and Winlow, 1981.)

Table 9.1. Observation of 1 : 1 p.s.p.s and latencies in follower cells of R.Pe.D.1.

p.s.p. type	Cell type	Mean latency		
		(*N*)	± *S.D.*	(*n*)
e.p.s.p.s	VD2/3	9	41·3 ± 6·3	(4)
	A gp	34	17·1 ± 4·9	(12)
	H cells	22	18·7 ± 4·9	(11)
	K cells	27	16·3 ± 4·0	(12)
b.p.s.p.s	A gp (I type)	6	19·5 ± 6·6	(6)
	G gp	13	44·0 ± 9·7	(8)
	I cells	12	38·8 ± 2·0	(4)
i.p.s.p.s	J cells	27	182·4 ± 44·0	(17)

(*N*) indicates the number of times that 1 : 1 p.s.p.s have been recorded between R.Pe.D.1 and any given follower cell. The mean p.s.p. latency is also shown, and here (*n*) indicates the number of preparations in which p.s.p. latency was accurately determined from superimposed oscilloscope sweeps. The p.s.p. latencies shown for each cell type were tested against those for all other cell types by means of the Student's *t*-test. The latencies to onset for e.p.s.p.s on cells of A gp, H cells and K cells and for b.p.s.p.s on A gp were not significantly different from one another. They were, however, significantly different (*P* <0·001 in most cases) from the latencies to onset of e.p.s.p.s on VD2/3 and b.p.s.p.s on G gp and I cells (which were not significantly different from one another). The J cell i.p.s.p. latencies were significantly different from all other p.s.p. latencies (*P* <0·001 in all cases). (From Winlow *et al.*, 1981.)

time of the presynaptic spike to its axon terminals would therefore be of the order of 8–10 ms. It is possible that the wide differences in p.s.p. latencies induced by R.Pe.D.1 are due to differences in 'the kinetics of the postsynaptic actions of the transmitter rather than conduction velocity or delay in release of transmitter' (Berry and Pentreath, 1976). I would speculate that second messengers such as cyclic nucleotides may mediate the very long latency i.p.s.p.s we have observed.

References

Benjamin, P. R. and Winlow, W. (1981). *Comp. Biochem. Physiol.* **70A**, 293–307.
Berry, M. S. and Pentreath, V. W. (1976). *Brain Res.* **105**, 1–20.
Haydon, P. G. and Winlow, W. (1981). *J. exp. Biol.* **94**, 149–57.
Winlow, W., Haydon, P. G. and Benjamin, P. R. (1981). *J. exp. Biol.* **94**, 137–48.

10

Modulation by dopamine and 5-hydroxytryptamine of rhythmic activity in buccal motoneurones of the pond snail, Lymnaea stagnalis

M. D. Tuersley and C. R. McCrohan

Department of Zoology, University of Manchester, Manchester M13 9PL, UK

10.1. Introduction

Molluscan nervous systems provide good preparations for studies of cellular actions of neuroactive amines such as dopamine and 5-hydroxytryptamine (5-HT) (Cottrell, 1977). These neurotransmitters have been located in identifiable neurones in the gastropod CNS and are known to affect neuronal activity. They act as neuromodulators for the control of rhythmic motor behaviour in a number of invertebrate species, including feeding in gastropods (Wieland and Gelperin, 1983). We have examined the effects of dopamine on a well described molluscan feeding system, using intracellular recording of synaptic activity in neurones responsible for the patterned motor output. The effects of dopamine are compared with those of 5-HT.

A rhythm-generating network of interneurones has been identified in the buccal ganglia of *Lymnaea stagnalis* (Benjamin, 1983). These neurones provide identifiable rhythmic synaptic inputs to buccal feeding moto-neurones which innervate the buccal musculature. The rhythm can be monitored by recording from the motoneurones. In this study, the whole isolated CNS, which is capable of generating rhythmic buccal motor output, was perfused with dopamine or 5-HT. Specific effects on the rhythm were recorded.

10.2. Methods

Spike activity and synaptic inputs in identified feeding motoneurones were monitored intracellularly for 10 min before and during perfusion. One-hundred millilitres of 10^{-5} M dopamine or 5-HT in buffered saline was allowed to flow into the 8 ml bath at a rate of 2.5 ml s^{-1}, gradually displacing the normal saline over a weir. This produced a ramp of drug concentration, culminating in a final concentration of 10^{-5} M, which remained until the end of the experiment. A control application of normal saline was given prior to perfusion.

10.3. **Results**

The effects of dopamine were determined using 24 isolated CNS preparations. A number of effects were identified. In 13 preparations, dopamine produced an increase in frequency of rhythmic synaptic inputs to buccal motoneurones, leading to more frequent burst activity (Figs. 10.1 and 10.2a). In addition, rhythmic motoneurone bursts were usually stronger (20 preparations; Fig. 10.1). This strengthening is probably due to a tonic depolarising effect of dopamine, rather than to increased amplitude of rhythmic synaptic inputs, since in non-rhythmic preparations dopamine has a tonic excitatory effect on feeding motoneurones. The increased frequency of synaptic inputs was often associated with a decrease in their duration (13 preparations) (Fig. 10.2a). The overall effect of dopamine was therefore to produce more frequent, stronger motoneuronal bursting without any significant effect on the type of bursts produced. Nine preparations showed a further response. Rhythmic bursting became more organised into bouts of bursting, interspersed with periods of silence (Fig. 10.1).

Like dopamine, 5-HT has a tonic depolarising effect on buccal motoneurones (McCrohan and Benjamin, 1980). However, our results with 5-HT were markedly different from those with dopamine. 5-HT affected the nature of rhythmic synaptic inputs to motoneurones by increasing their amplitude and duration. This produced motoneurone bursts completely different from those observed before perfusion (Fig. 10.2b). In addition, 5-HT often caused the rhythm generator to lock into a non-rhythmic state (Fig. 10.2b), thus switching off rhythmic motor output. This result was never obtained following dopamine perfusion.

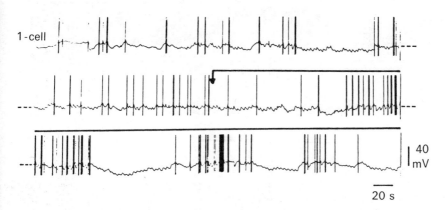

1-cell

40 mV

20 s

Fig. 10.1. Continuous intracellular recording from buccal motoneurone (1-cell) before and after application of 10^{-5} M dopamine (indicated by arrow and bar). In the presence of dopamine, rhythmic bursts of spikes are organised into bouts of about 50 s duration. Within these bouts, bursts of spikes are stronger and more frequent than those recorded before perfusion.

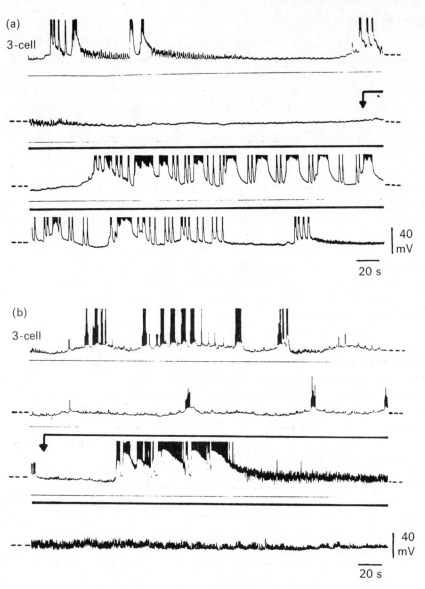

Fig. 10.2.   Effects of (a) 10^{-5} M dopamine, and (b) 10^{-5} M 5-HT perfusion (arrow and bar) on the same buccal motoneurone type (3-cell), recorded continuously in different preparations. (a) Dopamine leads to increased strength and frequency of motoneuronal bursting. Synaptic inputs leading to bursts are apparently similar to those recorded before perfusion. (b) 5-HT leads to rhythmic synaptic inputs with much larger amplitudes and durations than those recorded before perfusion, resulting in markedly stronger bursts. After about 160 s 5-HT perfusion, the rhythm generator locks into a non-rhythmic state, so that rhythmic synaptic inputs and bursting cease.

10.4. Discussion

These results indicate that dopamine and 5-HT have specific and distinguishable effects on the *Lymnaea* feeding system. Dopamine acts to modulate the timing of patterned output from the rhythm generator, while not significantly affecting its nature. 5-HT, on the other hand, seems to affect the rhythm generator itself, modifying or disrupting the rhythm. This suggests that *in vivo* the two amines may act independently to control different aspects of feeding motor output. Work on other gastropods has shown that the effects of dopamine and 5-HT on rhythmic output may be separated pharmacologically (Wieland and Gelperin, 1983; Trimble and Barker, 1984). The gastropod buccal ganglia present an ideal 'model' system for the study of modulatory effects of these amines on the coordinated output of identified networks of neurones at the synaptic level. Attention has already been paid to the role of 5-HT in the buccal ganglia, including the involvement of the giant serotonin-containing cerebral neurones (McCrohan and Benjamin, 1980). The physiological basis of dopamine's actions is unkown, though work on another pond snail, *Helisoma*, suggests that a dopaminergic system is intrinsic to the buccal ganglia themselves (Trimble and Barker, 1984).

Acknowledgements

M. D. Tuersley is supported by an SERC studentship.

References

Benjamin, P. R. (1983). In *Neural Origin of Rhythmic Movements*. A. Roberts and B. L. Roberts (eds), pp. 159–93. S.E.B. Symposium XXXVII.
Cottrell, G. A. (1977). *Neuroscience* **2**, 1–18.
McCrohan, C. R. and Benjamin, P. R. (1980). *J. exp. Biol.* **85**, 169–86.
Trimble, D. L. and Barker, D. L. (1984). *J. Neurobiol.* **15**, 37–49.
Wieland, S. J. and Gelperin, A. (1983). *J. Neurosci.* **3**, 1735–45.

Part II

Neuroendocrinology of dopamine

11
Dopaminergic control of anterior pituitary function

C. Dieguez, S. M. Foord, J. R. Peters, B. M. Lewis, P. Harris and M. F. Scanlon

Department of Medicine, University of Wales College of Medicine, Heath Park, Cardiff CF4 4XN, UK

11.1. Introduction

In 1968 Van Maanen and Smelik proposed that dopamine released into hypophyseal portal blood by tuberoinfundibular dopaminergic neurones was involved in the inhibition of PRL secretion by cells of the pituitary gland (Van Maanen and Smelik, 1968). This hypothesis was fully confirmed by Ben-Johnathan *et al.* (1977) when they found that dopamine concentrations in hypophyseal portal blood were significantly greater than in the systemic circulation. These data, together with the observations of MacLeod, Neill and colleagues, indicated that dopamine was the long-sought prolactin-inhibiting factor (PIF) and established dopamine as one of the major neuroregulators of anterior pituitary function. Subsequent studies involving the use of dopamine agonist and antagonist drugs have allowed a thorough analysis of the mechanisms involved in the dopaminergic inhibition of PRL and other anterior pituitary hormones.

In studying the role of dopamine in the control of anterior pituitary function, several factors must be considered. Dopamine can influence secretion of anterior pituitary hormones by acting at the level of the pituitary or the hypothalamus. These in turn will influence the secretion of anterior pituitary hormones. The action of dopamine at pituitary or hypothalamic levels can be modulated by target gland hormones (e.g. oestrogens and thyroid hormones), as well as other neuromodulators (e.g. opioid peptides). Furthermore, the activity of tuberoinfundibular dopaminergic neurones themselves can be altered by these factors.

In this review, we will discuss the available data on the actions of dopamine at pituitary and hypothalamic levels and the factors involved in the modulation of such actions. We will also review some of the methods that have been used for the study of the dopaminergic control of anterior pituitary function, highlighting their disadvantages and shortcomings.

11.2. Methods of assessing the dopamine control of anterior pituitary function

During the last 10 years, due to the availability of more specific dopaminergic agonists and antagonists, together with technical developments such as HPLC and improved cell-culture methodology, it is now possible to assess the interrelationship between tuberoinfundibular dopaminergic activity and anterior pituitary hormone secretion using several methods (see Table 11.1). Most studies have been performed on animals, and extrapolation to man is difficult, but, despite this, considerable headway has been made in our understanding of the processes involved in the dopaminergic control of anterior pituitary function in both animals and humans.

Most of the data obtained *in vivo* come from experiments involving the administration of dopamine agonists, antagonists and drugs that interfere with dopamine biosynthesis or catabolism. A major disadvantage of dopamine agonists is that they may exhibit other non-specific pharmacological actions, in addition to their dopamine-agonist activity (Scanlon *et al.*, 1979). However, the recent application of more specific D-2 receptor antagonists (e.g. metoclopramide and domperidone) has allowed us to gain further insight into the physiological role of dopamine as a neuroendocrine modulator of anterior pituitary function.

Table 11.1.   Methods of assessing dopamine control of anterior pituitary function.

In vivo. Intraventricular or intravenous administration of: dopamine or dopamine agonists; dopamine precursors (L-dopa); drugs that interfere with dopamine synthesis or action (antagonists, (dopamine antagonists, dopamine enzyme-synthesis blockers)

In vitro. (1) *Localisation*: fluorescence histochemistry; EM + autoradiography; radioligand binding studies

(2) *Functional*: quantification of the synthesis and/or release of a hormone, cAMP, adenylate cyclase, etc.
(a)   Acute studies: perfusion of hypothalamic or pituitary explants
(b)   long-term cell cultures (primary or tumoral) in monolayer or perfusion systems: organ culture; aggregates; dispersed cells; enriched pituitary cell types.

Estimation of the functional state of tuberoinfundibular dopaminergic neurones
    Non-steady-state techniques:
                            — decline in dopamine content in the ME after synthesis inhibition
                            — rate of accumulation of L-dopa after dopa decarboxylase (DDC)
                            — measurement of dopamine metabolites dihydroxyphenylacetic acid and homovanillic acid
                            — quantification of dopamine on anterior pituitary after synthesis inhibition by mass spectrometry
    Steady-state method: administration of [^{3}H]tyrosine and quantification of dopamine and trosine by HPLC or radioenzymatic methodology
                            quantification of dopamine on portal blood

Data acquired from *in vitro* studies relate to organ culture (in which the histiotypic organisation is maintained) or cell culture (in which cells are mechanically or enzymatically dispersed and maintained as monolayers). The major advantage of organ culture techniques is that a high degree of cellular integrity and the intrinsic cell-to-cell network is maintained. The major problem with this technique is the poor cell survival over time. In contrast, the use of primary or tumoral dispersed cell culture offers the advantage of study over long periods (usually between days and weeks). A potential drawback is the presence of changes in the phenotypic expression of these cells after culture. (For extensive reviews on *in vitro* methodologies, see Robbins and Reichlin, 1982, and Foord *et al.*, 1984*a*.)

Many techniques are also being used at present for the assessment of the neural activity of the tuberoinfundibular dopaminergic neurones. These can be divided into two groups. The non-steady-state methods consist of measuring the rate of decline of dopamine content, or accumulation of dopamine metabolites at median eminence or anterior pituitary, (measured by a sensitive radioenzymatic assay, fluorescence histochemistry, HPLC or mass spectrometry), after inhibition of tyrosine hydroxylase activity. Since the rate of dopamine decay follows first-order kinetics, an increased rate of dopamine decline represents an increased rate of dopamine turnover, whereas a decreased rate of decline is associated with a reduced dopamine turnover. Using this technique, it has been possible to obtain considerable information with regard to the mechanisms that regulate tubero-infundibular dopaminergic turnover. However, since the administration of enzyme synthesis inhibitors probably leads to a perturbation of the system, the physiological relevance of data gathered using this methodology is questionable. Recently, considerable attention has focused on the use of 'steady-state methods' that do not have the drawbacks of the non-steady-state methods. The technique is based on the quantification of the rate of formation of labelled dopamine after the previous administration of small quantities of the labelled precursor [^{3}H]tyrosine. The major drawback of this method is that it is technically more complex than enzyme inhibition methods. Other techniques being applied to the study of neuropeptide and neurotransmitter physiology include the use of synaptosomes and hypophyseal portal blood collection. The major drawback of the first is that it is difficult to differentiate what fraction of released product represents leakage from damaged membranes or a true response to a particular secretagogue. Hypophyseal portal blood collection, however, is technically complicated and depends on the use of anaesthetised animals. In summary, these data indicate the many experimental tools available for the study of the control of anterior pituitary function by dopamine. Each has advantages and drawbacks, but all are important since they provide us with information about a particular biochemical or physiological mechanism and some allow us to study the level of the phenotypic expression.

11.3. Dopamine control of neuropeptide release from the hypothalamus

Despite initial hopes, the measurement of hypothalamic peptides in body fluids and their use as indices of hypothalamic function have been largely unrewarding and the methodology is fraught with difficulties. The poor sensitivity of the radioimmunoassays used in relation to the low circulating concentrations of these peptides has been a major problem. Furthermore, many neuropeptides are also synthesised outside the CNS, and in consequence most of the circulating material (e.g. thyrotropin-releasing hormone, TRH, and somatostatin, SRIF) is not of neural origin. Therefore, most of the available data on the control of hypothalamic peptide synthesis and release by dopamine has been carried out using *in vitro* models. So far, the data reported (Table 11.2) have produced contradictory results, mainly derived

Table 11.2.   Actions of dopamine on hypothalamic peptide release *in vitro*

	Effect	*Experimental Model*
Luteinising hormone releasing factor	Stimulatory	Median eminence fragment
	No effect	Mediobasal hypothalamus and median eminence
Somatostatin	Stimulatory	Hypothalamic explants (intact); dispersed cells (adult)
	No effect	Hypothalamic fragments
Thyrotropin-releasing hormone	Stimulatory	Hypothalamic synaptosomes; hypothalamic fragments
	No effect	Hypothalamic explants (intact)
Corticotropin-releasing factor	Stimulatory	Hypothalamic synaptosomal preparation

from the wide variety of methodologies used. However, the overall pattern suggests that dopamine plays a stimulatory role in TRH, luteinising hormone releasing factor (LHRH), SRIF and corticotropin-releasing factor (CRF) secretion. Once again, however, different results may be found according to which experimental model is used. In perfused, whole, hypothalamic explants, for example, dopamine releases SRIF but has no action on TRH (Lewis *et al.*, 1984). In contrast, in acutely perfused hypothalamic fragments, dopamine can release TRH (Maeda and Frohman, 1980) without any action on SRIF (Terry *et al.*, 1980).

11.4. Dopamine control of anterior pituitary function

As shown in Table 11.3, the administration of high concentrations of dopamine to normal subjects produces a widespread inhibition of basal or stimulated anterior pituitary hormone secretion with the exception of

Table 11.3.   Effects of dopamine on anterior pituitary function in man

---

Dose 4 μg kg⁻¹ min⁻¹ infusion

(1)	PRL inhibition:	basal + thyrotropin-releasing hormone
(2)	Thyrotropin inhibition:	basal + thyrotropin-releasing hormone
(3)	Luteinising hormone inhibition:	basal + GnRH, maximal at midcyle
(4)	Follicle-stimulating hormone inhibition:	basal
(5)	Growth hormone release:	basal, transient
(6)	Growth hormone inhibition:	IST
(7)	No effect on basal cortisol or cortisol response to IST	

Dopamine antagonists (10 mg i.v. of metoclopramide or domperidone)
(1)   Increase in PRL
(2)   Increase in thyrotropin
(3)   Increase in luteinising hormone (only in tumoral hyperprolactinaemia)
(4)   Increase in growth hormone (only in hypogonadism)
(5)   No effect on follicle-stimulating hormone or cortisol

---

growth hormone (GH), where a transient stimulation occurs (Kaptein *et al.*, 1980). However, the specificity and physiological significance of such data must be questioned since a high dose was used and it is difficult to know whether this phenomenon is mediated via dopaminergic or noradrenergic mechanisms. With the recent use of more specific dopamine antagonists such as domperidone and metoclopramide, it has been possible to throw some light on this issue (see Table 11.2). From the data available with these drugs, the only clear-cut picture to emerge is the inhibitory action of dopamine on prolactin (PRL) and thyrotropin secretion by the lactotrophs and thyrotrophs, respectively.

11.4.1. *Dopamine control of prolactin*

The role of dopamine as the most important mediator of the hypothalamic inhibition of PRL release is now clearly established (Leong *et al.*, 1983) and the evidence is as follows:

(1)   Dopamine inhibits PRL release *in vivo*, *in vitro* and after infusion into hypophyseal portal vessels.
(2)   Endogenous dopamine antagonism leads to a rapid increase in PRL levels.
(3)   Dopamine is present in portal blood at concentrations which inhibit PRL release *in vivo*.
(4)   Specific high-affinity dopamine binding sites are present on anterior pituitary cells.

There is also good animal evidence that PRL may regulate its own secretion through a short-loop feedback action to increase hypothalamic dopaminergic activity:

(1) PRL administration to rats causes:
     (a) reduced PRL levels in intact but not in pituitary-transplanted rats;
     (b) increased dopamine turnover in the median eminence;
     (c) selective increase in the activity of tuberoinfundibular but not striatal dopamine neurones.
(2) Hypothalamic PRL implants reduce PRL release.
(3) Pituitary PRL content is reduced in animals with donor pituitaries under renal capsule.
(4) Hypophysial portal dopamine levels are reduced in rats with ectopically transplanted PRL-secreting adenomas.

The most direct evidence is the recent observation that dopamine concentrations in hypophyseal portal blood are increased in animals with ectopically transplanted PRL-secreting adenomas. Other studies have indicated that dopamine concentrations in rat hypophysial portal blood tend to be inversely related to physiological alterations in PRL levels. However, in an individual animal this relationship may not be sufficiently close to exclude the activity of several other factors now implicated in PRL control. Despite all these data, several questions still remain to be answered. Is dopamine the only PIF? What is the role of dopamine in the rise in PRL levels observed after suckling? What is the physiological role, if any, of the stimulatory action of dopamine on PRL release from rat anterior pituitary cells in culture? What is the role of dopamine transported via the posterior pituitary?

The major drawback to the investigation of PRL responses to dopamine antagonists as a marker of specific neuroendocrine dysfunction is that alterations in such responses are observed in a variety of different disease states, including metabolic and neuropsychiatric. This is in addition to the well known physiological fluctuations in PRL release (*vide infra*). In contrast, the introduction of dopamine agonist therapy in hyperprolactinaemia is of major importance and has totally changed the management and prognosis of patients with this condition (*vide infra*) (Scanlon *et al.*, 1979).

11.4.2. *Dopamine control of thyroid stimulating hormone*

Data derived from *in vitro* experiments using rat anterior pituitary cells in monolayer culture have shown that dopamine inhibits TSH release in a dose-related stereospecific manner and there is a striking parellelism between the inhibition of TSH and PRL by dopamine and dopamine agonists (Foord *et al.*, 1983). Preliminary data also indicate that TSH may specifically regulate its own release via the induction of dopamine receptors on thyrotroph cells. Release of TSH by cells from hypothyroid animals shows increased sensitivity to the inhibitory effects of dopamine, and parallel binding studies indicate that the number of dopamine receptors may be increased on these cells. These data suggest that thyroid hormones, or possibly TSH itself, may regulate directly this aspect of dopaminergic function (Foord *et al.*, 1984*b*).

Data from the use of dopamine and specific dopamine receptor blocking drugs such as domperidone, which do not penetrate the blood–brain barrier to any appreciable extent, are compatible with the view that this is a direct pituitary action and is probably mediated by dopamine receptors on the thyrotroph. The dopaminergic inhibition of TSH release varies according to sex (greater in females than in males), thyroid status (greater in patients with mild and subclinical hypothyroidism than in euthyroid or severely hypothyroid patients), time of day (greater at 11 p.m. than at 11 a.m.) and PRL status (increased in patients with microprolactinomas) (Peters *et al.*, 1982).

Generally, it can be seen that altered TSH responses to dopamine antagonism occur in a wide variety of hypothalamic–pituitary and neuropsychiatric disorders. Overlaps are wide, however, and although such alterations can be diagnostically helpful in some instances, they provide little information as to aetiology, site of disease or precise neuropeptide/ neurotransmitter imbalances. Perhaps one exception is the measurement of the combined TSH and PRL responses to domperidone in hyper- prolactinaemia. This has thrown some light on the nature of the disturbances which may occur in these patients and can be diagnostically helpful (Peters *et al.*, 1982).

11.4.3. *Dopamine control of luteinising hormone, growth hormone and adrenocorticotrophic hormone*

The role of dopamine in the control of LH secretion is complex and has been the subject of many conflicting reports in recent years. There have been variations in the dose of dopamine used, the specificity and pharmaco- logical activity of the various dopamine agonists and antagonists employed, and the gonadal status of the individual. It is now clear that infusion of high concentrations of dopamine (4 μg kg^{-1} min^{-1}) decreases both basal and stimulated LH levels in normal men and women and in patients with hyperprolactinaemia, whereas the administration of the potent dopamine agonist bromocriptine does not produce any changes in LH levels or has only a small stimulatory effect (Martin *et al.*, 1981). The reason for these discrepancies is unkown. It is possible that the inhibitory effect of dopamine is mediated by non-dopaminergic receptors such as α- and β- adrenoreceptors. Alternatively, the dopaminergic control of LH, in contrast to PRL and TSH, is not mediated by D-2 receptors. Finally, the reason for the lack of inhibition of LH after bromocriptine could be due to the α-adrenergic antagonist properties of this drug, having two different and opposite actions and one therefore masking the other. In contrast to the conflicting data obtained after administration of dopamine and dopamine agonists, there is a general consensus of opinion about the lack of action of dopamine antagonists on basal LH in normal men and women. However, some confusion appears to exist with regard to the LH responses to dopamine antagonists in patients with hyperprolactinaemia. The original observation of Quigley *et al.* (1979) of elevation in LH levels after metoclopramide in these patients has not been confirmed by others (Elli *et al.*, 1983). Similarly, after dopamine blockade with domperidone, no modification in LH was seen by Peters *et al.* (1984) and Elli *et al.* (1983). Interestingly, it has recently been reported (Elli *et al.*, 1983) that after prior enhancement of central dopaminergic tone by pretreatment with carbidopa and L-dopa, it has proved possible to observe an increase in LH levels following the administration of metoclopramide, in contrast to the lack of effect of domperidone under the same experimental conditions. This suggests that the dopaminergic control of LH is exerted at central level inside the blood-brain barrier.

The administration of dopamine agonists as well as dopamine infusion leads to a clear-cut acute release of GH in normal subjects (Kaptein *et al.*, 1980), and these responses can be inhibited by adrenergic blockade with phentolamine or cholinergic blockade (Delitala, 1982). These data, taken together with the inhibition of GH release by dopamine *in vitro* (Cronin *et al.*, 1984) suggest that the major effect of dopamine *in vivo* is exerted at central level via the stimulation of growth hormone releasing factor release and/or the inhibition of somatostatin.

The role of dopamine in the control of ACTH secretion is not yet settled. It appears at present that dopamine does not play a major role, since the administration of L-dopa or dopamine and dopamine agonist is not accompanied by any change in ACTH and/or cortisol levels (Delitala, 1982).

11.5. Dopamine control of pituitary adenomas

11.5.1 *Prolactinomas*

Hyperprolactinaemia with its consequent amenorrhoea/galactorrhoea and hypogonadism syndrome is a common clinical hypothalamic–pituitary disorder and it is now clear that prolactinomas are the commonest anterior pituitary tumours. The diagnosis of a prolactinoma is made easier when the sella turcica is abnormal on plain lateral skull X-ray. However, in many cases the sella is normal radiologically, and precise interpretation minor radiological change in the fossa is difficult and contentious. Despite these difficulties, the current view is that a significant number of hyperprolactinaemic patients who show no gross abnormalities of the pituitary fossa may well harbour a PRL-seicreting microadenoma.

The basal plasma concentration of PRL, when very elevated, points to the presence of prolactinoma. However, microprolactinomas can present with much lower basal PRL levels. It has been appreciated for some time that the PRL reponses to both TRH and dopamine antagonist drugs may be suppressed in patients with prolactinomas when studied as a group, and although reduced or absent PRL responses to TRH and metoclopramide correlate well with the presence of a prolactinoma, the presence of considerable PRL responses by no means excludes such a diagnosis (Peters *et al.*, 1983).

A novel approach to the diagnosis of PRL-secreting adenomas has been to examine the function of the normal pituitary tissue surrounding the adenoma. The sustained hyperprolactinaemia and reduced PRL responses to dopamine antagonism seen in patients with prolactinomas imply an underlying loss of dopaminergic inhibition of PRL release, and this is in accord with the recent demonstration of defective dopaminergic function in such patients. Since both PRL and TSH are regulated by inhibitory hypothalamic dopaminergic activity, we investigated whether the defective dopaminergic function was limited to the control of PRL release.

Patients with hyperprolactinaemia caused by presumed microprolactinomas show reduced dopaminergic inhibition of PRL release and increased dopaminergic inhibition of TSH release. Prolactinoma patients also show a normal circadian variation in the dopaminergic inhibition of TSH release (Rodriguez-Arnao *et al.*, 1983). The relationship between exaggerated TSH responses to dopaminergic blockade and microprolactinomas has been confirmed immunohistochemically (Peters *et al.*, 1984). Similar findings have been reported with regard to the dopaminergic inhibition of LH secretion (Quigley *et al.*, 1979), although others have failed to confirm these findings (Elli *et al.*, 1983). In contrast, patients with physiological hyperprolactinaemia associated with the puerperium show reduced TSH responses to acute dopamine antagonism and hence reduced dopaminergic inhibition of TSH release. Taken together, these data suggest that the autonomous PRL production by the prolactinoma leads to increased hypothalamic dopaminergic activity via the short-loop feedback mechanism, followed by an increase in the release of dopamine into hypophyseal portal blood leading to increased dopaminergic inhibition of TSH (and possibly LH) release. However, the increased dopamine concentrations are unable to control PRL release by the adenomatous cells, possibly because of reduced sensitivity to dopamine and/or because of defective delivery of dopamine caused by dissociation of the adenoma from the hypophysial portal blood supply.

In vitro studies on human PRL-secreting adenomas obtained after surgery have also yielded some interesting information. Radioligand binding assays performed using membranes from these tissues showed no changes in the affinity of dopamine receptors in comparison to normal tissue (Cronin and Evans, 1983). Unfortunately, quantification of the number of receptors proved inconclusive because of the difficulty of comparing tissues with very different numbers of lactotrophs. On the other hand, analysis of the dopamine inhibition of PRL *in vitro* showed an increase in the ED_{50} of dopamine in tumerous cells in comparison to normal tissue with no apparent change in the percentage of maximal inhibition of PRL (Bethea *et al.*, 1982). It was reported recently (Spada *et al.*, 1982) that micro- and macroprolactinomas behave quite differently *in vitro*, the former being much less sensitive to the inhibition by dopamine of both PRL and adenylate cyclase. This applied to both maximal inhibition (20% vs 72%, 10% vs 34%, respectively) and to the ED_{50} of such actions (8×10^{-7} vs 5×10^{-8} M dopamine and $1 \cdot 5 \times 10^{-6}$ M vs 3×10^{-5} M dopamine respectively).

In summary, these data suggest the presence of increased dopaminergic activity in patients with prolactinomas. The fact that the increased dopamine concentrations in hypophyseal portal blood are unable to control PRL release by the adenomatous cells could be due to two main factors: reduced sensitivity to dopamine and/or defective delivery of dopamine to the lactotrophs. It is possible that the relative importance of these two factors varies between micro- and macroprolactinomas.

11.5.2. *GH-secreting adenomas*

The commonest cause of increased GH secretion is a pituitary tumour secreting GH. Administration of dopamine or dopamine agonists produces a decrease in GH levels in a large number of these patients in contrast to the rise in GH usually seen in normal subjects. It is believed that this action is mediated via dopamine receptors since it is blocked by dopamine antagonists. This paradoxical response usually returns to normal after successful removal of the tumour.

Radioligand binding studies on membranes obtained from these tumours show the presence of dopamine receptors in most but not in all. Whether there is a change in the number or affinity of these receptors is still controversial (Cronin and Evans, 1983). On the other hand, *in vitro* studies looking at the action of dopamine on GH secretion and adenylate cyclase activity in adenomatous tissue have produced several interesting findings. Dopamine inhibits GH secretion in most but not all tumours. In fact, a stimulation of basal GH release after dopamine was observed in 3 out of 21 tumours studied (Spada *et al.*, 1982). Even more interesting was the fact that in most of the tumours that exhibited an inhibition of GH secretion by dopamine it was not possible to observe a parallel decrease in adenylate cyclase activity, suggesting that the dopaminergic inhibition of GH secretion in these tumours is exerted through a cyclic AMP-independent mechanism. Therefore, it is possible that dopamine receptors on GH-secreting adenomas are different from the D-2 receptors of PRL-secreting cells in which dopamine causes inhibition of adenylate cyclase.

11.5.3. *Other tumours*

Because of the low incidence of these tumours, information is scanty. A decrease in ACTH levels after bromocriptine administration to patients with Nelson's or Cushing's syndrome is usually seen in 40% of these subjects. Recent histological evidence suggests that the group of bromo-criptine responders are mostly the ones with a tumour originating in the intermediate lobe, whereas the group of bromocriptine non-responders usually harbour a tumour originating in the anterior pituitary. Similarly, *in vitro* studies have shown inhibition of ACTH by dopamine in some of these tumours (Lamberts, 1982). With regard to the role of dopamine in TSH-secreting tumours, even fewer data are available. In general, there is lack of response of TSH to the administration of dopamine agonists or antagonists. However, recent *in vitro* studies show the presence of dopamine receptors in one of these tumours (Chanson *et al.*, 1984). Also, after investigation of the action of dopamine on TSH secretion and adenylate cyclase activity in another case, a decrease in TSH and a concomitant increase in adenylate cyclic activity were found in response to dopamine (Spada *et al.*, 1983). Clearly this latter finding merits further investigation.

11.6. **Modulation of dopamine action on anterior pituitary function by target organs**

In addition to their actions on hypothalamic dopaminergic activity, oestrogens and thyroid hormones have actions directly at pituitary level. Understanding of the role of oestrogens is particularly difficulty because of marked interspecies variations (rodents vs primates) in their biological actions. An antidopaminergic activity of oestrogens on the dopaminergic inhibition of PRL in the rat has been clearly established *in vitro* (Raymond *et al.*, 1978). In contrast, recent data obtained in a similar experimental model using monkey anterior pituitaries have shown that this is not the case in primates (Bethea, 1984). In fact *in vivo* data obtained in primates, both monkeys and humans, suggest that oestrogens could play a facilitatory role in the dopamine inhibition of PRL secretion (for discussion on this subject see Ch. 19 by Valcavi *et al.*).

Recent data gathered using rat anterior pituitary cells in monolayer cultures of both euthyroid and hypothyroid rats suggest that thyroid hormones have opposite actions with regard to the dopamine control of PRL and TSH, namely a slight facilitation of dopaminergic inhibition of

PRL secretion and a powerful antagonism on the dopamine control of TSH. Preliminary data from these experiments suggest that these actions of thyroid hormones are mediated through modulation of the number of dopamine receptors on lactotrophs and thyrotrophs, respectively (Foord *et al.*, 1984*b*).

11.7. Factors affecting hypothalamic dopaminergic activity

Dopamine turnover in the median eminence is regulated by a variety of pharmacological and hormonal factors. One of the major regulators is dopamine itself, which can produce inhibition of dopamine synthesis because of the sensitivity of the enzyme tyrosine hydroxylase to the end-product feedback exerted by dopamine and related compounds. Of all the pharmacological agents known to influence dopamine turnover, probably the most important are dopamine agonists. In addition to these, several humoral factors can influence tuberoinfundibular turnover. Prolactin increases dopamine turnover whereas endorphins, enkephalins and thyroid hormones reduce it. Also, it has been shown that the administration of oestrogens to rats leads to a transient initial decrease but ultimately to an increase in dopamine turnover. This latter event may well be mediated by the higher circulating PRL levels in oestrogen-treated animals (Annunziato *et al.*, 1983), but caution should be exerted in the extrapolation of these data to the human situation for the reasons stated earlier.

References

Annunziato, L., Cerrito, F. and Balsamo, S. (1983). Dopamine biosynthesis regulation in the terminals of tuberoinfundibular neurons. In *Prolactin and Prolactinomas*. G. Tolis (ed.), pp. 57–70. Academic Press, New York.
Ben-Jonathan, N., Oliver, C., Weiner, H.J., Mical, R. S. and Porter, J. C. (1977). Dopamine in hypophysial portal blood of the rat during the estrous cycle and throughout pregnancy. *Endocrinology* **100**, 452–8.
Bethea, C. L. (1984). Effects of oestrogen on the dopaminergic inhibition of PRL in monkey anterior pituitary cells in culture. *Fourth International Congress on Prolactin*, Charlottesville (Abstract).
Bethea, C. L., Ramsdell, J., Jaffe, R., Wilson, C. and Weiner, R. (1982). Characterisation of the dopaminergic regulation of human-prolactin secreting cells cultured on extracellular matrix. *J. Clin. Endocrinol. Metab.* **54**, 893–902.
Chanson, P., Orgiazzi, J., Derome, P. J. *et al.* (1984). Paradoxical response of thyrotropin to L-dopa and presence of dopaminergic receptors in a thyrotropin-secreting pituitary adenoma. *J. Clin. Endocrinol. Metab.* **59**, 542–6.
Cronin, M. F. and Evans, W. S. (1983). Dopamine receptors in the normal and abnormal anterior pituitary gland. *Clin. Endocrinal. Metab.* **12**, 15–30.
Cronin, M. J., Thorner, M. O., Hellman, P. and Rogol, A. D. (1984). Bromocriptine inhibits growth hormone release from rat pituitary cells in primary culture. *Proc. Soc. Exp. Biol. Med.* **175**, 191–5.

Delitala, G. (1982). Neurotransmitter control of anterior pituitary hormone secretion and its clinical implications in man. In *Clinical Neuroendocrinology*. Besser and Martini (eds), pp. 67–139, Academic Press, New York.

Elli, R., Scaperrotta, C., Travaglini, P., Beck-Peccoz, P. and Faglia, G. (1983). Dopaminergic control of PRL, TSH and LH secretion in tumoral hyperprolactinaemia. *Third European Workshop on Pituitary Adenomas*. Abstract P.5.

Foord, S. M., Peters, J. R., Dieguez, C. *et al.* (1983). Dopamine receptors on rat anterior pituitary cells in culture: functional correlation with the inhibition of TSH and PRL secretion. *Endocrinology* 112, 1567–77.

Foord, S. M., Dieguez, C., Peters, J. R. *et al.* (1984a). Recent advances in the understanding of the catecholaminergic control of anterior pituitary function. In *Pituitary Adenomas*. S. W. Lamberts (ed.). Excerpta Medica, Amsterdam.

Foord, S. M., Peters, J. R., Dieguez, C. *et al.* (1984b). Hypothyroid pituitary cells in culture. An analysis of thyrotropin and prolactin responses to dopamine and dopamine receptor binding. *Endocrinology* 115, 407–15.

Kaptein, E. M., Kletzky, D. A., Spencer, C. A. and Nicoloff, J. T. (1980). Effects of prolonged dopamine infusion on anterior pituitary function in normal males. *J. Clin. Endocrinol. Metab.* 51, 488–91.

Lamberts, S. W. J. (1982). Cushing's disease is caused by adenomas originating in the anterior or the intermediate pituitary lobe. In *Pituitary Hyperfunction*. F. Camanni and E. Müller (eds), pp. 235–42. Raven Press, New York.

Leong, D. A., Frawley, L. S. and Neill, J. D. (1983). Neuroendocrine control of prolactin secretion. *Ann. Rev. Physiol.* 45, 109–27.

Lewis, B. M., Dieguez, C., Foord, S. M., Hall, R. and Scanlon, M. F. (1985). Dopaminergic control of somatostatin and thyrotropin-releasing hormone release from perfused adult rat hypothalami. *The Neurobiology of Dopamine Systems*. W. Winlow and R. Markstein (eds), ch. 17. Manchester University Press, Manchester.

Maeda, K. and Frohman, L. A. (1980). Release of somatostatin and thyrotropin-releasing hormone from rat hypothalamic fragments *in vitro*. *Endocrinology* 106, 1837–42.

Martin, W. H., Rogol, A. D., Kaiser, D. L. and Thorner, M. O. (1981). Dopaminergic mechanisms and luteinizing hormone (LH) secretion. Differential effects of dopamine and bromocriptine on LH release in normal women. *J. Clin. Endocrinol. Metab.* 52, 650–6.

Peters, J. R., Foord, S. M., Dieguez, C. *et al.* (1982). Microprolactinoma and functional hyperprolactinaemia: two clinical entities or two phases of the same disease? In *A Clinical Problem: Microprolactinoma. Intern. Congress Series* 568, 21–34.

Peters, J. R., Arnao, M. D. R., John, R. *et al.* (1983). Diurnal variation in basal TSH and PRL levels and in the dopaminergic inhibition of TSH and PRL in normals and in patients with immunohistochemical evidence of prolactinomas. *Third European Workshop on Pituitary Adenomas*. Abstract P. 23.

Peters, J. R., Rodriguez-Arnao, M. D., Foord, S. M. *et al.* (1984). Differential effects of acute DA receptor blockade with domperidone in TSH and LH release in hyperprolactinaemic patients. *J. Endocrinol. Invest.* (in press).

Quigley, M. E., Judd, S. J., Gilliland, G. B. and Yen, S. S. C. (1979). Effects of a dopamine antagonist on the release of gonadotrophin and prolactin in normal women and women with hyperprolactinaemic anovulation. *J. Clin. Endocrinol. Metab.* 48, 718–70.

Raymond, V., Beaulieu, M., Labrie, F. and Boissiez, J. (1978). Potent antidopaminergic activity of oestradiol at the pituitary level on prolactin release. *Science* 200, 1175–8.

Robbins, R. and Reichlin. R. (1982). *In vitro* systems for the study of secretion and synthesis of hypothalamic peptides. In *Neuroendocrinology Perspectives*, vol. 1. R. M. MacLeod and E. E. Müller (eds), pp. 111–35. Elsevier Biomedical Press, Amsterdam.

Rodriguez-Arnao, M. D., Peters, J. R., Foord, S. M. *et al.* (1983). Exaggerated circadian variation in basal thyrotropin (TSH) and in the dopaminergic inhibition of TSH release in pathological hyperprolactinaemia: evidence against a hypothalamic dopaminergic defect. *J. Clin. Endocrinol. Metab.* **57**, 975–80.

Scanlon, M. F., Pourmand, M., McGregor, A. M. *et al.* (1979). Some current aspects of clinical and experimental neuroendocrinology with particular reference to growth hormone, thyrotropin and prolactin. *J. Endocrinol. Invest.* **2**, 307–37.

Spada, A., Sartorio, A., Bassetti, M., Pezzo, G. and Giannattasio, G. (1982). *In vitro* effect of dopamine on growth hormone (GH) release from human GH-secreting pituitary adenomas. *J. Clin. Endocrinol. Metab.* **55**, 737–40.

Spada, A., Beck-Peccoz, P., Giannattasio, G. and Faglia, G. (1983). Effects of dopamine and somatostatin on TSH secretion and adenylate cyclase activity in one human TSH secreting pituitary adenoma. *Thirteenth Annual Meeting of the European Thyroid Association, Madrid.* Abstract 162.

Terry, L. C., Rorstad, O. P. and Martin, J. B. (1980). The release of biologically and immunologically reactive somatostatin from purified hypothalamic fragments. *Endocrinology* **107**, 659–63.

Van Maanen, J. H. and Smelik, P. G. (1968). Induction of pseudopregnancy in rats following local depletion of monoamines in the median eminence of the hypothalamus. *Neuroendocrinology* **3**, 177–86.

12
Dopaminergic control of growth hormone secretion

D. Cocchi

Department of Pharmacology, University of Milan, Milan, Italy

12.1. Introduction

It is now unequivocally established that in primate and subprimate species, growth hormone (GH) secretion is regulated by two distinct neurohormones, one inhibitory, somatostatin (Brazeau *et al.*, 1973) and the other stimulatory, the recently discovered growth hormone releasing factor (GRF) (Guillemin *et al.*, 1982). Activation of dopamine receptor sites is associated, at least in humans, with stimulation of GH release. Thus direct dopamine receptor agonists (e.g. apomorphine, bromocriptine: Lal *et al.*, 1972; Camanni *et al.*, 1975), or indirect dopamine agonists, viz. drugs which, to act, require the availability of an endogenous dopamine pool (e.g. nomifensine, amphetamine and metilphenidate: Besser *et al.*, 1969; Brown and William, 1976) all induce a rise in GH levels in man.

More conflicting results have been obtained in other species; for instance in the monkey, stimulation of GH release was induced by systemic infusion of dopamine (Steiner *et al.*, 1978). However, the dopamine-induced GH release was suppressed by concomitant infusion of FLA-63, an inhibitor of dopamine-β-hydroxylase, which interrupts norepinephrine synthesis, and by phentolamine, thus indicating adrenergic mediation of the dopamine-induced GH release (Steiner *et al.*, 1978). In addition, in the same species, subemetic doses of apomorphine failed to release GH (Jacoby *et al.*, 1974), and intraventricular infusion of dopamine suppressed plasma levels of the hormone (Toivola and Gale, 1970).

In the rat, the most commonly used species in the laboratory, studies on the effect of neurotransmitters on GH secretion are hampered by the particular kind of pulsatile secretion of GH characterised by striking variations from few nanograms per millilitre to some hundreds of nanograms per millilitre in a few minutes. Sampling animals in the absence of stress, such as through a catheter permanently inserted in the jugular vein, provided evidence for the physiological pulsatile secretion of GH.

Administration of graded doses of apomorphine into these unanaesthetised freely moving rats induced a biphasic effect on the secretion of GH. In fact, relatively low doses of the drug ($0 \cdot 1 – 1 \cdot 0$ mg kg^{-1} i.v.) induced an increase in plasma rat GH within 15–75 min, whereas administration of a higher dose of apomorphine ($5 \cdot 0$ mg kg^{-1}) induced a striking and sustained decrease of the hormone (Locatelli *et al.*, 1977).

12.2. Dopaminergic control of GH secretion in the dog

As opposed to the rat, the dog is a very suitable species for studies on the control of GH secretion since it behaves like humans in many aspects of GH regulation (Tsushima *et al.*, 1971; Cocola *et al.*, 1976). Administration of nomifensine, a dopamine-uptake inhibitor (Schacht *et al.*, 1977), caused stimulation of GH release in trained conscious dogs at doses of $1 \cdot 4$ and $2 \cdot 8$ mg kg^{-1} i.v. (Casanueva *et al.*, 1981). Pimozide, a specific blocker of dopamine receptors, significantly reduced the GH release evoked by nomifensine (Fig. 12.1), suggesting that the effect of the drug was due to

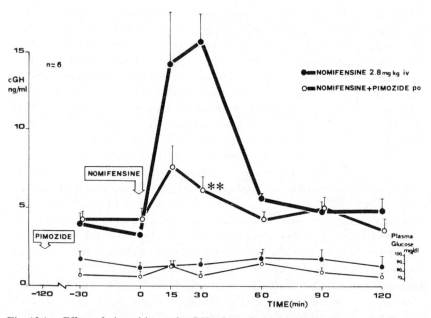

Fig. 12.1.   Effect of pimozide on the GH release induced by nomifensine in the dog. Nomifensine was administered at a dose of $2 \cdot 8$ mg kg^{-1} i.v.; pimozide was administered 1 mg orally, three times daily for 2 days, followed by 4 mg orally 120 min before testing. Each point represents mean GH levels in four dogs, and vertical lines represent the SEM. **$P < 0 \cdot 01$ vs nomifensine alone. The effect on plasma glucose of nomifensine alone or associated with pimozide is shown in the lower part of the graph. From Casanueva *et al.* (1981).

stimulation of dopamine receptors. Pretreatment with domperidone, an antagonist of dopamine receptors which does not cross the blood–brain barrier (BBB), failed to counteract the GH-releasing effect of nomifensine (Casanueva *et al.*, 1981). This finding and the observation that both pimozide (Fig. 12.1) and haloperidol (Casanueva *et al.*, 1981) instead prevent the action of nomifensine suggest that the dopamine receptors mediating GH stimulation lie within the BBB. Nomifensine, however, also potentiates noradrenergic neurotransmission by blocking norepinephrine re-uptake (Schacht *et al.*, 1977). That the GH-releasing effect of nomifensine in the dog was due in part also to activation of noradrenergic systems is demonstrated by the ability of phentolamine, an α-adrenoceptor antagonist, to abolish the nomifensine-induced rise of GH (Casanueva *et al.*, 1981).

12.3. Extra-hypothalamic regulation of GH secretion

In contrast to other hormones (e.g. prolactin), whose secretion is controlled mainly by the medial basal hypothalamus (Neill, 1980), subcortical areas, particularly the limbic system, appear to play an important role in regulating GH secretion. It has been reported that electric stimulation of the limbic system both in rats (Martin, 1972) and humans (Luedecke *et al.*, 1972) induces an increase of GH secretion. Most likely, dopamine plays a role in mediating this response since the limbic system receives important inputs from dopaminergic mesolimbic neurones originating in the mesencephalon, and electric stimulation of this area evokes a GH release in the rat (Martin, 1972). An alteration in the function of limbic dopamine has been related to the pathophysiology of schizophrenic illness and has led to the so-called 'dopamine hypothesis of schizophrenia' (Pandey *et al.*, 1984).

Administration of the direct dopamine agonist apomorphine to schizo-phrenic patients does not induce a GH response different from that in control subjects (Pandey *et al.*, 1977). However, the distribution of GH peaks in the schizophrenic population appears to be bimodal. The same authors in fact observed that when the schizophrenic population was subgrouped as acute and chronic, the mean GH value for patients with acute schizophrenia was significantly higher than that of chronic schizophrenics (Pandey *et al.*, 1977). The mean GH levels in the acute schizophrenic group was also significantly higher as compared with the normal control group.

These data would indicate that the population of schizophrenics is heterogeneous in nature. The possibility exists that the patients diagnosed as chronic might at one time have had an increased dopamine receptor sensitivity similar to that observed among acute patients, but that some adaptive processes associated with the chronicity of the illness or with the pharmacotherapy for the illness (chronic exposure to neuroleptic drugs) interfered with this sensitivity.

12.4. Dopamine in Huntington's chorea

Other human disorders with alteration of extra hypothalamic dopaminergic systems have been extensively investigated for the GH reponses to dopaminergic stimuli. Among them is Huntington's chorea, in which a functional prevalence of the striatal dopaminergic activity is assumed (Klawans, 1970).

If the dopaminergic tone is increased at the striatal level in Huntington's disease, a similar condition might be expected in other dopaminergic systems (hypothalamic, limbic) involved in the control of GH secretion. An enhanced GH responsiveness in choreic patients after administration of the dopamine agonist bromocriptine was first observed in our laboratory and later confirmed with other dopamine—mimetic drugs (Müller *et al.*, 1979). In contrast to our findings were the results of Chalmers *et al.*, (1978), who found that choreic patients responded poorly to bromocriptine; also, Levy *et al.* (1979) reported a poor GH release after administration of apomorphine. It is difficult to interpret such conflicting data. Different populations of patients could conceivably behave differently, especially considering the

Table 12.1. GH responses to lisuride and bromocriptine in seven untreated Huntington's chorea patients. (Reproduced from Caraceni *et al.*, 1983, with permission.)

Lisuride Case No.	Time (min)						
	0	30	60	90	120	150	180
1	0·1	0·2	0·2	0·2	0·3	0·5	0·6
2	2·0	0·5	0·4	0·3	0·3	0·4	5·3
3	0·7	0·5	0·5	0·6	0·7	0·6	0·7
4	4·5	8·7	7·9	6·2	5·0	1·1	1·1
5	0·5	1·4	9·0	21·0	10·0	—	—
6	0·3	2·8	15·5	18·0	8·0	3·1	1·3
7	1·7	0·7	18·4	15·3	9·1	—	5·2
Mean of 10 control subjects, $\bar{X} \pm$ SEM	1·9	1·4	2·2	5·4	5·5	4·6	1·9
	0·6	0·5	0·7	1·1	1·7	0·9	0·4

Bromocriptine Case No.	Time (min)								
	0	30	60	90	120	150	180	210	240
1	0·8	0·8	0·5	0·4	0·3	0·3	0·4	0·4	0·2
2	0·3	0·3	0·3	0·2	0·5	0·5	0·5	0·8	1·5
3	0·2	0·3	6·1	2·8	3·4	4·5	3·5	2·1	0·9
4	0·7	0·2	1·6	2·5	3·1	2·8	0·3	0·5	1·1
5	3·0	0·9	1·0	0·3	0·4	1·2	2·4	2·8	0·5
6	3·0	3·2	0·6	0·8	0·4	0·4	0·4	0·1	0·1
7	0·8	1·2	10·5	7·5	9·5	5·6	3·1	2·8	2·0
Mean of 12 control subjects, $\bar{X} \pm$ SEM	1·6	2·8	2·8	2·1	1·5	2·1	4·4	3·4	4·4
	0·4	0·1	1·1	0·6	0·3	0·6	1·3	1·1	0·8

small numbers of patients involved in the studies. Another point to bear in mind is the duration of previous neuroleptic therapy and the duration of the wash-out period, if any, before the stimulation tests.

In order to clarify the above-mentioned controversial data, Caraceni *et al.* (1983) evaluated the GH response to dopamine agonists in a small but selected population of choreic patients. These patients (two males and five females, with an average age of 48·5 years, an average history of the disease of 5 years and no previous exposure to neuroleptic therapy) received an acute administration of two dopamine agonists, bromocriptine and lisuride (2·5 mg and 200 μg orally, respectively). Different responses were obtained: three cases did not respond to lisuride and four cases exhibited a normal or exaggerated rise in plasma GH levels. The response to bromocriptine was almost identical. Cases 1 to 3 failed to respond, cases 5–7 responded normally, and only case 4 presented dissociated responses (no response to bromocriptine, normal response to lisuride, Table 12.1).

Due to the extreme homogeneity of the group, the different neuro-endocrine responses exhibited by the tested patients cannot be accounted for by differences in the severity and duration of the illness. In view of the complexity of the biochemical alterations which characterise Huntington's chorea, it may be hypothesised that not all the patients have the same biochemical defects. For instance, Bird and Iversen (1974) found a significantly reduced choline acetyltransferase activity in only 23 of 37 investigated brains from patients with Huntington's disease.

12.5. Dopaminergic control of growth hormone secretion in acromegaly

In contrast to their effect in normal subjects, dopaminergic drugs induce a paradoxical lowering of plasma GH levels in about 60% of acromegalic subjects. In addition to L-dopa, a large series of other dopamine agonists, such as apomorphine, piribedil, bromocriptine and lisuride, are able to suppress high resting levels of GH in a consistent population of acromegalic patients (Table 12.2). The susceptibility of some acromegalics to suppress plasma GH levels after administration of dopamine agonists formed the rationale for an effective pharmacologic treatment of the disease.

The ability of L-dopa and ergot derivates to suppress GH levels in acromegaly is not the result of an action exerted at central dopamine receptors mediating inhibition of GH release: a host of observations militates against this view. In contrast to direct dopamine agonists, indirect dopamine agonists such as amantadine, amphetamine and nomifensine, which act by releasing dopamine and/or by blocking dopamine re-uptake, do not affect GH levels in acromegaly (Table 12.2); L-dopa's ability to lower GH levels is not impaired when brain-permeable compounds such as phentolamine and isoproterenol are administered in advance (thus ruling out CNS-mediated actions of adrenergic receptor sites) (Cryer and

Table 12.2   Effect of dopamine agonists on GH release in acromegaly.

	Effect on GH levels
Drug	*References*
Direct dopamine agonists	
DA	Massara *et al.*, 1976; Verde *et al.*, 1976
Apomorphine	Chiodini *et al.*, 1974
Bromocriptine	Liuzzi *et al.*, 1974; Thorner *et al.*, 1975
Lisuride	Liuzzi *et al.*, 1974
Metergoline	Chiodini *et al.*, 1976
Piribedil	Camanni *et al.*, 1975*b*
L-dopa	Liuzzi *et al.*, 1972
Indirect dopamine agonists	
Amantadine	Camanni *et al.*, 1975*b*
D-145	Camanni *et al.*, 1975*b*
Amphetamine	Müller *et al.*, 1977
Nomifensine	Müller *et al.*, 1984

Daughaday, 1977). Finally, the action of L-dopa, bromocriptine and lisuride is antagonised by the specific dopamine receptor blocker, pimozide. These observations, together with the findings that L-dopa, dopamine and bromocriptine inhibit the release of GH from human pituitary GH-secreting adenomas *in vitro* (Mashiter *et al.*, 1977), suggest that dopaminergic drugs affect GH secretion in acromegaly by acting on dopamine receptors located on the somatotrophs. Biochemical evidence for such receptors has been provided recently (Bression *et al.*, 1984).

Further evidence for a pituitary site of action of dopaminergic drugs in acromegalic patients comes from a neuropharmacologic study (Camanni *et al.*, 1975*a*) on acromegalics, where L-dopa was given alone or preceded by carbidopa, an extracerebral inhibitor of dopa decarboxylase. Pretreatment with carbidopa, which allows the conversion of L-dopa to dopamine only in the brain, strikingly reduced the GH-lowering action of L-dopa. This suggests that the neuroendocrine effect of the latter in acromegaly is due to its peripheral conversion to dopamine, and hence to the direct action of dopamine at the level of the somatotrophs.

The ability of dopaminergic drugs to inhibit GH secretion seems to be a prerequisite only to tumoral somatotrophs which develop ectopic receptors for dopamine, reportedly absent on the normal cells. However, more recent *in vitro* data obtained in both animals and humans would indicate that dopamine receptors inhibiting GH secretion are present also on somatotrophs from normal pituitaries (Adams *et al.*, 1981; Cronin *et al.*, 1984). Other evidence also supports the possibility of dopamine receptors on somatotrophs (Goldsmith *et al.*, 1979) and internalisation of dopamine itself by rat somatotrophs (Rosenzweig and Kanwar, 1982). One possibility to explain these data is that a normal pituitary cell type, which may secrete both PRL and GH (Frawley and Neill, 1983), reminiscent of the so-called mammosomatotroph cell adenoma or a subclass of somatotrophs (Grindeland *et al.*, 1982), may also express the dopamine receptor phenotype.

From all these data it can be postulated that, by analogy with PRL, for GH also, dopaminomimetic compounds have two sites of action, at suprapituitary and at pituitary levels. In healthy subjects, the stimulatory action of these compounds on the brain overrides a direct effect on the pituitary. In acromegaly, due to a derangement in the normal secretion of hypophysiotropic hormones (i.e. GRF overproduction), the stimulatory CNS action on GH release would be completely lacking (responder acromegalics) or partially lacking (non-responder acromegalics); the inhibitory effect on the pituitary would thus predominate.

References

Adams, E. F., Brajkovich, I. E. and Mashiter, K. (1981). Growth hormone and prolactin secretion by dispersed cell cultures of a normal human pituitary. Effects of thyrotrophin releasing hormone, theophylline, somatostatin and 2-bromo-alpha-ergocriptine. *Acta Endocrinol.* **98**, 345–51.

Besser, G. M., Butler, P. W. P., Landon, J. and Rees, L. (1969). Influence of amphetamines on plasma corticosteroids and growth hormone levels in man. *Brit. Med. J.* **IV**, 528–30.

Bird, E. D. and Iversen, L. L. (1974). Huntington's Chorea: post-mortem measurement of glutamic acid decarboxylase, choline acetyltransferase and dopamine in basal ganglia. *Brain* **97**, 457–72.

Brazeau, P., Vale, W., Burgus, R., Ling, N., Butcher, M., Rivier, J. and Guillemin, R. (1973). Hypothalamic polypeptide that inhibits the secretion of immunoreactive pituitary growth hormone. *Science* **179**, 77–9.

Bression, D., Le Dafniet, M., Brandi, A. M., Racadot, J. and Peillon, F. (1984). Dopamine receptors in human PRL- and GH-secreting adenomas. In *Pituitary Hyperfunction. Physiopathology and Clinical Aspects.* F. Camanni and E. E. Müller (eds), pp. 111–24. Raven Press, New York.

Brown, W. A. and William, B. W. (1976). Methylphenidate increases serum growth hormone concentrations. *J. Clin. Endocrinol. Metab.* **43**, 937–9.

Camanni, F., Massara, F., Belforte, L. and Molinatti, G. M. (1975a). Changes in plasma GH following 2-Br-α-ergocriptine in normal and acromegalic subjects. *J. Clin. Endocrinol. Metab.* **40**, 363–6.

Camanni, F., Massara, F., Fassio, U., Molinatti, G. M. and Müller, E. E. (1975b). Effect of five dopaminergic drugs on plasma growth hormone levels in acromegalic subjects. *Neuroendocrinology* **19**, 227–40.

Camani, F., Picotti, G. B., Massara, F., Molinatti, G. B., Mantegazza, P. and Müller, E. E. (1978). Carbidopa inhibits the growth-hormone and prolactin suppressive effect of L-DOPA in acromegalic patients. *J. Clin. Endocrinol. Metab.* **47**, 647–52.

Caraceni, T., Giovannini, P. and Parati, E. (1983). Anterior pituitary function in extrapyramidal disorders. In *Neuroendocrine Perspectives*. E. E. Müller and R. M. MacLeod (eds), pp. 293–328. Elsevier, Amsterdam.

Casanueva, F., Betti, R., Cocchi, D., Zanardi, P., Motta, T. and Müller, E. E. (1981). A role for dopamine in growth hormone regulation in the dog. *Endocrinology* **108**, 1469–75.

Chalmers, R. J., Johnson R. H., Keogh, H. J. and Nanda, R. N. (1978). Growth hormone and prolactin response to bromocriptine in patients with Huntington's Chorea. *J. Neurol. Neurosurg. Psychiat.* **41**, 135–9.

Chiodini, P. G., Liuzzi, A., Botalla, L., Cremascoli, G. and Silvestrini, F. (1974). Inhibitory effect of dopaminergic stimulation on GH release in acromegaly. *J. Clin. Endocrinol. Metab.* **38**, 200–6.

Chiodini, P. G., Liuzzi, A., Müller, E. E., Botalla, L., Cremascoli, G., Oppizzi, G., Verde, G. and Silvestrini, F. (1976). Inhibitory effect of an ergoline derivative, methergoline, on growth hormone and prolactin levels in acromegalic patients. *J. Clin. Endocrinol. Metab.* **43**, 356–63.

Cocola, F., Udeschini, G., Panerai, A. E., Neri, F. and Müller, E. E. (1976). A rapid radioimmunoassay method for growth hormone in dog plasma. *Proc. Soc. Exp. Biol. Med.* **151**, 140–4.

Cronin, M. J., Thorner, M. O., Hellman, P. and Rogol, A. D. (1984). Bromocriptine inhibits growth hormone release from rat pituitary cells in primary culture. *Proc. Soc. Exp. Biol. Med.* **175**, 191–5.

Cryer, P. E. and Daughaday, M. (1977). Adrenergic modulation of growth hormone secretion in acromegaly: alpha- and beta-adrenergic blockade produce qualitatively normal response but no effect on L-dopa suppression. *J. Clin. Endocrinol. Metab.* **44**, 977–9.

Frawley, L. S. and Neill, J. D. (1983). Identification of a pituitary cell type that secretes both growth hormone and prolactin: detection by reverse hemolytic plaque assays. *Endocrinol. Soc. Ann. Meet*, **65**, 310.

Goldsmith, P. C., Cronin, M. J. and Weiner, R. I. (1979). Dopamine receptor sites in the anterior pituitary. *J. Histochem. Cytochem.* **27**, 1205–7.

Grindeland, R. E., Hymer, W. C., Lundgren, P. and Edwards, C. (1982). Differential secretion of bioassayable growth hormone by two types of somatotroph. *Physiologist* **25**, 262.

Guillemin, R., Brazeau, P., Bohlen, P., Esch, F., Ling, N. and Wehrenberg, W. B. (1982). Growth hormone-releasing factor from a human pancreatic tumor that caused acromegaly. *Science* **218**, 585–7.

Jacoby, J. H., Greenstein, M., Sassin, J. F. and Weitzman, E. D. (1974). The effect of monoamine precursors on the release of growth hormone in the Rhesus Monkey. *Neuroendocrinology* **14**, 95–102.

Klawans, H. L. (1970). A pharmacologic analysis of Huntington's Chorea. *Eur. Neurol.* **4**, 148–63.

Lal, S., de la Vega, C. E., Sourkes, T. L. and Friesen, H. G. (1972). Effect of apomorphine on human growth hormone secretion. *Lancet* ii: 661.

Levy, C. L., Carlson, H. E., Sowers, J. R., Goodlett, R. E., Tourtellotte, W. W. and Hersham, J. M. (1979). Growth hormone and prolactin secretion in Huntington's disease. *Life Sci.* **24**, 743–50.

Liuzzi, A., Chiodini, P. G., Botalla, L., Cremascoli, G. and Silvestrini, F. (1972). Inhibitory affect of L-dopa on GH release in acromegalic patients. *J. Clin. Endocrinol. Metab.* **35**, 941–3.

Liuzzi, A., Chiodini, P. G., Botalla, L., Cremascoli, G., Müller, E. E. and Silvestrini, F. (1974). Decreased plasma growth hormone (GH) levels in acromegalics following CB 154 (2-Br-α-ergocryptine). *J. Clin. Endocrinol. Metab.* **38**, 910–12.

Locatelli, V., Cocchi, D., Gil-Ad, I., Mantegazza, P. and Müller, E. E. (1977). Presynaptic dopamine receptors for the release of growth hormone in the rat. *Endocrinol. Soc. 56th Ann. Meet., Chicago*, p. 205.

Luedecke, G., Mueller, D. and Patino, J. (1972). Effects of stereotaxis in the human hypothalamus and limbic system on ACTH and GH secretion. *fifth Int. Congr. Endocrinol., Hamburg*, Abst., p. 272.

Martin, J. B. (1972). Plasma growth hormone (GH) response to hypothalamic or extrahypothalamic electrical stimulation. *Endocrinology* **91**, 107–15.

Mashiter, K., Adams, B., Beard, M. Holley, A. (1977). Bromocriptine inhibits prolactin and growth hormone release by human pituitary tumors in culture. *Lancet*, **ii**, 197–8.

Massara, F., Camanni, F., Belforte, L. and Molinatti, G. M. (1976). Dopamine-induced inhibition of prolactin and growth hormone secretion in acromegaly. *Lancet*, **i**, 485.

Müller, E. E., Liuzzi, A., Cocchi, D., Panerai, A. E., Oppizzi, G., Locatelli, V., Mantegazza, P., Silvestrini, F. and Chiodini, P. G. (1977). Role of dopaminergic receptors in the regulation of growth hormone secretion. In *Nonstriatal Dopaminergic Neurons*. E. Costa and G. L. Gessa (eds), pp. 127–39, Raven Press, New York.

Müller, E. E., Parati, E. A., Panerai, A. E., Cocchi, D. and Caraceni, T. (1979). Growth hormone hyperresponsiveness to dopaminergic stimulation in Huntington's Chorea. *Neuroendocrinology* **28**, 313–19.

Müller, E. E., Cavagnini, F., Martinez-Campos, A., Maraschini, C., Giovannini, P., Novelli, A. and De Leo, V. (1984). Dynamic testing of prolactin and growth hormone secretion in patients with neuroendocrine disorders. *Acta Endocrinol.* **107**, 155–63.

Neill, J. D. (1980). Neuroendocrine control of prolactin secretion. In *Frontiers in Neuroendocrinology*. L. Martini and W. F. Ganong (eds), pp. 126–55, Raven Press, New York.

Pandey, G. N., Garvey, D. L., Tamminga, C., Ericksen, S., Ali, S. I. and Davis, J. M. (1977). Post-synaptic supersensitivity in schizophrenia. *Am. J. Psychiat.* **134**, 518–22.

Pandey, G. N., Ali, S. I., Casper, R. and Davis, J. M. (1984). Neuroendocrine studies in schizophrenia In *Pscyhoneuroendocrine Dysfunction*, N. S. Shah and A. G. Donald (eds), pp. 313–19. Plenum, New York.

Rosenzweig, L. J. Kanwar, Y. S. (1982). Dopamine internalization by and intracellular distribution within prolactin cells and somatotrophs of the rat anterior pituitary as determined by quantitative electron microscope autoradiography. *Endocrinology* **111**, 1817–29.

Schacht, V., Leven, M. and Backer, G. (1977). Studies on brain metabolism of biogenic amines. *Brit. J. Clin. Pharmacol.* **4**, 772.

Steiner, R. A., Illner, P., Rolfs, A. D., Toivola, P. T. K. and Gale, C. C. (1978). Noradrenergic and dopaminergic regulation of GH and prolactin in baboons. *Neuroendocrinology* **26**, 15–31.

Thorner, M. O. Chait, A., Aitken, M., Benker, G., Bloom, S. R., Mortimer, C. H., Sanders, P., Stuart Mason, A. and Besser, G. M. (1975). Bromocriptine treatment of acromegaly. *Brit. Med. J.* **1**, 299–303.

Toivola, P. T. K. and Gale, C. C. (1970). Effect on temperature of biogenic amine infusion into hypothalamus of baboon. *Neuroendocrinology* **6**, 210–19.

Tsushima, T., Irie, M. and Sakuma, M. (1971). Radioimmunoassay for canine growth hormone. *Endocrinology* **89**, 685–93.

Verde, G., Oppizzi, G., Colussi, G., Botalla, L., Silvestrini, F., Chiodini, P. G., Liuzzi, A. and Müller, E. E. (1976). Effect of dopamine infusion on plasma levels of growth hormone in normal subjects and in acromegalic patients. *Clin. Endocrinol.* **5**, 419–23.

13

Dopamine and disorders of growth hormone secretion

S. M. Shalet

Department of Endocrinology, Christie and Withington Hospitals, Manchester, UK

13.1. Introduction

In normal man, dopamine has opposing actions on growth hormone (GH) release, by stimulating basal secretion but diminishing augmented release. Bansal *et al.* (1981) have shown that these dual actions of dopamine on GH secretion are mediated by a dopaminergic mechanism since they are mimicked by a specific dopamine agonist and prevented by a dopamine antagonist.

13.2. Acromegaly

In clinical practice the major disorder of GH secretion in which dopamine agonists have been used is acromegaly. In this condition there is an elevated circulating level of GH, due nearly always to the presence of a GH-secreting pituitary adenoma. Although some patients with acromegaly have very few symptoms, in others the disease can cause considerable distress, increasing disability and, eventually, early death. Treatment by surgery and radio-therapy, singly or combined, can certainly cure the disease but may take a long time to do so and is not always effective. Even in those who are cured eventually, symptoms may persist for several years after treatment. There is, therefore, a considerable need for a rapidly effective alternative treatment, and a number of pharmacological agents have been tried in the past. In 1972 Liuzzi *et al.* showed that L-dopa a short-acting dopamine agonist, which stimulates GH secretion in most normal subjects, would depress GH concentrations in certain acromegalic patients. They used bromocriptine, a long-acting dopamine agonist, in a group of seven untreated acromegalics, who were selected because they were responsive to L-dopa. It was found that a single dose of 2·5 mg bromocriptine lowered GH concentrations in all for up to 5 h (Liuzzi *et al.*, 1974a). Since that time bromocriptine has been used in a large number of patients, and although there remain some uncertainties as to its mode of action and its value in long-term treatment, it has come into widespread clinical use.

13.3. Effect of bromocriptine on GH concentrations and clinical features of acromegaly

Approximately 70 % of patients treated with bromocriptine show a significant decrease in GH concentration, which is sustained over the whole period of treatment (Vance *et al.*, 1984). However, in only a minority of patients does the GH concentration fall to near the normal range (< 10 mU litre^{-1}). One study that reported negative results with bromocriptine therapy deserves comment. Lindholm *et al.* (1981) compared the effects of placebo with those of bromocriptine in a short-term double-blind crossover study of 18 acromegalic patients. No difference was seen in the clinical symptoms or GH response to oral glucose during the placebo or bromocriptine treatment periods. These observations led the authors to suggest that it is doubtful if bromocriptine has a beneficial effect in acromegaly (Lindholm *et al.*, 1981). This view, although not generally accepted, has contributed to the development of a more critical appraisal of the effectiveness of therapy. Lindholm *et al.* (1981) provided no objective measures of clinical response, such as ring size, hand or foot volume or heel-pad thickness. Furthermore, they did not mention the time interval between the last dose of bromocriptine and the performance of an oral glucose tolerance test (OGTT) in their patients. However, Nortier *et al* (1984*a*) have shown that this time interval may be critical and that the GH level during OGTT may only reflect the mean plasma GH level during the day accurately if the OGTT is performed within 1 h of the last dose of bromocriptine.

The generally held view, which the author shares, is that bromocriptine is an effective drug for the treatment of acromegaly but it should be used as adjunctive not primary therapy. In most patients, surgery should be used initially, followed by radiotherapy if the postoperative GH levels remain elevated. While awaiting the beneficial effects of pituitary irradiation, which may take a number of years, a trial of bromocriptine therapy should be undertaken.

For most patients the effective daily dose of bromocriptine lies between 5 and 20 mg, with only very few showing any further decrease in GH concentration with increasing dose. The drug must be given in divided doses, preferably 6-hourly, and has been shown to be effective for up to 5 years of continuous treatment; however, escape from its suppressive effect has been reported in one case.

Clinical improvement occurs in 70–90 % of acromegalics treated with bromocriptine (Vance *et al.*, 1984). Many studies have reported improvement in or disappearance of: headache, excessive sweating, arthralgia and fatigue. A decrease in skin thickness and in the size of the hands and feet also occurs in a high proportion of patients treated.

An improvement in glucose tolerance was reported by Wass *et al.*, (1980), who showed that the proportion of acromegalics who could be defined as diabetic fell by about 60 % after 3 months' bromocriptine therapy, the exact figures varying a little depending on the criteria used. Glucose tolerance also improved in 38 patients who could not be defined as diabetic. Although no clear correlation could be defined between changes in glucose tolerance and in GH concentration, only those patients who showed a decrease in GH concentration showed improvement in glucose tolerance. Other metabolic changes, including a decrease (sometimes to normal) in urinary hydroxy-proline and urinary calcium excretion, have also been reported (Roelfsema *et al.*, 1979).

Objective confirmation of decrease in soft tissue thickness has been less commonly reported, but Wass *et al.* (1977) found a decrease in finger size, as assessed by jeweller's rings, in 30 of 34 patients, whereas Sachdev *et al.* (1975) found a decrease in heel-pad thickness of more than 5 mm in only two of 21 patients. Figure 13.1 shows the change in facial appearance of a patient after treatment for one-and-a-half years, and clearly shows reduction in skin thickness.

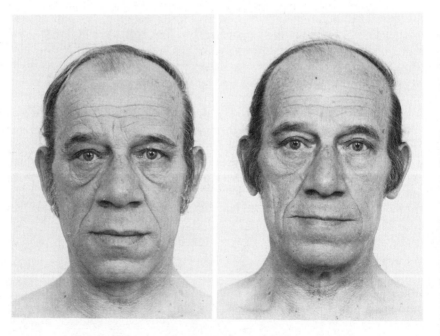

Fig. 13.1.   Changes in the facial appearance of an acromegalic patient before (left) and after (right) 18 months of treatment with bromocriptine.

13.4. Reduction in tumour size

In the majority of patients with large prolactin-secreting pituitary adenomas, dopamine agonist drugs inhibit prolactin secretion as well as reducing the size of the tumour. In acromegaly the results are less impressive. Oppizzi *et al.* (1984) studied 19 patients with large GH-secreting adenomas. GH secretion was inhibited by bromocriptine or lisuride (another dopamine agonist drug) in about 50 % of patients but there was a reduction in tumour size, demonstrated by computer tomography, in only two patients.

13.5. Selection of responsive patients

It has been suggested that patients who are likely to show a good response to long-term treatment can be differentiated from those who will not on the basis of their GH response to thyrotrophin-releasing hormone (TRH), the serum prolactin concentration and the acute change in GH concentration after a single oral dose of bromocriptine.

Liuzzi *et al.* (1974*b*) suggested that only those patients who showed an increase in GH concentration after TRH administration would respond to bromocriptine. However, subsequent experience has shown that this is not always true. Certainly the great majority have done so but a few have not, and a small number of those who showed no GH response to TRH responded to bromocriptine (Wass *et al.*, 1977). Werner *et al.* (1978) and Lamberts *et al.* (1983) have suggested that acromegalics with hyperprolactinaemia are more likely to respond to bromocriptine with an inhibition of GH secretion. Lamberts *et al.* (1983) showed that prolactin was present in the pituitary tumour tissue in 14 out of 35 consecutive acromegalics. Hyperprolactinaemia was found in 10 of these 14 patients. The GH levels after a single dose of bromocriptine were significantly more suppressed in the patients with mixed GH and prolactin-containing adenomas than in those with pure GH-containing pituitary adenomas. However, prolactin secretion by the adenoma is not the only cause of hyperprolactinaemia in acromegaly; the GH-producing tumour may interfere with the transport or secretion of hypothalamic prolactin-inhibiting factor (PIF) resulting in hypersecretion of prolactin by the normal lactotroph. Therefore, as hyperprolactinaemia in acromegaly may be due to very different aetiologies, it is not surprising that Roelfsema *et al.* (1979) found no correlation between the pretreatment prolactin level and the GH response to bromocriptine.

The predictive value of acute changes in GH levels following a single oral dose of 2·5 mg of bromocriptine was assessed by Belforte *et al.* (1977). Twenty-one of their 30 patients showed a greater than 50 % fall in GH concentration in the acute test, but only 18 of these showed a similar fall on chronic treatment. None of the nine who failed to respond in the acute test showed any response on long-term therapy.

Therefore these tests, particularly the GH response to a single dose of bromocriptine, allow selection of patients with a high chance of clinical and GH response on chronic therapy. However, it could be argued that, with a treatment which offers clinical improvement in 70–90% of patients, all patients who remain acromegalic after surgery, radiotherapy or both warrant a trial of bromocriptine. Thus, in most instances, a predictive test is unnecessary.

13.6. Mechanism of action

It has been clearly demonstrated that in acromegaly dopamine and dopaminergic drugs exert the inhibitory effect on GH release by a direct action at the pituitary level (Mashiter *et al.*, 1977; Peillon *et al.*, 1979). From experimental data using human GH-secreting and prolactin-secreting adenomatous cells in a perfusion column, Bression *et al.* (1982) have shown that GH-secreting adenomas are less sensitive to dopamine than prolactin-secreting adenomas. Moreover, less dopamine receptors are present on the membranes of pure GH-secreting adenomas compared with those releasing prolactin.

Ultrastructural studies on tumour cells removed from responsive patients after 3 weeks of bromocriptine treatment have shown that the GH-producing cells are densely granulated, suggesting that the predominant effect of the drug is to inhibit GH release rather than its synthesis (Fanghanel-S *et al.*, 1978).

Since Liuzzi *et al.* (1974*a*) showed that patients in whom GH secretion is decreased by L-dopa are also responsive to bromocriptine, it seems not unreasonable to assume that the two agents act at the same site *in vivo*. The effect of L-dopa can be almost completely inhibited by prior administration of carbidopa (Camanni *et al.*, 1978), which blocks conversion of L-dopa to dopamine outside the central nervous system but not within it, so that it would appear likely that its major effect is exerted directly on the anterior pituitary and not through activation of central dopaminergic mechanisms. However, the observation that GH concentrations in acromegalics who are responsive to bromocriptine tend to rise progressively during sleep (Chihara *et al.*, 1977) suggests that central influences may be partially able to overcome the blocking effects of the drug and could, perhaps, be involved in the resistance to its action seen in some patients.

Apart from its direct effect in inhibiting GH secretion, there is some evidence that bromocriptine may also act in other ways. Maneschi *et al.* (1978) showed a modest but significant increase in the metabolic clearance rate of GH in six acromegalic women treated with bromocriptine, and suggested that this might be due to an increase in hepatic and renal blood flow resulting from direct stimulation of visceral dopamine receptors.

A further effect on GH metabolism that has been proposed is alteration of the ratio of biologically active monomeric GH to less active high-molecular-weight forms (Besser *et al.*, 1976). Since the high-molecular-weight forms appear to have little or no biological activity but react equally with low-molecular-weight forms in radioimmunoassays, it is suggested that a disproportionate fall in biologically active GH may account for the disparity often observed between clinical response and change in total GH concentrations. Benker *et al.* (1979) approached this problem by applying gel chromatography to the sera of acromegalics in order to study the different molecular forms. After 4 weeks' treatment with bromocriptine, Benker *et al.* (1979) observed 'little' GH concentrations decreased to a greater extent than those of 'big big' GH or 'big' GH. However, they felt that the degree of change was insufficient to explain the clinical response of the disease to bromocriptine. Furthermore, Hizuka *et al.* (1984) were unable to show any effect of bromocriptine therapy on the elution profile of GH in eight acromegalic patients.

Since many of the biological effects of GH are mediated through somatomedins, assay of these substances might indicate whether a reduction in bioactive GH could occur in the absence of any change in immuno-reactive GH. Holdaway *et al.* (1978) could find no consistent relationship between change in total GH and somatomedin, and Shalet *et al.* (1980) found significant decreases in somatomedin activity only in patients who showed suppression of GH concentrations to near the normal range. In other patients, symptomatic relief was obtained even though total GH concentrations and somatomedin activity showed little or no significant change. However, both groups used a porcine costal cartilage bioassay technique for measurement of somatomedin activity. More recently, a highly specific radioimmunoassay for somatomedin C has been developed and utilised to compare the clinical and biochemical responses of acromegalic patients to bromocriptine. Wass *et al.* (1982) identified a subgroup of patients who showed clinical improvement with a decrease of somatomedin C levels but without reduction in GH values. However, Nortier *et al.* (1984*b*) were unable to substantiate these findings.

The proposed alteration of the ratio of biologically active monomeric GH to less active high-molecular-weight forms and its relevance to the clinical progress of the patient still need to be studied further. Nor is it yet clear which of the two hormone markers, GH or somatomedin C, provides the most useful biochemical assessment of disease activity in bromocriptine-treated acromegaly.

13.7. **Other drugs**

Of the known ergot derivatives with dopaminergic activity, the ergopeptine, bromocriptine, is the compound that has been studied most. However, the search has continued for new dopamine agonists which are more potent and longer acting than bromocriptine. For instance Thorner *et al.* (1978) studied the effects of lergotrile, an ergoline derivative, in eight acromegalics who had been shown to respond to bromocriptine with a fall in GH levels. Lergotrile appeared equally effective in reducing GH levels but there were more side-effects and it had a shorter duration of action than bromocriptine.

Lisuride is another ergoline derivative which appears as effective as bromocriptine in reducing GH levels in acromegaly but offers no particular advantages. The duration of action is similar to that of bromocriptine.

More recently, Kendall-Taylor *et al.* (1983) studied the effects of chronic administration of pergolide, another ergoline derivative with dopamine-agonistic effects. All eight acromegalic patients showed a statistically significant fall in GH levels; in three the decrease was rather small and did not parallel the change in prolactin; two patients noticed no clinical change. Two patients were troubled with side-effects and the drug was subsequently discontinued. In three patients ambulant GH fell to 5 mU litre^{-1} or less. For the whole group of eight patients there was a mean reduction of GH to 55% of intial value on 500 μg pergolide daily and to 40% on 1000 μg daily. Further increase to 1500 μg in two cases offered no advantage. It is clear that pergolide has a beneficial effect in acromegaly. The optimum dose is probably between 500 and 1000 μg daily but further studies are needed to confirm this. The drug is effective when given as a once-daily regime, suppressing GH for at least 22 h; this is in keeping with its long-acting *in vivo* effect on prolactin (Grossman *et al.*, 1980).

13.8. Growth hormone deficiency

Following the observation that dopamine stimulates the secretion of GH in normal man, L-dopa has been used as a test of GH reserve. In children with GH deficiency, the GH response to a single oral dose of L-dopa is blunted or absent, whereas in most of the other causes of short stature a normal GH response to L-dopa is observed.

Recently, Huseman and Hassing (1984) have suggested that GH deficiency in some children is due to a defect in the synthesis or release of endogenous L-dopa or dopamine. In a novel study they examined the effect of 6 months' therapy with either L-dopa (group I) or bromocriptine (group II) on GH secretion and growth in GH-deficiency children. The growth rate in group I increased to 5·7 cm year^{-1} during the 6 months from a pretreatment rate of 3·4 cm year^{-1}. Three of the four children had significantly increased height increments and two children achieved growth rates normal for their bone age. Similarly, the growth rate in group II increased to 4·8 cm year^{-1} from the pretreatment rate of 2·9 cm year^{-1}. Three children in group II had significantly increased height increments and two children had normal growth rates for bone age. The growth increments during L-dopa therapy occurred in the three children who had significant increases in GH and somatomedin C. Of the three children with significant growth increments during bromocriptine therapy, two had increases in somatomedin C and one achieved a normal peak GH value.

The results of the above study raise a number of questions concerning the mechanism of action and the selection of patients considered likely to respond. In the first instance further studies are necessary to refute or support the authors' conclusions.

References

Bansal, S. A., Lee, L. A. and Woolf, P. D. (1981). Dopaminergic stimulation and inhibition of growth hormone secretion in normal man: studies of the pharmacological specificity. *J. Clin. Endocrinol.* **53**, 1273–7.

Belforte, L., Camanni, F., Chiodini, P. G., Liuzzi, A., Massara, F., Molinatti, G. M., Müller, E. E. and Silvestrini, F. (1977). Long-term treatment with 2-Br-α-ergocryptine in acromegaly. *Acta Endocrinol. (Kbh)* **85**, 235–48.

Benker, G., Sandmann, K., Tharandt, L., Hackenberg, K. and Reinwein, D. (1979). Gel filtration studies of serum growth hormone in acromegaly following bromocriptine administration. *Hormone Res.* **11**, 151–60.

Besser, G. M., Thorner, M. O., Wass, J. A. H., Jones, A. E., Lowry, P. J., Rees, L. H. and Jones, A. (1976). Bromocriptine treatment of acromegaly. *Quart. J. Med.* **45**, 695.

Bression, D., Brandi, A. M., Nousbaum, A., LeDafniet, M., Racadot, J. and Peillon, F. (1982). Evidence of dopamine receptors in human growth hormone (GH)-secreting adenomas with concomitant study of dopamine inhibition of GH secretion in a perfusion system. *J. Clin. Endocrinol.* **55**, 589–93.

Camanni, F., Picotti, G. B., Massara, F., Molinatti, G. M., Mantegazza, P. and Müller, E. E. (1978) Carbidopa inhibits the growth hormone- and prolactin-suppressive effect of L-dopa in acromegalic patients. *J. clin. Endocrinol.* **7**, 647–51.

Chihara, K., Kato, Y., Abe, H., Furomoto, M., Maeda, K. and Imura, H. (1977). Sleep-related growth hormone release following 2-bromo-α-ergocryptine treatment in acromegalic patients. *J. Clin. Endocrinol.* **44**, 78–84.

Fanghanel-S, G., Larraza, O., Arauco, R., Esquivel, R., Martinez Campos, A. and Valverde, R. C. (1978). Serum growth hormone and ultrastructural studies of adenohypophysial tissue in bromocriptine-treated acromegalics. *Clin. Endocrinol.* **9**, 289–96.

Grossman, A., Yeo, T., Delitala, G., Hathway, N. R. and Besser, G. M. (1980). Two new dopamine agonists that are long-acting *in vivo* but short-acting *in vitro*. *Clin. Endocrinol.* **13**, 595–9.

Hizuka, N., Hendricks, C. M., Roth, J. and Gorden, P. (1984). Failure of bromocriptine to alter the qualitative characteristics of human growth hormone in acromegaly. *Metabolism* **33**, 582–4.

Holdaway, I. M., Frengley, P. A., Scott, D. J. and Ibbertson, H. K. (1978). Bromoergocryptine treatment of acromegaly persisting following conventional therapy. Clin. Endocrinol. **8**, 45–54.

Huseman, C. A. and Hassing, J. M. (1984). Evidence for dopaminergic stimulation of growth velocity in some hypopituitary children. *J. Clin. Endocrinol.* **58**, 419–25.

Kendall–Taylor, P., Upstill-Goddard, G. and Cook, D. (1983). Longterm pergolide treatment of acromegaly. *Clin. Endocrinol.* **19**, 711–19.

Lamberts, S. W. J., Klijn, J. G. M., van Vroonhoven, C. C. J., Stefanko, S. Z. and Liuzzi, A. (1983). The role of prolactin in the inhibitory action of bromocriptine on growth hormone secretion in acromegaly. *J. Clin. Endocrinol.* **103**, 446–50.

Lindholm, J., Rhshede, J., Vestergaard, S., Hummer, L., Faber, O. and Hagen, C. (1981). No effect of bromocriptine in acromegaly. *New Engl. J. Med.* **304**, 1450–4.

Liuzzi, A., Chiodini, P. G., Botalla, L., Cremascoli, G. and Silvestrini, F. (1972). Inhibitory effect of L-dopa on GH release in acromegalic patients. *J. Clin. Endocrinol.* **35**, 941–3.

Liuzzi, A., Chiodini, P. G., Botalla, L., Cremascoli, G., Müller, E. G. and Silvestrini, F. (1974a). Decreased plasma growth hormone (GH) levels in acromegalics following CB 154 (2-Br-α-ergocryptine) administration. *J. Clin. Endocrinol.* **38**, 910–12.

Liuzzi, A., Chiodini, P. G., Botalla, L., Silvestrini, F. and Müller, E. G. (1974b). Growth hormone (GH)-releasing activity of TRH and GH-lowering effect of dopaminergic drugs in acromegaly: homogeneity of the two responses. *J. Clin. Endocrinol.* **39**, 871–6.

Maneschi, F., Navalesi, R., Pilo, A. and Paci, A. (1978). (^{125}I)hGH metabolism in acromegaly: effects of chronic treatment with 2-Br-α-Ergocryptine. *J. Clin. Endocrinol.* **47**, 110–18.

Mashiter, K., Adams, E., Beard, M. and Holley, A. (1977). Bromocriptine inhibits prolactin and growth hormone release by human pituitary tumours in culture. *Lancet* **ii**, 197–8.

Nortier, J. W. R., Croughs, R. J. M., Thijssen, J. H. H. and Schwarz, F. (1984a). Plasma growth hormone suppressive effect of bromocriptine in acromegaly. Evaluation by plasma GH day profiles and plasma GH concentrations during oral glucose tolerance tests. *Clin. Endocrinol.* **20**, 565–71.

Nortier, J. W. R., Croughs, R. J. M., Thijssen, J. H. H. and Schwarz, F. (1984b). Bromocriptine therapy in acromegaly: Effects on plasma GH levels, somatomedin-C levels and clinical activity. *Clin. Endocrinol.* **22**, 209–17.

Oppizzi, G., Liuzzi, A., Chiodini, P., Dallabonzana, D., Spelta, B., Silvestrini, F., Borghi, G. and Tonon, C. (1984). Dopaminergic treatment of acromegaly: different effects on hormone secretion and tumour size. *J. Clin. Endocrinol.* **58**, 988–92.

Peillon, F., Cesselin, F., Bression, D., Zygelman, N., Brandi, A. M., Nousbaum, A. and Mauborgne, A. (1979). *In vitro* effect of dopamine and L-dopa on prolactin and growth hormone release from human pituitary adenomas. *J. Clin. Endocrinol.* **49**, 737–41.

Roelfsema, F., Goslings, B. M., Frolich, M., Moolenaar, A. J., Van Seters, A. P. and Van Slooten, H. (1979). The influence of bromocriptine on serum levels of growth hormone and other pituitary hormones and its metabolic effects in active acromegaly. *Clin. Endocrinol.* **11**, 235–44.

Sachdev, Y., Tunbridge, W. M. G., Weightman, D. R., Gomez-Pan, A., Hall, R. and Goolanicki, S. K. (1975). Bromocriptine therapy in acromegaly. *Lancet* **ii**, 1164–8.

Shalet, S. M., Price, D. A., Beardwell, C. G., Mindel, A. and MacFariane, I. A. (1980). Growth hormone and somatomedin levels in acromegalics treated with bromocriptine. *Hormone Res.* **12**, 121–9.

Thorner, M. O., Ryan, S. M., Wass, J. A. H., Jones, A., Williams, S. and Besser, G. M. (1978). Effect of the dopamine agonist, lergotrile mesylate on circulating anterior pituitary hormones in man. *J. Clin. Endocrinol.* **47**, 372–8.

Vance, M. L., Evans, W. S. and Thorner, M. O. (1984). Bromocriptine. *Ann. int. Med.* **100**, 78–91.

Wass, J. A. H., Thorner, M. O., Morris, D. V., Rees, L. H., Stuart Mason, A., Jones, A. E. and Besser, G. M. (1977). Long-term treatment of acromegaly with bromocriptine. *Brit. Med. J.* **2**, 875–8.

Wass, J. A. H., Cudworth, A. G., Bottazzo, G. F., Woodrow, J. C. and Besser, G. M. (1980). An assessment of glucose intolerance in acromegaly and its response to medical treatments. *Clin. Endocrinol.* **12**, 53–9.

Wass, J. A. H., Clemmons, D. R., Underwood, L. E., Barrow, J., Besser, G. M. and Van Wyk, J. J. (1982). Changes in circulating somatomedin-C levels in bromocriptine-treated acromegaly. *Clin. Endocrinol.* **17**, 369–77.

Werner, S., Hall, K. and Sjoberg, H. E. (1978). Bromocriptine therapy in patients with acromegaly: effect on growth hormone, somatomedin A and prolactin. *Acta Endocrinol. (Kbh)* **88**, 199–206.

14

Dopaminergic regulation of the prolactin-secreting cells: the role of calcium and cyclic AMP

G. Schettini[1], E. L. Hewlett[2], M. J. Cronin[3], T. Yasumoto[4] and R. M. MacLeod[2]

[1]*Department of Pharmacology, II School of Medicine, University of Naples, via S. Pansini 5, 80131 Naples, Italy;* [2]*Department of Internal Medicine,* [3]*Department of Physiology, University of Virginia, Charlottesville VA 22908, USA;* [4]*Faculty of Agriculture, Tohoku University, Tsutsumi-dori, Sendai 980, Japan*

14.1. Introduction

Prolactin (PRL) secretion by anterior pituitary glands is under the tonic inhibitory control of dopamine (MacLeod, 1976; Weiner and Ganong, 1978). The tubero-infundibular dopaminergic (TIDA) neurones, originating in the hypothalamus, produce and release dopamine into the hypophyseal portal circulation, which reaches the anterior pituitary gland (Ben-Jonathan et al., 1977; Gibbs and Neill, 1978), where the dopamine binds to specific receptors (Cronin, 1982), thereby initiating the inhibitory process. However, the post-receptor mechanisms responsible for mediating this inhibitory effect have not been completely characterised. Changes in cell Ca^{2+} metabolism, adenylate cyclase activity, cyclic AMP accumulation and phospholipid turnover in response to inhibitory or stimulatory activity mechanisms have been reported; moreover, these intracellular messengers do not seem to operate independently, but to interact and modulate each other's metabolism in determining the cellular response (Fig. 14.1) (Mitchell, 1975; Cheung, 1980; Means and Dedman, 1980; Berridge, 1981; Rasmussen, 1981; Canonico et al., 1983; Gautvik et al., 1983; Schettini et al., 1983b).

It is known that dopamine inhibits both cyclic AMP accumulation in primary cultures of anterior pituitary cells (Cronin and Thorner, 1982; Swennen and Denef, 1982; Cronin et al., 1983; Schettini et al., 1983b), and adenylate cyclase activity in anterior pituitary membrane preparations (De Camilli et al., 1979; Giannattasio et al., 1981; Onali et al., 1981). Dopamine also reduces phosphotidylinositol turnover (Canonico et al., 1983), which is often associated with Ca^{2+} mobilisation (Mitchell, 1975; Berridge, 1981). Although dopamine does not produce changes in basal or ionophore A23187-stimulated Ca^{2+} flux (Tam and Dannies, 1980), the biogenic amine

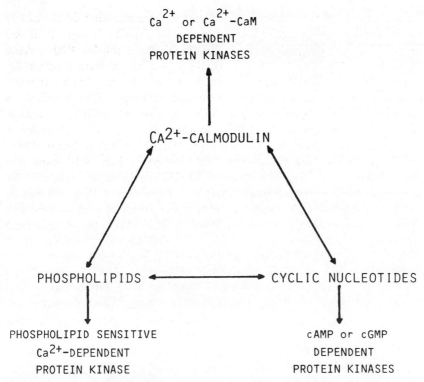

Fig. 14.1. Schematic representation of a possible general relationship among the major intracellular regulatory 'messengers'. Interrelationships between these messengers include their metabolism and functions, as well as their complementary roles with respect to their rate, distance and duration of action.

antagonises cyclic AMP accumulation as well as PRL release by ionophore A23187 (Tam and Dannies, 1980; Schettini *et al.*, 1983*b*). These observations suggest that dopamine acts at a site distal to Ca^{2+} mobilisation to cause the inhibition of PRL release (Tam and Dannies, 1980; Schettini *et al.*, 1983*b*).

Calcium serves as a regulator of cyclic nucleotide metabolism in some cellular systems and is thought to modulate both adenylate cyclase and phosphodiesterase activity (Rasmussen, 1981). The effectiveness of Ca^{2+} is often dependent upon the intracellular Ca^{2+} binding protein calmodulin (Cheung, 1980; Means and Dedman, 1980), which can regulate adenylate cyclase activity in brain (Brostrom *et al.*, 1977), pancreatic islets (Valverde *et al.*, 1979; Sharp *et al.*, 1980), neurohypophysis (Dratt *et al.*, 1981) and bovine lung tissue (Ofulue and Nijjar, 1981). Evidence also suggests the involvement of calmodulin in the regulation of PRL release and cyclic AMP formation, both in normal and clonal pituitary cells (Merritt *et al.*, 1981; Schettini *et al.*, 1983*a, b, c*).

In the present study we describe the role of Ca^{2+} and cyclic AMP, their interaction in the control of PRL secretion and their changes following a dopaminergic stimulus, in order to characterise the mechanism whereby dopamine inhibits PRL secretion. We used a variety of probes that either modify the Ca^{2+} status of the cells or interact with components of the adenylate cyclase holoenzyme.

14.2. Dopamine and the Ca^{2+} system

Many studies suggest a role for Ca^{2+} in the modulation of PRL release. For example, *in vitro* PRL secretion is enhanced by increased Ca^{2+} concentration (MacLeod and Fontham, 1970; Parsons, 1970) whereas perfusion of pituitary cells with Ca^{2+}-free medium or the addition of Ca^{2+} channel blockers suppresses PRL release (Thorner *et al.*, 1980). Depolarisation of pituitary cells with K^+ increases $^{45}Ca^{2+}$ uptake and induces Ca^{2+}-dependent PRL release (Milligan and Kraicer, 1971; Tashjian *et al.*, 1978). In addition, although acting via different mechanisms, the Ca^{2+} ionophore A 23187 and TRH (which enhance intracellular Ca^{2+} mobilisation) each stimulate PRL release from anterior pituitary cells in primary cultures (Tam and Dannies, 1980; Gershengorn, 1982) or various types of GH cells (Tashjian *et al.*, 1978; Gershengorn, 1980; Tan and Tashjian, 1981).

Recently, Yasumoto *et al.* (1979) reported the extraction and partial purification of maitotoxin (MTX), a Ca^{2+} channel activator, from a marine dinoflagellate, *Gambieridiscus toxicus*. We have used MTX to better characterise the role of Ca^{2+} and Ca^{2+} channels and their relation to the dopaminergic inhibition of PRL release. MTX stimulates the release of [^{3}H]noradrenaline and the influx of Ca^{2+} from a pheochromocytoma cell line PC12h (Takahashi *et al.*, 1982). Both effects of MTX were inhibited by Ca^{2+} channel blockers (verapamil or Mn^{2+}) or a local anaesthetic agent tetracaine, but not by tetrodotoxin or the absence of sodium in the external medium (Takahashi *et al.*, 1982). More recently the same authors, using mitochondrial and liposomal preparations, found that unlike Ca^{2+} ionophore A 23187 or the Na^+ ionophore, monensin, MTX exerted neither swelling of the mitochondria measured by changes in turbidity nor modifications in mitochondrial ion content examined by atomic absorption (Takahashi *et al.*, 1983), suggesting that MTX may be a Ca^{2+} channel activator with a mechanism of action different from that of a carboxylic ionophore such as A 23187. Moreover, only the omission of Ca^{2+} from the medium completely abolished the MTX effect.

When tested on PRL secretion, MTX dose-dependently stimulated PRL release. The data in Fig. 14. 2A show that incubation for 15 min with as little as 5 ng ml^{-1} MTX markedly increased hormonal release. Exposure of anterior pituitary cells to 50 ng ml^{-1} MTX stimulated PRL release within 1·5 min (Fig. 14.2B), which continued to be enhanced throughout the next

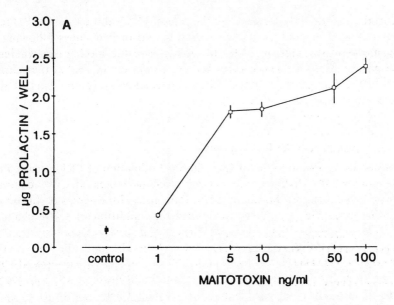

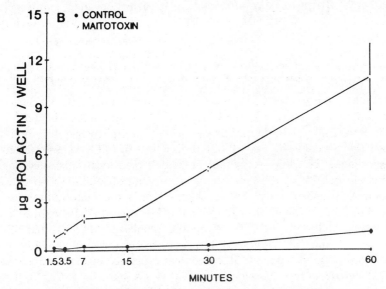

Fig. 14.2. (A) Effect of maitotoxin on prolactin release from primary cultures of anterior pituitary cells, prepared as described in Schettini *et al.*, 1983c. Cells were incubated for 30 min with increasing concentrations of maitotoxin (1 to 100 ng ml^{-1}). Maitotoxin, in a dose-related manner, significantly ($P<0.01$) stimulated the release of prolactin. (B) Time course of the maitotoxin stimulation of the release of prolactin from primary cultures of anterior pituitary cells. Maitotoxin (50 ng ml^{-1}) significantly ($P<0.05$) stimulated prolactin within 1·5 min of exposure. This enhanced hormonal release continued throughout the next 60 min of incubation ($P<0.01$).

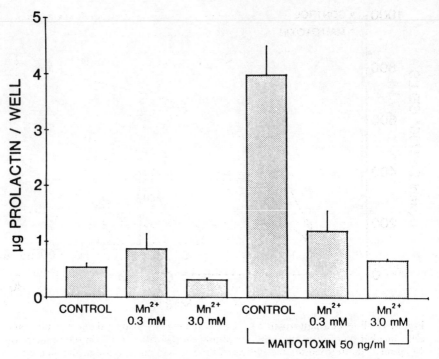

Fig. 14.3. Effect of manganese (Mn^{2+}) on maitotoxin stimulation of anterior pituitary prolactin release. The maitotoxin-induced release of prolactin was significantly reduced by 0·3 and 3·0 mM Mn^{2+} ($P<0.01$).

60 min of incubation. MTX, even at the higher concentration tested, had no effect on cell viability, as determined by the trypan blue exclusion test. Manganese (3 mM), a Ca^{2+} antagonist, significantly decreased basal PRL secretion. An MTX-induced increase in PRL release was reduced significantly by 0·3 and 3·0 mM Mn^{2+} (Fig. 14.3) (Schettini *et al.*, 1984).

Dispersed anterior pituitary cells in a column perfused with medium 199 containing 5 ng ml^{-1} MTX for 10 min acutely released PRL till the withdrawal of MTX and thereafter gradually returned to baseline.

Since in other studies the effect of MTX on $^{45}Ca^{2+}$ uptake was associated with neurotransmitter release (Takahashi *et al.*, 1982), we studied the effect of MTX on $^{45}Ca^{2+}$ flux in dispersed anterior pituitary cells. In the presence of 40 ng ml^{-1} of MTX, a greater level of $^{45}Ca^{2+}$ flux into the pituicytes was observed at every time studied (Fig. 14.4). The exchange of Ca^{2+} increased rapidly initially (30 s) and assumed an apparent equilibrium after 15 min. Over the time course studied, $^{45}Ca^{2+}$ exchange was distinctly different in the two groups. However, changes in Ca^{2+} uptake reflect a balance between influx, efflux or redistribution of intracellular Ca^{2+} pools. Therefore we measured net Ca^{2+} flux (which we assumed as uptake) at isotopic

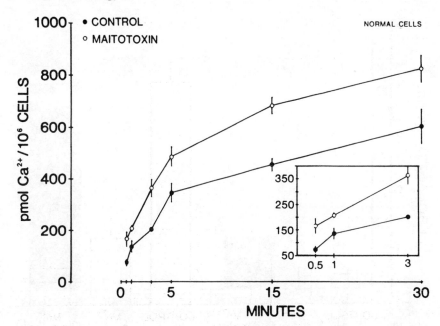

Fig. 14.4. Effect of maitotoxin on $^{45}Ca^{2+}$ exchange by dispersed anterior pituitary cells. Incubation of dispersed anterior pituitary cells with $^{45}Ca^{2+}$ in the presence of 40 ng ml^{-1} maitotoxin for the indicated time caused an increase in the amount of radioactive $^{45}Ca^{2+}$ associated with the cells in comparison with untreated cells. The effect of maitotoxin was evident at 30 s ($P < 0.05$) and continued over the time course studied ($P < 0.01$). The method for the $^{45}Ca^{2+}$ uptake studies is described by Schettini *et al.* (1985).

equilibrium, in GH$_3$ cells preincubated for 90 min with the isotope. Also in these conditions 50 ng ml^{-1} MTX stimulated net $^{45}Ca^{2+}$ flux, probably enhancing the isotope uptake. Thus it is possible to consider that MTX set in motion a series of cellular events starting from the stimulation of Ca^{2+} metabolism and ending in the secretory process (Fig. 14.5) (Schettini *et al.*, 1984, 1986).

Although the Ca^{2+} requirement for PRL secretion is clearly recognised, there is equivocal evidence that changes in pituitary Ca^{2+} flux represent a prime locus of action for dopamine. Electrophysiological studies suggest that fish pituitary lactotropes have spontaneous Ca^{2+}-dependent action potential that can be inhibited by dopamine (Taraskevich and Douglas, 1977, 1978). Surprisingly, Dufy *et al.* (1979) found that dopamine also inhibited action potential in GH$_3$/B$_6$ cells; however, these cells have no dopamine receptors and their PRL release is not affected by the catecholamine (Faure *et al.*, 1980). Nevertheless, alteration of Ca^{2+} flux through voltage-dependent channels could be an important mechanism whereby dopamine inhibits PRL release. In these elctrophysiological studies in the fish, a rapid

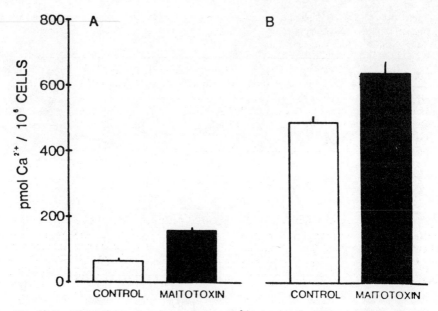

Fig. 14.5. The effect of maitotoxin on Ca^{2+} uptake by GH_3 tumour cells. (A) Maitotoxin significantly ($P<0.01$) increased the gross influx of Ca^{2+} into GH_3 cells. Cells were simultaneously exposed to $^{45}Ca^{2+}$ and maitotoxin or vehicle and the $^{45}Ca^{2+}$ content of the cells was determined at 20 s. (B) Maitotoxin significantly ($P<0.01$) increased the Ca^{2+} present in the GH_3 cells. Cells were equilibrated with $^{45}Ca^{2+}$ for 90 min and then exposed to maitotoxin or vehicle, and the $^{45}Ca^{2+}$ content of the cells was determined at 20 s.

suppression of pituitary electrical activity was produced by dopamine, and we found a significant reduction in PRL release within 3 min after perfusion of pituitary cells with dopamine (Thorner *et al.*, 1980). However, it is not established that these changes reflect a dopaminergic influence on pituitary Ca^{2+} flux.

Thorner *et al.* (1980) found that A 23187 partially reversed the dopaminergic inhibition of PRL release, suggesting that part of the action of dopamine was mediated by a Ca^{2+}-dependent mechanism. On the other hand, bromocriptine, a dopamine agonist, inhibited both basal and A 23187-induced PRL release without affecting $^{45}Ca^{2+}$ uptake in primary cultures of anterior pituitary cells (Tam and Dannies, 1980). Therefore it is likely that one of the events affected by dopamine occurs after Ca^{2+} mobilisation. To better characterise this aspect, since MTX-induced PRL release is Ca^{2+} mediated, we attempted to determine whether dopamine receptor activation influenced the enhanced PRL release produced by the toxin. In primary cultures of anterior pituitary cells, both dopamine and bromocriptine partially inhibited MTX-enhanced PRL release (Table 14.1).

Table 14.1 Effects of dopamine or bromocriptine on maitotoxin (MTX) stimulation of prolactin release by primary cultures of anterior pituitary cells. (Values are expressed as mean ± SEM of four samples per group.)

	Vehicle	Maitotoxin, 10 ng ml^{-1}
	Prolactin (ng well^{-1})	
Control	129 ± 30	529 ± 41*
Dopamine, 100 nM	53 ± 2*	290 ± 13†
Bromocriptine, 100 nM	61 ± 7*	204 ± 15†

* $P < 0.01$ vs respective control.
† $P < 0.01$ vs maitotoxin alone.

A 60% decrease in PRL release by cultured pituitary cells was produced by 100 nM dopamine; conversely a 400% increase was stimulated by 10 ng ml^{-1} MTX alone. A significant reduction in the MTX-induced release of PRL was achieved by incubating these cells with 100 nM dopamine. These findings are in agreement with our previously reported studies with A 23187 (Schettini *et al.*, 1983*b*) and suggest that at least part of the inhibitory action of dopamine on PRL release occurs after the influx of Ca^{2+}. Support for this assertion is also found in the observation that dopamine suppresses the TRH-mediated increase in PRL release (Matsushita *et al.*, 1983), a process thought to be dependently only upon intracellular Ca^{2+} and not enhanced Ca^{2+} influx (Gershengorn, 1982).

14.3. Dopamine and the cyclic AMP system

A large body of experimental evidence indicates that various physiological and pharmacological agents that enhance the accumulation of cyclic AMP in the anterior pituitary also stimulate the release of PRL. Phospho-diesterase inhibitors, e.g. theophylline and isobutylmethylxanthine (IBMX), which increase cyclic AMP accumulation in both normal and clonal pituitary cells, also stimulate PRL release (Dannies *et al.*, 1976; Hill *et al.*, 1976; Ray and Wallis, 1981, 1982; Swennen and Denef, 1982; Schettini *et al.*, 1983*c*). Cholera toxin (Rappaport and Grant, 1974; Tam and Dannies, 1981; Cronin and Thorner, 1982; Ray and Wallis, 1982), a potent stimulator of adenylate cyclase activity, VIP (Enjalbert *et al.*, 1980; Onali *et al.*, 1981) and forskolin, which stimulates pituitary (Seamon *et al.*, 1981; Schettini *et al.*, 1986) adenylate cyclase activity, each enhance PRL secretion. The role of dopamine in this system was suggested by several reports indicating that dopamine inhibits cyclic AMP accumulation in primary cultures of adenohypophyseal cells (Labrie *et al.*, 1980; Cronin and Thorner, 1982; Swennen and Denef, 1982; Schettini *et al.*, 1983*b*), and reduces adenylate cyclase activity in normal anterior pituitary cells (Giannattasio *et al.*, 1981).

Furthermore, adenylate cyclase activity is more sensitive to dopaminergic inhibition in female than in male rat pituitary glands (Giannattasio *et al.*, 1981; Enjalbert and Bokjaert, 1983). Dopamine was recently shown to reduce the activation of adenylate cyclase induced by VIP but not the activation induced by prostaglandin E_1 (Onali *et al.*, 1981). De Camilli *et al.* (1979) showed that dopamine reduces adenylate cyclase activity in human prolactinoma tissue, and Spada *et al.* (1983) extended this observation by showing that adenylate cyclase activity was more sensitive to dopamine in macro- than in microadenomatous tissue. Therefore it appears likely that pituitary dopamine receptors can inhibit the adenylate cyclase system and this inhibition may be an important post-receptor mechanism in the dopaminergic inhibition of PRL release.

To characterise further the mechanism whereby dopamine interacts with the various components of the pituitary adenylate cyclase system, we studied the effects of forskolin, a diterpene derivative, which directly activates the catalytic subunit of adenylate cyclase (Seamon and Daly, 1981; Ross, 1982). This agent stimulated in a dose-dependent manner cyclic AMP accumulation in anterior pituitary cells within 30 min. Co-incubation of the cells with 10 μM dopamine partially reduced the forskolin-enhanced activity of adenylate cyclase (Fig. 14.6). However, 100μM forskolin overcame the dopaminergic inhibition of cyclic AMP levels.

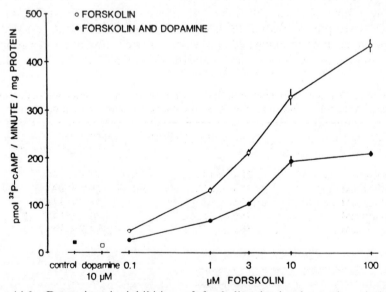

Fig. 14.6. Dopaminergic inhibition of forskolin-stimulated anterior pituitary andenylate cyclase activity. The enzyme assay was performed following the method described by Onali *et al.* (1981) and Schettini *et al.* (1985). Forskolin (1–100 μM) significantly ($P<0.01$) stimulated adenylate cyclase activity. Dopamine (10 μM) significantly ($P<0.01$) inhibited forskolin-stimulated enzyme activity.

In another study conducted over a 4-h period, forskolin enhanced PRL release from the cells, and simultaneous co-incubation with 10 μM dopamine produced a reduction of stimulated PRL secretion. We then studied the effect of forskolin, sodium fluoride (NaF) and Mg^{2+} on pituitary adenylase cyclase activity. NaF requires the GTP binding protein to enhance the enzyme activity (Ross and Gilman, 1980). Although Mg^{2+} is known to affect the adenylate cyclase system at several levels and to act as a cofactor rather than a direct stimulus, its effect was compared with the two other 'stimulators' of the enzyme, forskolin and NaF. The data presented in Fig. 14.6 and Table 14.2 show that all three agents stimulated adenylate cyclase activity progressively, forskolin being the most potent. Dopamine significantly attenuated the enzyme stimulation induced by the lower concentrations of NaF and Mg^{2+}; however, higher concentrations of both agents overcame the dopaminergic inhibition (Table 14.2), and enzyme activation by all concentrations of forskolin was always reduced by approximately 50 % by dopamine (Fig. 14.6).

Although it is now well established that dopamine receptor stimulation decreases adenylate cyclase activity, other components of this mechanism in the mammotroph plasma membrane have been obscure until recently. It is now suggested that this dopamine receptor–adenylate cyclase interaction is mediated by a guanine nucleotide dependent complex named G_i or N_i. This coupling protein is apparently comprised of a 41 000 and a 35 000 Da protein dimer (Bokoch, 1983) that is believed to dissociate upon inhibitory agonist receptor activation (Bokoch, 1983; Hildebrandt *et al.*, 1983). This concept has been promulgated on the basis of the action of a toxin secreted by *Bordetella pertussis*, namely pertussis toxin. This protein ADP-

Table 14.2. Dopaminergic inhibition of NaF or Mg^{++} activation of anterior pituitary adenylate cyclase. (Values are expressed as mean $\pm$ SEM of four determinations per group.)

| | *Cyclic AMP* (pmol mg protein^{-1} min^{-1}) | |
	Agent alone	*Agent plus* 10 μM *dopamine*
Control	21·4 $\pm$ 0·3	15·3 $\pm$ 0·1†
NaF, 1 mM	37·3 $\pm$ 1·1*	26·4 $\pm$ 1·4†
NaF, 3 mM	75·1 $\pm$ 4·7*	54·5 $\pm$ 0·3†
NaF, 10 mM	91·0 $\pm$ 2·5*	80·9 $\pm$ 1·5**
NaF, 30 mM	86·5 $\pm$ 2·3*	80·3 $\pm$ 3·8
Control	19·8 $\pm$ 1·9	12·4 $\pm$ 0·4†
Mg^{++} 3 mM	39·3 $\pm$ 1·6*	29·5 $\pm$ 0·6†
Mg^{++} 10 mM	56·4 $\pm$ 3·6*	48·9 $\pm$ 3·5
Mg^{++} 30 mM	78·6 $\pm$ 5·6*	80·4 $\pm$ 2·0

* $P < 0·01$ vs respective control.
† $P < 0·01$ vs respective agent alone.
** $P < 0·05$ vs respective agent alone.

ribosylates the 41 000 Da moiety (Bokoch, 1983) presumably resulting in an increased affinity for the 35 000 Da protein (Katada and Ui, 1982; Burns *et al.*, 1983). The ultimate effect of pertussis toxin fragment in the mamotroph is to uncouple functionally the dopamine receptor such that inhibition of cyclic AMP accumulation and PRL release by dopamine agonists are attenuated or abolished. Whether this inhibitory coupling protein is necessary to alter Ca^{2+} flux or phosphatidylinositol turnover after dopamine receptor activation is a fundamental question.

However, in our experimental conditions, cells preincubated with pertussis toxin for 24 h showed an adenylate cyclase activity refractory to the dopamine inhibition (Table 14.3). All together, these results, along with other literature reports, consistently show that anterior pituitary dopamine receptors are negatively coupled with pituitary adenylate cyclase activity, functionally regulating the inhibition of PRL secretion.

Table 14.3. Effect of pertussis toxin pretreatment on dopamine inhibition of anterior pituitary adenylate cyclase activity. (Each value represents the mean $\pm$ SEM of four determinations per group.)

| | Cyclic AMP (pmol mg^{-1} protein min^{-1}) | |
	No toxin pretreatment	Toxin pretreatment
Control	$7\cdot2 \pm 0\cdot2$	$4\cdot9 \pm 0\cdot5$
Dopamine, 10 μM	$5\cdot9 \pm 0\cdot1*$	$4\cdot9 \pm 0\cdot2$

* $P < 0\cdot05$ versus respective control.

14.4. Interaction between Ca^{2+}–calmodulin and cyclic AMP generating system: relation to dopaminergic inhibition

Ca^{2+} and cyclic AMP have integrated actions for controlling cellular function (Rasmussen, 1981) such as stimulus-secretion coupling. Prolactin release induced by cyclic AMP analogues or phosphodiesterase inhibitors requires the presence of Ca^{2+} (Lemay and Labrie, 1972; Tam and Dannies, 1981), and in the absence of Ca^{2+} these agents stimulate cyclic AMP accumulation but not hormonal release (Spence *et al.*, 1980). These studies, however, do not directly address the issue of Ca^{2+} and the formation of cyclic AMP as they relate to the secretory process.

The Ca^{2+} ionophore A 23187 was shown to enhance $^{45}Ca^{2+}$ uptake by primary cultures of anterior pituitary cells and to stimulate PRL release (Tam and Dannies, 1980). A 23187 is thought to promote Ca^{2+} uptake via a mechanism unrelated to normal Ca^{2+} channels (Ray and Wallis, 1982). We investigated the dopaminergic influence on the relationship between Ca^{2+} and cyclic AMP in the control of PRL release using A 23187. Those studies showed that the stimulation of PRL release by A 23187 was accompanied by increased cellular cyclic AMP accumulation (Schettini *et al.*, 1983*b*). A time-course study of this effect showed that 10 μM A 23187 maximally stimulated pituitary cyclic AMP accumulation ($+200\%$) within 1·5 min, whereas an enhanced PRL release reached its maximum at 15–30 min (Schettini *et al.*, 1983*c*). Therefore, the increased Ca^{2+} and the early activation of the pituitary cyclic AMP system may act synergistically to stimulate the release of PRL, although a definite association between these events cannot be made at this time.

These studies were extended by using the Ca^{2+} channel activator MTX in order to better characterise the interaction between Ca^{2+} and cyclic AMP system with regard to the PRL secretory process and their relation to dopaminergic inhibition.

The effects of MTX were measured in both untreated pituitary cells and in cells pretreated for 2 h with 0·2 mM IBMX. In the absence of IBMX, MTX not only stimulated the release of PRL at 3 min, a statistically significant effect that continued for at least 30 min, but also significantly prevented the

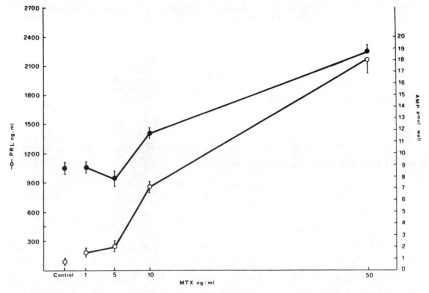

Fig. 14.7. Effect of a 15-min maitotoxin treatment on cyclic AMP accumulation and prolactin release by primary cultures of anterior pituitary cells pretreated with isobutylmethylxanthine (IBMX). Cells were washed twice with warmed (37°C) serum-free RPMI-1640 with antibiotic and preincubated for 2 h with 0·2 mM IBMX. The medium was then removed and replaced with fresh medium containing IBMX plus maitotoxin at the respective concentrations for 15 min. At the termination of the experiment the medium was quickly removed for prolactin assay and cyclic AMP was extracted by rapidly adding 0·5 ml of 0·1 N HCl for 10 min at 37°C. Samples were stored at -20°C until assayed.

progressive time-related decrease in pituitary cyclic AMP content. In IBMX-treated cells, MTX produced a stimulation of both cyclic AMP accumulation and PRL release (Fig. 14.7). Maitotoxin significantly stimulated PRL release at 30 s and cyclic AMP accumulation was increased after 1 min of exposure to the agent. The effect of graded doses of MTX on cyclic AMP accumulation and PRL release in IBMX-pretreated cells was studied. Maitotoxin progressively stimulated the accumulation of cyclic AMP and the release of PRL during a 15 min exposure (Fig. 14.7). These data show that MTX concentrations of 1 and 5 ng ml^{-1} were ineffective to enhance either of the parameters; however, 10 and 50 ng ml^{-1} increase both pituitary cyclic AMP accumulation and PRL release.

We subsequently studied whether MTX modified anterior pituitary adenylate cyclase activity. Hemipituitary glands were exposed for 10 min to an MTX concentration of 100 ng ml^{-1} and then homogenised, and their membranes were assayed for adenylate cyclase activity. The data in Fig. 14.8 show that adenylate cyclase activity from MTX-pretreated glands is increased compared with non-treated membranes. The addition of MTX directly to the membrane preparation, however, did not alter the enzyme activity, thus excluding a direct effect of the agent on the enzyme. These data

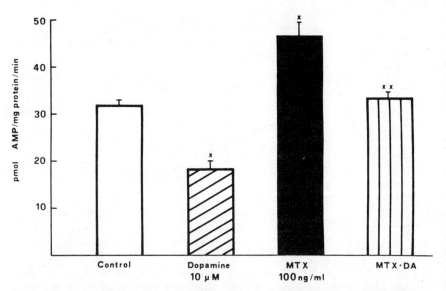

Fig. 14.8. Dopaminergic inhibition of basal and maitotoxin-stimulated anterior pituitary adenylate cyclase activity. Hemipituitary glands were incubated for 10 min with MTX, 100 ng ml^{-1}, then homogenised and their membranes prepared for adenylate cyclase activity evaluation. MTX significantly stimulated ($P < 0.01$) adenylate cyclase activity in comparison with control. Dopamine directly added to membrane preparations inhibited both basal ($P < 0.01$) and MTX-stimulated adenylate cyclase ($P < 0.01$).

show that the stimulatory effect of MTX on adenylate cyclase activity appears to be dependent on its Ca^{2+} channel activating properties, requiring intact, viable cells.

The hypothesis that enhanced pituitary cyclase activity is produced by MTX through its Ca^{2+}-activating property prompted us to investigate whether a Ca^{2+} signal is sufficient to produce this effect. The data in Fig. 14.9 show that increasing concentrations of $CaCl_2$, up to a 100 μM (Ca^{2+}–EGTA), significantly stimulated anterior pituitary adenylate cyclase activity, whereas higher concentrations were either ineffective or inhibitory to the enzyme activity. These results suggest that Ca^{2+} regulates the anterior pituitary cyclic AMP generating system metabolism, thereby suggesting a cooperative and integrated role between the two messenger systems in the control of PRL secretion.

Ca^{2+} regulation of adenylate cyclase activity in many cellular systems is modulated by calmodulin (Cheung, 1980; Means and Dedman, 1980; Rasmussen, 1981). To study whether calmodulin affects anterior pituitary adenylate cyclase activity, we added calmodulin to anterior pituitary membrane preparations, both in the presence and absence of added Ca^{2+}. Membranes were prepared as described (see legend to Fig. 14.6) or washed thoroughly

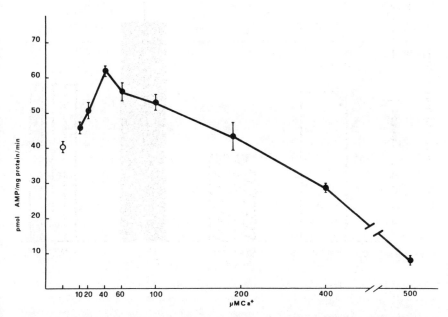

Fig. 14.9.   Effect of Ca^{2+} on anterior pituitary adenylate cyclase activity. Adenylate cyclase activity was significantly stimulated by 10 μM $CaCl_2$ ($P<0.05$) to 100 μM $CaCl_2$ ($P<0.01$); 200 μM $CaCl_2$ was not significantly different from control; and 300 and 500 μM $CaCl_2$ were inhibitory ($P<0.01$) to the enzyme's activity.

(three times) with 2 mM EGTA to reduce endogenous Ca^{2+} and calmodulin content (Gnegy and Treisman, 1981). The addition of 10 μg to normal membrane preparations, both with and without exogenously added Ca^{2+}, modestly enhanced anterior pituitary adenylate cyclase activity, whereas the membranes carefully washed with EGTA showed a significantly lower basal enzymatic activity (Table 14.4) and were less responsive to exogenous Ca^{2+} and calmodulin added alone. The addition of calmodulin was more effective in enhancing pituitary adenylate cyclase activity in the presence Ca^{2+} than without the cation. In contrast the Ca^{2+} binding protein parvalbumin (Schettini *et al.*, 1986) did not modify adenylate cyclase activity under the conditions studied (Table 14.4).

Table 14.4. Effect of calmodulin (CaM) and parvalbumin (PVA) on anterior pituitary adenylate cyclase activity. The enzyme activity was measured by the standard assay procedure in membranes thoroughly washed with 2 mM EGTA, both in the absence and in the presence of 300 μM EGTA. Each value represents the mean $\pm$ SEM of four determinations per group.

| | *Cyclic AMP* (pmol mg protein^{-1} min^{-1}) | |
	Agent alone	*Agent* + 40 μM CaCl$_2$
Control	42·3 $\pm$ 0·5	51·5 $\pm$ 3·2
CaM, 10 μg	49·2 $\pm$ 2·1	66·6 $\pm$ 1·8†
CaM, 20 μg	51·1 $\pm$ 0·6*	81·8 $\pm$ 1·4†
PVA, 20 μg	41·7 $\pm$ 1·7	43·5 $\pm$ 3·9

* $P < 0.01$ vs control.
† $P < 0.01$ vs agent alone.

The effects of dopamine on the MTX-induced increase in cyclic AMP formation and PRL release was studied by incubating IBMX-treated anterior pituitary cells for 15 min with dopamine or bromocriptine. The cells were then exposed to the same concentrations of dopamine or bromocriptine for 15 min with or without MTX, 5 or 10 ng ml^{-1}. Dopamine (1 and 10 μM) attenuated both cyclic AMP accumulation and PRL release promoted by an MTX concentration of 5 ng ml^{-1} whereas only 10 μM dopamine reduced PRL release brought about by 10 ng ml^{-1} and neither concentration of dopamine had an effect on the enhanced cyclic AMP accumulation. In the absence of IBMX, dopamine significantly reduced both cyclic AMP accumulation and PRL release induced by MTX at 10 ng ml^{-1} (Table 14.5).

It is important to note that MTX is a very effective secretagogue and is presumed to stimulate cyclic AMP accumulation in all types of anterior pituitary cells, an effect amplified by IBMX pretreatment. In contrast dopamine is thought to affect predominantly pituitary cyclic AMP

Table 14.5. Dopamine inhibition of maitotoxin-stimulated cyclic AMP accumulation and prolactin release by anterior pituitary cells in the absence of isobutylmethylxanthine. Each value represents the mean ± SEM of four determinations per group.

	Cyclic AMP (pmol well^{-1})	PRL(ng well^{-1})
Control	0·73 ± 0·16	132 ± 23
MTX, 10 ng ml^{-1}	1·47 ± 0·04*	529 ± 41*
Dopamine 0·1 μM + MTX	1·04 ± 0·03†	290 ± 13†
Dopamine 1·0 μM + MTX	1·09 ± 0·08†	275 ± 34†

* $P<0·01$ vs respective control.
† $P<0·01$ vs respective MTX-treated group.

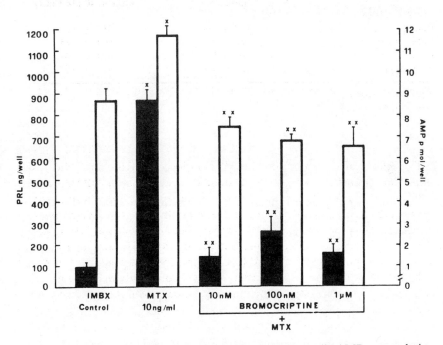

Fig. 14.10. Effect of bromocriptine on MTX-stimulated cyclic AMP accumulation and prolactin release in IBMX-pretreated cells. MTX stimulated ($P<0·01$) both cyclic accumulation and prolactin secretion as compared with control. Bromocriptine (10, 100 and 1000 nM) significantly ($P<0·01$) inhibited MTX stimulation of both cyclic AMP and prolactin secretion.

production in PRL-secreting cells. Since the concentration of dopamine necessary to inhibit pituitary adenylate cyclase activity is more than one order of magnitude greater than that required to inhibit PRL release (Enjalbert and Bokjaert, 1983), it is likely that its inhibitory action on lactotroph cyclic AMP accumulation is masked by the MTX-enhanced production of the nucleotide in other pituitary cell types. This hypothesis is supported by the observation that bromocriptine, whose IC_{50} to inhibit adenylate cyclase activity and PRL secretion is more than one order of magnitude lower than that of dopamine (Enjalbert and Bokjaert, 1983), greatly inhibited these parameters. The findings in Fig. 14.10 show that bromocriptine, in the presence of IBMX, significantly reduced the accumulation of pituitary cyclic AMP and the release of PRL induced by MTX at 10 ng ml^{-1}. It should be noted that 10 nM bromocriptine was nearly maximally effective to reduce both cyclic AMP accumulation and PRL release under these conditions. Similar results were obtained with bromocriptine in the absence of IBMX (data not presented).

The effect of dopamine on basal pituitary adenylate cyclase activity and that previously activated by MTX was also studied. The data in Fig. 14.9 demonstrate that 10 μM dopamine reduced basal pituitary adenylate cyclase activity by 43 % and that MTX-activated adenylate cyclase activity was diminished by 28 %.

Since no clear and definitive evidence emerges as regards the effect of dopamine on Ca^{2+} mobilisation, these data, together with our work and other previous reports in the literature, consistently support the concept of a site of action for the dopaminergic inhibition of PRL release after Ca^{2+} mobilisation, with adenylate cyclase a likely target for dopaminergic inhibition of PRL secretion.

14.5. Conclusions

Basal and secretagogue-stimulated PRL secretion is certainly affected by Ca^{2+} and cyclic AMP system, which possibly interact at various levels in regulating PRL release. In fact anterior pituitary adenylate cyclase appears to be regulated by Ca^{2+}–calmodulin, as well as by direct receptor coupling. However, the fact that Ca^{2+}–calmodulin enhances anterior pituitary adenylate cyclase activity does not mean that its biological role in the control of the PRL secretory process occurs solely through the cyclic AMP system. Indeed, cyclic AMP, Ca^{2+}–calmodulin dependent kinases and C-kinase exist in the pituitary and may activate distinctive pathways of phosphorilation, as reported in GH_3 cells, but also may respond in a co-ordinated integrated manner to a receptor stimulus to control the secretory process (Fig. 14.11).

Dopamine, regarded as a physiological inhibitor of PRL secretion, is certain to affect the pituitary D-2 receptor coupled with the cyclic AMP

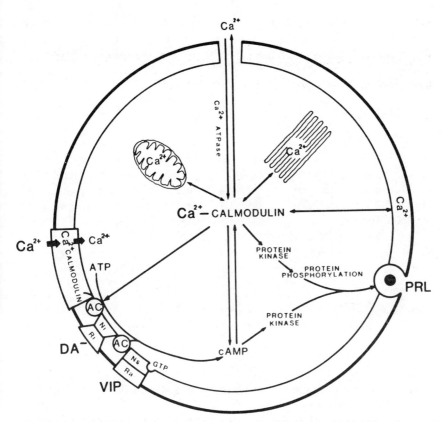

Fig. 14.11. Diagramatic representation of the interrelationship between Ca^{2+}–calmodulin, cyclic AMP and dopamine in the control of prolactin secretion. The binding of Ca^{2+} to calmodulin, at the level of the membrane or cytosol, may activate adenylate cyclase (AC) activity and, therefore, increase cyclic AMP formation. By stimulating cyclic AMP-dependent protein kinase(s), cyclic AMP may induce the phosphorylation of specific proteins, allowing prolactin release. It is also possible that a Ca^{2+}–calmodulin-dependent protein kinase (independent of cyclic AMP) stimulates the release of prolactin. The action of dopamine occurs at the level of a dopamine receptor D-2, which inhibits adenylate cyclase activity. The stimulation of these receptors prevents the formation of cyclic AMP and prolactin release, both basally and after Ca^{2+}–calmodulin stimulation. VIP, vasoactive intestinal polipeptide; R_i, inhibitory receptor; R_a, stimulating receptor; N_i, inhibitory nucleotide binding protein; N_s, stimulating nucleotide binding protein; GTP, guanosine triphosphate.

generating system. As has been summarised in this chapter, we and others have demonstrated that dopamine decreases adenylate cyclase activity and cyclic AMP accumulation: these effects are consistently linked with reduction in PRL secretion. On the other hand, until now, there is little or no evidence to support the idea of any effect of dopamine on cellular Ca^{2+}

movements. Moreover, dopamine, although not affecting Ca^{2+} fluxes, decreases Ca^{2+} ionophore or MTX-stimulated adenylate cyclase activity, cyclic AMP accumulation and PRL secretion. Therefore a mechanism whereby dopamine inhibits PRL secretion appears to be distal or lateral to Ca^{2+} mobilisation, relating to a Ca^{2+}-dependent or independent cyclic AMP production.

Acknowledgements

We gratefully acknowledge the technical assistance of S. B. O'Dell, G. Baber, C. Valdenegro, M. MacQueen and G. Meyers. We also wish to thank Dr R. H. Kretsinger for the generous gift of calmodulin and parvalbumin, and the Diabetes Core Laboratory of the University of Virginia for providing ^{32}P-ATP and for assaying cyclic AMP. This research was supported by USPHS Grant CA-07535 (R.M.M.); RCDA 1KO4NS00601, NS 18409, AM 22125 (M.J.C.); AI 18000, AM 22125 (E.L.H.); and 1 F05 TWO 3267 (G.S.).

References

Ben-Jonathan, N., Oliver, C., Weiner, H. J., Mical, R. S. and Porter, J. C. (1977). Dopamine in hypophyseal portal plasma of the rat during the estrous cycle and throughout pregnancy. *Endocrinology* **100**, 452.

Berridge, M. J. (1981). Phosphatidylinositol hydrolysis: a multifunctional transducing mechanism. *Mol. Cell. Endocrinol.* **24**, 115.

Bokoch, G. M., Katada, T., Northrup, J. K., Hewlett, E. L. and Gilman, A. G. (1983). Identification of a predominant substrate for ADP-ribosylation by islet-activating protein. *J. Biol. Chem.* **258**, 2072.

Brostrom, C. O., Brostrom, M. A. and Wolff, D. J. (1977). Calcium-dependent adenylate cyclase from rat cerebral cortex. *J. Biol. Chem.* **252**, 5677.

Burns, D. L., Hewlett, E. L., Moss, J. and Voughan, M. (1983). Pertussis toxin inhibits enkephalin stimulation of GTPase of NG 108–15 cells. *J. Biol. Chem.* **258**, 1435.

Canonico, P. L., Valdenegro, C. A. and MacLeod, R. M. (1983). The inhibition of phosphatidylinositol turnover: a possible post-receptor mechanism for the prolactin secretion inhibiting effect of dopamine. *Endocrinology* **113**, 7.

Cheung, W. Y. (1980). Calmodulin plays a pivotal role in cellular regulation. *Science* **207**, 19.

Cronin, M. J. (1982). The role and direct measurement of dopamine receptor(s) of the anterior pituitary. In *Neuroendocrine Perspectives*, vol. 1. E. E. Müller and R. M. Macleod (eds), pp. 167–210, Elsevier, Amsterdam.

Cronin, M. J. and Thorner, M. O. (1982). Dopamine and bromocriptine inhibit cyclic AMP accumulation in the anterior pituitary: the effect of cholera toxin. *J. Cyclic Nucleotide Res.* **8**, 267.

Cronin, M. J., Mayers, G. A., MacLeod, R. M. and Hewlett, E. L. (1983). Pertussis toxin uncouples dopamine agonist inhibition of prolactin release. *Am. J. Physiol.* **244**, E499.

Dannies, P. S., Gautvik, K. M. and Tashjian, A. H. Jr. (1976). A possible role of cyclic AMP in mediating the effects of thyrotropin-releasing hormone on prolactin release and on prolactin and growth hormone synthesis in pituitary cells in culture. *Endocrinology* **98**, 1147.

De Camilli, P., Macconi, D. and Spada, A. (1979). Dopamine inhibits adenylate cyclase in human prolactin secreting adenomas. *Nature (Lond.)* **278**, 252.

Dratt, D. A., Torp-Pedersen, C. and Thorn, N. A. (1981). Effects of Ca^{2+} and calmodulin on cyclic nucleotide metabolism in neurosecretosomes isolated from ox neurohypophyses. *Brain Res.* **204**, 121.

Dufy, B., Vincent, J. D., Fleury, H., DuPasquier, P. D., Gourdji, D. and Tixier-Vidal, A. (1979). Dopamine inhibition of action potentials in prolactin secreting cell line is modulated by estrogen. *Nature (Lond.)* **282**, 855.

Enjalbert, A. and Bokjaert, J. (1983). Pharmacological characterisation of the D_2 dopamine receptor negatively coupled with adenylate cyclase in rat anterior pituitary. *Mol. Pharmacol.* **23**, 576.

Enjalbert, A., Arancibia, S., Ruberg, M., Priam, M., Bluet-Pajot, M. T., Rotsztejn, W. H. and Kordon, C. (1980). Stimulation of *in vitro* prolactin release by vasoactive intestinal peptide. *Neuroendocrinology* **31**, 200.

Faure, N., Cronin, M. J., Martial, J. A. and Weiner, R. I. (1980). Decreased responsiveness of GH_3 cells to the dopaminergic inhibition of prolactin. *Endocrinology* **107**, 1022.

Gautvik, K. M., Gozdeladre, J. O., Jahnsen, T., Haug, E., Hausson, V. and Lystad, E. (1983). Thyrotropin receptor binding and adenylate cyclase activation in cultured prolactin producing rat pituitary tumour cells (GH cells). *J. Biol. Chem.* **258**, 10304.

Gershengorn, M. C. (1980). Thyrotropin releasing hormone stimulation of prolactin release: evidence for a membrane potential independent Ca^{2+}-dependent mechanism of action. *J. Biol. Chem.* **255**, 1801.

Gershengorn, M. C. (1982). Thyrotropin releasing hormone. Review of the mechanisms of acute stimulation of pituitary hormone release. *Mol. Cell Biochem.* **45**, 163.

Giannattasio, G., DeFerrari, M. E. and Spada, A. (1981). Dopamine inhibited adenylate cyclase in female rat adenohypophysis. *Life Sci.* **28**, 1605.

Gibbs, D. M. and Neill, J. D. (1978). Dopamine levels in hypophyseal stalk blood in the rat are sufficient to inhibit prolactin secretion *in vivo*. *Endocrinology* **102**, 1895.

Gnegy, M. and Treisman, G. (1981). Effect of calmodulin on dopamine-sensitive adenylate cyclase activity in rat striatal membranes. *Mol. Pharmacol.* **19**, 256.

Hildebrand, J. D., Sekura, R. D., Codina, J., Iyengar, R., Manclark, C. R. and Birnmaumer, L. (1983). Stimulation and inhibition of adenyl cyclases mediated by distinct regulatory proteins. *Nature (Lond.)* **302**, 706.

Hill, M. K., MacLeod, R. M. and Orcutt, P. (1976). Dibutyl cyclic AMP, adenosine and guanosine blockade of the dopamine, ergocryptine and apomorphine inhibition of prolactin release *in vitro*. *Endocrinology* **99**, 1612.

Katada, T. and Ui, M. (1982). Direct modification of the membrane adenylate cyclase system by islet-activating protein due to ADP-ribosylation of a membrane protein. *Proc. nat. Acad. Sci. USA* **79**, 3129.

Labrie, F., Borgeat, P., Barden, N., Godbout, M., Beaulieu, M., Ferland, L. and Lavoie, M. (1980). Mechanisms of action of hypothalamic hormones in the anterior pituitary. In *Polypeptide Hormones*. R. F. Beers Jr. and E. G. Basset (eds), pp. 235–51. Raven Press, New York.

Lemay, A. and Labrie, F. (1972). Calcium-dependent stimulation of prolactin release in rat anterior pituitary *in vitro* by N^6-monobutyryl adenosine 3′, 5′-monophosphate. *FEBS Lett.* **20**, 7.

MacLeod, R. M. (1976). Regulation of prolactin secretion. In *Frontiers in Neuroendocrinology*, vol. 4. L. Martini and W. F. Ganong (eds), p. 169.

MacLeod, R. M. and Fontham, E. H. (1970). Influence of ionic environment on the *in vitro* synthesis and release of pituitary hormones. *Endocrinology* **86**, 863.

Matsushita, N., Yazura, K., Shimatsu, A., Katakomi, H., Yahaihara, N. and Imura, H. (1983). Effects of VIP, TRH, GABA and dopamine on prolactin release from superfused rat anterior pituitary cells. *Life Sci.* **32**, 1263.

Means, A. R. and Dedman, J. R. (1980). Calmodulin: an intracellular calcium receptor. *Nature (Lond.)* **285**, 73.

Merritt, J. E., Tomlinson, S. and Brown, B. L. (1981). Phenothiazines inhibit prolactin secretion *in vitro*. A possible role for calmodulin in stimulus-secretion coupling in the pituitary. *FEBS Lett.* **35**, 107.

Milligan, J. V. and Kraicer, J. (1971). ^{45}Ca uptake during the *in vitro* release of hormone from the rat adenohypophysis. *Endocrinology* **89**, 766.

Mitchell, R. H. (1975). Inositol phospholipids and cell surface receptor function. *Biochem. Biophys. Acta* **415**, 81.

Ofulue, A. F. and Nijjar, M. S. (1981). Calmodulin activation of rat lung adenylate cyclase is independent of the cytoplasmic factors modulating the enzyme. *Biochem. J.* **200**, 475.

Onali, P. L., Schwartz, J. P. and Costa, E. (1981). Dopaminergic modulation of adenylate cyclase stimulation by vasoactive intestinal peptide in anterior pituitary. *Proc. nat. Acad. Sci. USA* **78**, 6531.

Parsons, J. A. (1970). Effect of cations on prolactin and growth hormone secretion by rat adenohypophysis *in vitro*. *J. Physiol. (Lond.)* **210**, 973.

Rappaport, R. S. and Grant, N. H. (1974). Growth hormone releasing factor of microbial origin. *Nature (Lond.)* **248**, 73.

Rasmussen, H. (1981). Calcium and cyclic AMP as synarchic messengers. Wiley, New York.

Ray, K. P. and Wallis, M. (1981). Effects of dopamine on prolactin secretion and cyclic AMP accumulation in the rat anterior pituitary gland. *Biochem. J.* **194**, 119.

Ray, K. P. and Wallis, M. (1982). Actions of dopamine on prolactin secretion and cyclic AMP metabolism in ovine pituitary cells. *Mol. Cell. Endocrinol.* **27**, 139.

Ross, E. M. (1982). Phosphatidylcholine-promoted interaction of the catalytic and regulatory protein adenylate cyclase. *J. Biol. Chem.* **257**, 10751.

Ross, E. M. and Gilman, A. G. (1980). Biochemical properties of hormone sensitive adenylate cyclase. *Ann. Rev. Biochem.* **49**, 533.

Schettini, G., Judd, A. M. and MacLeod, R. M. (1983*a*). *In vitro* studies on basal and stimulated prolactin release by rat anterior pituitary: a possible role for calmodulin. *Endocrinology* **112**, 64.

Schettini, G., Cronin, M. J. and MacLeod, R. M. (1983*b*). Adenosine 3', 5'-monophosphate (cAMP) and calcium–calmodulin interrelation in the control of prolactin secretion: evidence for dopamine inhibition of cAMP accumulation and prolactin release after calcium mobilization. *Endocrinology* **112**, 1801.

Schettini, G., Cronin, M. J. and MacLeod, R. M. (1983*c*). Calmodulin regulation of prolactin secretion and cyclic AMP metabolism in the anterior pituitary. In *Neuromodulation and Brain Function. Advances in the Bioscience*. G. Biggio, P. F. Spano, G. Toffano and G. L. Gessa (eds). pp. 281–92. Pergamon Press, Oxford.

Schettini, G., Koike, K., Login, I. S., Judd, A. M. Cronin, M. J. Yasumoto, T. and MacLeod, R. M. (1984). Maitotoxin, a Ca^{2+} channel activator, stimulates hormonal release and calcium flux in rat anterior pituitary cells '*in vitro*'. *Am. J. Physiol.* **247**, E520.

Schettini, G., Hewlett, E. L., Cronin, M. J., Koike, K., Yasumoto, T. and MacLeod, R. M. (1986) Dopaminergic inhibition of anterior pituitary adenylate cyclase activity and prolactin release: The effects of perturbing calcium or catalytic adenylate cyclase activity. *Neuroendocrinology* (in press).

Seamon, K. and Daly, J. W. (1981). Activation of adenylate cyclase by the diterpene forskolin does not require the guanine nucleotide regulatory protein. *J. Biol.*

Chem. **256**, 9799.

Seamon, K., Padgett, W. and Daly, J. W. (1981). Forskolin: unique diterpene activator of adenylate cyclase in membranes and intact cells. *Biochemistry* **78**, 3363.

Sharp, G. W. G., Wiedenkeller, D. E., Kaelin, D., Siegel, E. G. and Wollheim, C. B. (1980). Stimulation of adenylate cyclase by Ca^{2+} and calmodulin in rat islets of Langerhans. *Diabetes* **29**, 74.

Spada, A., Nicosia, S., Cortellazzi, L., Pezzo, G., Bassetti, M., Sartorio, A. and Giannattasio, G. (1983). *In vitro* studies on prolactin release and adenylate cyclase activity in human prolactin secreting pituitary adenomas. Different sensitivity of macro- and microadenomas to dopamine and vasoactive intestinal polipeptide. *J. Clin. Endocrinol. Metab.* **56**, 1.

Spence, J. W., Sheppard, M. S. and Kraicer, J. (1980). Release of growth hormone from purified somatotrophs: interrelation between Ca^{2+} and adenosine 3′, 5′-monophosphate. *Endocrinology* **106**, 764.

Swennen, L. and Denef, C. (1982). Physiological concentrations of dopamine decrease adenosine 3′, 5′-monophosphate levels in cultured rat anterior pituitary cells and enriched populations of lactotroph: evidence for causal relationship to inhibition of prolactin release. *Endocrinology* **111**, 398.

Takahashi, M., Ohizumi, Y. and Yasumoto, T. (1982). Maitotoxin, a Ca^{2+} channel activator candidate. *J. Biol. Chem.* **257**, 7287.

Takahashi, M., Tatsumi, M., Ohizumi, Y. and Yasumoto, T. (1983). Ca^{2+} channel activating function of maitotoxin, the most potent marine toxin known, in clonal rat pheochromocytoma cells. *J. Biol. Chem.* **258**, 10944.

Tam, S. W. and Dannies, P. S. (1980). Dopaminergic inhibition of ionophore A23187-stimulated release of prolactin from rat anterior pituitary cells. *J. Biol. Chem.* **255**, 6595.

Tam, S. W. and Dannies, P. S. (1981). The role of adenosine 3′, 5′-monophosphate in dopaminergic inhibition of prolactin release in anterior pituitary cells. *Endocrinology* **109**, 403.

Tan, K. T. and Tashjian, A. H. Jr. (1981). Receptor-mediated release of plasma membrane-associated calcium and stimulation of calcium uptake by thyrotropin releasing hormone in pituitary cells in culture. *J. Biol. Chem.* **256**, 8994.

Taraskevich, P. S. and Douglas, W. W. (1977). Action potentials occur in cells of the normal anterior pituitary gland and are stimulated by the hypophysiotropic peptide thyrotropin-releasing hormone. *Proc. nat. Acad. Sci. USA* **74**, 4064.

Taraskevich, P. S. and Douglas, W. W. (1978). Catecholamines of supposed inhibitory hypophysiotrophic function suppress action potential in prolactin cells. *Nature (Lond.)* **276**, 837.

Tashjian, A. H. Jr., Lomedico, M. E. and Maina, D. (1978). Role of calcium in the thyrotropin-releasing hormone-stimulated release of prolactin from pituitary cells in culture. *Biochem. Biophys. Res. Commun.* **81**, 798.

Thorner, M. O., Hakett, J. T., Murad, F. and MacLeod, R. M. (1980). Calcium rather than cyclic AMP as the physiological intracellular regulator of prolactin release. *Neuroendocrinology* **31**, 390.

Valverde, J., Vanderpieers, A., Anjaneyulu, R. and Malaisse, W. J. (1979). Calmodulin activation of adenylate cyclase in pancreatic islets. *Science* **206**, 225.

Weiner, R. I. and Ganong, W. F. (1978). Role of brain monoamines and histamine in regulation of anterior pituitary secretion. *Physiol. Rev.* **58**, 905.

Yasumoto, T., Nakajima, I., Oshima, Y. and Bagnis, R. (1979). A new toxic dinoflagellate found in association with ciguatera. In *Toxic Dinoflagellate* Blooms. D. L. Taylor and H. Seliger (eds), pp. 65–70. Elsevier/North Holland, New York.

15

Post-receptor mechanisms mediating dopaminergic inhibition of prolactin secretion

B. L. Brown, J. G. Baird, J. E. Merritt and P. R. M. Dobson

Department of Human Metabolism & Clinical Biochemistry, University of Sheffield Medical School, Beech Hill Road, Sheffield S10 2RX, UK.

15.1. Introduction

Dopamine, released from terminals of tuberoinfundibular neurones into the hypophyseal portal system, is recognised to be the primary inhibitory regulator of prolactin secretion (MacLeod, 1976; Weiner and Ganong, 1978; Cronin, 1982). Dopamine is a potent inhibitor of prolactin secretion both *in vivo* and *in vitro* (Cronin, 1982), and the amounts present in portal blood are sufficient to fulfil this role (Neill *et al.*, 1981; Arita and Porter, 1984). Moreover, high-affinity stereoselective dopamine receptors have been characterised on anterior pituitary cell membranes using dopamine agonists (Calabro and MacLeod, 1978; Caron *et al.*, 1978) and antagonists (Creese *et al.*, 1977; Cronin and Weiner, 1979). The binding characteristics of different dopamine agonists are closely correlated with their ability to inhibit prolactin release. Immunostaining has also revealed large numbers of dopamine receptors on the prolactin-secreting cells (Goldsmith *et al.*, 1978). These receptors have been proposed as being of the D-2 type of receptor (Kebabian and Calne, 1979).

Despite the considerable knowledge about dopamine receptors in the pituitary, the postreceptor mechanisms mediating the effects of dopamine on secretion remain unresolved. Indeed, dopamine has been reported to induce changes in adenylate cyclase activity, cyclic AMP accumulation, phosphoinositide metabolism and cellular Ca^{2+} levels. Generally, the two major intracellular signalling systems which are thought to be involved in mediating extracellular signals are cyclic AMP (acting via cyclic AMP dependent protein kinases) and Ca^{2+} (acting via either Ca^{2+}/calmodulin protein kinases or phospholipid and Ca^{2+} kinases). The Ca^{2+}-signalling system is intimately linked to the hydrolysis of the membrane-associated inositol phospholipids. Indeed, current evidence suggests that those agonists whose action involves Ca^{2+} as the intracellular mediator cause rapid hydrolysis of phosphatidylinositol 4,5-bisphosphate (PIP$_2$) (and possibly also phosphatidylinositol 4-phosphate (PIP)) leading to the formation of inositol 1,4,5-trisphosphate (IP_3) and 1,2-diacylglycerol (DAG). IP_3 has been shown to cause the mobilisation of intracellular Ca^{2+} (Streb et al., 1983; Berridge, 1984) and diacylglycerol activates the phospholipid-dependent protein kinase C (Castagna et al., 1982).

The aim of this short review is to analyse the respective roles, of these two major second-messenger systems in mediating the effect of dopamine on prolactin secretion.

15.2. Adenylate cyclase/cyclic AMP system

There is considerable evidence to suggest that changes in the intracellular concentration of cyclic AMP are associated with alteration in prolactin secretion. For example, agents that raise the intracellular concentration of cyclic AMP, such as VIP, which stimulates adenylate cyclase, and IBMX, which inhibits phosphodiesterase, cause increases in prolactin secretion (Guild and Drummond, 1983). Moreover, the non-specific adenylate cyclase stimulators, cholera toxin and forskolin, also stimulate secretion (Thorner et al., 1980; Guild and Drummond, 1983; Delbecke et al., 1984). In addition, muscarinic cholinergic agonists inhibit both adenylate cyclase activity and prolactin secretion (Onali et al., 1983; Wojcikiewicz et al., 1984), an effect which is blocked by pre-incubation with pertussis toxin (Brown et al., 1984; Wojcikiewicz et al., 1984).

In 1978, we reported that dopamine inhibited cyclic AMP accumulation in a partially purified preparation of rat lactotrophs (Barnes et al., 1978). This effect was not observed using a thyrotroph-enriched fraction. There have since been a number of reports showing an inhibitory effect of dopamine on cyclic AMP accumulation (Ray and Wallis, 1981, 1982; Swennen and Denef, 1982; Cronin et al., 1983; Schettini et al., 1983b, 1985). Furthermore, inhibitory effects of dopamine on adenylate cyclase activity have been reported (De Camilli et al., 1979; Giannattasio et al., 1981; Onali et al., 1981; Cronin et al., 1983; Spada et al., 1983; Schettini et al., 1985).

Three related points need to be considered. First, the concentrations of dopamine required for this inhibitory effect on adenylate cyclase are often high. Secondly, it is not always clear whether the effects on cyclic AMP metabolism is direct or indirect. Thirdly, changes in cyclic AMP concentrations and the secretory rate may not be related. In support of a direct effect, Cronin et al. (1983) observed that preincubation with pertussis toxin (which causes ADP-ribosylation of the inhibitory (N_i) unit of adenylate cyclase) blocked the effect of dopamine on both prolactin secretion and cyclic AMP accumulation. This suggests that the inhibitory action of dopaminergic agonists on prolactin release is analogous to the action of other inhibitory ligands coupled to adenylate cyclase. Schettini and colleagues (1985) report that, although dopamine inhibited adenylate cyclase activity stimulated by fluoride, Mg^{++} and forskolin, it also inhibited activity stimulated by maitotoxin, a calcium channel agonist. This group have also reported that changes in Ca^{2+} and calmodulin stimulate adenylate cyclase (Schettini *et al.*, 1983*a*) and that A23187 (the bivalent cation ionophore) also increased cyclic AMP accumulation. Thus, the possibility exists that dopamine-induced changes in Ca^{2+} may also influence intracellular cyclic AMP. Moreover, there are reports that dopamine can inhibit prolactin secretion in the face of elevated intracellular concentrations of cyclic AMP (e.g. Ray and Wallis, 1981), suggesting that one locus for dopamine action is subsequent to the generation of cyclic AMP.

15.3. Calcium

There is a considerable body of evidence indicating that Ca^{2+} plays a major role in the control of prolactin secretion. Indeed, reduction in the extracellular concentration of Ca^{2+} results in attenuation of the rate of prolactin release (Thorner *et al.*, 1980; Ray and Wallis, 1982; Merritt and Brown, 1984*a*). Conversely, increased extracellular Ca^{2+} enhances prolactin secretion (MacLeod and Fontham, 1970; Parsons, 1970). Agents that increase intracellular Ca^{2+}, such as depolarisation with K^+ or the addition of Ca^{2+} ionophores, are associated with increased secretory rate (Tashjian *et al.*, 1978; Tam and Dannies, 1980; Thorner *et al.*, 1980; Gershengorn, 1982; Merritt and Brown, 1984*a*). Moreover, TRH increases the frequency of spontaneous action potentials (Taraskevich and Douglas, 1977), and evokes a transient increase in intracellular Ca^{2+} in normal (Schofield, 1983) and tumour (Gershengorn, 1982) anterior pituitary cells. Recent evidence suggests that TRH may stimulate prolactin secretion via release of Ca^{2+} from intracellular stores (Albert and Tashjian, 1984; Gershengorn *et al.*, 1984). This is most likely to be mediated via TRH-stimulated hydrolysis of phosphoinositides (Leung *et al.*, 1982; Baird *et al.*, 1983; Martin, 1983; Rebecchi and Gershengorn, 1983; Macphee and Drummond, 1984). Finally, agents which bind to and inhibit the action of the calcium-binding protein, calmodulin, also inhibit prolactin secretion (Merritt *et al.*, 1983; Schettini *et al.*, 1983*a,b*). However, the influence of dopamine on Ca^{2+}-dependent processes remains controversial.

We have shown that dopamine inhibits prolactin secretion evoked by IBMX, TRH and A23187, but that the effect of dopamine on K^+-stimulated secretion is not so pronounced (Merritt and Brown, 1984*b*). This may indicate that voltage-dependent channels are normally relatively insensitive to inhibition by dopamine whereas secretion stimulated via agonist-dependent Ca^{2+} mobilisation as a result of IP_3 production (from PIP_2 hydrolysis), or ionophore-induced increases in Ca^{2+}, are susceptible to inhibition. The latter observation would point to one locus for dopamine inhibition being subsequent to a rise in Ca^{2+} concentrations. Ray and

Wallis (1982) concluded that dopamine affected Ca^{2+} mobilisation and not calcium flux. Moreover, Schofield (1983), using the calcium indicator Quin 2, showed that, in bovine pituitary cells, dopamine (10 μM) could prevent a rise in intracellular Ca^{2+} concentrations induced by TRH (1 nM).

If TRH-induced Ca^{2+} mobilisation is affected by dopamine, then this catecholamine may be acting by inhibiting phosphoinositide breakdown to IP_3 and DAG. Indeed, Canonico and co-workers (1983) have reported that labelling of phosphatidylinositol (PI) using $[^{32}P]$phosphate was attenuated in the presence of dopamine. This inhibition was reversed by the dopamine antagonist haloperidol. However, since labelling of PI occurs as a secondary event to the primary effect of PI breakdown, we have measured the accumulation of the water-soluble inositol phosphates (the initial products of PI breakdown). Our initial studies indicated that dopamine had little or no effect on inositol phosphate accumulation in rat anterior pituitary cells, either in the absence or the presence of TRH (Brown *et al.*, 1985). However, Simmonds and Strange (1985) have reported an inhibition of inositol phosphate accumulation by dopamine and apomorphine in bovine pituitary cells, but only in the presence of TRH. Nevertheless, on the basis of the current hypothesis, it will be important to measure changes in IP_3 (rather than total inositol phosphates), since only inositol 1,4,5-trisphosphate appears to be capable of mobilising intracellular Ca^{2+}. If dopamine only attenuates stimulated phosphoinositide breakdown, the possibility exists that dopaminergic inhibition of basal secretion is mediated by a different mechanism.

Thus, in summary, it appears that dopamine may act, at least in part, by inhibition of either adenylate cyclase or of phosphoinositide turnover. However, it is not yet clear whether either is a direct effect. Nevertheless, although alterations in the intracellular concentration of $Ca^{2+,}$ and cyclic AMP seem to play a role in the mechanism of action of dopamine, interactions between the Ca^{2+} system (raised intracellular Ca^{2+} and protein kinase activation) and the cyclic AMP system may be important and warrant further investigation.

References

Albert, P. R. and Tashjian, A. H. Jr (1984). Thyrotropin-releasing hormone-induced spike and plateau in cytosolic free Ca^{2+} concentrations in pituitary cells. *J. Biol. Chem.* **259**, 5827–32.

Arita, J. and Porter, J. C. (1984). Relationship between dopamine release into hypophyseal portal blood and prolactin release after morphine treatment in rats. *Neuroendocrinology* **38**, 62–7.

Baird, J. G., Dobson, P. R. M., Wojcikiewicz, R. J. H. and Brown, B. L. (1983). Thyrotropin-releasing hormone stimulates inositol phosphate production in normal anterior pituitary cells and GH_3 tumour cells in the presence of lithium. *Bioscience Reports* **3**, 1091–9.

Barnes, G. D., Brown, B. L., Gard, T. G., Atkinson, D. and Ekins, R. P. (1978). Effect of TRH and dopamine on cyclic AMP levels in enriched mammotroph and thyrotroph cells. *Mol. Cell Endocrinol.* **12**, 273–84.

Berridge, M. J. (1984). Inositol trisphosphate and diacylglycerol as second messengers. *Biochem. J.* **220**, 345–60.

Brown, B. L., Wojcikiewicz, R. J. H., Dobson, P. R. M., Robinson, A. and Irons, L. I. (1984). Pertussis toxin blocks the inhibitory effect of muscarinic cholinergic agonists on cyclic AMP accumulation and prolactin secretion in GH_3 anterior-pituitary cells. *Biochem. J.* **233**, 145–9.

Brown, B. L., Baird, J. G., Quilliam, L. A., Merritt, J. E. and Dobson, P. R. M. (1985). Calcium mediated intracellular signalling in the control of prolactin secretion from rat anterior pituitary cells. In *Fidia Research Series, vol. I. Prolactin—basic and clinical correlates*. R. M. MacLeod, M. O. Thorner and T. D. Scapagnini (eds). Liviana Press.

Calabro, M. A. and MacLeod, R. M. (1978). Binding of dopamine to bovine anterior pituitary gland membranes. *Neuroendocrinology* **25**, 32–46.

Canonico, P. L., Valdenegro, C. A. and MacLeod, R. M. (1983). The inhibition of phosphatidylinositol turnover: a possible post-receptor mechanism for the prolactin-inhibiting effect of dopamine. *Endocrinology (Baltimore)* **113**, 7–14.

Caron, M. G., Beaulieu, M., Raymond, V., Gagne, B., Drouin, J., Lefkowitz, R. J. and Labrie, F. (1978). Dopaminergic receptors in the anterior pituitary gland. *J. Biol. Chem.* **253**, 2244–53.

Castagna, M., Takai, Y., Kaibuchi, K., Sano, K., Kikkawa, U. and Nishizuka, Y. (1982). Direct activation of calcium-activated, phospholipid-dependent protein kinase by tumor-promoting phorbol esters. *J. Biol. Chem.* **257**, 7847–51.

Creese, I., Schneider, R. and Snyder, S. H. (1977). ^{3}H-spiroperidol labels dopamine receptors in pituitary and brain. *Eur. J. Pharmacol.* **46**, 377–81.

Cronin, M. J. (1982). Some calcium and lysosome antagonists inhibit ^{3}H-spiperone binding to the porcine anterior pituitary. *Life Sci.* **30**, 1385–9.

Cronin, M. J. and Weiner, R. I. (1979). ^{3}H-spiroperidol (spiperone) binding to a putative dopamine receptor in sheep and steer pituitary and stalk median eminence. *Endocrinology (Baltimore)* **104**, 307–12.

Cronin, M. J., Myers, G. A., MacLeod, R. M. and Hewlett, E. L. (1983). Pertussis toxin uncouples dopamine agonist inhibition of prolactin release. *Am. J. Physiol.* **244**, E499–504.

De Camilli, P., Macconi, D. and Spada, A. (1979). Dopamine inhibits adenylate cyclase in human prolactin-secreting pituitary adenomas. *Nature (Lond.)* **278**, 252–4.

Delbecke, D., Scammell, J. G. and Dannies, P. S. (1984). Difference in calcium requirements for forskolin-induced release of prolactin from normal pituitary and $GH_4 C_1$ cells in culture. *Endocrinology (Baltimore)* **114**, 1433–40.

Gershengorn, M. C. (1982). Thyrotropin releasing hormone. A review of the mechanisms of acute stimulation of pituitary hormone release. *Mol. Cell. Biochem.* **45**, 163–79.

Gershengorn, M. C., Geras, E., Purrello, V. S. and Rebecchi, M. J. (1984). Inositol trisphosphate mediates thyrotropin-releasing hormone mobilisation of non-mitochondrial calcium in rat mammotropic pituitary cells. *J. Biol. Chem.* **259**, 10675–81.

Giannattasio, G., De Ferrari, M. E. and Spada, A. (1981). Dopamine inhibited adenylate cyclase in female rat adenohypophysis. *Life Sci.* **28**, 1605–12.

Goldsmith, P. C., Cronin, M. J., Rubin, R. J. and Weiner, R. I. (1978). Immunocytochemical staining of dopamine receptors on mammotrophs. *Program of the Eighth Annual Meeting of the Society for Neuroscience, St. Louis, MO* (Abstract 1098).

Guild, S. and Drummond, A. H. (1983). Adenosine 3'5'monophosphate-dependent release of prolactin from GH_3 pituitary tumour cells. *Biochem. J.* **216**, 551–7.

Kebabian, J. W. and Calne, D. B. (1979). Multiple receptors for dopamine. *Nature (Lond.)* **277**, 93–6.

Leung, P. C. K., Raymond, V. and Labrie, F. (1982). Mechanism of action of TRH: involvement of the phosphatidylinositol (PI) response in the action of TRH in rat anterior pituitary cells. *Life Sci.* **31**, 3037–41.

MacLeod, R. M. (1976). Regulation of prolactin secretion. In *Frontiers in Neuroendocrinology, vol. I.* L. Martini and W. F. Ganong (eds), p. 169. Raven Press, New York.

MacLeod, R. M. and Fontham, E. H. (1970). Influence of the ionic environment on the *in vitro* synthesis and release of pituitary hormones. *Endocrinology (Baltimore)* **86**, 863–9.

Macphee, C. H. and Drummond, A. H. (1984). Thyrotropin-releasing hormone stimulates rapid breakdown of phosphatidylinositol 4,5-bisphosphonate and phosphatidylinositol 4-phosphate in GH_3 pituitary tumour cells. *Mol. Pharmacol.* **25**, 193–200.

Martin, T. F. J. (1983). Thyrotropin-releasing hormone rapidly activates the phosphodiester hydrolysis of polyphosphoinositides in GH_3 pituitary cells. *J. Biol. Chem.* **258**, 14816–22.

Merritt, J. E. and Brown, B. L. (1984a). An investigation of the involvement of calcium in the control of prolactin secretion: studies with low calcium, methoxyverapamil, cobalt and manganese. *J. Endocrinol.* **101**, 319–25.

Merritt, J. E. and Brown, B. L. (1984b). The possible involvement of both calcium and cyclic AMP in the dopaminergic inhibition of prolactin secretion. *Life Sci.* **35**, 707–11.

Merritt, J. E., MacNeil, S., Tomlinson, S. and Brown, B. L. (1983). The relationship between prolactin secretion and calmodulin activity. *J. Endocrinol.* **98**, 423–9.

Neill, J. D., Frawley, S., Plotsky, P. M. and Tindall, G. T. (1981). Dopamine in hypophyseal stalk blood of the rhesus monkey and its role in regulating prolactin secretion. *Endocrinology (Baltimore)* **108**, 489–94.

Onali, P., Schwarz, J. P. and Costa, E. (1981). Dopaminergic modulation of adenylate cyclase stimulation by vasoactive intestinal peptide in anterior pituitary. *Proc. nat. Acad. Sci. USA* **78**, 6531–4.

Onali, P., Eva, C., Olianas, M. C., Schwartz, J. P. and Costa, E. (1983). In GH_3 pituitary cells, acetylcholine and vasoactive intestinal peptide antagonistically modulate cyclase, cyclic AMP content, and prolactin release. *Mol. Pharmacol.* **24**, 189–94.

Parsons, J. A. (1970). Effects of cations on prolactin and growth hormone secretion by rat adenohypophyses *in vitro*. *J. Physiol (London)* **210**, 973–87.

Ray, K. P. and Wallis, M. (1981). Effect of dopamine on prolactin secretion and cyclic AMP accumulation in the rat anterior pituitary gland. *Biochem. J.* **194**, 119–28.

Ray, K. P. and Wallis, M. (1982). Actions of dopamine on prolactin secretion and cyclic AMP metabolism in ovine pituitary cells. *Mol. Cell. Endocrinol.* **27**, 139–55.

Rebecchi, M. J. and Gershengorn, M. C. (1983). Thyroliberin stimulates rapid hydrolysis of phosphatidylinositol 4,5-bisphosphate by a phosphodiesterase in rat mammotropic pituitary cells. *Biochem. J.* **216**, 299–308.

Schettini, G., Judd, A. M. and MacLeod, R. M. (1983a). *In vitro* studies on basal and stimulated prolactin release by rat anterior pituitary: a possible role for calmodulin. *Endocrinology* **122**, 64–70.

Schettini, G., Cronin, M. J. and MacLeod, R. M. (1983*b*). Cyclic AMP and calcium–calmodulin interrelation in the control of prolactin secretion: evidence for dopamine inhibition of cyclic AMP accumulation and prolactin release after calcium mobilisation. *Endocrinology (Baltimore)* **112**, 1801–7.

Schettini, G., Hewlett, E. L., Cronin, M. J., Schiassi, A., Yasumoto, T. and MacLeod, R. M. (1985). Dopaminergic regulation of the prolactin-secreting cells: the role of calcium and cyclic AMP. *Neurobiology of Dopamine*. W. Winlow and R. Markstein (eds), ch. 14. Manchester University Press, Manchester.

Schofield, J. G. (1983). Use of a trapped fluorescent indicator to demonstrate effects of thyroliberin and dopamine on cytoplasmic calcium concentrations in bovine pituitary cells. *FEBS Lett.* **159**, 79–82.

Simmonds, S. H. and Strange, P. G. (1985). Effect of D_2 dopamine receptor on inositol phospholipid metabolism in anterior pituitary. *Proceedings of the British Pharmacological Society* (in press).

Spada, A., Nicosia, S., Cortelazzi, L., Pezzo, G., Bassetti, M., Sartorio, A. and Giannattasio, G. (1983). *In vitro* studies on prolactin release and adenylate cyclase activity in human prolactin secreting pituitary adenomas. Different sensitivity of macro- and microadenomas to dopamine and vasoactive intestinal peptide. *J. Clin. Endocrinol. Metab.* **56**, 1-10.

Streb, H., Irvine, R. F., Berridge, M. J. and Schulz, I. (1983). Release of Ca^{2+} from a non-mitochondrial intracellular store in pancreatic acinar cells by inositol 1,4,5-triphosphate. *Nature (Lond.)* **306**, 67–69.

Swennen, L. and Denef, C. (1982). Physiological concentrations of dopamine decrease adenosine 3′5′-monophosphate levels in cultured rat anterior pituitary cells and enriched populations of lactotrophs. *Endocrinology (Baltimore)* **111**, 398–405.

Tam, S. W. and Dannies, P. S. (1980). Dopaminergic inhibition of ionophore A23187-stimulated release of prolactin from rat anterior pituitary cells. *J. Biol. Chem.* **255**, 6595–9.

Taraskevich, P. S. and Douglas, W. W. (1977). Action potentials occur in cells of the normal anterior pituitary gland and are stimulated by the hypophysiotropic peptide thyrotropin-releasing hormone. *Proc. Nat. Acad. Sci. USA* **74**, 4064–7.

Tashjian, A. H., Lomedico, M. E. and Mains, D. (1978). Role of calcium in the thyrotropin-releasing hormone stimulated release of prolactin from pituitary cells in culture. *Biochem. Biophys. Res. Commun.* **81**, 798–806.

Thorner, M. O., Hackett, J. T., Murad, F. MacLeod, R. M. (1980). Calcium rather than cyclic AMP as the physiological intracellular regulator of prolactin release. *Neuroendocrinology* **31**, 390–402.

Weiner, R. I. and Ganong, W. F. (1978). Role of brain monoamines and histamine in regulation of anterior pituitary secretion. *Physiol. Rev.* **58**, 905–76.

Wojcikiewicz, R. J. H., Dobson, P. R. M. and Brown, B. L. (1984). Muscarinic acetylcholine receptor activation causes inhibition of cyclic AMP accumulation, prolactin and growth hormone secretion in GH_3 rat anterior pituitary cells. *Biochem. Biophys. Acta.* **805**, 25–9.

16

Recent dopaminergic agents

E. Flückiger and R. Markstein
Preclinical Research, Sandoz Ltd, CH-4002 Basel, Switzerland

16.1. Introduction

When speaking of 'recent dopaminergic agents' we are faced with the products of a fast-expanding field of research. The number of chemical structures which are carriers of a dopamine-like information seems nearly unlimited. In addition, behavioural, biochemical and then physicochemical methods produced results suggesting that not all dopaminomimetic drugs have the same profile of actions, and that differences exist in the dopamine receptor populations. Thus, as with other biogenic amine transmitters, the subclassification of dopamine receptors became a necessity.

For this subclassification, anatomical, biochemical, physicochemical or behavioural criteria were used. As dopamine has functions not only in the brain and some of its appendages but also in the peripheral nervous system, the cardiovascular system, the gastrointestinal system, and certain peripheral endocrine and reproductive organs, it cannot be expected that studies are similarly advanced in the different organ systems. Indeed, detailed information is available only about some parts of the central nervous system, the anterior and intermediate lobe of the pituitary, and the peripheral cardiovascular system.

Pharmacological characterisation by agonists and antagonists suggests that the dopamine receptors in the cardiovascular system are probably different from those in the central nervous system (Goldberg and Kohli, 1983).

16.2. Selective drugs for dopamine receptors

In recent years great efforts have been undertaken towards finding new drugs selective for dopamine-receptor subtypes. In Fig. 16.1 a number of such structures are shown which stand out from the mass on the basis of qualitative criteria. It is evident that selective drugs are needed to characterise receptors and to elucidate their function in the neuronal network of the central nervous system or in the periphery. On the other hand it is difficult to develop selective drugs if the receptors involved cannot be studied directly. Some of the drugs shown in Fig. 16.1 are examples of such difficulties. For instance, the claim for four of these drugs to be autoreceptor specific is not undisputed (Markstein and Lahaye, 1983; Williams *et al.*, 1983). Thus, 3-PPP and TL-99 were found to inhibit prolactin release from the rat pituitary by a direct action on dopamine receptors (Eriksson *et al.*, 1983; Gudelsky *et al.*, 1983; Meltzer *et al.*, 1983), and to induce emesis in dogs (Martin *et al.*, 1982). Although such properties make the claim for autoreceptor selectivity impossible, the suggestion made by Eriksson *et al.* (1983) is interesting that 'non-synaptic' dopamine receptors may form a distinct subgroup embracing autoreceptors as well as receptors on prolactin cells and in the area postrema.

Compounds 1–4 of Fig. 16.1 are not the only drugs claimed to be autoreceptor selective. Comparing the effects of TL-99 and 3-PPP in various functional tests with those of other dopaminomimetics revealed that classical ergolines may also possess this property (Martin *et al.*, 1982). Thus, when using an 'autoreceptor selectivity index', pergolide and CF 25-397 unexpectedly proved to be as selective as TL-99 or 3-PPP. Does this have consequences for the clinical profile of such drugs as compared with a drug lacking autoreceptor selectivity, such as bromocriptine? Present knowledge does not suggest such differences.

Another receptor which is changing its image is the D-1 receptor (Kebabian and Calne, 1979), which until quite recently was in search of a (behavioural) function. Recent observations strongly indicate that D-1 agonistic activity may actually contribute in an important way to the motor effects of apomorphine and of certain other, dual, dopamine agonists (pergolide, CQ 32-084). This explains the unexpected motor activity of the D-1 selective compound SK & F 38393 (Gershanik *et al.*, 1983; Herrera-Marschitz and Ungerstedt, 1984).

Just as dopamine itself is not selective for any dopamine-receptor subtype, so many other dopaminomimetic compounds also are not really selective but in addition interact with other receptors, e.g. adrenoceptors and serotonin receptors. This pluripotency is usually recognised with ergot derivatives but it is not restricted to this chemical class. Comparing the ability of dopaminomimetic ergot derivatives to inhibit the specific binding of [^{3}H]-rauwolscine, [^{3}H]-prazosine and [^{3}H]-spiperone to membranes of rat brain (Beart *et al.*, 1983; McPherson and Beart, 1983), it was found that bromocriptine, CQ 32-084 and pergolide could claim selectivity for D-2 binding sites over adrenoceptors, and several others 'might rather be considered as α-adrenergic than dopaminergic drugs'. As adrenergic mechanisms may also be involved in the regulation of DOPA biosynthesis (Goldstein *et al.*, 1983), it is clear that such drugs are not very suitable for studying dopaminergic functions.

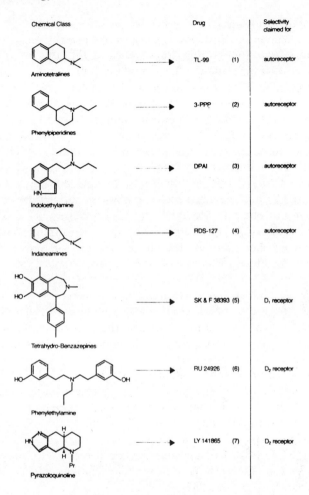

Fig. 16.1. General chemical structures of drugs with claimed selectivity for dopamine receptor subtypes. Literature: (1) Goodale *et al.* (1980); (2) Hjorth *et al.* (1981); (3) Clemens *et al.* (1984); (4) Arnerič *et al.* (1983); (5) Setler *et al.* (1978); (6) Euvrard *et al.* (1980); (7) Tsuruta *et al.* (1981).

In recent years attempts to define the pharmacophore of ergot compounds for dopaminomimetic activity revealed that structural changes do not produce parallel changes in D-1 and D-2 agonistic activity. For instance, bromocriptine is an agonist at D-2 receptors and an antagonist at D-1 receptors. Substitution of the methyl group at N_6 by an ethyl group converted the D-1 antagonistic activity into D-1 agonistic activity without greatly influencing the agonistic activity at D-2 receptors (Flückiger and Markstein, 1983).

Another structure–activity relationship was established with 8α-aminoergolines with which human pharmacological studies were also performed: CH 29-717, CQ 32-084 and CQP 201-403. Their chemical structures are presented in Fig. 16.2 and selected pharmacological activities are summarised in Table. 16.1, together with data for bromocriptine and apomorphine. It is interesting to note that elongation of the terminal N-dimethyl groups in the side chain in position 8α of CH 29-717 to N-diethyl groups (to produce CQ 32-084) reduced the *in vivo* potency to inhibit prolactin secretion but increased the central *in vivo* and *in vitro* potency and switched the D-1 antagonist activity into D-1 agonist activity. In the next step the methyl group on N_6 of ring D was replaced by *n*-propyl (to produce CQP 201-403) which enhanced all effects on the dopaminergic activities presented in Table 16.1. Methylation of N_1 in ring A of CH 29-717 produced CU 32-085 (=mesulergine, DCl prop.) (Flückiger *et al.*, 1979). This compound has an interesting pharmacodynamic profile in experimental animals (Flückiger *et al.*, 1983). This is probably explained by the

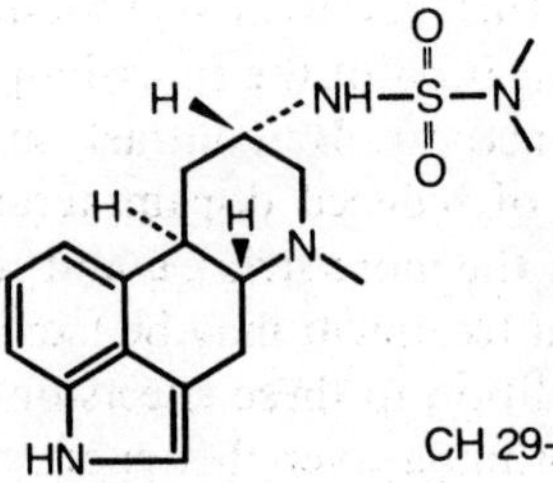

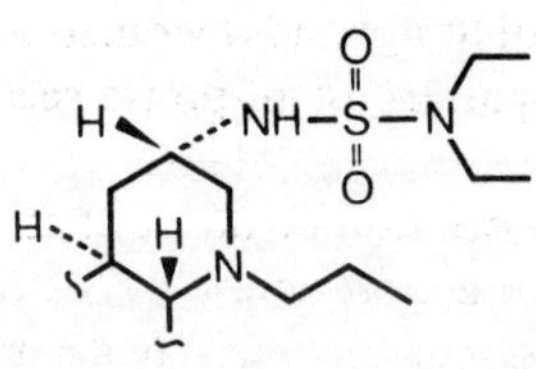

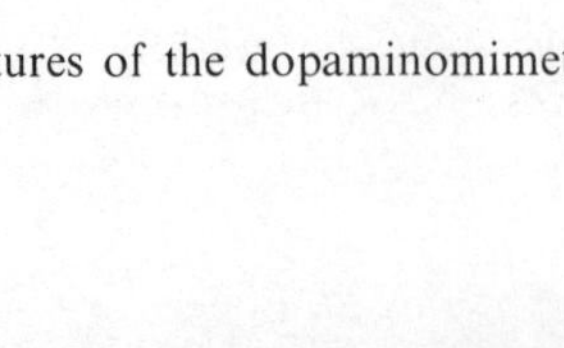

Fig. 16.2. Chemical structures of the dopaminomimetics CH 29-717, CQ 32-084 and CQP 201-403.

Table 16.1. Dopaminergic activities *in vivo* and *in vitro* of 8α-amino-ergolines (for chemical structures see Fig. 16.2) compared with dopamine and apomorphine.

	(1) PRL ↓ rat (2 h) ID$_{50}$, μg kg^{-1} s.c.	(2) Nidation ↓ rat ED$_{50}$, μg kg^{-1} s.c.	(3) Emesis dog ED$_{50}$, μg kg^{-1} i.v.	(4) Ungerstedt rat MED, μg kg^{-1} s.c.	(5) ACH release ↓ rat striatum EC$_{50}$ (nM)	(6) DA-AC ↑ bovine retina E$_r$	EC$_{50}$ μM
CH 29-717	0·9	13	<5	50	5	0	$K_i=6$
CQ 32-084	5	28	<5	10	2·5	10	35
CQP 201-403	0·3	7	0·7	<10	0·4	38	2·2
Bromocriptine	8	750	7·5	100	5	0	K_i/0·4
Apomorphine	1800	00	—	10	25	38	0·73

Methods used: (1, 2) Flückiger *et al.* (1978, 1979). (3) ED$_{50}$ values were evaluated by log probit analysis from the number of animals vomiting within 5 min after i.v. administration of different doses of the test drug. For each dose, at least 10 animals were used. (4) Vigouret *et al.* (1978). MED = minimal effective dose. (5, 6) Markstein (1981); E$_r$ = maximal stimulation of adenylate cyclase activity as a percentage of that of dopamine.

fact that mesulergine (*in vitro*) is an antagonist at D-1 and D-2 receptors (Markstein, 1983). D-2 antagonism *in vivo* is exemplified by the observation that CU 32-085 does not attenuate prolactin secretion in primary cultures of rat pituitary cells but antagonises dose-dependently the inhibitory action of CH 29-717 on prolactin secretion (see Fig. 16.3) (Marko, 1984). The compound is converted metabolically into metabolites having dopamine agonistic activity (Markstein, 1983; Enz *et al.*, 1984). With such a drug, which does not produce a burst of dopamine agonistic action when entering the systemic circulation, typical side-effects in man such as nausea or postural hypotension should theoretically become less prominent (see Goldstein *et al.*, 1984).

Starting from lisuride, saturation of the double bond 9,10 produced dihydrolisuride (= VUFB 6638 m SPOFA; TDHL), which unexpectedly showed partial dopamine-antagonistic potency (Wachtel, 1983*a*; Wachtel and Darrow, 1983). This partial dopamine antagonism is also expected to produce a favourable shift in the dopaminergic profile of actions. The

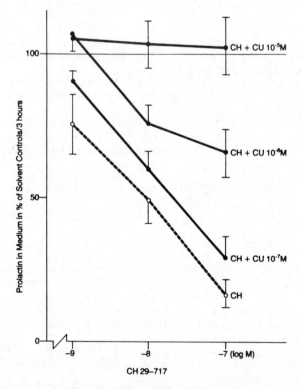

Fig. 16.3. The effects of CU 32-085 and CH 29-717 alone and in combination on prolactin release into the medium of primary cultures of dissociated cells from rat anterior pituitaries. (After Marko, 1984, with permission.)

unexpected production of a dopamine antagonist, by introducing bromine in position 2 of lisuride (2-Br-lisuride), should also be mentioned (Wachtel, 1983*b*). The profiles of actions of dopamine agonist/antagonist other than of TDHL have not yet been systematically studied and compared, but such studies are underway in several laboratories.

16.3. Summary

When reviewing 'recent dopaminomimetic agents' one is faced with the products and the problems of a fast-expanding field. Results from behavioural, biochemical and physicochemical investigations show that only a few of the presently available dopaminomimetic drugs are really selective for dopamine-receptor subtypes.

Ergot derivatives in particular interact with several other monoaminergic receptor types. For the characterisation of dopamine receptors, only selective drugs should be used.

References

Arnerič, S. P., Long, J. P., Williams, M., Goodale, D. B., Mott, J., Lakoski, J. M. and Gebhart, G. F. (1983). RDS-127 (2-di-*n*-propylamino-4,7-dimethoxyindane): central effects. *Therap.* **224**, 161–70.

Beart, P. M., Gundlach, A. L. and McPherson, G. M. (1983). Selectivity of some ergot derivatives for α_1- and α_2-adrenoceptors of rat cerebral cortex. *Brit. J. Pharmacol.* **29** (*Proc. Suppl.*), 196P.

Clemens, J.-A., Fuller, R. W., Phebus, L. A., Smalstig, E. B., Hynes, M. D., Cassady, J. M., Nichols, D. E., Kelly, E. and Persons, P. (1984). Stimulation of presynaptic dopamine autoreceptors by 4-(2-di-*n*-propylaminoethyl)indole (DPAI). *Life Sci.* **34**, 1015–22.

Enz, A., Donatsch, P., Briner, U., Jaton, A. L., Markstein, R. and Palacios, J. M. (1984). Dopaminergic profile of mesulergine. In *Proc. 1st ENEA Meeting*, Basel, March 1984. X. Lataste, (ed.). Academic Press.

Eriksson, E., Modigh, K., Carlsson, A. and Wikström, H. (1983). Dopamine receptors involved in prolactin secretion pharmacologically characterized by means of 3-PPP enantiomers. *Eur. J. Pharmacol.* **96**, 29–36.

Euvrard, C., Ferland, L., Di Paolo, T., Beaulieu, M., Labrie, F., Oberlander, C., Raynaud, J. P. and Boissier, J. R. (1980). Activity of two new potent dopaminergic agonists at the striatal and anterior pituitary levels. Neuropharmacology **19**, 379–86.

Flückiger, E. and Markstein, R. (1983). Receptor pharmacology of ergot compounds. In *Prolactin and Prolactinomas*. G. Tolis, C. Stefanis, T. Mountokalakis and F. Labrie (eds), pp. 105–13, Raven Press, New York.

Flückiger, E., Briner, U., Doepfner, W., Kovacs, E., Marbach, P. and Wagner, H. R. (1978). Prolactin secretion inhibition by a new 8α-aminoergoline, CH 29-717. *Experientia* **34**, 1130–1331.

Flückiger, E., Briner, U., Bürki, H. R., Marbach, P., Wagner, H. R. and Doepfner, W. (1979). Two novel prolactin release-inhibiting 8α-amino-ergolines. *Experientia* **35**, 1677–8.

Flückiger, E., Briner, U., Enz, A., Markstein, R. and Vigouret, J. M. (1983). Dopaminergic ergot compounds, an overview. In _Lisuride and Other Dopamine Agonists._ D. B. Calne, R. Horowski, R. J. McDonald and W. Wuttke (eds), pp. 1–9. Raven Press, New York.

Gershanik, O., Heikkila, R. E. and Duvoisin, R. C. (1983). Effects of dopamine depletion on rotational behaviour to dopamine agonists. _Brain Res._ **261**, 358–69.

Goldberg, L. I. and Kohli, J. D. (1983). Differentiation of dopamine receptors in the periphery. In _Dopamine Receptors._ C. Kaiser and J. W. Kebabian (eds), pp. 101–13, ACS Symposium Series 224, American Chemical Society, Washington, D.C.

Goldstein, M., Lew, J. Y., Regev, I., Lieberman, A. and Engel, J. (1983). The effects of dopamine agonists and of D-2-adrenoceptor antagonists on dopaminergic neurotransmission. In _Dopamine Receptor Agonists 2._ A. Carlsson and J. L. G. Nilsson (eds), pp. 168–74, Swedish Pharmaceutical Press, Stockholm.

Goldstein, M., Lieberman, A. and Battista, A. F. (1984). The therapeutic potential of centrally acting dopamine agonists. _TIPS_ **5**, 227–30.

Goodale, D. B., Rusterholz, D. B., Long, J. P., Tlynn, B., Cannon, J. G. and Lee, T. (1980). Neurochemical and behavioural evidence for a selective presynaptic dopamine receptor agonist. _Science_ **210**, 1141–3.

Gudelsky, G. A., Passaro, E. and Meltzer, H. Y. (1983). Effect of two dopamine agonists, TL-99 and 3-PPP, on prolactin secretion in the rat. _Eur. J. Pharmacol._ **90**, 423–5.

Herrera-Marschitz, M. and Ungerstedt, U. (1984). Evidence that apomorphine and pergolide induce rotation in rats by different actions on D-1 and D-2 receptor sites. _Eur. J. Pharmacol._ **98**, 165–76.

Hjorth, S., Carlsson, A., Wikström, H., Lindberg, P., Sanchez, P., Hacksell, U., Arvidsson, L.-E., Svensson, U. and Nilsson, J. L. G. (1981). 3-PPP, a new centrally acting DA receptor agonist with selectivity for autoreceptors. _Life Sci._ **28**, 1225–38.

Kebabian, J. W. and Calne, D. B. (1979). Multiple receptors for dopamine _Nature (Lond.)_ **277**, 93–6.

Marko, M. (1984). Dopamine agonistic potency of two novel prolactin release-inhibiting ergolines. _Eur. J. Pharmacol._ **00**, 000–00.

Markstein, R. (1981). Neurochemical effects of some ergot derivatives: a basis for their antiparkinsonian actions. _J. Neural Transm._ **51**, 39–59.

Markstein, R. (1983). Mesulergine and its 1,20-N,N-bidemethylated metabolite interact directly with D-1 and D-2 receptors. _Eur. J. Pharmacol._ **95**, 101–7.

Markstein, R. and Lahaye, D. (1983). _In vitro_ effect of the racemic mixture and the (−)enantiomer of _N-n_-propyl-3-(3-hydroxyphenyl) piperidine (3-PPP) on postsynaptic dopamine receptors and on presynaptic dopamine autoreceptors. _J. Neural Transm._ **58**, 43–53.

Martin, G. E., Williams, M. and Haubrich, D. R. (1982). A pharmacological comparison of 6,7-dihydroxy-2-dimethylaminotetraline (TL-99) and _N-n_-propyl-3-(3-hydroxyphenyl) piperidine (3-PPP) with selected dopamine agonists. _J. Pharmacol. Exp. Therap._ **223**, 298–304.

McPherson, G. A. and Beart, P. M. (1983). The selectivity of some ergot derivatives for α_1- and α_2-adrenoceptors of rat cerebral cortex. _Eur. J. Pharmacol._ **91**, 363–9.

Meltzer, H. Y., Gudelsky, G. A., Simonovic, M. and Fang, V. S. (1983). Effect of dopamine agonists and antagonists on prolactin and growth hormone secretion. In _Dopamine Receptor Agonists 1_, A. Carlsson and J. L. G. Nilsson (eds), pp. 200–18, Swedish Pharmaceutical Press, Stockholm.

Setler, P. E., Saran, H. M., Zirkle, C. L. and Saunders, H. L. (1978). The central effects of a novel dopamine agonist. _Eur. J. Pharmacol._ **50**, 419–30.

Tsuruta, K., Frey, E. A., Grewe, C. W., Cote, T. E., Eskay, R. L. and Kebabian, J. W. (1981). Evidence that LY-141865 specifically stimulates D-2 receptors. *Nature (Lond.)* **292**, 463–5.

Vigouret, J.-M., Bürki, H. R., Jaton, A. L., Züger, P. E. and Loew, D. M. (1978). Neurochemical and neuropharmacological investigations with four ergot derivatives: bromocriptine, dihydroergotoxine, CF 25-397 and CM 29-712. *Pharmacology* (Suppl. 1) **16**, 156–73.

Wachtel, H. (1983*a*). Central dopaminergic and antidopaminergic effects of ergot derivatives structurally related to lisuride. In *Lisuride and Other Dopamine Agonists*, D. B. Calne, R. Horowski, R. J. MacDonald and W. Wuttke (eds), pp. 109–25. Raven Press, New York.

Wachtel, H. (1983*b*). 2-Bromo-lisuride, an ergot derivative with potential neuroleptic activity. *Arch. Pharmacol.* **322** (Suppl.), R90.

Wachtel, H. and Darrow, R. (1983). Dual action of central dopamine function of transdihydrolisuride, a 9,10-dihydrogenated analogue of the ergot dopamine agonist lisuride. *Life Sci.* **32**, 421–32.

Williams, M., Martin, G. E., McClure, D. E., Baldwin, J. J. and Wathing, K. J. (1983). Interaction of the component enantiomers of the putative dopamine autoreceptor agonist, TL-99 (6,7-dihydroxy-2-dimethylamino tetraline) with dopaminergic systems in mammalian brain and teleost retina. *Naunyn-Schmiedeberg's Arch. Pharmacol.* **324**, 275–80.

17

Dopaminergic control of somatostatin and thyrotropin-releasing hormone release from perfused adult rat hypothalami

B. M. Lewis, C. Dieguez, S. M. Foord, R. Hall and M. F. Scanlon

Department of Medicine, University of Wales College of Medicine, Heath Park, Cardiff CF4 4XN, UK

17.1. Introduction

The role of hypothalamic dopaminergic pathways in the control of somatostatin (SRIF) and thyrotropin-releasing hormone (TRH) release has not been fully clarified. Some workers have reported an inhibitory role whereas others have demonstrated that dopamine stimulates both SRIF and TRH release from rat hypothalamic fragments or hypothalamic synaptosomes (Robbins and Reichlin, 1982). This stimulatory effect of dopamine is further supported by the demonstration of an increase in secretion rate and concentration of SRIF in rat hypophyseal portal blood after stimulation with dopamine (Chihara *et al.*, 1979).

However, very little is known about the specificity and the receptor type that mediates such actions. Dopamine receptors have been divided into two general categories. D-1 receptors are responsible for stimulating dopamine-sensitive adenylate cyclase activity upon agonist activation. D-2 receptors either inhibit or do not affect adenylate cyclase activity (Creese *et al.*, 1983). The aim of this study was to establish the dose-response relationships of dopamine and the dopamine agonist bromocriptine on TRH and SRIF secretion, and to determine whether these actions could be reduced by selective D-2 receptor blockade with metoclopramide. We used intact hypothalami in a superfusion system which has the advantage over static incubations that metabolic products are constantly removed. In consequence, the dynamic response of the tissue can be observed using the tissue as its own control.

17.2. Methods

Hypothalami were rapidly removed from male Wistar rats weighing 200–250 g and placed in a Gelman 25 mm filter holder. Each filter holder contained six hypothalami which were perfused at 37°C with Krebs Ringer bicarbonate (KRB) solution (pH 7·4). To reduce non-specific peptide absorption, 0·1 % BSA was added; bacitracin (300 μg ml^{-1}) was added to prevent enzymatic degradation. The perfusion system was maintained by a Treonic IP3 Digital Syringe Pump (Vickers Medical) at a flow rate of 200 μl min^{-1}. Ten-minute fractions were collected into acetic acid (final concentration 1 N).

Experimental treatments were applied when steady-state basal TRH and SRIF levels were reached following a 50-min prewash. On each occasion a control perfusion was carried out. Following each treatment the hypothalami were perfused with 60 mM K$^+$.

Aliquots of the perfusate (200 μl) were lyophilised and reconstituted in distilled water. Well described radioimmunoassays were employed for SRIF and TRH quantitation. The SRIF-14 antibody used was specific for the mid-portion of the molecule (residues 5–8) and showed complete cross-reactivity with SRIF-28 on an equimolar basis. Standards were made up in tubes containing 200 μl of lyophilised KRB solution.

17.3. Results

Administration of varying concentrations of dopamine (10^{-8} M to 10^{-5} M) produced a clear, dose-related release of SRIF from perfused hypothalami (Fig. 17.1). The lowest concentration which produced a

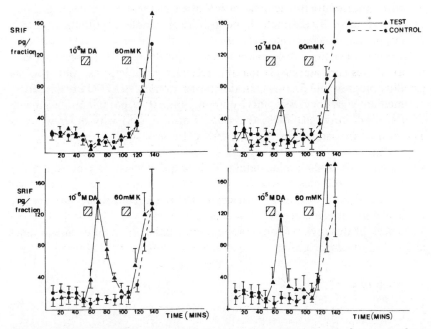

Fig. 17.1. Effect of different concentrations of dopamine (DA) (10^{-8} M to 10^{-5} M) and 60 mM K$^+$ on SRIF release.

significant effect was 10^{-7} M. In contrast, TRH release was unaltered by these concentrations of dopamine. However, exposure to 60 mM K^+ produced striking release of both neuropeptides.

Further experiments using the D-2 receptor agonist bromocriptine showed a similar pattern. Bromocriptine (10^{-6} M) produced a seven-fold increase in SRIF release with no effect on TRH release. The administration of 10^{-8} M metoclopramide 10 min prior to the infusion of 10^{-6} M dopamine abolished the increase in SRIF secretion.

17.4. Discussion

These data demonstrate that dopamine produces consistent dose-related release of SRIF from perfused adult rat hypothalami with no action on TRH release. Furthermore, this dopaminergic effect could be mimicked by the D-2 agonist bromocriptine and antagonised by D-2 receptor blockade with metoclopramide, suggesting mediation via a D-2 receptor.

Our results are in agreement with those of Richardson *et al.* (1983) but differ from those of Terry *et al.* (1980) who found that dopamine had no effect on SRIF release from perfused hypothalamic fragments.

The conflicting evidence which is available may be due to several factors:

(1) A variety of approaches both '*in vivo*' and '*in vitro*' have been utilised, with the majority of the work being carried out on isolated preparations. These include the culture of whole hypothalami, slices, fragments, mechanically or enzymatically dispersed cells and also synaptosomal and other subcellular preparations. It is not possible to make a direct comparison between such diverse methods and this is illustrated by the finding that dopamine stimulates LHRH release from median eminence but not from mediobasal hypothalami fragments (Negro-Vilar *et al.*, 1979). This could also explain why other workers have demonstrated an increase in TRH release with dopamine but we did not.

(2) The use of different antibodies with affinities for varying molecular forms of SRIF could be a reason for contradictory results as the release of the different molecular forms of SRIF may be selectively mediated by a variety of mechanisms.

(3) Many of the earlier studies exposed the tissue to a single and usually high dose of dopamine. There is therefore the possibility of non-specific interactions with other receptor types (viz. noradrenaline) which could cause misleading results.

In conclusion, our data show that dopamine plays a stimulatory role on SRIF release with no changes on TRH release from perfused intact hypothalami. With further studies using more specific drugs with greater selectivity for different neurotransmitter types, together with HPLC of the immunoreactive material, it will be possible to form a clearer picture of the factors controlling neuropeptide synthesis and release from the hypothalamus.

References

Chihara, K., Arimura, A. and Schally, A. V. (1979). *Endocrinology* **104**, 1656–62.

Creese, I., Sibley, D. R., Hamblin, M. W. and Leff, S. E. (1983). *Ann. Rev. Neurosci.* **6**, 43–71.

Negro-Vilar, A., Ojeda, S. R. and McCann, S. M. (1979). *Endocrinology* **104**, 1749–57.

Richardson, S. B., Nguyen, T. and Hollander, C. S. (1983). *Am. J. Phys.* (*Endocrinol. Metab.*) **244**, E560–6.

Robbins, R. and Reichlin, S. (1982). In *Neuroendocrinology Perspectives*, Vol. 1. E. E. Müller and R. M. MacLeod (eds), pp. 111–34. Elsevier Biomedical Press, Amsterdam.

Terry, L. C., Rastad, O. P. and Martin, J. B. (1980). *Endocrinology* **107**, 794–800.

18

Lack of influence of endogenous cortisol on the dopaminergic inhibition of thyrotropin and prolactin

J. Salvador, I. Cano, R. Rodriguez, J. J.Barberia and E. Moncada

Department of Endocrinology, University Hospital of Navarra, University of Navarra, Pamplona, Spain

18.1. Introduction

It has been demonstrated that the inhibition of thyrotropin (TSH) release by dopamine as assessed by its response to dopamine antagonists, exhibits a circadian rhythm in normal subjects, showing lower values in the morning (Scanlon *et al.*, 1980) just when cortisol levels are at their highest point. Since pharmacological doses of glucocorticoids are able to inhibit TSH release and the prolactin (PRL) response to dopamine antagonism (Sowers *et al.*, 1977), the endogenous level of cortisol might play a role in the modulation of the TSH increment which follows administration of dopamine antagonists.

We have tested this hypothesis by studying the TSH and PRL responses to dopamine antagonists under basal conditions and after abolishing the morning cortisol level by administration of the adrenal synthesis blocker, metyrapone. If our hypothesis is correct, one might expect a bigger TSH response to a dopamine antagonist when the 08.00 cortisol level is lowered by treatment with metyrapone. This would support a physiological influence of endogenously secreted cortisol on the dopaminergic inhibition of TSH.

18.2. Subjects and methods

Six normal euthyroid females aged 22-38 years were studied. Informed consent was obtained from all of them. TSH and PRL responses to domperidone (DOM, 10 mg i.v.) were determined at 08.00 before and on the third day of treatment with metyrapone (750 mg 6-hourly). Growth hormone and gonadotrophins were estimated in only three subjects. Blood was sampled at 0, 10, 20, 30, 45, 60, 90 and 120 min. Both tests were separated by a week. Cortisol levels were measured at 08.00, 12.00, 16.00, 20.00 and 24.00 during the days both DOM tests were carried out. Serum was collected and stored at $-20°C$ until assayed. All samples from a single subject were run in one assay. Cortisol, growth hormone, gonadotrophins, TSH and PRL were determined by radioimmunoassay.

Results were analysed by using paired Student's *t*-test. All data are expressed as mean ± standard error (SE).

18.3. Results

No significant side-effects were noticed when subjects were on metyrapone. All females showed a normal circadian variation of cortisol levels, which was significantly blunted during treatment with metyrapone, especially at 08.00 (control 17.48 ± 2.60 vs metyrapone 5.15 ± 0.50 μg 100 ml^{-1}, $P <$ 0.005; Fig. 18.1).

As expected, a clear increase in urinary 17-oxygenic steroids was found when the adrenal blocker was given (control 10.76 ± 1.10 vs metyrapone

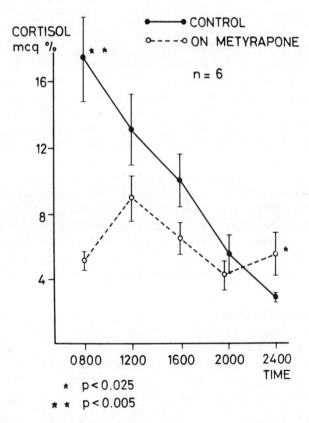

Fig. 18.1. Daily variation of cortisol levels in six normal females on basal conditions (●——●) and during metyrapone administration (O----O). Data are means ± SE.

$36\cdot38 \pm 7\cdot20$ mg 24 h^{-1}, $P<0\cdot01$), confirming the efficacy of the cortisol synthesis blockade. Injection of DOM did not lead to any change in growth hormone or gonadotrophins, either basally or during metyrapone adminis-tration in three subjects.

Baseline TSH value was slightly higher in the control test $(3\cdot30\pm0\cdot34$ mU litre$^{-1})$ than during metyrapone administration $(3\cdot05 \pm0\cdot25$ mU litre$^{-1})$ but no statistical differences could be found. A clear increase in TSH levels after DOM was observed in the control test as well as when subjects were on metyrapone $(P<0\cdot05$ at 30 min in both tests). As shown in Fig. 18.2, when females received metyrapone, their TSH response to DOM was slightly smaller than that in pretreatment conditions, but was

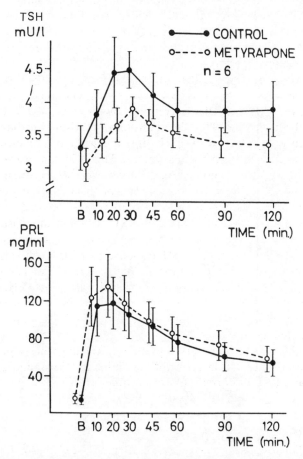

Fig. 18.2. TSH and PRL responses to domperidone in six normal females on basal conditions (●———●) and during metyrapone administration (O————O). Data are means ± SE.

not statistically significant. Similar results are obtained when both curves are analysed in terms of maximal percentage changes over the baseline value (control: 44·83 ± 19·35 vs metyrapone 33 ± 11·89 %, P = NS).

Baseline as well as stimulated PRL levels were almost identical in both conditions (Fig. 18.2), and the percentage changes over the basal value observed during metyrapone administration (765 ± 211 %) did not differ from those found before the adrenal blocker was given (744 ± 163 %).

18.4. Discussion

This study shows that the decline of morning cortisol levels induced by metyrapone administration does not result in greater TSH or PRL release after administration of domperidone at that time. This indicates that the dopaminergic inhibition of TSH and PRL released at anterior pituitary/median eminence level (DOM does not cross the blood–brain barrier) is not dependent on circulating concentrations of endogenous cortisol in normal subjects.

There is considerable evidence that high levels of adrenal steroids, either exogenously administered or endogenously hypersecreted, have a negative influence on TSH and PRL secretion (Sowers *et al.*, 1977). In a previous study (Salvador *et al.*, 1983), the abolition of cortisol circadian rhythm did not produce any change in TSH diurnal variation, suggesting a lack of interdependence between both hormones. This is in contrast to the findings of Patel *et al.* (1974) who reported the absence of nocturnal elevation of TSH following administration of pharmacological doses of glucocorticoids. The most likely explanation for these controversial data is that the effects of steroid administration on TSH and PRL secretion are dose related and represent pharmacological rather than physiological interactions.

The observation that the TSH response to DOM is slightly decreased when subjects received metyrapone might suggest the possibility of an increase in somatostatin, which has inhibitory effects on TSH, secondary to an elevation of growth hormone levels caused by metyrapone (Kunita *et al.*, 1970), but no changes in this hormone were found after administration of the adrenal blocker.

In summary, these results show that our initial hypothesis, suggesting that cortisol diurnal variation does not determine the dopaminergic inhibition of TSH and PRL in normal subjects, is not correct.

References

Kunita, H., Takebe, K., Nakagawa, K., Sawano, S. and Horiuchi, Y. (1970). *J. Clin. Endocrinol. Metab.* **31**, 301–6.

Patel, Y. C., Baker, H. W. G., Burger, H. G., Johns, M. W. and Ledinek, J. E. (1974). *J. Endocrinol. Metab.* **62**, 421–2.

Salvador, J., Wilson, D., Edwards, K., Peters, J. R., Foord, S., Dieguez, C., Hall, R. and Scanlon, M. F. (1983). In *Proceedings of the 3rd European Workshop on Pituitary Adenomas*, Amsterdam, p. 13.

Scanlon, M. F., Weetman, A. P., Lewis, M., Pourmand, M., Rodriguez Arnao, M. D., Weightman, D. R. and Hall, R. (1980). *J. Clin. Endocrinol. Metab.* **51**, 1251–6.

Sowers, J. R., Carlson, H. E., Brautbar, N. and Hershman, J. M. (1977). *J. Clin. Endocrinol. Metab.* **44**, 237–41.

19

Oestrogens increase the sensitivity of prolactin to the inhibitory actions of dopamine in hyperprolactinaemic patients

R. Valcavi, R. Elli, P. E. Harris, C. Dieguez, S. M. Foord, J. R. Peters, R. Hall and M. F. Scanlon

Department of Medicine, University of Wales College of Medicine, Heath Park, Cardiff CF4 4XN, UK

19.1. Introduction

It has generally been considered that oestrogens and dopamine are antagonistic with respect to lactotroph function. Histologically it is well established that oestrogens stimulate lactotroph hypertrophy and mitosis. At the functional level, oestrogens stimulate the synthesis of prolactin (PRL) mRNA, and in the rat they antagonise the effects of dopamine directly at the level of the lactotroph, leading to sustained hyperprolactinaemia and ultimately prolactinoma formation (see reviews by Franks, 1983, and Leong *et al.*, 1983). Although there is no clear evidence in man that oestrogen-containing compounds lead to prolactinoma formation, it is generally felt that oestrogens should be used with caution in patients with pathological hyperprolactinaemia.

In normal human females, PRL sensitivity to the inhibitory effect of dopamine is maximal at midcycle following the pre-ovulatory oestradiol surge, and the suppressive effect of dopamine infusion on PRL is reduced in ovariectomised women (Judd *et al.*, 1978; Franks, 1983). These data suggest that in man the situation is different from that in the rat, oestrogens possibly increasing PRL sensitivity to the inhibitory actions of dopamine. We have tested this hypothesis in hyperprolactinaemic women by studying the effects of short-term low-dose oestrogen pretreatment on the PRL response to dopamine infusion.

19.2. Patients and methods

Seven female patients (aged 22–57 years) with hyperprolactinaemia due to presumed PRL-secreting microadenomas and otherwise intact anterior pituitary function were investigated. Each subject underwent three tests on separate days following an overnight fast. Between 08.30 and 09.30, intravenous cannulae were inserted into both antecubital veins. After 30 min at rest, the submaximal infusion of dopamine ($0.06\ \mu g\ kg^{-1}\ min^{-1}$) or normal saline (NS) commenced and samples were collected every 15 min over the 3-hour infusion period. On day 1, patients received saline; on day 2 and on day 5 the same dopamine infusion was repeated after the administration of ethinyl oestradiol 100 μg day^{-1} for 3 days. Prolactin was measured by radioimmunoassay as previously described. Statistical analysis was performed using the paired Student's t-test.

19.3. Results

Following oestrogen treatment, basal PRL levels rose in all subjects (mean $\pm$ SE, mU litre^{-1}; 2903 ± 761 vs 2293 ± 684 ($P < 0.05$). As expected, submaximal dopamine infusion decreased the serum PRL concentration in all patients (Fig. 19.1). The mean $\pm$ SE sum of decrements from the baseline in serum PRL levels over the 3 h of infusion (8013 ± 1902 mU litre^{-1}) was

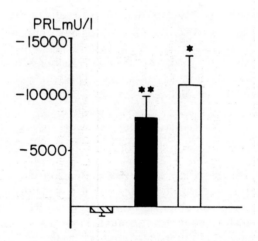

Fig. 19.1.

significantly ($P < 0.001$) greater than saline controls (618 ± 850 mU litre^{-1}). The repeated dopamine infusion after oestrogens produced a greater decrement in PRL levels (sum of decrements 10998 ± 2489 mU litre^{-1}) than obtained with dopamine alone, with a difference of 2971 $\pm$ 880 mU litre^{-1} ($P < 0.02$). The mean $\pm$ SE percentage decrease in basal PRL levels during dopamine infusion was slightly but not significantly greater following oestrogen pretreatment ($38.4 \pm 2.4\%$ vs $33.4 \pm 2.1\%$, NS) depending upon the increase in serum PRL concentration after oestrogen priming.

Although the numbers were small, the increments in basal serum PRL levels after oestrogen treatment showed a significant correlation ($r = 0.93$; $P < 0.001$), with the greater decrement in PRL levels during the oestrogen-primed dopamine infusion.

19.4. Discussion

In normal females only minimal hyperprolactin may occasionally be seen following oestrogen treatment, and indeed a lowering of basal PRL levels has been reported in some women treated with 17-B-oestradiol (Helgason *et al.*, 1982). It is possible that PRL levels in such oestrogen-treated normal subjects are controlled by activation of the short-loop feedback of PRL on hypothalamic dopaminergic activity. By contrast, in hyperprolactinaemic patients who show evidence of reduced dopaminergic inhibition of PRL release at the lactotroph level, the more striking and significant rise in PRL levels which follows oestrogen treatment may be facilitated by this reduced dopaminergic activity. In addition to these effects on basal PRL levels, we have found that oestrogen priming over this short time period increased PRL sensitivity to the inhibitory action of dopamine. Furthermore, the degree of increase in the dopaminergic inhibition of PRL is directly related to the degree of rise in basal PRL levels induced by oestrogen.

This finding, which differs from the situation in rat studies, is supported by previous data from both primate and human studies. In ovariectomised stalk-sectioned rhesus monkeys, hyperprolactinaemia occurs because of the disruption of normal inhibitory hypothalamic–pituitary dopaminergic connections in these animals. Prolactin sensitivity to the inhibitory effect of infused dopamine is increased by oestrogen replacement therapy. Administration of ethinyl oestradiol to agonadal women increases the dopaminergic inhibition of PRL to the pattern of response seen on day two of normally cycling women (Yen, 1979).

In summary, these data provide evidence that in prolactinoma patients, oestrogens may have a dual action on the lactotroph: to stimulate synthesis and release of PRL and to increase the sensitivity of the lactotroph to the inhibitory actions of dopamine. It should be emphasised that the duration of this study was short. This may account for the considerable inter-individual variation in responses. It is possible that longer periods of oestrogen pretreatment may produce a greater uniformity of response and further increases in PRL sensitivity to the inhibitory action of dopamine. It should also be noted that although the decremental changes in PRL were increased during dopamine infusion following oestrogen priming, the absolute final values achieved were not lower than during dopamine infusion in the untreated state. Therefore any potential therapeutic developments must await the results of further long-term studies with both oestrogens and anti-oestrogens.

References

Franks, S. (1983). *Clin. Sci.* **65**, 757–62.
Helgason, S., Wieking, N., Carlstrom, K., Damber, M. G. and Von Schonetz, B. (1982). *J. Clin. Endocrinol. Metab.* **54**, 404–8.
Judd, S. J., Rakoff, J. S. and Yen, S. S. C. (1978). *J. Clin. Endocrinol. Metab.* **47**, 494–8.
Leong, D. A., Frawley, S. and Neill, J. D. (1983). *Ann. Rev. Physiol.* **45**, 109–27.
Yen, S. S. C. (1979). *Central Regulation of the Endocrine System*, pp. 387–416. Plenum, New York.

20

CU 32085 is a dopamine agonist *in vivo* and a dopamine antagonist *in vitro*

C. A. Edwards, G. Shewring, C. Dieguez, S. M. Foord, R. Hall, J. R. Peters and M. F. Scanlon

Department of Medicine, University of Wales College of Medicine, Heath Park, Cardiff CF4 4XN, UK

20.1. Introduction

CU 32085 is an 8α-amino ergoline that has been suggested to be a dopamine agonist on the basis of its ability to inhibit prolactin secretion in the rat (Fluckiger *et al.*, 1979). However, the kinetics of this inhibition are slower than those of the other dopamine agonists (Fluckiger *et al.*, 1979). In addition, it has also been reported that CU 32085 has a biphasic influence on striatal dopamine synthesis and turnover *in vivo* in the rat (Enz, 1981). This observation also argues against CU 32085 being a direct dopamine agonist. In this paper we report that CU 32085, similar to other dopamine agonists, binds to D-2 receptors in particulate preparations of bovine anterior pituitary cells, but that its ability to counteract the inhibitory effect of dopamine on prolactin (**PRL**) secretion *in vitro* is characteristic of a dopamine antagonist.

20.2. Materials and methods

Three-day monolayer cultures of adult male Wistar rat anterior pituitary cells were prepared by a conventional methodology as previously described (Foord *et al.*, 1983). The culture medium was α-modified minimum essential medium supplemented with 10% fetal calf serum and antibiotics (all from Gibco, Paisley, Scotland). Functional experiments were performed over a 3-h period and rat prolactin was measured using NIADDK materials.

Particulate preparations of bovine anterior pituitary membranes were prepared by conventional homogenisation/high-speed centrifugation techniques. [^{3}H]Dihydroergocriptine (1–2 nM) was incubated with anterior pituitary membranes ($+/-$ competing drug) for 60 min at room temperature before filtration over glass fibre filters under vacuum.

20.3. **Results and discussion**

Figure 20.1. shows the effect of CU 32085 on prolactin secretion from monolayer cultures of rat anterior pituitary cells in the presence or absence of dopamine. Although CU 32085 was able to inhibit prolactin secretion, its maximum effect was not more than 40% and its ED_{50} was approximately 500 nM. Dopamine is recognised as being a relatively low affinity agonist at its own receptor but it inhibits PRL secretion by 55% with an ED_{50} of 10 nM in the same system (Foord *et al.*, 1983). When CU 32085 is incubated with 10^{-7}, 10^{-6} or 10^{-5} M dopamine and anterior pituitary cells, it is able to prevent the dopaminergic inhibition of prolactin secretion (Fig. 20.1.) with an ED_{50} of approximately 500 nM against 10^{-7} M dopamine. Given that 10^{-7} M dopamine is a maximally effective dose and that CU 32085 has some ability to inhibit PRL secretion by itself, it is clear the CU 32085 has appreciable dopaminergic antagonist activity.

Figure 20.2. illustrates competition curves for [³H]dihydroergocriptine binding to bovine anterior pituitary membranes. The top panel shows monophasic displacement curves (typical of dopamine antagonists), the middle panel biphasic displacement curves (typical of dopamine agonists) and the lower panel control displacement curves obtained with non-dopaminergic compounds. It can be seen that CU 32085 competes for [³H]dihydroergocriptine with high affinity and with a biphasic profile typical of a dopamine receptor agonist in this system. It should also be noted that the other compound that behaves anomalously in this series is the ergot derivative bromocriptine, which is an extremely potent inhibitor of prolactin secretion *in vitro* and *in vivo* yet has a monophasic antagonist-

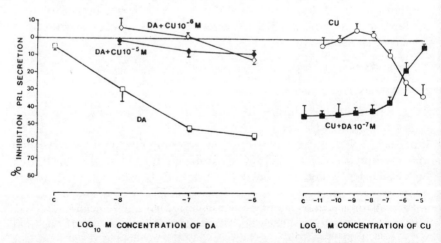

Fig. 20.1. The effect of CU 32085 on the secretion of prolactin by rat anterior pituitary cells when administered with dopamine (10^{-7}, 10^{-6} and 10^{-5} M) or alone. The data are the means of three individual experiments.

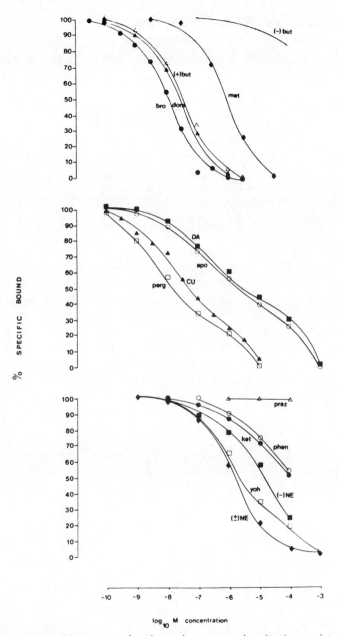

Fig. 20.2. Competition curves for dopamine antagonists (top), agonists (middle) and non-dopaminergic compounds (bottom) for [³H]dihydroergokryptine binding to bovine anterior pituitary membranes. but = Butaclamol; met = metoclopramide; dom = domperidone; bro = bromocriptine; apo = apomorphine; perg = pergolide; praz = prazosin; phen = phentolamine; ket = ketanserin; NE = norepinephrine; yoh = yohimbine. *n* = at least 5 for each curve.

type binding profile. The anomalous behaviour of ergopeptine drugs in dopamine binding assays has been reported previously (Sibley and Creese, 1983) but the converse behaviour of an ergoline-derivative, dopamine-antagonist activity with an agonist-type binding profile is a novel observation. CU 32085 has an unusual pharmacology since it appears to show mixed agonist/antagonist activity acutely *in vitro* and *in vivo* but agonist activity after chronic *in vivo* administration.

References

Enz, A. (1981). *Life Sci.* **29**, 2227.
Fluckiger, E., Briner, U., Burki, H. R., Marbach, P., Wagner, H. R. and Doepfner, W. (1979). *Experientia* **35**, 1677.
Foord, S. M., Peters, J. R., Dieguez, C., Scanlon, M. F. and Hall, R. (1983). *Endocrinology* **112**, 1567.
Sibley, D. R. and Creese, I. (1983). *Mol. Pharmacol.* **23**, 585.

21

The influence of dopamine neurones in the zona incerta on sexual behaviour in the rat

Jeremy P. Grierson and Catherine A. Wilson

Department of Obstetrics & Gynaecology, St. George's Hospital Medical School, Cranmer Terrace, London SW17 ORE, UK

21.1. Introduction

The similarity of the steroidal control of gonadotrophin release and sexual behaviour in the female rat has led to the hypothesis that a single hypothalamic neural site exists that integrates these two reproductive functions (Foreman and Moss, 1979).

Dopamine has a modulatory effect on both functions, although there is confusion in the literature as to whether it is stimulatory or inhibitory (Wilson, 1979; Crowley and Zemplan, 1981). There are three main dopaminergic tracts in the CNS, and two of these innervate the hypothalamus: the tubero-infundibular tract (originating in the arcuate nucleus (A12) and terminating in the median eminence) and the incerto-hypothalamic tract (with cell bodies in the thalamus (A11), zona incerta (A13) and the periventricular nucleus (A14) and with nerve terminals in the zona incerta (ZI), anterior hypothalamus and preoptic area).

There is substantial evidence that the preoptic area and the anterior hypothalamus are concerned with reproductive functions and as such implicate the incerto-hypothalamic tract and the ZI as part of the same control mechanism. Recently we have shown that the medial portion of the ZI has a stimulatory role in the control of gonadotrophin release (MacKenzie et al., 1984) and now we are investigating its role in the control of sexual behaviour.

21.2. Methods

21.2.1. *Experiment 1*

Female Wistar rats (225–275 g, Bantin and Kingman) were ovariectomised 3 weeks before the start of the experiment and housed in pairs under reverse lighting (lights off 11.00–21.00). After 2 weeks a unilateral cannula (22 gauge,

Clark Electromedical Instruments, Pangbourne) was stereotaxically positioned in the ZI (coordinates: A5·4, lat. 0·3, vent. 7·5, DeGroot atlas). A steroid priming of 0·25 μg oestradiol benzoate (OB) daily, for 3 days, followed by 0·2 mg progesterone on the fourth day, 4 h before testing, was employed. This regime allows multiple testing in one afternoon without significant changes in sensitivity. Receptivity was measured by calculating the lordosis quotient (LQ): the number of lordotic reflexes divided by the number of mounts (in this case 15), as a percentage. From the results of the preinjection test the animals were divided into one of two groups depending on whether they exhibited an LQ of 50 % or over, receptive animals; or under 50 %, non-receptive animals. In this way potential stimulatory and inhibitory effects could be observed. Straight after the first test, 2 or 4 μg dopamine per rat was infused into the ZI in 2 min, using a micrometer-controlled syringe with a flexible tube attached to a 28 gauge internal cannula. Dopamine was dissolved in 0·4 μl of 5 % glucose/saline solution. This vehicle alone was given to 12 animals. Another group of rats received either 0·4 μl saline or 2 μl sulpiride (L.E.D.C, Paris) dissolved in 0·4 μl saline. Receptivity testing was repeated 30 min after the injection.

21.2.2. Experiment 2

Female Wistar rats, ovariectomised for 3 weeks and kept in reverse lighting, were primed with 10 μg OB and their receptivity level was tested 48 h later. From these results the animals were divided into receptive and non-receptive groups. Lesions were then made bilaterally in the ZI using a monopolar electrode (Clark Electromedical Instruments) and passing a current of 0·25 mA for 30 s using a Grass Lesionmaker. The operations were carried out under sodium pentobarbitol anaesthetic. One week later these animals were primed again with the same dose of steroid and the behavioural test was repeated.

At the end of both experiments a histological examination of the brain, using thionin-stained sections, was made to check the position of the cannula or lesions.

21.3. Results

21.3.1. Experiment 1

Dopamine infused into the ZI of non-receptive rats had a slight stimulatory effect on sexual behaviour, significant at $P < 0.05$ (Table 21.1). There was no difference in response to the two doses, 2 and 4 μg. Sulpiride had no effect on receptivity. Dopamine, at the two concentrations used, had no effect on the behaviour of receptive animals, whereas sulpiride, a D-2 antagonist, significantly inhibited receptivity (Table 21.2).

Table 21.1.   The effect of dopamine (DA) in the ZI on sexual behaviour in non-receptive female rats.

Treatment	Number of rats showing increased receptivity	Mean LQ, % (range in brackets)	
		Preinjection	Postinjection
0·4 μl glucose/saline	4/12	12 (0–40)	17 (0–70)
2–4 μg DA in ZI	8/10	13 (0–40)	29* (10–80)
2–4 μg DA outside ZI	5/14	14 (0–40)	18 (0–90)

*P<0·05.

Table 21.2.   The effects of dopamine (DA) in the ZI on sexual behaviour in receptive female rats.

Treatment	Number of rats showing decreased receptivity	Mean LQ, % (range in brackets)	
		Preinjection	Postinjection
0·4 μl glucose/saline	1/4	100	98 (90–100)
2–4 μg DA in ZI	2/5	84 (60–100)	88 (70–100)
2 μg sulpiride	6/6	83 (55–100)	53* (15–80)

* P<0·05.

21.3.2. Experiment 2

Lesioning the ZI significantly enhanced lordosis in the non-receptive rat, whereas sham lesions and lesions outside the ZI had little or no effect (Table 21.3). The lesions in these animals were approximately 1 mm in diameter. Lesions in the ZI had no significant effects in receptive animals (Table 21.4).

Table 21.3.   The effect of lesions in the ZI on sexual behaviour of non-receptive female rats.

Treatment	Number of rats showing increased receptivity	Mean LQ, % (range in brackets)	
		Preinjection	Postinjection
Sham lesions	0/3	18 (0–45)	13 (0–40)
Bilateral lesions in ZI	8/8	3 (0–25)	48* (5–100)
Bilateral lesions outside ZI	2/5	10 (0–45)	34 (0–85)

* P<0·01.

Table 21.4.   The effect of lesions in the ZI on sexual behaviour in receptive female rats.

Treatment	Number of rats showing decreased receptivity	Mean LQ, % (range in brackets)	
		Preinjection	Postinjection
Sham lesions	5/7	91 (70–100)	71 (60–95)
Bilateral lesions in ZI	4/8	86 (65–100)	77 (10–100)
Bilateral lesions	6/8	88 (75–100)	75 (50–100)

21.4. Discussion

From these results it would appear that dopamine, injected into the ZI, has a limited facilitatory effect on receptive behaviour in the female rat. This supports evidence produced by Foreman and Moss (1979), who demonstrated the stimulatory effects of dopamine and apomorphine when injected into the preoptic area and the ventromedial hypothalamus. The behavioural changes these workers reported were greater than those presented here and required a dopamine dose of 200 ng per rat (a dose which had no effect on our animals). The effect of sulpiride injected into the ZI was consistent with the above results, i.e. a dopamine antagonist reduces receptivity in the receptive rat. Therefore the data suggest that reduced dopaminergic transmission in the incerto-hypothalamic tract inhibits sexual behaviour and increased transmission stimulates it, with perhaps lesser role for the zona incerta.

It is a paradox that electrolytic lesions in the same area also stimulate receptivity; according to our theory above, if the dopamine-containing cell bodies and nerve terminals are destroyed, behaviour should be inhibited. One suggestion is that the lesions may be ablating the GABAergic cells, known to be present in the ZI (Oertel et al, 1982), more effectively than the dopamine-containing cells. We propose that the GABA neurones have a tonic inhibitory effect on dopaminergic transmission and, therefore, sexual behaviour. Dopamine neurones from the thalamus exert an inhibitory effect on the GABAergic tract, resulting in the stimulation of behaviour. This hypothesis is currently under investigation.

References

Crowley, W. R. and Zemplan, F. P. (1981). In *Neuroendocrinology of Reproduction*. N. T. Adler (ed.), pp. 451–84. Plenum, New York.

Foreman, M. M. and Moss, R. L. (1979). *Physiol. Behav.* **22**, 283–9.

MacKenzie, F. J., Hunter, A. J., Daly, C. and Wilson, C. A. (1984). *Neuroendocrinology* **39**, 289–95.

Oertel, W. H., Tappaz, M. L., Berod, A. and Mugnaini, E. (1982). *Brain Res. Bull.* **9,** 463–74.
Wilson, C. A. (1979). In *Hypothalamic Neurotransmitters and Gonadotrophin Release,* vol. 1. C. A. Finn (ed.), pp. 383–473. Oxford Sci. Publ., Oxford.

Part III
Schizophrenia

22

Clinical aspects of schizophrenia

R. H. S. Mindham
Department of Psychiatry, University of Leeds, 15 Hyde Terrace,
Leeds LS2 9LT, UK

22.1. Introduction

In a section devoted to biological research in schizophrenia, it is appropriate to give some consideration to clinical aspects of the disorder. I shall deal with three major issues: the nature of schizophrenia and its identification, the measurement of symptoms, and the progress which has been made in its management. Possibly the most important matter is the nature of schizophrenia; is it a disease, or a disorder, or a reaction? Is it a single disorder or a common manifestation of several distinct physical changes? Can it be reliably recognised, and do clinicians agree on the criteria to be applied? Perhaps most important in the context of this book, has it a tangible identity, such that it might be studied without the results losing meaning? Even though the precise nature of schizophrenia is unknown, it is still possible to study the severity of symptoms, to examine the effects of treatment, and to assess the outcome of the disorder long term. Methods have been developed for the measurement of severity of symptoms in schizophrenia. From a clinical viewpoint, treatment and management are of paramount importance. Although great advances have been made in these fields, how satisfactory are they, what effects have they had, and what further developments are needed?

22.2. The nature of schizophrenia and its recognition

The recognition that groups of symptoms which regularly occur together may be manifestations of a common underlying cause we owe to Sydenham, the seventeenth-century physician (McHugh and Slavney, 1983). He realised that illnesses with similar clinical features often had the same underlying cause. This view contrasted with earlier approaches, originating from Hippocrates, which emphasised the unique nature of illness in the individual patient rather than drawing attention to the similarities between illnesses in different subjects. Although Sydenham's thinking was an important conceptual advance, its application led only to the gradual recognition of diseases and the processes that underlie them. A great variety of 'syndromes' came to be recognised, and gradually a coherent body of knowledge developed of the causes, symptoms, course and outcome of many diseases. Knowledge of some disorders took several generations to accrue.

The application of Sydenham's ideas to psychiatric disorders has been slower than in physical disease. General paralysis of the insane is the psychiatric disorder which has most closely followed the model established in medicine generally. This condition was recognised in 1822 by Calmeil and Bayle independently (Ackernecht, 1968), and by the turn of the century symptoms had been linked with a specific neuropathology; but it was not until the discovery and introduction of penicillin in the 1930s and 1940s, over a hundred years after the disease was first recognised, that a truly effective and widely applicable treatment was developed. No other psychiatric illnesses have become so thoroughly understood or so effectively treated.

The syndromes known collectively as schizophrenia have not travelled far along this path. Individual syndromes were described in the second half of the nineteenth century; Morel in 1856 described *démence précoce*; Kahlbaum in 1868 described '*katatonie*'; Hecker in 1870 described '*hébephrénie*', but it was Kraepelin who in 1896 brought these disorders and *dementia paranoides* together in a single group which he called *dementia praecox*. These disorders had the common characteristics of following a chronic or relapsing course with deterioration; in contrast to manic depressive insanity in which there was full recovery after each episode. In 1911 Bleuler coined the name '*schizophrenia*' and wrote of dementia praecox as 'the group of schizophrenias' (Bleuler, 1911). Since that time, in spite of research in many areas, no definite pathology for schizophrenia has been established, and the integrity of the syndrome has depended upon the consistency of its clinical features. Interestingly, although the symptomatology of schizophrenia may vary in its content, the form of the syndrome is the same all over the world. Schizophrenia is widespread, and probably occurs in similar proportions in all populations, whatever their cultural or racial origins.

The uncertainty surrounding the nature of schizophrenia has important implications for research. Is schizophrenia a single disorder, or a group of disorders with similar clinical features? In the case of the toxic confusional states, essentially the same clinical picture can be produced by a remarkable variety of pathological causes (Wolff and Curran, 1935). Can the disorder be recognised with sufficient certainty and consistency, for the purposes of both clinical work and research? Research has shown clearly that psychiatrists with different backgrounds and training, and working in different countries, use the term in ways which are different in very important respects. The US–UK diagnostic project, for example, showed that the disorder was viewed more narrowly in the United Kingdom, and that psychiatrists in the USA, with a broader concept of the disorder, included subjects who would have been regarded as suffering from manic-depressive psychosis or personality disorder in the United Kingdom (Cooper *et al.*, 1972). The differences found between the UK and the USA clearly require the results of studies from these two countries to be interpreted in different terms. More often, insufficient information is available to allow such a judgement to be made. The subtypes of schizophrenia, paranoid, simple, hebephrenic and catatonic, are no more readily or reliably recognised and individual cases need not have a single common feature.

Originally the disorder was defined in terms of certain psycho-pathological features. This approach stems largely from Bleuler, who attempted to define the features central to the disorder and shared by all patients; these he defined as: loosening of associations; blunting or incongruity of affect; autism or self-absorption; and ambivalence. Although these features ring true for the clinician, they do not define the condition sufficiently clearly for the purposes of clinical work, let alone research.

The diagnostic criteria suggested by Schneider (1959) are more precise. These are as follows: audible thoughts; voices arguing; voices commenting on subject's actions; influences playing on the body, somatic passivity (includes hallucinations); thought withdrawal; diffusion or broadcasting of thought; made feelings; made impulses; made volitional acts; delusional perception. Schneider chose certain common features of the changed mental state in schizophrenic illness which are more readily defined and recognised. These are not necessarily the most conspicuous, disabling or otherwise important symptoms, but those which have the greatest utility in dianosis. These criteria have been a great advance in recognising schizophrenia, even though there are weaknesses here too. None of the criteria is specific to the disorder; all may occur in coarse brain disease and in some other disorders; some schizophrenics have none of the first-rank symptoms; and, most important of all, diagnosis is based upon examination of the mental state, possibly on a single occasion when symptoms may not be present by chance, and does not consider the long-term progression and character of the disorder.

Other attempts have been made to improve the reliability and validity of the methods by which schizophrenia could be recognised. The development of international classification dates back to the introduction of the International Classification of Diseases for Psychiatric Disorders in 1889 (Kendell, 1975). This kind of procedure gives rise to great problems of acceptability. Only in the period after the Second World War did the

International Classification of Diseases become at all widely recognised and used in mental health. Following a commissioned report by Professor Erwin Stengel on the reasons for the lack of success of the psychiatric section of the ICD (Stengel, 1960), a glossary of terms was added to the classification, greatly increasing the ease with which it could be used. In spite of this, the ICD for Psychiatric Disorders has had much greater utility in epidemiological research than in the identification and study of individual cases. A similar approach in the United States, the Diagnostic and Statistical Manual of the American Psychiatric Association (1980), although more generally acceptable to clinicians, presents many conceptual problems and by no means meets the needs of biological research (Klerman, 1984; Vaillant, 1984).

The development of the present state examination (PSE) (Wing *et al.*, 1974), is based on clinical examination utilising criteria derived from those of Schneider; this approach led to an improvement in the reliability in which information is gathered but has the same weaknesses as Schneider's first rank criteria themselves. The data from the PSE can be processed using the CATEGO computer program to provide diagnoses, with advantages in consistency.

Table 22.1.   Feighner's criteria for diagnosis of schizophrenia.

For a diagnosis of schizophrenia, A through C are required

A. *Both of the following are necessary:*
   1. Chronic illness with at least six months of symptoms prior to the index evaluation without return to the premorbid level of psychosocial adjustment.
   2. Absence of a period of depressive or manic symptoms sufficient to qualify for affective disorder or probable affective disorder.

B. *The patient must have at least one of the following:*
   1. Delusions or hallucinations without significantly perplexity or disorientation associated with them.
   2. Verbal production that makes communication difficult because of a lack of logical or understandable organisation. (In the presence of muteness the diagnostic decision must be deferred.)

(We recognise that many patients with schizophrenia have a characteristic blunted or inappropriate affect; however, when it occurs in mild form, inter-rater agreement is difficult to achieve. We believe that on the basis of presently available information, blunted affect occurs rarely or not at all in the absence of B-1 or B-2.)

C.   *At least three of the following manifestations must be present for a diagnosis of 'definite' schizophrenia, two for a diagnosis of 'probable' schizophrenia:*
   1. Single.
   2. Poor premorbid social adjustment or work history.
   3. Family history of schizophrenia.
   4. Absence of alcoholism or drug abuse within one year of onset of psychosis.
   5. Onset of illness prior to age 40.

The weakness of all these approaches has led to the widespread use of operational criteria in psychiatric research. These criteria have no theoretical basis or heuristic value, but simply define criteria which may lead to the reliable recognition of psychiatric disorders. Widely used examples include those developed by Spitzer *et al.* (1974) and by Feighner *et al.* (1972) (Table 22.1). These methods use mixed features of the disorder to aid its recognition, and include items concerned with age of onset, clinical features, cause, disability, family and personal history, etc. This approach has proved to be of great value in research in that reliability is improved and communication between research workers is promoted, but has obvious limitations in furthering knowledge of the conditions themselves. No laboratory methods for recognising the disorder are available.

The very need to resort to such methods for identifying patients starkly illustrates the extent of our ignorance of the disorder known as schizophrenia. At best the present status of the syndrome is provisional. Only future work will show whether the condition is one disorder or several, whether there are nuclear features, or indeed whether the concept is valid at all. Any disscusions later in the book are qualified by the weaknesses of our methods of recognising the disorder in clinical practice, however precise the scientific procedures themselves are.

22.3. Measuring the severity of symptoms

For the purposes of research it is necessary not only to recognise disorders accurately and reliably, but also to measure their severity. This is particularly necessary for the evaluation of methods of treatment.

My predecessor in Leeds, Emeritus Professor Hamilton, was a pioneer in the development of scales for rating the severity of symptoms and signs in depressive disorders. Early work in this field began 25 years ago and has been both extensively developed and used. Similar, but probably less successful, developments have occurred in relation to schizophrenic disorders. As in the case of depressive disorders, these have included methods of measuring the severity of physical symptoms such as sleep, appetite, energy and activity, and of measuring more strictly psycho-pathological symptoms, such as changes in mood, thought content, perception, formal though disorder, and cognition. In addition, ratings of behaviour have been made including items on sociability, activity, relationships with friends, relatives and staff, output in occupational therapies and so on, and these have often been based on observations made by nursing staff, occupational therapists and others.

It is apparent that procedures of this kind present serious methodological problems. To what extent is a mentally ill subject able to cooperate in such tasks; what status can be given to a rating of a patient's subjective experiences which cannot be further evaluated more objectively; how can imprecise concepts be sufficiently clearly defined for reliable reporting by observers? These are some of the major problems of the practice of clinical psychiatry itself, brought into greater prominence by the requirements of research.

The problems do not end there. Where data have been collected, how can the various measures be drawn together, compared and weighted for importance? For example, how many hallucinations are equivalent to a delusional belief, or passivity experiences to incongruence of mood, and how do these relate to overactivity? In practice, ratings are simply added together, and scores between and within subjects are compared as indicators of severity and of change. The shortcomings of such procedures are obvious, but what else can be put in their place in the present stage of knowledge?

Ratings of symptoms have been extensively used in psychiatric research, and especially in the evaluation of treatments. Similar procedures have been employed in the measurement of unwanted effects of treatments, in particular the drug-induced movement disorders, and their management, with comparable problems (Mindham, 1976).

The physiological characteristics of subjects with schizophrenia have been extensively investigated (Venables and Wing, 1962) but the findings are not specific to schizophrenia and have no general application in diagnosis or in measurement of the severity and progression of the disorder at this current stage of their development.

22.4. Management of schizophrenia

Although a great deal of effort and much money have been spent on the management and care of patients suffering from schizophrenia for two centuries and more, effective treatment has only been available during the last thirty years. As the disease affects about 1 % of the entire population of the world, and in a substantial proportion of those affected the illness is severe and its effects prolonged, its moribidity is high (World Health Organisation, 1973).

The need to care for these large numbers of severely disabled people was a major factor in the development of asylums for the insane in the last century. In the United Kingdom, provision was minimal before the nineteenth cedntury, but following the Lunacy Act of 1808, several county asylums were built, and following the Act of 1845, local authorities were compelled to care for the insane. Large hospitals, often distant from centres of population, sprang up all over the United Kingdom. Similar patterns of provision have been seen in Europe and in the USA.

These hospitals offered only care and containment of the patient. Specific treatments were quite unkown and many methods were employed for no good reason and to little effect. Morphia and paraldehyde were used to sedate violent patients and the solitary confinement of patients in secure wards and rooms was often necessary, for the protection both of patients themselves and of others. Matters somewhat improved when the barbiturates were introduced in the 1920s, but the control of violence was then achieved at the expense of inducing heavy sedation or even sleep. In the 1930s a period of 'shock treatments' followed, with drug-induced convulsions, electro-convulsive therapy and insulin coma therapy being among the more popular methods. The introduction of prefrontal lobotomy by Moniz (1937) led to its widespread use in long-standing schizophrenia, especially where there were intractable behavioural problems. The treatment of patients suffering from psychotic illness, with no prospect of cure, and probably showing violence and other forms of socially unacceptable behaviour, had led to a concentration on security and containment and for the occupation and care of patients rather than on their rehabilitation. The mental hospitals became villages with their own cultures, rules and problems, and have been vividly described by Goffman (1961).

It was only the introduction of the major tranquillisers or neuroleptics in the early 1950s which changed this most unsatisfactory situation. The effects of both the phenothiazine, chlorpromazine, and the extract of *Rauwolfia serpentina*, reserpine, were startling; over a period of 3 to 8 weeks, many patients showed amelioration in many of their symptoms which included hallucinations, delusions, thought disorder and disturbances of motility and volition. It was not long before the phenothiazines were in widespread use and were shown to be effective in a wide range of schizophrenic symptoms in controlled studies (National Institute of Mental Health, 1964). At a later date, they were also proved to be effective in preventing recurrence of symptoms (Leff and Wing, 1971; Hirsch *et al.*, 1973).

The effects on institutions have been slower but no less dramatic. From a peak of population in the late 1950s, the mental hospitals have steadily shown a reduction in the number of patients resident there; as long-term patients have died, they have not been replaced (Fig. 22.1). Other changes have been just as striking; locked wards are retained only for the most disturbed patients (at a local mental hospital 20 beds out of 1020). 'Seclusion' of patients is uncommon and occurs only in restricted parts of hospitals rather than in every ward. Violence is a rarity. But possibly most important of all, patients have become accessible to other methods of treatment, of which the most important is rehabilitation for life *outside* the institution. One of the costs of these advances has been the occurrence of movement disorders induced by the drugs used. These include drug-induced parkinsonism, dystonias, akathisia, and, in the longer term, tardive dyskinesias (Ayd, 1961; Marsden *et al.*, 1984)—see also chapters 26, 35 and 36.

Modern management consists of the use of drugs in controlling acute and chronic symptoms, combined with supportive measures carried out in day hospitals, out-patient clinics, hostels, and sheltered workshops. Social

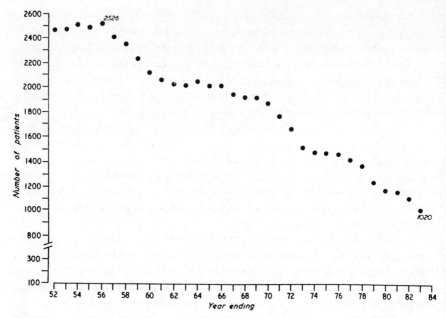

Fig. 22.1.   Numbers of patients resident on 31 December 1952–1984 at High Royds Hospital, West Yorkshire, UK.

measures play an important part in management and many argue that these have been as important as drugs in bringing about improvement in treatment. However, it can be argued, and I would certainly do so, that the introduction of neuroleptic drugs has been the essential factor which has changed the course of events and allowed other treatments and methods of management to be used effectively.

In the early 1960s it seemed that schizophrenia might be cured, but that has not proved to be the case. Neuroleptic drugs are capable of suppressing acute symptoms of schizophrenia in a large proportion of patients. Some of these patients recover and never experience a further attack. The majority, however, have further discrete episodes from which recovery is rarely complete, or develop a chronic illness in which the psychotic process continues. These patients require continued medication, either to prevent relapse or to keep symptoms in check, and a proportion of patients continue to have symptoms in spite of a large dosage of medication. Furthermore, many patients show symptoms such as a blunting of affect, lack of initiative and drive, and emotional incontinence or incongruity, which do not respond well to medication at any stage of treatment. Such clinical observations have led Crow and his colleagues to postulate the existence of two varieties of schizophrenia, one with 'positive' florid symptoms, such as hallucinations, delusions and thought disorder, which respond to drug treatment and are not associated with deficits of cognitive function in the long term; and another with 'negative' symptoms, such as social withdrawal, emotional blunting, and lack of initiative, associated with cognitive deterioration in the longer term, and a demonstrable shrinking of the brain on computer-assisted tomography.

Many patients who show acceptable control of symptoms are by no means normal: they may show important unwanted effects of medication; they may lack initiative, interest and drive; they may be socially isolated and unemployable; drug treatment and social measures may allow them to live outside hospitals but not to lead normal domestic lives. Thus the dramatic advances in treatment brought about by drugs are by no means completely satisfactory. In spite of early optimism, patients requiring long-term institutional care continue to accumulate, although at a much lower rate than previously (Mann and Cree, 1976).

We are still in need of further treatments for schizophrenic patients. We require drugs with a wider range of effects; drugs free of serious unwanted effects; and drugs which will prevent the long-term deterioration seen in some patients. In spite of the wide variety of chemical substances used in the treatment of schizophrenia, all appear to exert their effect by blocking the action of dopaminergic neurotransmission (Snyder *et al.*, 1974). It seems likely that drugs with other pharmacological effects are required to make a further advance in treatment. Where should we look? It is indeed sobering to review the discovery of the neuroleptic drugs and their introduction to clinical practice in the treatment of schizophrenia and to recognise that they were discovered by chance with little contribution from fundamental research. Such is the state of our ignorance of psychiatric disorders that chance discovery may still provide the best opportunity of new treatments in the field.

22.5. Summary

In the knowledge that much of this book is devoted to the functioning of particular groups of cells in particular parts of the brain, I have deliberately drawn attention to what I see as important clinical aspects of schizophrenia affecting both individual patients and groups of patients. I have emphasised the uncertainty as to the nature and status of schizophrenia; the difficulties in recognising it and the relevance this has to the interpretation of neurobiological information; and the methodological problems in measurement of symptoms of the illness. In the area of management of the disorder I have sketched the development of treatments for schizophrenia, the remarkable impact of the neuroleptic drugs, and the problems and deficiencies in the treatments at present available. I hope that I have demonstrated the enormous gap that exists between the clinician and the neurobiologist, and that I have underlined the vital necessity of bridging it if research and its application to clinical practice are to be properly conducted.

References

Ackernecht, E. H. (1968). *A Short History of Psychiatry*. Transl. S. Wolff. Hafner Publishing Company, New York.

American Psychiatric Association (1980). *Diagnostic and Statistical Manual of Mental Disorders* (3rd edition) DSM III. The American Psychiatric Association, Washington DC.

Ayd, F. J. (1961). A survey of drug-induced extrapyramidal reactions. *J. Amer. Med. Ass.* **175**, 1054.

Bleuler, E. (1911). Dementia praecox or the group of schizophrenics. Transl. J. Zinkin, 1950. International University Press, New York.

Cooper, J. E., Kendell, R. E., Gurland, B. J., Sharpe, L., Copeland, J. R. M. and Simon, R. (1972). *Psychiatric Diagnosis in New York and London*. Maudsley Monograph No. 20. Oxford University Press, London.

Crow, T. J. (1983). Schizophrenic deterioration. *Brit. J. Psychiat.* **143**, 80.

Feighner, J. P., Robins, E., Guze, S. B., Woodruff, R. A., Winokur, G. and Munoz, R. (1972). Diagnostic criteria for use in psychiatric research. *Arch. Gen. Psychiat.* **26**, 57.

Goffman, E. (1961). *Asylums: Essays on the Social Situations of Mental Patients and Other Inmates*. Penguin Books, Harmondsworth.

Hirsch, S. R., Gaind, R., Rohde, P. D., Stevens, B. C. and Wing, J. K. (1973). Outpatient maintenance of chronic schizophrenic patients with long-acting fluphenazine: double blind placebo trial. *Brit. Med. J.* **1**, 633.

Kendell, R. E. (1975). *The Role of Diagnosis in Psychiatry*. Blackwell Scientific Publications, Oxford.

Klerman, G. L. (1984). A debate on DSM III. *Amer. J. Psychiat.* **141**, 539.

Leff, J. P. and Wing, J. K. (1971). Trial of maintenance therapy in schizophrenia. *Brit. Med. J.* **3**, 599.

McHugh, P. R. and Slavney, P. R. (1983). *The Perspectives of Psychiatry*. Johns Hopkins University Press, Baltimore.

Mann, S. and Cree, W. (1976). New long-stay psychiatric patients: a national sample of 15 mental hospitals in England and Wales. *Psychol. Med.* **6**, 603.

Marsden, C. D., Mindham, R. H. S. and Mackay, A. V. P. (1986). Drug induced extrapyramidal syndromes. In *The Psychopharmacology of Schizophrenia*. S. R. Hirsch and P. Bradley. Oxford University Press, Oxford.

Mindham, R. H. S. (1976). The assessment of drug-induced extrapyramidal reactions and of drugs given for their control. *Brit. J. Clin. Pharmacol.* **3**, 345.

Moniz, E. (1937). Prefrontal leucotomy in the treatment of mental disorders. *Amer. J. Psychiat.* **93**, 1379.

National Institute of Mental Health Psychopharmacology Service Center Collaborative Study Group (1964). Phenothiazine treatment in acute schizophrenia. *Arch. Gen.Psychiat.* **10**, 246.

Schneider, K. (1959). *Clinical Psychopathology*. Transl. M. W. Hamilton. Grune and Stratton, London and New York.

Snyder, S. H., Bannerjee, S. P., Yamamura, H. I. and Greenberg, D. (1974). Drugs, neurotransmitters and schizophrenia. *Science* **184**, 1234.

Spitzer, R. L., Endicott, J. and Robins, E. (1974). Research diagnostic criteria. Biometrics Research, New York State Department of Mental Hygiene, New York.

Stengel, E. (1960). Classification of mental disorders. *Bull. World Health Organisation* **21**, 601.

Vaillant, G. E. (1984). The disadvantages of DSM III outweigh its advantages. *Amer. J. Psychiat.* **141**, 542.

Venables, P. H. and Wing, J. K. (1962). Level of arousal and the sub-classification of schizophrenia. *Arch. Gen. Psychiat.* **7**, 114.

Wing, J. K., Cooper, J. E. and Sartorius, N. (1974). *Measurement and Classification of Psychiatric Symptoms. An Instruction Manual for the PSE and Catego Programme.* Cambridge University Press, London.

Wolff, H. G. and Curran, D. (1935). Nature of delirium and allied states: the dysergasic reaction. *Arch. Neurol. Psychiat.* **33**, 1175.

World Health Organisation (1973). *The International Pilot Study of Schizophrenia.* World Health Organisation, Geneva.

World Health Organisation (1979). *Mental disorders: glossary and guide to their classification in accordance with the ninth revision of the International Classification of Diseases.* World Health Organisation, Geneva.

23

Some philosophical reservations on the neurobiology of schizophrenia

F. A. Jenner and S. S. Johl

University of Sheffield, Royal Hallamshire Hospital, Glossop Road, Sheffield S10 2JF, UK

In the previous chapter, Professor Mindham has given very eloquently a historical and classical account of psychiatry and the problems its practitioners recognise in their attempts to delineate the syndrome of schizophrenia. One requires such information before one criticises it or tries to improve on it. From the biochemist's point of view, it must be realised that in any study of schizophrenia one relies on the work of clinicians. Without their categories there is no subject nor any word with a meaning. We feel and would argue that biochemists should study brains, not psychoses. If they do happen to contribute to the understanding of the latter, that is a great bonus. Successful science is, after all, that which gets good results, and exactly how to do that is what is so unpredictable.

Biochemists, though, should remember that the language of clinical psychiatry and that of cerebral chemistry are quite different. Translation from one to the other is across the really mysterious brain–mind barrier. To put the situation into its context and to point out the difficulties, one can hardly do better than quote Kraepelin (1904) (see Jaspers, 1963). Kraepelin was the most influential writer on the classification of psychiatric conditions in the whole history of the subject. Nevertheless, he was himself influenced by Antoine Bayle's (1822) great discovery of dementia paralytica, cerebral syphilis, the so-called general paralysis of the insane (GPI). Whether or not Kraepelin actually thought that Bayle had made his discovery by his clinical acumen we do not know. The fact is, though, that the advancing field of bacteriology of that period influenced the whole of medical science, and GPI patients were numerous in the mental hospitals. What could have been more reasonable than to assume that some analogous diseases were also present? What, however, could the psychiatrist, Kraepelin, do but try to delineate the other disease entities by the behaviour and speech of his patients? In a valiant attempt to do this, Kraepelin produced the concept of manic-depressive illness and dementia praecox, the latter to be subsequently renamed schizophrenia.

However, it seems important to note that Bayle's (1822) diagnosis was based on thickened meninges, a so-called arachnoiditis, and not on psychological symptoms. Further, Kraepelin's own studies of dementia paralytica showed that the diagnosis of general paralysis of the insane could not be made with any confidence by studying the patient's psychological symptoms. Indeed, the incidence in his clinic dropped from 30% to 8% when lumber punctures were introduced. Kraepelin openly acknowledged these striking factors.

There is no implication here that schizophrenic symptoms cannot be produced by cerebral changes; indeed Kraepelin's difficulties with GPI demonstrate the opposite, as do the symptoms often seen in, for example, systemic lupus erythematosis. The point being made is that, when we use these 'clues', we might be in danger of extrapolating in ways which might 'distort' psychiatric knowledge in general. The difficulty of diagnosing the nature of cerebral changes from functional psychological changes has to be emphasised.

In more recent years, clinical acumen has been combined with numerical taxonomies based on statistical technologies. Kendell (1975) is the most important and most assiduous applier of such approaches to clinical material. Unfortunately, such attempts have failed to demonstrate any natural articulation in nature, that is, clear lines of demarcation of any psychiatric disease entities, and, what is more, they are ill conceived and were bound to fail. Classification of things in the world cannot be objective; they are always for a purpose. Mathematics cannot give us purposes; if we have already differentiated things, discriminant function analysis can help us maintain most efficiently the differentiation we have made. Cluster analysis as a technique can obviously enable us to make the most efficient clusterings of what we have included and already weighted, and it will give us any number of clusters we request as long as it is less than N, the number of objects to be clustered. These problems and others in a range of techniques are alluded to in an elementary way by Dunn and Everitt (1982), and in a more thorough-going and difficult way by Sneath and Sokal (1973) and Jardine and Sibson (1971).

It is, of course, important for us to agree that one can use simple statistical techniques to assess who responds to various drugs. However, one has then to define 'responds', and that is going to be the nub of our argument. Certainly a statistically significant number of so-called schizo-phrenic patients respond to treatment with phenothiazines and other major tranquillisers, or, as some French writers put it, to *camisoles chemiques* (chemical straitjackets), as do so-called manic patients and many other people.

Here we are anxious to consider what, if anything, happens when we become very critical in this field. We want to encourage, but we also want to warn biochemists and other brain research workers of hidden aspects—or more accurately consequences—of what they are doing. Indeed, as ourselves apostates from the church of organic psychiatrists, we feel that being invited to contribute to this book is like a generous invitation to speak at an important diocesan meeting. Our object, then, should be to try to be helpful to liberal-minded colleagues.

In a sense this makes us very conscious that it can be argued that philosophy never contributed much to scientific progress, though that depends on what one understands by the word. Clear thinking is obviously required. If one asks what would have happened had Maxwell asked questions about the existence of magnetic lines of force, one must conclude that he might have achieved much less than he did. Indeed, asking fundamental questions about language and meaning probably impedes the development of science. On the other hand, no one quite knows what is and what is not to be asked, nor how happy we can be to have a science that limits what is to be asked. Where, in fact, is the border between fundamental science and more superficial technology? Aristotle viewed science as 'seeing and approaching horizons to see all manner of vistas beyond'. Technology was finding a means to achieve an end in mind, for example repairing a burst pipe.

The danger persists in studies of schizophrenia that the intention is to make the individual fit in with our outlook and behaviour, and to use whatever means we can find to do this. No one can necessarily argue that that is bad—indeed if you can't beat them, join them—but such a view would situate psychochemistry differently from a non-culturally relative science to something of a political science.

People without experience in clinical psychiatry, and many with much experience, will, however, argue that the schizophrenic is so bizarre in behaviour and thought that he cannot be understood in simple human terms; he must, therefore, be finally explained in terms of cerebral pathology. That view, which Karl Jaspers (1963) pioneered in psychiatry and acquired from the philosophy of Dilthey (1914) and the sociology of Max Weber (1922), permeated much of the history of clinical psychiatry. Certainly, indirectly basing psychiatry on philosophy and sociology, the biochemist must be warned that that aspect of the diagnosis depends on the imaginative understanding of the psychiatrist, hardly the concrete type of operational definition the biochemist himself depends on and looks for in his laboratory, but one on which he must depend if he orders supplies of schizophrenics' brains or urine. Perhaps we can then be forgiven for stating what we believe we have gradually learnt: the profound odds which some people will play for in life.

For some Indian people, to have broken bones is desirable in order to beg. We have recently been dealing with a woman who, almost immediately after a clear if belated confrontation with a son she had tried for too long to manipulate, ate her own faeces, smeared her walls and convinced expert opinion of her dementia. Many people kill themselves, even set fire to themselves; certainly many young girls starve themselves, etc., etc. Szasz (1971) reports that in the Middle Ages many confessed to being witches. Any statement confidently asserting what is beyond human understanding is peculiarly parochial and necessarily precarious. We have learnt to suspect that in fields so difficult to recheck, unlike synthetic chemistry, much that appears to be hard science is bogus for a whole spectrum of reasons, from Cyril Burts' writings to the unconscious selectivity of discarding some results of deludedly confident scientists. In psychoanalytical language, the spectrum ranges from hysteria, with, by definition, unconscious mechanisms, to fraud, with, by definition, the too conscious activities.

In its pure form, despite Freud's own materialist philosophy, psychoanalysis attempts to describe how one mental state arises from another. To some extent, discussions are independent of any information about the brain, which might as well be cotton wool. The struggle is to understand the other person in terms of his background and human context. Its danger is illustrated by Jerome Frank's remark that psychoanalysis is good because even if the patient doesn't get better, the doctor does know he is doing the right thing. One day the patient may change.

In the neurophysiology laboratory, cerebral activity can be discussed as though mind does not exist. Indeed the readings and other recordings are such that no mental process will or can be found. But even though the work is almost completely useless for clinical practice, those workers too know they are doing the right thing. One day we will really be able to explain things physically.

Karl Jaspers, although anti-Freudian in outlook, saw the psychiatrist as a circus artiste, standing with one leg on each of the galloping horses of understanding and explaining. This founding father of psychopathology did not envisage any significant rapprochement, and criticised those riders, who in the middle of the act, fell into one or other saddle. However, Jaspers thought one horse was adequate for the neurotic and a different one for the psychotic. He was less aware of more modern philosophies of science and their concern with fashions of scientific theories.

Much can often still be more easily explained by an old theory than by the new one. An important question involves when and why we change our minds. Indeed, what is the overwhelming evidence which textbooks report from history and extol, and experiments so often fail to produce? The struggle within science is to be an important scientist, but, because one is civilised and adult, to do so within the rules of the game. A goal is scored by overwhelming evidence but who or what does the evidence overwhelm? In this game, points are to be scored that are probably more like those in a boxing match than in, for example, rugby. Someone's judgement is colouring the scores even if the competitors struggle to put things beyond doubt, indeed to reach the knock-out.

To be accused of choosing truths or ways or languages for approaching them is also for the accuser to assert that one can do so. Of course, we would concede that a completely deterministic world is conceivable, one in which we are puppets and one in which, from Popper's point of view, it is undeniable in any experiment because the errors in observation are reasonable explanations of the failure to clinch matters absolutely.

Having become something of psychodynamicists and relativists rather than realists, we might be asked: why do you now avoid considering the organic basis of problems? We must answer that (a) we don't totally do that, but (b) we recognise that to tend to do so involves a decision. Recognition of that, however, also makes our case, namely that much of what we call knowledge is based on choices we have made about how to look at things, and a similar question can be put to colleagues. It is, as Popper would presumably say, ridiculous to deny the truths revealed by scientific method, a method probably based on accepting conjectures (guesses) which resist serious attempts at refutation (testing). The psychodynamicist nevertheless wants to look at the language in which the guessing and testing are embedded, and then to measure the resistance to denial of the Kuhnian (1970) paradigms or axioms that may be involved.

Surely there is some truth in declaring that many do believe 'it is all biochemical really', when much can be explained otherwise. In a sense those people may not be denying a value for sociology, psychology and physiology as well as biochemistry, and would indeed concede that biochemistry is all physics and chemistry really. There are levels which we work at for convenience, they might add, but reduction to more fundamental levels is a central aim of science (for some scientists). Nothing is really greater than theoretical science. Those people are at least the Aunt Sallies we have set up in order to knock them down. Whether that cap fits our readers or not, they must judge for themselves. The aunt we have in mind might not want to deny that schizophrenia, which she asserts is really due to an inherited cerebral enzymatic disorder, will present itself in ways influenced by the environment. Without the rights events it might never be noticed in the really schizophrenic person. Perhaps it would be added that defects in almost any complex machine—certainly a computer or other involved electronic gadget, for example a TV set on too long—will only become apparent under certain circumstances.

A defect in a computer, though, is something about the machine that makes it unsuitable for a purpose for which it was constructed. Human beings may be nothing more than computers, but that line of analogy makes one look for the purpose of our construction to define our deviance. On the other hand, we must notice that all computers will fail if they are misused or badly programmed; robustness might then be a measure we could develop, but the least robust may also be the best and most delicate instrument needing and rewarding more care and attention, as is true of chemical balances. Objective sciences cannot answer any question containing 'should': like the question of how much attention we should pay to robustness versus sensitivity towards conditions we might change.

But our interrogator may still ask: but what is really more fundamental, the hardware or the software of the computer? To answer his question, we always refer to rickets, Sheffield and vitamin D. The great discovery was made by Edward Mellanby who did not live in the fog with the children he returned to it, while at least half aware of what some would call the real problem, slum conditions, which caused rickets. What, then, was more fundamental; the sociology that explained the scientists who chose to study why children's bones bend in slums, or the biochemistry? What happens to ill-nourished children living with sunlight excluded for much of their lives would have been of little interest and a wicked experiment if no one had been willing, half-unconsciously, to do it to them. Even separating and classifying rickets and TB, tuberculous osteomyelitis would certainly have been less important if the social context had not made them important. The political question is always; what is or is not important to whom?

We do not know the purpose of life, but we do know something about the conditions in which some of us, indeed most, seem to like to breathe and live, even if, as Count Leo Tolstoy apparently wrote: there is nothing human beings will not tolerate as long as those around them have the same conditions.

The classical psychiatrist, another Aunt Sally who hardly exists, considers schizophrenia now as a condition diagnosable by certain specific symptoms. The essentially required symptoms include delusions, the social and personal origins of which the psychiatrist cannot understand; reports of influences on the mind from outside forces as well as lack of privacy of one's own thoughts, which others, it is claimed, can hear or know; and auditory hallucinations, typically hearing other people talking about one's moral behaviour and worth. Rather classically, the patient can declare himself to be very important, even God-like, can accuse others of persecuting him, and later still he develops his own vocabulary and so-called private language.

The anti-psychiatrists, or some of them, suggest the patient is essentially alienated. His delusions are only designated schizophrenic if they are essentially his own. His alienation arises from the impossibility he perceives of a reasonable contract with his social environment. It is a response to emotional entrapment. Under such conditions, mental processes, which always require a social context to modulate and control them, are set free from the very process that enables us to talk about accepted realities. Alone in the world, perception changes, as in transatlantic yacht races. Further, all human beings want to be the centre of the universe, indeed to be God; in early childhood they are. Subsequently they learn about others and the fact that, to begin with, their power base in negligible. Complicity with mother is essential to learn language and gain favours, and her language and moral outlook and later that of others become an essential part of the adult. Here we cannot go into the social basis of knowledge and construction of reality. We can, of course, allude to the problems, even in foundation studies of mathematics and logic (see, for example, Kline, 1980; Putnam, 1981). However, we do not need to be such radical relativists because schizophrenics remain good at arithmetic. The problems they present are more about the more obvious social realities, indeed about fears of other people. Their fear is expressed by Shakespeare, probably mistakenly, that there is no art which can read the mind written on the face of another person. To trust and feel secure in a human context is a problem. Unfortunately for them, with such a difficulty there is the fact that, just as people need vitamins for bodily health, they need human contact for mental health. We live in our perception of other people's perception of us, and our language and world are created by consensus: the science of a period is most obviously so and we must be to some extent complicit.

The anti-psychiatrist himself, of course, has to be complicit when he finds himself forced to use the word 'schizophrenia' which he is rejecting. He rejects it in one sense and yet talks about it, otherwise he cannot communicate in the modern world at all. Psychiatrists like us, writing to a physician, would know to some extent what the physician would understand by this, and without his being given a lecture he is forced to answer the question: is the patient schizophrenic?

Some light can be thrown on the sort of view that some would take by the poetic concept of Ortega y Gasset (1921). Ortega suggested that every generation sends a lot of caravans across the deserts of life and some of the turbulence of adolescence is really getting from one caravan to another. To some extent, if you join the band of drug addicts, punk-rockers, students or some other group, you are OK. You can obviously wander from the caravan in the desert a little way, but if you go right off by yourself, you are alone in the desert, and then you alienation leads to your hallucination and your own destruction.

The above can only be a glimpse of the type of alternative view. In no way does it deny that a person wandering as it were, in the desert wants to come back to the warmth of the group. His problem is fear in doing so. In no way does it deny that phenothiazines and related drugs probably act on the dopaminergic system and make that fear less, and in no way does one want to imply that that may not be the best thing one can do, in the same way as giving a gin and tonic is often the best thing one can do. One does, however, want to point out that adolescent crises are very difficult for psychiatrists to separate from schizophrenia, and that a jump from successful use of the antidopaminergic agent tò concepts about the aetiology can be very crude. Indeed, because there are choices in what one might do in approaching the problem, then these are projections of one's values. We are arguing that the degree to which one can humanise the thing rather than technicalise is congenial to human beings, and should by and large be chosen, other things being equal. We are certainly not pretending that such a path is easy, nor indeed likely to be followed when the person who is trying to help will get so fed up with the other person that drugs have an immense convenience. We are, however, defending the role of academics who continue to fear the Brave New World of such an approach.

Perhaps it is not totally wrong of us to say a little about how one of us misspent his youth and how he came to this position. For a long time he thought that the predictable psychotic was the obvious model to study in order to work out physiological mechanisms. It is well known that some psychotic syndromes recur to a timetable which cannot be explained by social events, and as this can continue year in and year out in some people's lives, and was so well documented in the past by Gjessing (1976), it seemed that an understanding of it would be very illuminating in terms of psychochemistry. At the time he started, of course, the leading textbooks of psychiatry talked about periodic catatonia as that type of schizophrenia in which the chemistry was to some extent known. It was due to the retention of nitrogen-containing toxic factors which could be washed out with massive doses of thyroxine. It was generally considered that Gjessing had proved this point and it is therefore obvious as to why it was attractive to try to find out what were the factors, what was the nitrogen-containing material, etc. Of course, as happens in many projects in clinical medicine, once the research began, the disease disappeared and an entity which had formerly been common became almost unkown. Nevertheless, it still existed and there were remarkable examples in Gjessing's clinic in Norway. However, we also began to notice, and Gjessing had also demonstrated, that the precision of the periodicity depended to some extent on the constancy of the environment. When we moved people in meaningful ways to strange sociological environments, the course of the illness changed totally. For scientific reasons, of course, one avoided doing this without, in a sense, realising what one was doing. There was evidence that meaningful events could ameliorate the condition, and later evidence showed that in some of these patients lithium carbonate and phenothiazines also stopped these conditions. It was really those observations, which are reported in Jenner and Damas-Mora (1983), that made us realise there were quite massive choices in what one went on studying, and what one emphasised of course would produce a hypertrophy of that part of one's knowledge.

We do now know, and certainly classical psychiatrists would agree, that highly emotionally expressive families are detrimental to patients who have 'recovered' from schizophrenia. To some extent the scientific mind wants to say that $x\%$ is due to the genetics and $y\%$ due to the environment, in the same way as such formulae must have been tried for intelligence (in the latter case, of course, not uninfluenced by the work of Cyril Burt). Unfortunately, it is not quite as easy as that, because the influences cannot be measured in the same units, and obviously genetic propensity requires an environment to do anything. We have not here attempted to go into criticism of the current genetic views, but they are not as firm as the textbooks suggest. There are problems, and perhaps the best review is by Lidz *et al.* (1981) on the most important Danish–American studies, where the correlation between half-siblings is actually higher than between real siblings. The DSM III type of diagnostic formulation (American Psychiatric Association, 1980) was influenced by the genetic studies. Previous concepts of the diagnosis of schizophrenia do not give such significant biological correlations. Further, the genetic factor is not like that in Huntington's chorea or phenylketonuria. It is at best a tendency. However, to close, the problem with the approach we are advocating is of course that understanding of a human being is always precarious. It cannot be objectified like statistics, even though we should draw attention to the very obvious fact of how much of British statistics was developed. In terms of techniques, choices were made in favour of those that would be most likely to give statistical results, particularly in, for example, Pearson's correlation coefficient (see Mackenzie, 1980).

Obviously, man's security comes from confident knowledge. Understanding another person is knowledge not held with the same sort of confidence that is often acquired in scientific testing, but it is that human understanding and intuition by which we live and which we need. Without it we know very little about our relations, wives, husbands, friends, etc., and we are suggesting that there is a case for dissidents like ourselves to go on feeling that those whom others think are quite mad are also understandable. All that we are saying, of course, could be quite wrong, but we hope with Voltaire, indeed we know with Voltaire, that you have gone a great deal of the way while possibly disagreeing with us, by reading what we have been allowed to write, for which again we thank the Editors very much.

References

American Psychiatric Association (1980). *Diagnostic and Statistical Manual of Mental Disorders, III.* American Psychiatric Association, Washington, DC.
Bayle, A. L. J. (1822). Recherches sur les maladies mentales. Thèse présentée et soutenue a la Faculté de Medicine de Paris. Centenaire de la Thèse de Bayle, Paris. Masson, Paris.
Dilthey, W. (1914). *Gesammelte Schriften.* Leipzig.
Dunn, G. and Everitt, B. S. (1982). *An Introduction to Mathematical Taxonomy.* Cambridge University Press, Cambridge.
Gjessing, R. L. (1976). *Contribution to the Somatology of Periodic Catatonia.* L. Gjessing and F. A. Jenner (eds). Pergamon, Oxford.
Jardine, N. and Sibson, R. (1971). *Mathematical Taxonomy.* Wiley, London.
Jaspers, K. (1963). General Pscyhopathology. Translated from the 7th Edn by J. Hoenig and M. W. Hamilton. Manchester University Press, Manchester.

Jenner, F. A. and Damas-Mora, J. (1983). Philosophical reflections on and neurobiological studies of some periodic psychoses. In *Neurobiology of Periodic Psychoses*. N. Hatotani and J. Nomura (eds). Igaku-Shoin, Tokyo and New York.

Kendell, R. E. (1975). *The Role of Diagnosis in Psychiatry*. Blackwell, Oxford.

Kline, M. (1980). *Mathematics. The Loss of Certainty*. Oxford University Press, New York.

Kraepelin, E. (1904). *Psychiatrie*, vol. I. *Allgemeine Psychiatrie*. Barth, Leipzig. (See Jaspers, 1963, p. 568.)

Kuhn, T. S. (1970). *The Structure of Scientific Revolutions*, 2nd Edn Chicago: University of Chicago Press.

Lidz, T., Blatt, S. and Cook, B. (1981). Critique of Danish–American studies of the adopted-away offsprings of schizophrenic parents. *Amer. J. Psychiat.*, **138**, 1063–8.

Mackenzie, D. (1980). *Statistics in Britain 1865–1930. The Social Construction of Scientific Knowledge*. Edinburgh University Press, Edinburgh.

Ortega y Gasset, J. (1921). *Meditaciones del Quijote*. Calpe, Madrid.

Putnam, H. (1981). *Reason, Truth and History*. Cambridge University Press, Cambridge.

Sneath, P. H. A. and Sokal, R. R. (1973). *Numerical Taxonomy*. Freeman, London.

Szasz, T. S. (1971). *The Manufacture of Madness*. Routledge & Kegan Paul, London.

Weber, M. (1922). *Gesammelte Beitrage zur Wissenschaftslehre*. Mohr, Tübingen.

Site of action of anti-schizophrenics

A. El-Sobky

Department of Psychiatry, University of Leicester, Leicester Royal Infirmary, Leicester LE2 7LX, UK

24.1. Definition of terms

The appellation of certain groups of drugs as 'anti-schizophrenic' is based on the fact that these drugs have in common the ability to reverse or arrest the course of schizophrenic illness. Implied in this categorisation is a measure of specificity of these drugs to schizophrenia inasmuch as terms like 'anti-tuberculous' or 'anti-diabetic' indicate that certain therapeutic agents exert a compensatory or curative effect in clinical conditions for which the drugs are specific and often exclusive. This is not the case with regard to anti-schizophrenics, hence the lack of accuracy in that term's usage.

24.1.1. *Neuroleptics (NL)*

Soon after the discovery of chlorpromazine (CPZ) by Laborit, and its antipsychotic effects by Delay and Deniker in 1952, reserpine was introduced by Kline in 1954. Delay and Deniker coined the term 'neuroleptic' (NL) to describe the range of pharmacological actions shared by reserpine and CPZ (Carlsson, 1981). The term 'neuroleptic' has, however, been used to define the characteristics of compounds with potential clinical use as dopaminergic antagonists. Such characteristics were often tested in the experimental animal on the early assumption that certain behavioural and pharmacological findings were essential for a drug to qualify as a neuroleptic drug. This has led to the situation where some compounds, e.g. metoclopramide, which act as typical neuroleptics in the experimental setting, lack antipsychotic effects, whereas compounds that do not fulfil the experimental criteria of neuroleptics yet possess a potent antipsychotic effect, e.g. clozapine (CZP).

24.1.2. *Antipsychotics*

Perhaps the least controversial name that has been given to these drugs is 'antipsychotics', since it clearly makes no claims to specificity nor does it require the fulfilment of certain experimental criteria. It does not, however, take account of the frequent clinical use of the drugs in non-psychotic conditions. For example, moderate or severe anxiety as a primary condition or in conjunction with depressive illness often responds to low to moderate doses of antipsychotic drugs.

24.1.3. *Major tranquillisers*

As an alternative, the term 'major tranquilliser' is used to describe the sustained tranquillisation and consequent amelioration of excitement or agitation with decreasing sedation over the first three to four weeks of treatment. This is often referred to as tranquillisation without sedation, in contrast to 'minor tranquillisers' which are predominantly sedative, e.g. barbiturates and benzodiazepines.

Other names, such as 'ataractic', 'anti-hallucinatory' and 'anti-delusional', have not resolved the difficulty. For the present, and until we develop a clear understanding of the pathophysiology of schizophrenia, we have to accept the terms 'antipsychotic', 'neuroleptic' and, to a lesser extent, 'anti-schizophrenics' on the same level as 'anti-pyretic', since these drugs are not specific to schizophrenia nor do they seem to alter a certain aetiological mechanism. Like analgesics and anti-pyretics, they seem to act on a secondary site as compensatory agents, and the reversal of the schizophrenic process takes effect indirectly, probably via other as yet unclarified mechanisms.

24.2. Schizophrenia

In spite of the pathoplastic effects of culture and the immediate context of the individual patient, both the genesis and natural history of the illness seem to be universal phenomena, with an incidence of 15 to 20 new cases per 100 000 per annum, multi-factorial aetiology and prominent genetic contribution. In general there is a tendency towards chronicity or recurrence that may be averted by appropriate pharmacological intervention and social management, particularly in patients with favourable prognostic indices.

Several sets of diagnostic criteria have developed, notably those of Carpenter and Strauss, 'WHO', Feighner, Schneider, the New Haven Schizophrenic Index, Taylor and Abrams, and the Research Diagnostic Criteria (Spitzer *et al.*, 1981). Despite some disagreement on the diagnostic value or weight given to periods of chronicity and the presence or absence of essential symptoms, these diagnostic criteria overlap to a reasonable extent. They tend to exclude general non-specific psychotic features as well as symptoms and signs suggestive of affective or cognitive disorders and thus converge on the core symptoms of schizophrenia, providing consistent and valid tools for diagnosis as well as measurement of change in the clinical picture with a high rate of inter-rater reliability. Furthermore, they make possible the sub-classification into schizophrenic types with different clinical profiles, degrees of certainty and prognostic outlooks.

24.2.1. *The Dopamine hypothesis of schizophrenia*

The hypothesis that hyperdopaminergic activity underlies schizophrenia is based on observations that dopamine agonists precipitate a schizophrenic-like state whereas dopamine antagonists may have antipsychotic effects (Carlsson, 1977). Crow (1980) proposed a two-syndrome model of schizophrenia. *Type I* is an acute florid clinical picture with positive symptoms. This type is likely to be the result of an increased dopaminergic transmission, is amphetamine sensitive, is ameliormed by NL and carries a good prognosis.

Type II, on the other hand, may arise *de novo* (as ? simple schizophrenia on the *International Classification of Disease*, 9th edition), or as a result of a chronic course of the acute variety. The pathological changes here are probably structural in nature, with a prominent cell loss. This type does not respond to NL and is not sensitive to dopamine agonists. The clinical picture is that of a defect state with negative symptoms.

24.2.2. *Drug treatment of schizophrenia*

Until the early 1950s and prior to the introduction of chlorpromazine, 70 % of patients admitted to mental hospitals with a diagnosis of schizophrenia were expected to spend two or more years as in-patients. This is in; contrast to the present figure of 10 % (Cooper, 1983; and see also Chapter 22).

The most important single factor in this dramatic change in the psychiatric landscape was undoubtedly the advent of the antipsychotic drugs. This is clearly reflected in the occupancy of mental hospital beds (Fig. 24.1) and is largely responsible for the change of orientation of psychiatric services towards community care.

24.3. **Antipsychotic drugs**

24.3.1. *Classification*

Chemical formulae of antipsychotic drugs clearly show that they are heterogeneous groups of compounds that can be separated into relatively large subgroups (see Table 24.1). Although such classification is of limited clinical significance, it has a considerable value in the development of new drugs; chemical variation of certain molecules may prove useful in increasing potency or reducing unwanted effects.

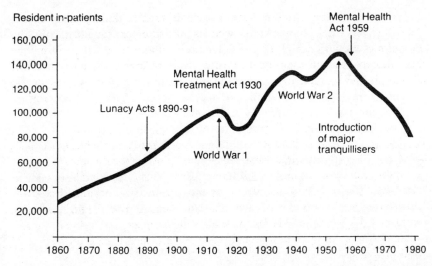

Fig. 24.1. In-patients resident in mental illness hospitals, England and Wales, 1860–1978.

Table 24.1. Major antipsychotic groups and representing members.

Group	Representing member
I. Phenothiazines	
aliphatic side chain	Chlorpromazine
piperidine side chain	Thioridazine
piperazine side chain	Trifluoperazine
II. Thioxanthines	Flupenthixol
III. Butyrophenones	Haloperidol
IV. Diphenylbutylpiperidine	Pimozide
V. Phenylpiperazines	Oxypertine
VI. Dibenzodiazepine	Clozapine
VII. Rauwolfia alkaloids	Reserpine

24.3.2. Clinical profile

The available evidence does not support claims of preference or specificity of one member of the antipsychotic drugs to schizophrenia or to any other psychotic conditions (Sulser and Robinson, 1981). They all exert similar effects with variations in their potency, side-effects and degrees of sedation (Hirsch, 1982).

The antipsychotic effects of neuroleptics are as follows:

(1) Amelioration of excitement and agitation. This is possibly due to reduction of the affective colouring of the perceived environmental

events via their effect on the amygdala and limbic system. This is achieved without marked sleepiness, ataxia or clouding of consciousness.

(2) Filtering out of the irrelevant data and reducing their significance, possibly via their effect on the reticular activating system of the midbrain as well as the limbic system and cortex. This may account for the reduction in the intensity and severity of thought disorder, delusions and hallucinations, and the state of hyperarousal, presumed in acute schizophrenia.

(3) General antipsychotic effects are seen in their control of states of delirium, behavioural disorders, particularly of the aggressive–destructive type, states of excitement and stupor, repetitive and stereotyped movement in cases of brain damage, and the flight of ideas and the pressure of thought characteristics of manic states.

24.3.3. *Site of action*

The site of action of the antipsychotic drugs has been gradually elucidated over the last twenty years. The possible mechanisms of action have also been largely outlined, but the link between the drugs, action and the clinical therapeutic effect remains speculative. This is not unreasonable since the pharmacological profile, and findings from animal experiments, as well as neurological and endocrinological side-effects, may well indicate which system(s) of the brain are influenced by antipsychotics. These observations cannot, however, explain how the schizophrenic process is reversed or arrested. An often-quoted example of this is the use of diuretics in cases of heart failure where the primary site of action of the drug is quite remote from the primary pathology and its therapeutic effect is achieved through a series of secondary physiological changes.

24.3.4. *Reserpine*

Reserpine exerts its antipsychotic effect through depletion of the neurotransmitters dopamine, noradrenaline and serotonin, possibly with some selectivity for dopamine. Reserpine is unique in this respect (Carlsson, 1981).

24.3.5. *Neuroleptics*

Apart from reserpine, all other antipsychotics are competitive antagonists of dopamine at different receptor sites. For instance, they block the postsynaptic receptors involved in stimulation of cyclic AMP formation. The blockade of presynaptic receptors results in stimulation of synthesis, release and metabolism of dopamine (Carlsson and Lindquist, 1963; Carlsson, 1977, 1981; El-Sobky, 1983).

Neuroleptics do vary in their differential effects on the dopaminergic subsystems in the brain, their inhibitory effect on dopamine-sensitive adenylate cyclase, and their effect on other central transmitter systems, particularly those using acetylcholine, noradrenaline and serotonin, and consequently in their side-effects.

The early assumption that extra-pyramidal symptoms (EPS) were essential characteristics of the antipsychotic profile is now shown to be false since some neuroleptics are not universal dopamine blockers and are more selective in their target site of the dopaminergic system outside the nigrostriatal pathways (El-Sobky, 1983). Thus the available neuroleptics could be broadly classified on functional or clinical grounds as typical or atypical.

24.3.6. *Dopamine inhibition and stimulation*

Evidence for dopamine antagonist effects of NL was derived from the observations that in man amphetamine, an indirect dopamine agonist which causes its release into the synaptic cleft, can induce a clinical picture closely akin to paranoid schizophrenia. Other dopamine agonists, e.g. methylphenidate, L-dopa and bromocriptine could precipitate similar clinical pictures. The drug-induced psychotic picture can be reversed by NL. Furthermore, α-methyl paratyrosine (α-MPT), a specific inhibitor of tyrosine hydroxylase (an enzyme necessary for the conversion of tyrosine to dopa), does potentiate antipsychotic effects of NLs.

Clinically, therapy with typical neuroleptics is usually associated with the incidence of extrapyramidal symptoms and hyperprolactinaemia resulting from blockade of dopamine receptors.

Stimulation of the dopaminergic activity in the dopamine-rich regions of the animal brain by microinjections of dopamine and dopamine agonists such as amphetamine or apomorphine results in increased motor activity and stereotyped behaviour. These two well established dopamine agonist effects depend on intact mesolimbic and mesocortical pathways.

In addition to the stereotyped behaviour and motor hyperactivity, apomorphine induces continuous climbing behaviour in the mouse. This is a serotonin-antagonist effect which is enhanced by methysergide, a specific serotonin antagonist. Climbing is reduced by the specific serotonin agonist quizapine, as well as by metoclopramide which is a non-antipsychotic dopamine agonist (Costall and Naylor, 1977).

24.4. **The general characteristics of neuroleptics**

Neuroleptics abolish stereotypy, overactivity and apomorphine-induced behaviour in doses sufficient to produce catalepsy in the untreated animal. This is related to their relative potency and their ability to stimulate the synthesis, release and metabolism of dopamine (see Fig. 24.2). Their affinity

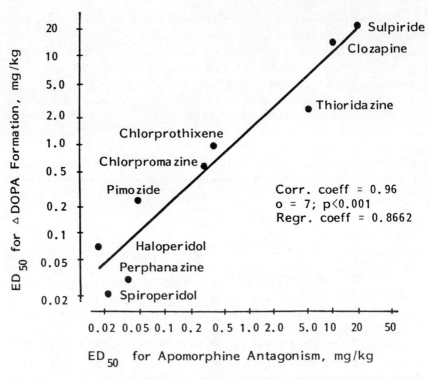

Fig. 24.2. Relationship between apomorphine antagonism and dopa formation (Sulser and Robinson, 1981).

for the dopamine receptor shown by their ability to displace [^{3}H]haloperidol from the binding site is also related to their potency as antipsychotics (see Fig. 24.3).

The postsynaptic dopamine receptor blockade is widespread in the case of typical neuroleptics like fluphenazine, chlorpromazine and haloperidol. With the exception of haloperidol, typical NLs inhibit the dopamine-sensitive adenylate cyclase.

High correlations have been reported between cataleptic effects, ED_{50} for dopa formation, ED_{50} for apomorphine antagonism, their clinical potency, the side-effects, particularly EPS, and their affinity to displace [^{3}H]haloperidol.

Atypical neuroleptics depart from these general observations individually. Thioridazine, sulpiride and clozapine act more selectively on the meso-limbic and mesocortical pathways. Their ratio of striatal to limbic increase of dopamine turnover is 1·2 compared with 1·8 for the typical neuroleptic.

These atypical compounds are non-cataleptic and are not associated with extrapyramidal side-effects. Thioridazine is a potent α-adrenergic receptor

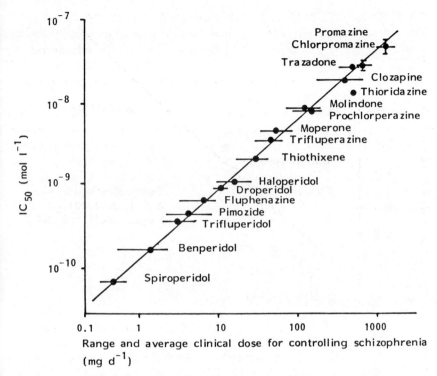

Fig. 24.3. Relationship between clinical potency and [³H]haloperidol displacement (Carlsson, 1981).

antagonist, and that may account for its noted peripheral effects such as hypotension and its central sedative effect. Its lack of EPS is attributed to an in-built, potent, anticholinergic effect, although it is possible that its α-adrenergic blocking nature may initiate a series of feedback mechanisms which correct the relative cholinergic–dopaminergic imbalance.

Specific dopamine receptors have been postulated for CZP (Burki *et al.*, 1980), thus accounting for its virtual lack of cataleptic and EPS effect. Like thioridazine, clozapine is associated with an increased noradrenaline turnover in the brain.

24.4.1. *Side-effects of neuroleptics*

There is evidence that in the striatum dopamine exerts tonic inhibitory effects on cholinergic interneurones, thus contributing to maintaining a balance between cholinergic and dopaminergic functions (Hattori *et al.*, 1976).

With the dopamine blockade of the NL therapy, these interneurones release more acetylcholine, and a functional imbalance follows, resulting in a relative underactivity in dopaminergic and relative overactivity in cholinergic transmission. This is possibly achieved via a long loop involving GABA-containing neurones. Early extrapyramidal side-effects are often noticed within hours or days of the beginning of neuroleptic therapy and they are responsive to anticholinergic drugs. These include acute dystonic reactions, akathesia and hypokinesia. Within two to three weeks of therapy, patients can develop a clinical picture of Parkinsonism, with masked facial expressions, hypersalivation, rigidity and tremors. Again these patients respond to anticholinergic drugs, although there is evidence that over a long period such drugs may contribute to the onset of tardive dyskinesia.

Long-term use of neuroleptics results in an increase in the density and sensitivity of postsynaptic dopamine receptors. This is thought to cause the buccolingual masticatory hyperkinesia known as tardive dyskinesia, which eventually progresses to the trunk and limbs. It is an irreversible picture and is not known to respond to any specific therapy. Anticholinergics often exacerbate the clinical picture.

24.4.2. *The neuroleptic malignant syndrome*

This is a recently identified adverse reaction to neuroleptic therapy in which akinesia, hyperpyrexia and altered consciousness may develop at any time of the treatment with NL (Trent Drug Information Services, 1984). Haloperidol and fluphenazine have been implicated most, but other neuroleptics have also been reported to cause NMS. This condition may progress to autonomic and respiratory failure. The mortality rate is reported to be between 20 and 30% and the syndrome is not dose related (Neppe, 1984).

Neuroleptic malignant syndrome is attributed to the increased synthesis of dopamine and increased sensitivity of the dopaminergic receptor due to long-term therapy. These two factors combined would allow relatively small increases in neuroleptic doses to cause a profound receptor blockade which is not compensated for by adequate dopamine synthesis. This is supported by the reports that L-dopa and bromocriptine have proved beneficial in the treatment.

24.4.3. *Endocrinological side-effects*

Dopamine reaches the anterior pituitary gland through the hypothalamic–hypophyseal circulation to act as PIF (prolactin inhibitory factor). It also controls the release of growth hormone, R thyrotropin-releasing hormone and luteinising hormone. With the blockade of dopaminergic receptors, hyperprolactinaemia follows. That side-effect does not seem to show tolerance to long-term therapy. Hyperprolactinaemia may lead to galactorrhoea, infertility, gynaecomastia and hypogonadism.

24.4.4. *Presynaptic receptors*

Presynaptic dopamine autoreceptors have been identified (Bunney and Aghajanian, 1976.) If stimulated by micro-injection of dopamine or by dopaminergic agonists, these autoreceptors cause inhibition of the firing rate of the dopaminergic neurones, an effect that is reversed by neuroleptics. They may also be responsible for the feedback leading to an increase of the dopaminergic synthesis consequent on the neuroleptic effect.

24.4.5. *Postsynaptic receptors*

There is ample evidence to demonstrate that the dopamine-sensitive adenylate cyclase is stimulated by dopamine and antagonised by neuroleptics. It is claimed that there are two types of dopaminergic receptor, i.e. D-1 and D-2, with different sensitivities to dopamine agonists and antagonists, or that there is only one receptor which may take agonistic or antagonistic configuration depending on the normal physiological role that it performs (Laduron, 1981; El-Sobky, 1983).

Recent work (Calne, 1981; Kebabian and Cote, 1981) suggests two types of DA receptors, D-1, which is adenylate cyclase dependent, and D-2, which is adenylate cyclase independent (see also ch. 2). Some of the new-generation NLs, e.g. sulpiride, act specifically on D-2 receptors.

24.5. Conclusion

It is likely that the antipsychotic effect of the neuroleptics is subserved through their action on the mesolimbic and mesocortical dopaminergic systems. This antipsychotic effect is not specific to schizophrenia, and the dopamine theory of schizophrenia should therefore be seen as a likely hypothesis of the biochemistry of psychosis in the most general terms.

We are still unable to identify the intervening processes between dopamine blockade and the reduction in psychotic scores. Neuroleptics can be divided into typical and atypical types according to their individual profiles (see Fig. 24.4).

Apart from the clinical indices of likelihood of recovery and relapse, no reliable pharmacological measures have yet been identified for the prediction of therapeutic response. The serious and long-lasting side-effects of the available neuroleptics, as well as their lack of specificity, make it essential to search for further, newer, safer and more specific compounds.

References

Bunney, B. S. and Aghajanian, G. K. (1976). In *Antipsychotic Drugs: Pharmacodynamics and Pharmacokinetics*. G. Sedvall, B. Uvans and Y. Zotterman (eds), Wenner-Gren Symp. Pergamon Press, Oxford.

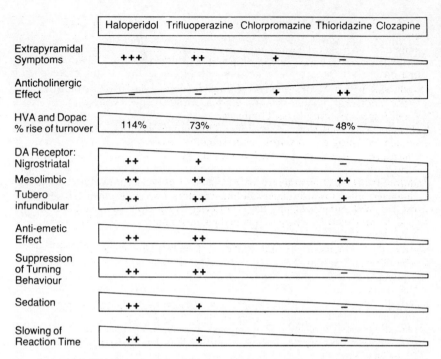

Fig. 24.4. Model for classification of antipsychotics according to their clinical profile (El-Sobky, 1983).

Burki, H. W., Alder, J. and Asper, H. (1980). Binding of thioridazine, haloperidol and clozapine to 3H-haloperidol and 3H-clozapine receptors in the striatum and limbic system of the rat phenothiazines and structurally related drugs: basic and clinical studies. In *Proceedings of the Fourth International Symposium, New York.* E. Usdin, H. Eckert and I. S. Forest (eds), pp. 71–8. Elsevier/North Holland, New York.

Calne, D. B. (1981). Clinical relevance of dopamine receptor classification towards understanding receptors. *Current Reviews in Biomedicine*, vol. 1. J. W. Lamble (ed.). Elsevier/North Holland, New York.

Carlsson, A. (1977). Does dopamine play a role in schizophrenia? *Psychol. Med.* 583–93.

Carlsson, A. (1981). Mechanism of action of neuroleptic drugs *Psychopharmacology: a Generation of Progress.* M. A. Lipton, A. Morris, A. DiMascio and K. F. Killam (eds), pp. 1057–70. Raven Press, New York.

Carlsson, A. and Lindquist, M. (1963). Effect of chlorpromazine and haloperidol on formation of 3-methoxy-tyramine and normetanephrine in mouse brain. *Acta Pharmacol. Toxicol.* **20**, 140–4.

Cooper, J. A. (1983). Schizophrenia and allied conditions. *Med. Int.* 1 (33), 1546–50.

Costall, B. and Naylor, R. J. (1977). Behavioural characterisation of neuroleptic properties in the rodent. *Proc. R. Soc. Med.* **70** (Suppl. 10), 5–14.

Crow, T. J. (1980). Molecular pathology of schizophrenia: more than one disease process. *Brit. Med. J.* **280**, 66–8.

El-Sobky, A. (1983). *In Search of Profile*. Publ. Sandoz Products Ltd., Basle.

Hattori, T., Singh, V. K., McGeer, P. L. and McGeer, E. G. (1976). Immunohistochemical localisation of choline acetyltransferase containing neo-striatal neurons and their relationship with dopaminergic synapse. *Brain Res.* **102**, 164–73.

Hirsch, S. R. (1982). Medication and physical treatment of schizophrenia. *Handbook of Psychiatry, vol. 3: Psychoses of Uncertain Aetiology*. J. K. Wing (ed.), pp. 74–87. Cambridge University Press, Cambridge.

Kebabian, J. W. and Cote, T. E. (1981). Dopamine receptors and cyclic AMP: a decade of progress towards understanding receptors. *Current Reviews in Biomedicine*, New York. 1. J. W. Lamble (ed.), pp. 112–17. Elsevier/North Holland, New York.

Laduron, P. (1981). Dopamine receptor from an *in vivo* concept towards a molecular characterisation towards understanding receptors. *Current Reviews in Biomedicine*, vol. 1. J. W. Lamble (ed.), pp. 105–11. Elsevier/North Holland, New York.

Neppe, V. M. (1984). The neuroleptic malignant syndrome. A priority system. *S. A. Med. J.* **65**, 523–5.

Spitzer, R. L., Fleiss, J. L. and Endicott, J. (1981). Problems of classification: reliability and validity. *Psychopharmacology: a Generation of Progress*. M. A. Lipton, A. Morris, A. DiMascio and K. F. Killam (eds), pp. 857–69. Raven Press, New York.

Sulser, F. and Robinson, S. E. (1981). Clinical implications of pharmacological differences among psychotic drugs (with particular emphasis on biochemical central synaptic adrenergic mechanisms). *Psychopharmacology: a Generation of Progress*. M. A. Lipton, A. DiMascio and K. F. Killam (eds), pp. 943–60. Raven Press, New York.

Trent Drug Information Services (1984). Malignant Neuroleptic Syndrome, Editorial. *Psychiatric alert*. Trent Drug Information Services No. 28.

25

Acute and chronic consequences of persistently disturbing cerebral dopamine function

B. Costall, A. M. Domeney and R. J. Naylor

Postgraduate School of Studies in Pharmacology, University of Bradford, Bradford BD7 1DP, UK

25.1. Introduction

For many years, following the introduction of the neuroleptic agents, their behavioural and biochemical profile was essentially defined by effects observed after acute administration of a single dose. These studies showed the distinguishing feature of such compounds to be their ability to inhibit cerebral dopamine function. However, in man such compounds are given repeatedly, and analogous studies were subsequently carried out in experimental animals which were challenged for several weeks or months. Such studies confirmed a dopamine involvement with antipsychotic drug activity but have widened this concept in questioning the persistence of dopamine receptor blockade on chronic treatment (see review by Rupniak *et al.*, 1983).

A second divergence between the experimental and clinical situation has been in selection of the 'subject'. Schizophrenic patients are not 'normal', and yet 'normal' animals have been used to assess antipsychotic drug activity. A most simplistic interpretation from the clinic would be that schizophrenia involves a dopamine 'overactivity' (see review by Haracz, 1983), and some attempts have been made to mimic this in animals by using dopamine agonists. Yet the brief disturbance in cerebral dopamine function initiated by an agonist drug is unlikely to mimic the neurotransmitter disturbance of schizophrenia. Nevertheless, the use of dopamine agonists to detect dopamine antagonist activity, combined with other simple behavioural tests, have detected compounds that have revolutionised the treatment of schizophrenia. However, those compounds which have been detected have virtual uniformity of action, having the same benefits and disadvantages in use, and similar acute and long-term side effects. The need for new developments is great. New animal models need to be developed which reflect the clinical situation as closely as possible, not only to allow the detection of novel therapeutic agents but also to shed new light on the aetiology of schizophrenia, to detail the ways in which discrete brain systems respond during neurotransmitter dysfunction and subsequently to acute or long-term challenge with a neuroleptic or dopamine antagonist drug.

In reassessing the clinical situation for extrapolation to an animal model, two key observations remain. First, dopamine agonists can induce or precipitate psychotic behaviour, and, secondly, substances which are able to reduce or antagonise dopamine function, directly or indirectly, provide the only realistic therapy (see Van Kammen *et al.*, 1982*a,b*; Donaldson *et al.*, 1983). These two points have formed the cornerstones of the dopamine hypothesis of schizophrenia (see Stevens, 1978). Yet this hypothesis has become self-limiting: it does not, for example, account for the marked differences in responses of schizophrenic patients to dopamine agonist challenge; some show a worsening, some an amelioration and others no change at all in the psychoses (Van Kammen *et al.*, 1982*a*). Further, administration of the classic dopamine antagonists, the neuroleptics, to man can effect a cerebral dopamine receptor blockade within minutes of administration, and yet the antipsychotic action is rarely evidenced within 14 days of starting treatment. Then, even accepting these anomalies, one must also consider that the neuroleptic agents are not uniformly effective, but are more able to control the positive symptoms than the negative symptoms, with marked individual variations in sensitivities and resistance to neuroleptic therapy. The conclusions must inevitably be drawn that the concept of a raised dopamine function may apply to only a proportion of patients, or that a dopamine dysfunction causes complex changes and a variety of symptomatologies and responsiveness to neuroleptic drugs.

Accepting that a disturbance in dopamine function does constitute the most important clue to our understanding of the biology of schizophrenia, we decided to re-evaluate the role of cerebral dopamine in the control of motor behaviour and as a site of antipsychotic drug activity, and to develop a test system which may aid the detection of novel drug activity. We began with a consideration of three decades of clinical experience with dopamine agonists and antagonists:

(1)   Schizophrenic patients show marked individual differences in their responsiveness to dopamine agonists and dopamine antagonists.

(2)   Since there is no evidence of a generalised disruption of all cerebral dopamine systems in schizophrenia, it is likely that any change would be occurring within a discrete brain system.

(3)   It is unlikely that a transitory disturbance in dopamine function could result in schizophrenia: a persistent neurotransmitter change would seem the more likely cause.

(4)   It is likely that forebrain dopamine-containing areas are the site of disturbance in schizophrenia.

25.2. Persistent stimulation of dopamine systems in experimental animals

We have therefore commenced a series of experiments assessing the acute and chronic consequences of discretely and persistently stimulating dopamine systems in experimental animals. First, we considered that the animals should be pre-selected. Since schizophrenic patients show great diversity in both sensitivity and response to dopamine agonists and antagonists, animals were selected on the basis of their response to a dopamine agonist. We selected $(-)N$-n-propylnorapomorphine $[(-)NPA]$ as a reference, directly acting, dopamine agonist able to influence limbic function and cause hyperactivity responses. Within a group of rats challenged with $(-)NPA$, a proportion will respond with a high-intensity hyperactivity and a proportion will exhibit a very low-activity response. There is always an intermediate band of moderate-activity animals, and usually a small group of non-responders. The proportion of animals separating into the different subgroups can vary markedly (Fig. 25.1). A very important point is that these groups of animals cannot be distinguished on the basis of their spontaneous activity: like man, the rat can show marked variation in responsiveness to dopamine agonist challenge, and we have incorporated these differences into our test system by pre-selecting animals.

It is generally accepted that in the rodent the function of the forebrain dopamine systems contributes to control of motor activity or psychomotor drive, and therefore locomotor activity was taken as a behavioural measure

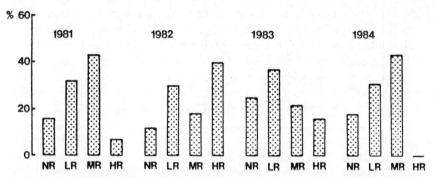

Fig. 25.1. Selection of rats (male Sprague-Dawley) into different activity groups according to their responsiveness to $(-)NPA$, 0.05 mg kg^{-1} s.c. assessed over a 4-year period. Representative data are given, derived in October of each year. It should be noted that there are seasonal variations. Locomotor activity was measured in individual perspex cages, each fitted with one photocell unit placed off-centre: interruptions of the beams were recorded electromechanically in counts per 5 min. Data on this figure indicate the number of rats in each subgroup as a percentage of the total tested. NR, non-responders (less than five counts per 5-min period); LR, low responders (10–25 counts per 5 min); MR, moderate responders (30–55 counts per 5 min); HR, high responders (60–85 counts per 5 min period). For each set of data presented, 200–300 rats were screened.

of change in the functional capacity of these systems during persistent neurotransmitter change. This persistent neurotransmitter change was effected via Alzet osmotic minipumps located subcutaneously in the back neck region and connected to permanently indwelling intracerebral cannulae stereotaxically located. These pumps allowed a solution containing dopamine to be delivered persistently, 0.48 μl h^{-1}, 24 h day^{-1} for up to 13 days into any selected area discretely located by stereotaxic manipulation.

The first brain area to be selected for study was the nucleus accumbens, primarily because the function of this area had previously been subject to extensive analysis following acute administration. The slow, persistent infusion of dopamine discretely into the nucleus accumbens caused dose-related (4.34–34.72 ng min^{-1}) enhancements in locomotor activity which occurred in phases during the 13-day infusion period such that peaks of responding occurred between days 4 to 5 and 10 to 11 (Fig. 25.2). These

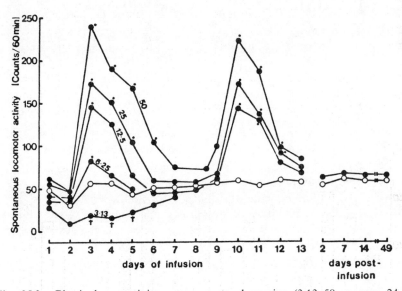

Fig. 25.2. Phasic hyperactivity responses to dopamine (3·13–50 μg over 24 h) infused into the centre of the rat nucleus accumbens for 13 days (ant. 9·4, lat. ±1·6, vert. 0·0; De Groot, 1959). ●—● indicates the dopamine responses, ○—○ the responses of animals receiving an intra-accumbens infusion of the vehicle for dopamine. Spontaneous locomotor activity was measured in photocell cages and is presented in counts per 60 min. Measurements were taken both during infusion and for 49 days post-infusion. Animals used in this study were initially selected as low-activity responders to (−)NPA, although similar data (no significant differences) were derived for animals pre-selected as high-activity responders. $n = 6$–12. SEMs < 11·9%. Significant enhancement of spontaneous locomotion indicated as *$P <$ 0·05–$P < 0·005$ (two-way ANOVA followed by Dunnett's test for multiple comparisons). It should also be noted that a low dose of dopamine can significantly reduce spontaneous locomotion (†$P < 0·01$).

responses to a dopamine infusion into the mesolimbic nucleus accumbens were identical in animals pre-selected as 'low' and 'high' responders to a dopamine agonist, and the timing of the peaks and troughs of activity occurred with remarkable precision as experiments were repeated many times over during a 4-year period.

The response to dopamine infusion could not have been predicted on the basis of responding to a single challenge with dopamine, since 100 000 ng is the minimum dose to secure only a brief period of increased motor activity (Costall and Naylor, 1976). Since locomotor hyperactivity was never observed during the first or second day of infusion, it became clear that the persistence of the dopamine challenge is a critical determinant of the motor response. It also became clear that brain compensatory mechanisms may come into play rapidly to adjust changed motor functioning to 'normal', and thus lead to phasic responding to a continued dopamine challenge.

However, perhaps the most dramatic changes caused by the dopamine infusion were revealed some 2 to 3 weeks after the infusion was withdrawn, at a time when levels of spontaneous activity had returned to control values to give no indication of the fundamental changes that had occurred in sensitivity to a dopamine agonist challenge. Throughout the dopamine infusion both the 'high' and 'low' responsive animals showed a decreased responsiveness to ($-$)NPA, possibly as a result of a 'desensitisation' of dopamine receptors (Fig. 25.3). However, 2 to 3 weeks after discontinuing the infusion, those animals which had been pre-selected as low responders to ($-$)NPA developed rapidly increasing responses to ($-$)NPA such that their level of responding equated to that of high-active animals; indeed such animals exhibited a 200 % increase in response to ($-$)NPA (Fig. 25.3). This increased responsiveness, which could persist for many months, was also shown to other dopamine agonists such as pergolide and LY141865 (Fig. 25.4). Of particular importance, these initially low-responsive animals also exhibited markedly exaggerated responses to L-dopa (in the order of 600 %) during the post-infusion period (Fig. 25.4). This would strongly suggest that those dopamine receptor mechanisms which increase their sensitivities as a consequence of dopamine infusion are sensitive to dopamine itself. It is also a most important observation that exceptionally large doses of neuroleptic agents are required to antagonise the L-dopa response (Costall *et al.*, 1984*a*).

Whereas 'low activity' animals, consequent on an intra-accumbens dopamine infusion, are rendered 'high active', the converse holds for the initially 'high active' animals which maintain a 'low active' response not only during infusion but for many weeks post-infusion. This reversal of animal activity status is highly dependent on the dose of dopamine infused, with doses lower than 4.34–8.68 ng min^{-1} failing to trigger the change, notwithstanding that lower doses of dopamine can trigger the phasic hyperactivity reponse during infusion (Costall *et al.*, 1983). It is concluded that an overt behavioural change to dopamine receptor stimulation requires that stimulation to be maintained persistently over a definite time period, that dopamine receptor stimulation of a precise intensity is required in order that long-term changes in dopamine receptor sensitivity may be triggered, and, most unexpectedly, the nature of the long-term consequence of a dopamine infusion into the centre of the nucleus accumbens of rats is

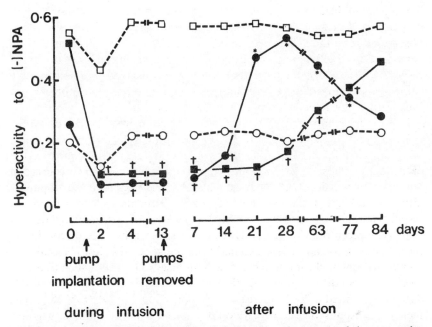

Fig. 25.3. Consequences of dopamine infusion into the centre of the rat nucleus accumbens on locomotor activity status as defined by responsiveness to $(-)$NPA, 0.05 mg kg^{-1} s.c. Hyperactivity to $(-)$NPA was measured in individual photocell cages; the data plotted intensity (counts per 5 min) versus times; and the area 'under the curve' was determined: hyperactivity to $(-)$NPA is therefore presented in arbitrary units derived from 'area under the hyperactivity curve'. Dopamine was infused at a rate of 0.48 μlitre h^{-1}, 25 μg per 24 h for 13 days. Responses to $(-)$NPA are shown during this 13-day infusion period and for 84 days post-infusion. Two groups of animals were used in these studies, low responders to $(-)$NPA (●, ○) and high responders (■, □). Each group was further divided into two, one half receiving intra-accumbens dopamine, the other an infusion of the vehicle for dopamine. The vehicle-infused animals (□----□ high active, ○----○ low active) maintained their activity status during infusion and post-infusion. Dopamine-infused animals (■—■ high active, ●—● low active) showed significantly depressed responding to $(-)$NPA during infusion (†$P<0.01$ to $P<0.001$) and the high-active animals continued to exhibit this reduced responding during the post-infusion period (†$P<0.05$ to $P<0.001$). In contrast, the initially low-active animals receiving dopamine (●—●) showed high-active responding following a 2- to 3-week delay after infusion withdrawal (*$P<0.01$ to $P<0.001$, two-way ANOVA followed by Dunnett's test for multiple comparisons). $n=6$–12. SEMs<12.3%.

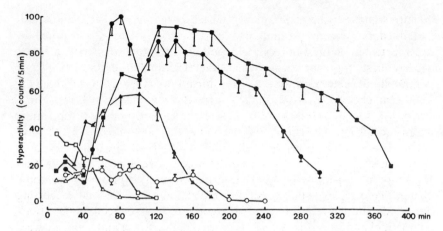

Fig. 25.4. Enhanced hyperactivity responding to L-DOPA, pergolide and LY141865 by rats initially selected as low-activity responders to (−)NPA but rendered high active to (−)NPA as a consequence of withdrawing a 13-day infusion of dopamine into the centre of the nucleus accumbens (25 μg per 24 h, 0·48 μlitre h^{-1}. Hyperactivity was measured in photocell cages and is given in counts per 5 min. Data shown were determined 9–12 weeks following withdrawal of the dopamine infusion when animal status showed maximum conversion from low to high activity. ■—■ 0·5 mg kg^{-1} s.c. pergolide, ●—○ 200 mg kg^{-1} i.p. L-DOPA given after a 50 mg kg^{-1} i.p. treatment with benserazide, ▲—▲ 0·25 mg kg^{-1} s.c. LY141865. Open symbols show control data for each dopamine agonist determined in rats which had received vehicle infusion into the nucleus accumbens (data indistinguishable from that of normal, untreated rats, □—□ pergolide, ○—○ L-DOPA (+benserazide), △—△ LY141865). *n* = 6–12. Representative SEMs shown. The hyperactivity in response to L-DOPA, pergolide and LY141865 is significantly greater in the dopamine-infused animals as compared with the control data (*P* < 0·001 for areas under curves, Student's *t* test).

critically dependent on the initial locomotor status of the rat as defined by a dopamine agonist.

In associated experiments we have shown the behavioural data to have clear biochemical correlates. Thus, rats of 'high' and 'low' responsiveness to (−)NPA have relatively 'high' and 'low' numbers of [³H]-NPA binding sites within the nucleus accumbens, and this situation is reversed as the long-term consequences of a dopamine infusion become apparent (Costall *et al.*, 1982).

The specificity of dopamine's action to cause the described changes both during and post-infusion was assessed by infusing other neurotransmitter substances (acetylcholine, GABA, serotonin and noradrenaline) into the centre of the nucleus accumbens. These treatments failed to institute the changes caused by dopamine, either during or after infusion (Costall *et al.*, 1984b). It is concluded that the changes caused by dopamine are specific for dopamine. However, whether dopamine exerts its actions via dopamine receptors can only be assessed by the use of antagonist drugs, and this prompted a series of experiments to determine the consequence of neuroleptic antagonism during the period of dopamine infusion.

The question posed was whether the behavioural expression of hyperactivity during an intra-accumbens dopamine infusion is critical for the long-term changes post-infusion. The possibility that a relatively brief period of dopamine excess may dictate changed responses to dopamine agonist challenge persisting for many months post-infusion could have considerable implications for our understanding of the aetiology of a disease associated with neurotransmitter excess, such as schizophrenia. Can the action of dopamine be blocked during infusion to prevent the chronic change? To investigate this possibility, the phases of hyperactivity occurring during dopamine infusion were antagonised by the neuroleptic agents haloperidol and sulpiride. Very low doses of haloperidol and sulpiride were found to prevent the hyperactivity responses very effectively during infusion. However, most unexpectedly, rather than reducing or preventing the long-term effects, these could actually be exacerbated by the neuroleptic intervention (Costall *et al.*, 1984c). Although this clearly shows that the behavioural expression of hyperactivity during a mesolimbic dopamine infusion is not essential for the subsequent development of changed sensitivity to dopamine agonist challenge, the actual exacerbation of response remains to be explained.

The implications are important: a raised dopamine activity in limbic brain areas was indicated to institute long-term changes in dopamine receptors via non-neuroleptic sensitive sites. Further studies were therefore initiated to determine whether higher dosage sulpiride, administered three times daily or by continuous intraperitoneal infusion, could provide adequate neuroleptic 'cover' to prevent the dopamine response and its consequences. The three-times-daily approach, as with haloperidol, again led to an exacerbation of the long-term change (Fig. 25.5) but sulpiride given as a continuous infusion most successfully prevented the acute and chronic consequences of a mesolimbic dopamine infusion (Fig. 25.5; Costall *et al.*, 1985). It is therefore concluded that the long-term changes induced by dopamine infusion are mediated via neuroleptic-sensitive sites but that the precise manner of neuroleptic administration, whether this be to provide fluctuating blood levels after repeated daily treatments or continuous and constant blood levels after infusion, can dramatically modify the consequences of neuroleptic action. That daily treatment with clozapine can also prevent the consequences of dopamine infusion (Fig. 25.6) is of obvious interest, and the effectiveness of other neuroleptic agents is currently being assessed.

At this stage of the studies it was clearly of interest to determine whether the spectrum or motor change seen during and after infusing dopamine into the nucleus accumbens could also be initiated from other limbic and cortical

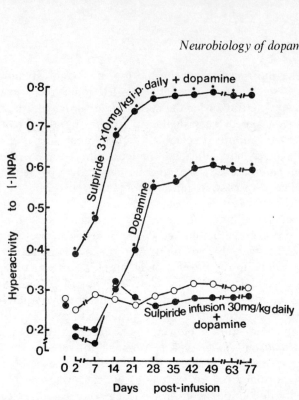

Fig. 25.5. Hyperactivity responding to (−)NPA following the infusion of dopamine into the centre of the nucleus accumbens, withdrawn for 2–77 days: influence of sulpiride. Hyperactivity to (−)NPA was derived as described previously, and is presented in arbitrary units. Day 0 indicates the initial hyperactivity responses of the animals (two representative groups given, ○, ●, low-activity responders to (−)NPA). Animals were then subject to an intra-accumbens infusion of dopamine alone, dopamine infusion with a concomitant sulpiride treatment given three times daily (with one priming dose the day before infusion commenced), 10 mg kg^{-1} i.p. at each dose (given at 7 a.m., 2 p.m., 11 p.m.), or dopamine infusion with a concomitant treatment with sulpiride delivered by intraperitoneal infusion (30 mg kg^{-1} day^{-1}, again with a priming dose of 10 mg kg^{-1} i.p. on the day before infusion commenced). ○—○, responses of animals receiving intra-accumbens infusion of vehicle. $n = 5$–10. SEMs $< 12.6\%$. Significant increase in the (—)NPA activity response during the post-infusion period from low to high responding) indicated as *$P < 0.001$ (two-way ANOVA followed by Dunnett's test for multiple comparisons).

areas. These studies have been limited to the use of a single dose of dopamine (25 μg daily for 13 days) infused into a posterior region of the nucleus accumbens, the anterior olfactory nucleus, the tuberculum olfactorium and areas of the cortical anteromedial dopamine system. The different spectra of responses which can be obtained range from no effect during infusion but with a long-term consequence of changed activity status (posterior region of the nucleus accumbens (Fig. 25.7) and also the anteromedial dopamine system (data not shown)) to biphasic hyperactivity

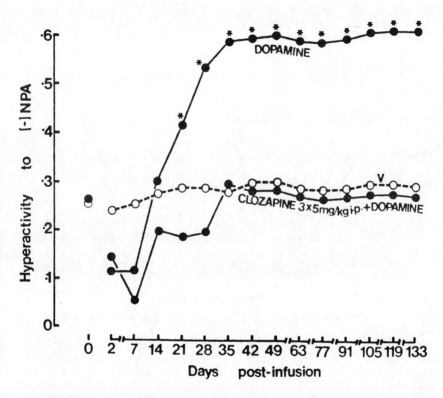

Fig. 25.6. Hyperactivity responding to (−)NPA following the infusion of dopamine into the centre of the nucleus accumbens, withdrawn for 2–133 days: influence of clozapine. Hyeractivity values for (−)NPA were derived as previously described (presented in arbitrary units). Day 0 indicates the initial hyperactivity responding of the animals (low-activity responders to (−)NPA). Animals were subject to an intra-accumbens infusion of dopamine alone (●—●), an intra-accumbens infusion of dopamine combined with a three-times-daily treatment with clozapine (●—●, 3×5 mg kg^{-1} day^{-1}, given at 7 a.m., 2 p.m., 11 p.m.). A priming dose of 5 mg kg^{-1} i.p. clozapine was given the day before dopamine infusion commenced. ○–○ indicates the responses of control-activity animals receiving intra-accumbens vehicle for dopamine and the vehicle for clozapine i.p. $n = 5$–10. SEMs < 11·2%. Significant increase in the hyperactivity responding to (−)NPA during the post-infusion period (a change from low- to high-activity status) is indicated as *$P < 0.001$ (two-way ANOVA followed by Dunnett's test for multiple comparisons).

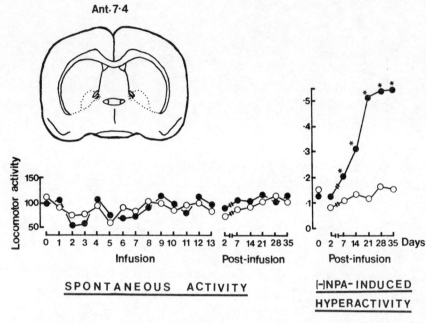

Fig. 25.7. Effects of infusing dopamine, 25 μg day^{-1}, 0·48 μlitre h^{-1}, into the posterior region of the nucleus accumbens of the rat (ant. 7·4, vert. 0·0, lat. $\pm$1·6). Spontaneous locomotor activity (counts per 60 min measured in photocell cages) is shown for the 13 days of infusion and for days 2–35 post-infusion, ●—● indicating the responses of rats receiving the dopamine infusion for 13 days, and ○—○ indicating the responses of rats receiving an infusion of the vehicle for dopamine. There are no significant differences between the spontaneous locomotor activity responses of the dopamine- and vehicle-infused animals ($P < 0.05$). Data are also given for the responsiveness of these two groups of animals 2–35 days post-infusion to (−)NPA 0·05 mg kg^{-1} s.c. All animals were pre-selected as low-activity responders to (−)NPA (shown on day 0, ●, ○). Subsequent to withdrawing infusion, the vehicle-control animals maintained the low-activity response whereas the dopamine-infused rats exhibited increasing hyperactivity responses to (−)NPA such that they acquired and maintained a high-activity response 21–35 days post-infusion (*$P < 0.01$ to $P < 0.001$, two-way ANOVA followed by Dunnett's test for multiple comparisons). $n = 6$–12. SEMs < 12·1 %.

responding during infusion without a long-term consequences of motor change (the anterior olfactory nucleus, Fig. 25.8). Thus, the behavioural expressions of acute and chronic change to a persistent dopamine overstimulation of a forebrain limbic or cortical area are not necessarily coincident, and one effect may occur independently of the other. The nature of the mechanisms mediating such changes remains to be elucidated, and this serves again to emphasise the complexities of the regulatory capacities of forebrain dopamine systems to control motor activity.

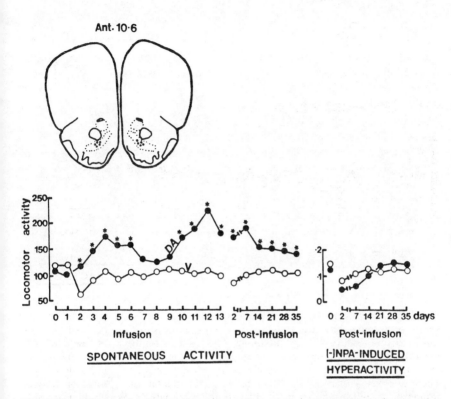

Fig. 25.8.   Effects of infusing dopamine, 25 μg day^{-1}, 0·48 μlitre h^{-1}, into the anterior olfactory nucleus of the rat (ant. 10·6, vert. 0·0, lat. $\pm$1·6). Spontaneous locomotor activity (counts per 60 min, measured in photocell cages) is shown for the 13 days of infusion and for days 2–35 post-infusion, ●—● indicating the responses of rats receiving the dopamine (DA) infusion for 13 days, and O—O indicating the responses of rats receiving an infusion of the vehicle (V) for dopamine. Dopamine infusion into the anterior olfactory nucleus caused significant increases in spontaneous locomotion above vehicle control levels, both during the 13-day period of infusion and during a 2- to 35-day post-infusion period (*$P<0·01$ to $P<0·001$, two-way ANOVA followed by Dunnett's test for multiple comparisons). Data are also given for the responsiveness of these two groups of animals 2–35 days post-infusion to (−)NPA, 0·05 mg kg^{-1} s.c. All animals were pre-selected as low-activity responders to (−)NPA (shown on day 0, O, ●). There were no significant ($P>0·05$) changes in the responsiveness of the dopamine-infused animals to (−)NPA on days 2–35 post-infusion: all animals maintained their low-activity status to (−)NPA. $n=6$–12. SEMs $<12·6\%$.

25.3. Relevance of dopamine infusion experiments to pathophysiology of schizophrenia

The final question is whether the changes in behaviour that follow the infusion of dopamine into the rodent forebrain are relevant to the pathophysiology of schizophrenia. Is it possible that the overt appearance of a psychotic disorder may reflect a persistently raised dopamine activity in limbic brain areas? The animal data would indicate that an overt behavioural disturbance may not necessarily be the consequence of a limbic dopamine excess, but that this excess, expressed behaviourally or not, may dictate the subsequent response of the patient to a dopamine agonist and, if the process is repeated, one may predict a situation developing where the limbic dopamine mechanisms may become excessively sensitive to endogenous dopamine. Certainly, in the animal model, L-dopa can cause a marked response following only one period of dopamine disturbance. It could be hypothesised that, in the human suffering repeated episodes of dopamine excess, the limbic mechanisms may become increasingly sensitive to dopamine such that the periods of remission become ever reduced in duration. Finally, there may become a point where 'normal' dopamine function, not even an excess, is sufficient to trigger a disturbance of limbic function: at this stage a patient would become chronically schizophrenic.

Two further points require consideration: first, the consequence of neuroleptic intervention, and, secondly, that one group of animals can be rendered chronically less responsive to dopamine agonist action as a consequence of the limbic dopamine infusion. For reasons not as yet entirely clear, neuroleptic challenge by repeated daily dosing can exacerbate the chronic effects of a limbic dopamine excess, and the maintenance of constant blood levels of neuroleptic may be critical for the prevention of the long-term consequences. Could this observation contribute to the success in application of depot medication in schizophrenia? The resistance of the behavioural expression of a limbic dopamine overstimulation in animals may also be extrapolated meaningfully to the clinical situation where, dependent on the degree of dopamine disturbance, patients may be expected to require anything from modest to heroic doses of neuroleptic agent to control the overt behavioural expression.

The recognition of clear subgroups of schizophrenic patients now finds counterparts in the rodent world. The relationship of the oversensitivity of rats to dopamine agonists to the clinical situation has already been discussed. There remains the group of rats which, as a consequence of an identical limbic dopamine excess, become less responsive to dopamine agonist challenge. The analogy in the clinic would be a schizophrenic patient of low sensitivity to dopamine agonists having a behavioural status which would be helped little by neuroleptic intervention. It would obviously be incautious to interpret the animal findings in terms of Crow's type I and type II schizophrenia (Crow, 1982), but the present data do, nevertheless, indicate that subgroups can occur in animal populations as well as in man when limbic function becomes disturbed.

References

Costall, B. and Naylor, R. J. (1976). The behavioural effects of dopamine applied intracerebrally to areas of the mesolimbic system. *Eur. J. Pharmacol.* **32**, 87–92.

Costall, B., Domeney, A. M. and Naylor, R. J. (1982). Behavioural and biochemical consequences of persistent overstimulation of mesolimbic dopamine systems in the rat. *Neuropharmacology* **21**, 327–35.

Costall, B., Domeney, A. M. and Naylor, R. J. (1983). A comparison of the behavioural consequences of chronic stimulation of dopamine receptors in the nucleus accumbens of rat brain effected by a continuous infusion or by single daily injections. *Naunyn-Schmiedeberg's Arch. Pharmacol.* **324**, 27–33.

Costall, B., Domeney, A. M. and Naylor, R. J. (1984a). Chronic enhancement of dopamine agonist action after intra-accumbens dopamine infusion. *Brit. J. Pharmacol.* **82**, 308P.

Costall, B., Domeney, A. M. and Naylor, R. J. (1984b). Locomotor hyperactivity caused by dopamine infusion into the nucleus accumbens of rat brain: specificity of action. *Pscyhopharmacology* **82**, 174–80.

Costall, B., Domeney, A. M. and Naylor, R. J. (1984c). Long-term consequences of antagonism by neuroleptics of behavioural events occurring during mesolimbic dopamine infusion. *Neuropharmacology* **23**, 287–94.

Costall, B., Domeney, A. M. and Naylor, R. J. (1985). The continuity of dopamine receptor antagonism can dictate the long-term behavioural consequences of a mesolimbic dopamine infusion. *Neuropharmacology* **24**, 193–7.

Crow, T. J. (1982). The biology of schizophrenia. *Experientia* **38**, 1275–82.

De Groot, J. (1959). The rat forebrain in stereotaxic coordinates. *Verh. K. Ned. Akad. Wet.* **52**, 14–39.

Donaldson, S. R., Gellenberg, A. J. and Baldessarini, R. J. (1983). The pharmacological treatment of schizophrenia: a progress report. *Schizophrenia Bull.* **9**, 504–27.

Haracz, J. L. (1983). The dopamine hypothesis: an overview of studies with schizophrenia patients. *Schizophrenia Bull.* **8**, 438–69.

Rupniak, N. M. J., Jenner, P. and Marsden, C. D. (1983). The effect of chronic neuroleptic administration on cerebral dopamine receptor function. *Life Sci.* **32**, 2289–311.

Stevens, J. R. (1978). Research in schizophrenia: regulation of dopamine in the mesolimbic system. *Impact of Science on Society* **28**, 39–56.

Van Kammen, D. P., Docherty, J. P., Marder, S. R., Schulz, S. C., Dalton, L. and Bunney, W. E. (1982a). Antipsychotic effects of pimozide in schizophrenia. *Arch. Gen. Psychiat.* **39**, 261–6.

Van Kammen, D. P., Docherty, J. P., Marder, S. R., Rayner, J. N. and Bunney, W. E. (1982b). Long-term pimozide pretreatment differentially affects behavioural responses to dextroamphetamine in schizophrenia. *Arch. Gen. Psychiat.* **39**, 275–81.

A reappraisal of abnormal, involuntary movements (tardive dyskinesia) in schizophrenia and other disorders: animal models and alternative hypotheses

J. L. Waddington†, H. A. Youssef[1], K. M. O'Boyle and A. G. Molloy

Department of Clinical Pharmacology, Royal College of Surgeons in Ireland, St. Stephen's Green, Dublin 2, Ireland
and
[1]*St. Davnet's Hospital, Monaghan, Ireland*

26.1. Introduction

There is still no satisfactory reason as to why some schizophrenic and other patients receiving long-term neuroleptic treatment develop a syndrome of abnormal, involuntary movements whereas the majority do not. These choreoathetoid movements involve principally the orobuccolingual and facial area (orofacial or buccolingual–masticatory dyskinesia), although they sometimes involve the limbs and trunk. Because of the common association with long-term exposure to such medication, the label of 'tardive' dyskinesia is usually applied. Use of this term implies a direct causal role of neuroleptic drugs. Striatal dopamine receptor supersensitivity, occurring as an overcompensatory response to neuroleptic-induced dopamine receptor blockade, is the most commonly accepted cause of hyperfunction of dopaminergic systems in the basal ganglia. These powerful heuristic hypotheses have proved disappointing in identifying risk factors and in generating new therapeutic strategies either to limit or to suppress the development of such abnormal movements. Comprehensive reviews covering the literature up to 1980 confirm this (Jeste and Wyatt, 1982; Kane and Smith, 1982). However, although it appears that these basic hypotheses require modification or even substantial revision, there are few indicators of how this process should proceed.

It is often overlooked that certain recent studies have questioned some of the commonly accepted notions that involuntary movements in schizophrenic and other patients may be caused by neuroleptics (see Waddington, 1984, and below). These studies have important implications for the issues of why the syndrome is seen in some but not all neuroleptic-treated patients

†To whom correspondence should be addressed.

(i.e. those of vulnerability or 'high risk' status) and of why indistinguishable movements can be seen to varying degrees in patients who have never been treated with neuroleptics. It seems that in the absence of any consistent evidence that patients with the syndrome have had longer or more vigorous treatment with neuroleptics (Kane and Smith, 1982), the need is to identify predisposing factors within individual affected patients rather than in particular modes of treatment. Only one characteristic, that of ageing, has been implicated in any consistent way (Smith and Baldessarini, 1980). However, we have recently reported preliminary data indicating that older chronic schizophrenic patients with abnormal movements are much more likely to show intellectual impairment/cognitive dysfunction and negative symptoms than similar patients without abnormal movements (Waddington *et al.*, 1985). These clinical characteristics are those of the 'defect state' of schizophrenia, or the Type 2 syndrome (Crow, 1980), where structural brain changes, especially ventricular dilatation, appear to be demonstrable by pneumoencephalography and CT scan (see Crow, 1982; Seidman, 1983).

It is provocative that these features of dementia and other organic brain disorders characterise so many of those patients who show involuntary movements in the absence of a history of exposure to neuroleptics (see Waddington, 1984). However, the urge to reinterpret such data in terms of simple organic predisposing factors for the syndrome must be tempered by the inconsistencies of the earlier literature (Kane and Smith, 1982). Also, any new or revised proposals must be able to accommodate and account for the association between neuroleptic treatment and the emergence of abnormal movements in younger schizophrenic patients with positive, florid symptomatology and in those non-schizophrenic patients receiving neuroleptic treatment for affective disorders or gastrointestinal dysfunction. Our purpose is to give a more extended account of some of our recent clinical investigations, and of related animal studies; then, drawing on the most recent literature, we develop a series of arguments and speculations as to how we might reconsider the syndrome of 'tardive' dyskinesia.

26.2. Clinical studies in schizophrenia

Our study population (Waddington *et al.*, 1985) consisted of 68 schizophrenic patients who have required long-term hospital care in St. Davnet's Hospital, Monaghan. For the diagnosis of schizophrenia the criteria of Feighner *et al.* (1972) were used. The patients were assessed for movement disorder using the Abnormal, Involuntary Movement Scale (AIMS) and their intellectual/cognitive function was evaluated using a modified abbreviated mental test of orientation and memory (see Qureshi and Hodkinson, 1974). After movement and neuropsychological evaluation, case notes and histories were reviewed to extract demographic variables together with details of the clinical course of the illness and of drug treatment. Using standard tables (Davis, 1976), drug doses were converted to equivalents of chlorpromazine (CPZ). In agreement with previous reports (Owens *et al.*, 1982; Karson *et al.*, 1983) we found buccolingual–masticatory movements to be much more common than those of the limbs and trunk, and therefore all subsequent references to abnormal, involuntary movements in our patients relate to these cardinal orofacial signs. Their prevalence in this population was 41% using the criterion of an AIMS score of $\geqslant 2$ on at least one orofacial region, with no significant sex differences over the groups as a whole.

Comparisons between patients with and without such abnormal movements are shown in Table 26.1. Patients with involuntary movements were significantly older, and tended to have a longer duration of illness and to be older when first given neuroleptics. Regarding treatment variables, patients with involuntary movements had been treated with neuroleptics for the same time as their counterparts without such movement disorder. Surprisingly, however, they had received significantly smaller average daily

Table 26.1.   Characteristics of schizophrenic patients with and without abnormal, involuntary movements (AIM)

	$-AIM(N = 40)$	$+AIM(N = 28)$
General		
Age	$61 \cdot 1 \pm 2 \cdot 1$	$68 \cdot 5 \pm 2 \cdot 7^*$
Duration of illness	$33 \cdot 4 \pm 1 \cdot 8$	$38 \cdot 5 \pm 2 \cdot 5$
Age at first neuroleptic	$44 \cdot 4 \pm 2 \cdot 2$	$50 \cdot 6 \pm 2 \cdot 6$
Treatment		
Duration of neuroleptics (years)	$13 \cdot 8 \pm 1 \cdot 0$	$13 \cdot 0 \pm 1 \cdot 3$
Average daily dose (CPZ, mg)	449 ± 69	$259 \pm 49^*$
Lifetime intake (CPZ, g)	2233 ± 386	1180 ± 231
Duration drug free	$2 \cdot 4 \pm 0 \cdot 6$	$3 \cdot 5 \pm 0 \cdot 8$
Drug free > 1 month	$17/37(45 \cdot 9\%)$	$18/25(72 \cdot 0\%)$
Duration of anticholinergics (years)	$5 \cdot 9 \pm 0 \cdot 8$	$4 \cdot 9 \pm 0 \cdot 9$
Negative symptoms		
Muteness	$6/40(15 \cdot 0\%)$	$11/28(39 \cdot 3\%)^*$
Blunted affect	$7/40(17 \cdot 5\%)$	$11/28(39 \cdot 3\%)$
Either or both of above	$13/40(32 \cdot 5\%)$	$20/28(71 \cdot 4\%)^{**}$
Intellectual impairment		
Mean score	$6 \cdot 7 \pm 0 \cdot 3$	$5 \cdot 3 \pm 0 \cdot 6^*$
Score $< 50\%$	$2/32(6 \cdot 2\%)$	$6/16(37 \cdot 5\%)$
Defect state/Type 2 signs		
One or more of muteness, blunted affect, intellectual impairment	$14/40(35 \cdot 0\%)$	$25/28(89 \cdot 3\%)^{***}$

Means $\pm$ SEM*, $P < 0 \cdot 05$;**, $P < 0 \cdot 01$;***, $P < 0 \cdot 001$.

doses and total lifetime intakes of neuroleptics. Interestingly, they also appeared to have spent a greater time drug-free since the first administration of neuroleptics. These two groups of patients had received anticholinergic drugs for indistinguishable periods of time. Regarding clinical features of their illness, patients with abnormal movements were significantly more likely to have shown the negative symptoms of muteness and/or blunted affect. In patients who were mute, it was not possible to assess intellectual/cognitive function. However, in those patients who were not mute, a significantly lower mean score on neuropsychological assessment (i.e. greater intellectual impairment) was found in the group with involuntary movements, and they were significantly more likely to score less than half correct. Patients with abnormal movements were distinguished by their highly significant association with at least one of the above signs of muteness, blunted affect or an intellectual function score of less than half (Table 26.1).

26.2.1. *General factors*

Our study population of schizophrenic patients appeared typical of others investigated for involuntary movements. The prevalence estimate and the greater age of patients with such a movement disorder is entirely consistent with results from other investigations in similar long-term hospitalised populations (Smith and Baldessarini, 1980; Kane and Smith, 1982).

26.2.2. *Treatment factors*

Very few controlled studies have shown a relationship between cumulative drug intake and abnormal movements (Kane and Smith, 1982). The present study further attests to the absence of any such simple relationship. Our data are also in agreement with the prevailing view that there is no consistent relationship between length of drug exposure and abnormal movements. However, the recent study of Toenniessen *et al.* (1985) indicates that in older patients an association between length of treatment and abnormal movements is demonstrable, but is only apparent early in exposure over the first two years. Their data are consistent with the suggestion of Kane and Smith (1982) that associations between treatment variables and abnormal movements may be confined to those patients with high vulnerability and may occur during shorter exposures than have been generally assumed. This emphasises again the crucial importance of identifying predisposing factors characterising the vulnerable or 'at-risk' patient, distinct from particular modes of treatment.

Our finding that patients with involuntary movements had received smaller average daily doses and reduced total lifetime intakes of neuroleptics may at first seem paradoxical. Certainly their drug treatment appeared to have been less vigorous, consistent with the report by Smith *et al.* (1978) of a negative relationship between total neuroleptic intake during hospitalisation and involuntary movements. The present findings might superficially suggest that patients with abnormal movements were less severely ill than their more vigorously medicated counterparts. However, an alternative explanation would be that movement-disordered patients were not less severely ill *per se* but, rather, manifested fewer or less intense, positive, florid schizophrenic symptoms, i.e. less prominent delusions, hallucinations and thought disorder. Their greater age would be consistent with a greater likelihood of deterioration to the 'defect state' where negative symptoms such as poverty of speech and flattening of affect might be more prominent. The case notes reveal that, over the years, the psychiatrists in charge of our patients have from time to time commendably attempted to reduce neuroleptic dosages to match the extent of productive symptoms, and to minimise medication of patients in the 'defect state' where the least benefit is likely to be derived. This practice supports the interpretation that we favour.

'Drug holidays' have been advocated on theoretical grounds by some authors to reduce the likelihood of emergence of involuntary movements, but there have been a number of provocative reports suggesting that a greater frequency of drug-free episodes during long-term neuroleptic treatment may in fact be associated with an increased risk of the syndrome (Kane and Smith, 1982; Branchey and Branchey, 1984). In our study, patients with abnormal movements were more likely to have spent at least 1 month drug-free than those without the syndrome, and tended to have spent a greater total period drug-free since their first exposure to neuroleptics; in agreement with the other reports in the literature, and with the difficulties of analysis and interpretation inherent in such studies, these results must be considered to be similarly provocative but inconclusive. In these other studies, drug-free episodes occurred often as a result of out-patient non-compliance (see, for example, Branchey and Branchey, 1984). In our own in-patient population, such drug-free periods were more usually a consequence of a positive clinical decision that neuroleptic treatment was not justified (temporarily or foreseeably). Thus these extended drug-free periods in our patients with involuntary movements probably reflect, at least in part, the greater prominence of negative symptoms in this group.

Although there has been speculation that the use of anticholinergic drugs may increase the risk of development of abnormal movements, the literature as a whole fails to substantiate this (Kane and Smith, 1982). We found that patients with and without involuntary movements did not differ in terms of exposure to anticholinergics. Because of the consistent use of only a small number of drugs over narrow and equivalent daily dose ranges (usually benztropine 2–4 mg, occasionally orphenadrine 50–150 mg), systematic dose comparisons were not helpful. There was no association between abnormal movements and the presence or absence of either current neuroleptic or anticholinergic medication.

26.2.3. *Negative symptoms*

Our finding of a highly significant excess of muteness and/or blunted affect in chronic schizophrenic in-patients with abnormal movements was striking. Such associations have been noted previously in predominantly psychogeriatric (Itil *et al.*, 1981) and out-patient (McCreadie *et al.*, 1982; Csernansky *et al.*, 1983) populations. This is of particular interest as such negative symptoms have been found to be over-represented in schizophrenic patients with structural brain changes as revealed by CT scan (Andreasen *et al.*, 1982; Crow, 1982; Kling *et al.*, 1983). It is those same clinical features which may be resistant to the influence of neuroleptic drugs (Johnstone *et al.*, 1978; Angrist *et al.*, 1980), and resistance to neuroleptic treatment in schizophrenia has itself been associated with such structural brain changes (see Smith and Maser, 1983).

Because of the continuing debate as to the nature, significance, specificity and assessment of negative symptoms in schizophrenia (see Lewine *et al.*, 1983) it would be unwise to overgeneralise from the present clinical findings in isolation. However, we feel they may be the first indicators of a consistent theme that will be further developed below.

26.2.4. *Intellectual impairment/cognitive dysfunction*

Over the past several years a substantial body of evidence has accumulated to the effect that, contrary to previous general assumptions, a proportion of chronic schizophrenic patients show evidence of intellectual impairment and cognitive dysfunction (see Crow, 1982; Seidman, 1983; Taylor and Abrams, 1984). We find chronic schizophrenic patients with involuntary movements to be significantly more intellectually impaired than those without such movement disorder; in some patients with abnormal movements there was a profound loss of cognitive function. The earlier literature is equivocal on this issue (Kane and Smith, 1982) though Itil *et al.* (1981) have since reported a greater impairment of orientation and memory in psychogeriatric patients with involuntary movements than in similar patients without such movements. Two very recent studies have also addressed this issue. Wolf *et al.* (1983) failed to find significant differences in particular aspects of cognitive function between schizophrenic patients with and without abnormal movements. However, Struve and Willner (1983) reported that schizophrenic patients with involuntary movements showed greater cognitive impairment than those without involuntary movements. The finding of the latter authors that this association is confined to patients with 'persistent' abnormal movements is similar to our own, namely that there is no such association in patients with 'transient' or 'withdrawal-emergent' abnormal movements.

A number of recent CT scan studies have reported intellectual impairment and cognitive dysfunction in schizophrenia to be associated with structural brain changes (see Crow, 1982; Seidman, 1983; Smith and Maser, 1983). The most recent eclectric review of the many controversies in this area of CT studies in schizophrenia (Maser and Keith, 1983) concludes that 'On balance, the weight of evidence is clearly on the side of a neuroanatomical abnormality' (page 281). We suggest that chronic schizophrenic patients with persistent involuntary movements are more intellectually impaired, as well as more likely to show negative symptoms, than similar patients without such movement disorder, and that both of these characteristics are associated with structural brain changes.

26.2.5. *CT scan studies in patients with abnormal, involuntary movements*

This issue of structural abnormalities in the brains of neuroleptic-treated patients with abnormal, involuntary movements has been addressed in a number of recent quantitative studies. Famuyiwa *et al.* (1979) found schizophrenic patients with involuntary movements to be significantly more likely to have a pathological rating on the ventricular index and to be more impaired as shown by a cognitive learning test than patients without such movements. In elderly, predominantly schizophrenic patients with abnormal movements, Jeste *et al.* (1980) found measures of caudate atrophy to be intermediate between similar patients without abnormal movements and a group with Huntington's disease, though only the latter group differed significantly from the others. Brainin *et al.* (1983) could not establish a significant relationship between severity of involuntary movements and CT scan abnormalities, but signs of structural abnormality were noted in some patients where movements were pronounced.

The most extensive published study is that of Bartels and Themelis (1983). CT scans in a large number of predominantly schizophrenic patients, most of whom were in the 'defect' state, revealed significant radiological abnormalities on several measures in the group with abnormal movements when compared with matched patients without such movements. These abnormalities included increased ventricular width and atrophy of the head of the caudate, and were associated with significantly greater cognitive deficits in the movement-disordered group on neuropsychological assessment. Interestingly, Pandurangi *et al.* (1980) have reported pneumoencephalographic evidence of caudate atrophy in schizophrenic patients with involuntary movements when compared with similar patients without such movements; this effect was confined to patients with 'persistent' involuntary movements, and was not seen in two cases where movements progressively decreased after withdrawal of medication.

Involuntary movements are most prevalent in older chronic schizophrenic populations, and it is in such patients that the majority of the above studies have been carried out. However, the syndrome can be seen in younger patients, and it is important to consider the extent to which the above radiological and clinical signs can be seen in the age range from the twenties up to, say, 45 years. The literature on this issue is more controversial; however, CT scan abnormalities have indeed been reported in schizophrenic patients in their twenties (Dewan *et al.*, 1983; Woods and Wolf, 1983) and in teenage patients with schizophrenia spectrum disorder (Schulz *et al.*, 1983*a*).

Even in such young patients, structural brain changes have been associated with poor response to neuroleptic treatment (Schulz *et al.*, 1983*b*), intellectual impairment (Rieder *et al.*, 1979; Golden *et al.*, 1980), and differences in clinical state in relation to negative vs positive symptoms (Luchins *et al.*, 1983); negative symptoms can indeed be seen in some young patients (Lewine *et al.*, 1983). Few have systematically compared these signs in younger patients with and without involuntary movements. However, the patients of Struve and Willner (1983) who showed involuntary movements and were cognitively impaired had a mean age only in their early thirties. Also, in our own patients under 45 years, those with involuntary movements ($N = 2$) had exhibited negative symptoms whereas those without such movements ($N = 5$) had not.

These studies on CT scan changes should be considered alongside those on the association of abnormal movements with negative symptoms and intellectual impairment, two clinical signs which are themselves associated with structural brain changes. We suggest that it is difficult to avoid the conclusion that abnormal movements in chronic schizophrenia, at least those that are 'persistent', are associated with an excess of objective evidence and clinical signs of structural abnormalities of the brain, and that this association is not the sole prerogative of the older chronic patient.

26.2.6. Abnormal, involuntary movements in the absence of exposure to neuroleptics

Two recent and extensive studies have confirmed that abnormal orofacial movements are a rather infrequent phenomenon in normal subjects for whom a history of neuroleptic exposure and of neurological or psychiatric disorder has been excluded; the prevalence of such signs was less than 8 % even in individuals in the eighth decade of life (Kane *et al.*, 1982; Klawans and Barr, 1982). Such data are usually interpreted as indicating a sound basis for assuming the neuroleptic dependence of the syndrome in high-prevalence populations receiving such treatment. Interestingly, they also indicate that such movements can, albeit rarely, be seen as a consequence of CNS changes associated with ageing and senescence and unrelated to medication.

A new significance must be attached to such studies on normal subjects following the crucial investigation of Owens *et al.* (1982). These authors were able to assess the prevalence and severity of abnormal movements in a rare and very important population of 47 older chronic schizophrenic patients who had never been exposed to neuroleptics because of the therapeutic community and psychodynamic orientation of the psychiatrist in charge of their care. In comparison with 364 similar patients in the same hospital who had received typical long-term neuroleptic treatment, those 47 that had remained neuroleptic-free showed a high and indistinguishable prevalence of abnormal, predominantly orofacial, movements; these were similar to the movements seen in their neuroleptic-treated counterparts both qualitatively and in terms of their severity. This striking and provocative finding requires careful analysis. The patients studied by Owens *et al.* (1982) were an older, institutionalised chronic schizophrenic population, characterised by behavioural deterioration, negative symptoms and intellectual impairment (Owens and Johnstone, 1980) indicative of schizophrenia in its most severe 'demented' form. These drug-free patients were slightly but significantly older than their medicated counterparts, and after age correction abnormal movements were found to be slightly but significantly more prominent in the neuroleptic-exposed group (Crow *et al.*, 1982). This study seems to suggest that the baseline of abnormal movements can change, depending upon the neurological and psychiatric status of the patient; when the clinical picture is characterised by intellectual impairment and negative symptoms, as is especially likely in older patients, the neural substrate of this state can itself be associated with abnormal movements even in the absence of exposure to neuroleptics. In the patients of Owens *et al.* (1982), there is a preliminary report of an association between abnormal movements and CT scan abnormalities, particularly lateral ventricular enlargement (Crow *et al.*, 1983).

We ourselves have been able to obtain data on a small number of chronic schizophrenic patients who had remained free from neuroleptic treatment. These elderly patients, all in their ninth decade of life, had not been exposed to neuroleptics because of a number of associated factors, rather different from those applying to the study of Owens and his colleagues; the clinical picture in our subgroup was characteristic of the 'defect state' of schizophrenia, florid positive symptoms having abated in the era immediately prior to the introduction of neuroleptics, and the clinical judgement was that the potential dangers of neuroleptic treatment outweighed the small likelihood of any therapeutic improvement being derived. These patients were compared with a similar schizophrenic group who had a history of long-term neuroleptic treatment and who were further subdivided on the basis of whether they were or were not currently receiving neuroleptics; because of the problem of age matching, these two comparison subgroups were composed of patients in the same (ninth) decade of life, hence the small size of numbers concerned. Abnormal movements were

Table 26.2. Scores on the Abnormal, Involuntary Movement Scale (AIMS) in elderly (ninth decade) schizophrenic patients in relation to presence or absence of history of neuroleptic treatment

	Neuroleptic treatment		
	Never (N = 4; 3F,1M)	Past (N = 4; 2M,2F)	Present (N = 3; 2F,1M)
Neuroleptic exposure	—	11·4 ± 3·6	20·0 ± 2·0
Withdrawal period	—	4·4 ± 2·1	—
AIMS score	5·0 ± 0·9	5·0 ± 1·3	4·7 ± 1·5

Data are means ± SEM.

observed in all groups; there were no significant differences in scores on the Abnormal, Involuntary Movement Scale between patients who had and who had not been exposed to neuroleptics, and this lack of difference could not be accounted for in terms of current neuroleptic treatment (Table 26.2).

We have recently reviewed the historical literature on abnormal movements in psychiatric patients prior to the advent of neuroleptics (Waddington, 1984). When these usually qualitative descriptions of abnormal movements are compared with the criteria of modern rating scales for the assessment of such movements, it is clear that the early literature is not inconsistent with the results of Owens and his colleagues and of ourselves. As neuroleptics have been used so ubiquitously for over 25 years, opportunities to assess schizophrenic patients never exposed to such drugs now occur only in the most exceptional circumstances. It is therefore interesting that in one recent group of 36 schizophrenic patients with abnormal movements who also showed an excess of flattened affect and social withdrawal (McCreadie *et al.*, 1982), there was no record of two patients ever having received neuroleptics.

A crucial question is whether these phenomena are specific for schizophrenia or else reflect the proven or assumed structural brain changes that can also occur in classically organic, usually dementing, illnesses. The literature on the prevalence of abnormal movements in various patient groups not exposed to neuroleptics is usually considered to be highly inconsistent (range 0–39%; see Kane and Smith, 1982). However, this literature becomes more consistent when one subdivides such studies on the basis of whether the relevant populations did or did not include cases of dementia. Several recent studies exemplify the importance of this distinction. Thus Martinelli and Gabellini (1982) found abnormal orofacial movements in 7·7% of 104 patients up to 84 years old who showed essential tremor, indistinguishable from the prevalence in the normal elderly population (Klawans and Barr, 1982). Conversely, four studies have reported high prevalences of abnormal orofacial movements (18·0–36·7%) in populations of elderly patients (age range 59–102; total $N = 829$) for whom a history of exposure to neuroleptics had been excluded but in which an (unspecified) proportion of cases showed clinical evidence of dementia, among other mixed neurological signs (Delwaide and Desseilles, 1977; Bourgeois *et al.*, 1980; Blowers, 1981). Mölsa *et al.* (1982) have recently reported abnormal orofacial movements in 16·8% of 143 patients with Alzheimer's disease.

It should be emphasised that when abnormal movements have been compared in the same study in patients who have and have not been exposed to neuroleptics, prevalences are higher in the neuroleptic-treated group (Bourgeois *et al.*, 1980; Blowers, 1981); this agrees with the age-corrected study of Owens and his colleagues (see Crow *et al.*, 1982). An earlier extensive study of Brandon *et al.* (1971) is often misinterpreted as simply showing no excess of abnormal movements in patients who have received neuroleptics over those who have not. Careful reappraisal of these data reveals that such movements were significantly more prevalent in neuroleptic-treated patients over 50, and that a substantial majority of the group showing abnormal movements in the absence of neuroleptic exposure showed clinical signs of dementia (see Kane and Smith, 1982; Toenniessen *et al.*, 1985; Waddington, 1984). The most recent CT scan studies have confirmed the presence of structural abnormalities of the brain in patients with a variety of dementing illnesses (see Damasio *et al.*, 1983).

26.2.7. Interim conclusion

It may be that the phenomenon of abnormal movements in patients not exposed to neuroleptics is an extreme manifestation of an organic (often structurally determined) vulnerability in schizophrenia and classically organic brain disease(s). Neuroleptics may enhance such processes, elevating to a varying extent the baseline prevalence of abnormal movements.

26.3. Studies in animals

Although there have been no entirely satisfactory animal models of abnormal, involuntary movements arising after very prolonged neuroleptic administration, some elements of consistency have recently appeared in a literature once characterised by abject contradiction (see Waddington, 1984). We describe here some of our recent studies of neuroleptic effects on orofacial movements in rats, extending our earlier series of studies (Waddington *et al.*, 1983a, b). Our idea was to utilise a subject population in which organic brain changes would be more likely to be present to investigate some of the ideas presented earlier in this chapter. By analogy with the human situation, we have studied similar phenomena in aged/senescent animals in comparison with conventional young counterparts.

Male Sprague–Dawley rats (young, 3 months; aged/senescent, 22 months) were given monthly i.m. injections of the depot neuroleptics haloperidol decanoate (25 mg kg^{-1}) and fluphenazine decanoate (2·5 mg kg^{-1}) or of oil vehicle alone. Before and periodically after beginning treatment they were assessed for orofacial movements using a procedure similar to that applied to the clinical population. Thus movements of various orofacial regions were rated on a 0 (absent) to 4 (severe) scale by an observer unaware of the treatment history of each animal. Movements were assessed in various groups 30 min after an acute challenge with vehicle, or with haloperidol (0·75 mg kg^{-1} s. c.) or the anticholinergic procyclidine (7·5 mg kg^{-1} s. c.) in a crossover design. Representatives of each treatment group were taken at the 3-month point for assay of striatal D-2 dopamine receptors using [^{3}H]-spiperone (O'Boyle and Waddington, 1984).

Orofacial movements, most prominently vacuous chewing but with occasional tongue protrusions, occurred to a small extent in young animals before initiating treatments; they were not exacerbated by acute haloperidol challenge, a trend towards attenuation being noted (vehicle scores: 0·41 ± 0·12; haloperidol scores: 0·24 ± 0·10; means ± SEM, $N = 17$). These movements were significantly more prominent ($P < 0·01$) in aged animals before initiating treatment; they were also not exacerbated by acute haloperidol challenge, and again a trend towards attenuation was noted (vehicle scores: 1·53 ± 0·25; haloperidol scores: 1·26 ± 0·24; $N = 19$). This age-related increase in spontaneous mouth movements was not associated with an increase in spontaneous or apomorphine-induced stereotyped behaviour when assessed using conventional techniques (Molloy and Waddington, 1984). Stereotypy scores (young vs old) were as follows: spontaneous, 0·6 ± 0·4 vs 0·8 ± 0·3; apomorphine 0·5 mg kg^{-1}, 2·8 ± 0· 3 vs 2·8 ± 0·5; apomorphine 4·0 mg kg^{-1}, 3·0 ± 0·4 vs 3·0 ± 0·4, $N = 4$–5). We have recently reported a significant decrease in the density of striatal D-2 receptors over this age range, with no change in the characteristics of the D-1 dopamine receptor labelled with [^{3}H]-piflutixol; increased non-specific binding of this ligand in the aged striatum suggested pathophysiological changes that are unrelated to changes in dopamine receptors (Waddington *et al.*, 1985).

After beginning long-term neuroleptic treatment in young animals, there was very little influence on either neuroleptic or orofacial movements over the first 1·5 months; between 1·5 and 3 months, however, a significant excess ($P < 0·01$) of orofacial movements (vacuous chewing) emerged in both treatment groups (Fig. 26.1). The baseline of orofacial movements in aged animals was significantly elevated before initiating neuroleptic treatment, but drug-associated effects were minimal in this group. Figure 26.1 illustrates the outcome after 3 months of identical neuroleptic treatments in

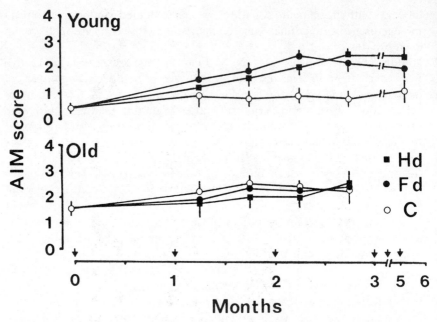

Fig. 26.1.   Orofacial movements as abnormal, involuntary movement (AIM) scores. Assessments were made before and after beginning monthly treatments (arrowed) with haloperidol decanoate (Hd), fluphenazine decanoate (Fd), or oil vehicle controls (C), in parallel young and old animals. Significant differences: young animals, Hd and Fd vs C, $P < 0.05$ at 1·75 months and $P < 0.01$ at 2·25 months; controls, young vs old, $P < 0.01$. $N = 14$–15 up to 3 months; $N = 7$–9 thereafter.

these aged animals, compared with the parallel young groups. Although orofacial movements were more prominent in each aged group than in their matched young counterparts, the only significant difference was that ($P < 0.05$) between aged and young vehicle-injected animals.

The effects of acute pharmacological challenges in those young animals continuing to receive neuroleptics beyond the third month are shown in Fig. 26.2. Challenge with the anticholinergic procyclidine had little consistent effect on mouth movements, though in the majority of animals these were increased or unaltered. Challenge with haloperidol tended to produce somewhat more consistent results; in two-thirds of animals, mouth movements were decreased or unaltered. It was notable that in animals given acute challenge with vehicle, orofacial movements in the control group receiving monthly oil injections were somewhat elevated in comparison with ratings made in these same animals before and immediately after the acute challenge period (months 3–5; compare Figs 26.1 and 26.2). In those animals taken for receptor binding studies at the 3-month point, striatal D–2 receptor density was elevated in young animals by

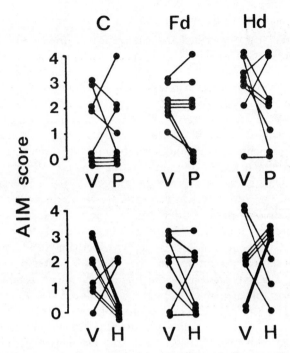

Fig. 26.2. Orofacial movements 30 min after acute challenge with procyclidine 7·5 mg kg^{-1} (P), haloperidol 0·75 mg kg^{-1} (H), or appropriate vehicles (V), in the three treatment groups of Fig. 26.1. Assessments were made in a crossover design for each acute drug challenge, between 3 and 5 months after beginning long-term treatments in young animals.

haloperidol decanoate 25 mg kg^{-1} month^{-1} ($P < 0·01$) but not by fluphenazine decanoate 2·5 mg kg^{-1} month^{-1} (Table 26.3); there was a significant overall effect of ageing ($P < 0·01$), both to decrease the density of striatal D-2 receptors and to diminish the degree of elevation induced by prolonged haloperidol decanoate treatment.

Table 26.3. Characteristics of striatal [^{3}H]-spiperone binding to D-2 dopamine receptors after 3-month neuroleptic treatments with haloperidol decanoate or fluphenazine decanoate. Affinity (K_d, nM) and receptor density (B_{max}, pmol g^{-1}) were determined in young and old animals receiving parallel treatments

| | Young | | Old | |
	K_d	B_{max}	K_d	B_{max}**,a
Control	0·07±0·01	13·8±1·4	0·09±0·02	11·5±1·4
Haloperidol	0·08±0·01	20·8±1·7**	0·09±0·02	14·5±1·2
Fluphenazine	0·09±0·01	12·7±0·6	0·09±0·01	10·1±0·8

Means±SEM, $n = 5$–6; **,$P < 0·01$;^{a}, significant overall effect of age.

26.3.1. Pharmacology of the animal model

We have previously described the emergence of an excess of late-onset orofacial movements in young rats during 6 months of treatment with representatives from each of the major classes of neuroleptic drugs. These movements usually emerged to excess while drug administration continued; an exception was haloperidol, where excess orofacial movements emerged only after neuroleptic withdrawal. The excess of orofacial movements was an enhancement of a predisposition towards such movements, and it persisted after neuroleptic withdrawal beyond the decline in dopamine receptor supersensitivity (Waddington *et al.*, 1983*a, b*; Waddington, 1984). We now replicate the emergence of an excess of late-onset orofacial movements during prolonged neuroleptic treatment in young animals. In agreement with the clinical literature in general, these movements were usually suppressed or unaltered by an acute haloperidol challenge (see Jeste and Wyatt, 1982), being only occasionally exacerbated (Casey and Denney, 1977). The clinical literature on the effects of previous or current anticholinergic agents on abnormal orofacial movements is now recognised to be more complicated than is commonly believed (see Jeste and Wyatt, 1982; Kane and Smith, 1982; Gardos and Cole, 1983). Thus, contrary to the usual assumption that anticholinergics are almost invariably detrimental to involuntary movements in general, Casey and Denney (1977) and Perenyi *et al.* (1983) have described cases in which opposite responses were found; also, both acute anticholinergic challenge and withdrawal of current anticholinergics have been reported to influence involuntary movements of the trunk, extremities and other somatic areas more than they influence the orofacial region (Chouinard *et al.*, 1979; Perenyi *et al.*, 1983). The finding that these orofacial movements in rats were more commonly exacerbated or else unaltered by anticholinergic drugs, but were sometimes attenuated, does not therefore seem inconsistent with an eclectic review of the clinical literature.

Those orofacial movements which occurred to excess either spontaneously in aged animals or after long-term neuroleptic treatment in young animals were again consistently dissociable from dopamine receptor supersensitivity and from dopaminergic hyperfunction in general; there was indirect evidence for a non-dopaminergic pathophysiology. It is important to emphasise that the ability of acute neuroleptics or anticholinergics to sometimes influence orofacial movements does not require that abnormal dopamine–acetylcholine function underlies the syndrome, though it is consistent with these processes exerting a modulatory action.

26.3.2. Implications of the animal model

Our animal studies confirm that an excess of late-onset orofacial movements emerges in younger animals given prolonged neuroleptic treatment;

neuroleptics again failed to 'create' such movements, the process being one of exacerbation of oral behaviours already within the animals (and in man's) behavioural repertoire to a variable but usually minimal extent. Indistinguishable movements were also seen to occur to excess in animals as a consequence of CNS changes associated with ageing and senescence. The prominence of such movements in aged animals given prolonged neuroleptic treatment had its basis more in these CNS changes than in such drug administration. These data suggest that such phenomena in animals not exposed to neuroleptics are an extreme manifestation of an organically determined vulnerability associated with maturation, ageing and senescence. Neuroleptics may enhance such processes, elevating to a varying extent the baseline prevalence of orofacial movements. The present animal data will thus support a proposition that is strikingly similar to the interim conclusion (section 26.2.7) derived from clinical data.

26.4. Conclusion

Our interim conclusion was that abnormal movements arise in neuroleptic-treated patients with schizophrenia and organic psychosyndromes because of drug interaction with an organic (often structurally determined) covert vulnerability which, in its most extreme form, can be manifested as overt abnormal movements without the necessity of exposure to neuroleptics. If such arguments are to form the basis of a more general hypothesis relating to abnormal movements, they must be able to accommodate the emergence of the syndrome in neuroleptic-treated patients with diagnoses other than schizophrenia or an organic psychosyndrome. There are reports of the syndrome in patients who have received neuroleptics for a wide variety of psychiatric, neurological or medical illnesses (see Kane and Smith, 1982).

Clearly, the emergence of abnormal movements in patients treated with neuroleptics for affective or gastrointestinal disorders appears to present the greatest problem for a general hypothesis of organic/structural vulnerability. However, following the demonstration of structural abnormalities of the brain in schizophrenia, several recent CT scan studies have addressed the issue of the specificity of such changes by investigating other diagnostic groups. Similar structural abnormalities have been identified in the brains of patients with depression, mania and bipolar affective disorder (Luchins *et al.*, 1983; Rieder *et al.*, 1983; Targum *et al.*, 1983). Also, Rush *et al.* (1983) have confirmed earlier reports that cognitive dysfunction can be a concomitant of affective disorder. These findings are provocative in the light of the recent report that patients with affective disorders who show abnormal movements are more cognitively impaired than those affective patients without such movements (Wolf *et al.*, 1983). It thus appears that our arguments, derived from studies in schizophrenia and in organic psychosyndromes, are applicable to affective disorders.

In non-psychiatric populations, neuroleptic drugs are sometimes prescribed for gastrointestinal disorders. Metoclopramide is the most widely used agent, and recent studies have confirmed more extensively the original case reports of abnormal orofacial movements in such patients (Grimes *et al.*, 1982; Wiholm *et al.*, 1984). However, these recent studies report that all patients with abnormal movements were aged 65 or over, with the great majority older than 70 years. It should be noted that ventricular enlargement and other structural abnormalities can be seen in CT scans of some elderly subjects, and that elderly subjects can show mild intellectual impairment, insufficient to warrant a diagnosis of dementia (see Damasio *et al.*, 1983). Thus, our arguments are not disproved by the emergence of abnormal movements in patients given neuroleptics for gastrointestinal disorders; the same vulnerability factors may still apply.

Some insight into the nature of this organic/structural vulnerability may be found in the report by Weinhold *et al.* (1981). They describe the emergence of abnormal movements in two young schizophrenic brothers treated with neuroleptics, and in each there was evidence of cognitive deficits. Such familial occurrence of the syndrome in younger patients, both with intellectual impairment, strongly supports an organic (and genetic) factor in vulnerability.

Regarding the pharmacological and neurochemical basis of the interaction between neuroleptic drugs and a vulnerability factor, the inadequacies of the dopamine receptor supersensitivity hypothesis are becoming increasingly recognised. Our animal studies, which are in such accord with clinical investigations, consistently suggest that the basis for these effects of neuroleptics may be found in non-dopaminergic systems of the brain; certainly dopamine receptor supersensitivity appears neither sufficient nor necessary for the emergence of persistent orofacial movements. It cannot be overlooked that two recent receptor binding studies have independently reported that post-mortem striatal tissue from schizophrenic patients rated in life for the presence of such abnormal movements does not show dopamine receptor supersensitivity in comparison with tissue from schizophrenic patients without these movements (Crow *et al.*, 1982; Riederer *et al.*, 1983). Though other weaknesses of the dopamine receptor supersensitivity hypothesis are discussed elsewhere, it must be acknowledged that no simple alternative is yet apparent to replace it (see Waddington, 1984). The neurochemical basis of any interaction between neuroleptic drugs and vulnerability factors remains obscure.

The report of Campbell *et al.* (1983) may indicate how one could consider this interface between organic vulnerability and neuroleptic treatment. They report an association between long-standing obstructive pulmonary disease and both the emergence of spontaneous abnormal movements and the exacerbation of neuroleptic-associated abnormal movements. Such chronic hypoxia has been shown to result in reduced activity in the synthetic pathway for GABA; also, Gunne *et al.* (1984) have recently reported GABAergic deficits in extrapyramidal areas such as the substantia nigra in primates with neuroleptic-associated abnormal movements. Therefore, hypoxic states may acquire a new significance in considerations of organic factors, including structural brain changes, which we suggest confer vulnerability to the emergence of abnormal, involuntary movements.

Acknowledgements

We thank the Medical Research Council of Ireland, the Royal College of Surgeons in Ireland, the Royal College of Physicians of Ireland, Sanity, the Mason Medical Research Foundation, Squibb (UK) and Janssen Pharmaceutica for financial support, and Wellcome Ireland Ltd for gifts of drugs.

References

Andreasen, N. C., Olsen, S. A., Dennert, J. W. and Smith M. R. (1982). Ventricular enlargement in schizophrenia: relationship to positive and negative symptoms. *Amer. J. Psychiat.* **139**, 297–302.

Angrist, B. J., Rotrosen, J. and Gershon, S. (1980). Different effects of amphetamine and neuroleptics on negative vs positive symptoms in schizophrenia. *Psychopharmacology* **72**, 17–19.

Bartels, M. and Themelis, J. (1983). Computerised tomography in tardive dyskinesia. *Arch. Psychiat. Nervenkr.* **233**, 371–9.

Blowers, A. J. (1981). Epidemiology of tardive dyskinesia in the elderly. *Neuropharmacology* **20**, 1339–40.

Bourgeois, M., Bouilh, P., Tignol, J. and Yesavage, J. (1980). Spontaneous dyskinesia vs neuroleptic-induced dyskinesia in 270 elderly subjects. *J. Nerv. Ment. Dis.* **168**, 177–8.

Brainin, M., Reisner, Th. and Zeitlhofer, J. (1983). Tardive dyskinesia: clinical correlation with computed tomography in patients aged less than 60 years. *J. Neurol. Neurosurg. Psychiat.* **46**, 1037–40.

Branchey, M. and Branchey, L. (1984). Patterns of psychotropic drug use and tardive dyskinesia. *J. Clin. Psychopharmacol.* **4**, 41–5.

Brandon, S., McClelland, H. A. and Protheroe, C. (1971). A study of facial dyskinesia in a mental hospital population. *Brit. J. Psychiat.* **118**, 171–84.

Campbell, R. J., Fann, W. E. and Thornby, J. I. (1983). Abnormal involuntary movements and chronic obstructive pulmonary disease. *J. Clin. Psychopharmacol.* **3**, 179–82.

Casey, D. E. and Denney, D. (1977). Pharmacological characterisation of tardive dyskinesia. *Psychopharmacology* **54**, 1–8.

Chouinard, G., De Montigny, C. and Annable, L. (1979). Tardive dyskinesia and antiparkinsonian medication. *Amer. J. Psychiat.* **136**, 228–9.

Crow, T. J. (1980). Molecular pathology of schizophrenia: more than one disease process? *Brit. Med. J.* **280**, 66–8.

Crow, T. J. (1982). Schizophrenia. In *Disorders of Neurohumoral Transmission.* T. J. Crow (ed.), pp. 287–340. Academic Press, London.

Crow, T. J., Cross, A. J., Johnstone, E. C., Owen, F., Owens, D. G. C. and Waddington, J. T. (1982). Abnormal involuntary movements in schizophrenia: are they related to the disease process or its treatment, are they related to changes in dopamine receptors? *J. Clin. Psychopharmacol.* **2**, 336–40.

Crow, T. J., Cross, A. J., Ferrier, I. N., Johnstone, E. C., Owen, F., Owens, D. G. C., Poulter, M. and Roberts, G. W. (1983). Dopamine receptors and other neurochemical parameters assessed in post-mortem brain in patients with and without abnormal involuntary movements ('tardive dyskinesia'). *Progr. Neuro-Psychopharmacol. Suppl.* 51–2.

Csernansky, J. G., Kaplan, J., Holman, C. A. and Hollister, L. E. (1983). Serum neuroleptic activity, prolactin and tardive dyskinesia in schizophrenic outpatients. *Psychopharmacology* **81**, 115–18.

Damasio, H., Eslinger, P., Damasio, A. R., Rizzo, M., Huang, H. K. and Demeter, S. (1983). Quantitative computed tomographic analysis in the diagnosis of dementia. *Arch. Neurol.* **40**, 715–19.

Davis, J. M. (1976). Comparative doses and costs of antipsychotic medication. *Arch. Gen. Psychiat.* **33**, 858–61.

Delwaide, P. J. and Desseilles, M. (1977). Spontaneous buccolinguofacial dyskinesia in the elderly. *Acta Neurol. Scand.* **56**, 256–62.

Dewan, M. J., Pandurangi, A. K., Lee, S. H., Ramachandran, T., Levy, B., Boucher, M., Yozawitz, A. and Major, L. F. (1983). Central brain morphology in chronic schizophrenic patients: a controlled CT study. *Biol. Psychiat.* **18**, 1133–40.

Famuyiwa, O. O., Eccleston, D., Donaldson, A. A. and Garside, R. F. (1979). Tardive dyskinesia and dementia. *Brit. J. Psychiat.* **135**, 500–4.

Feighner, J. P., Robins, E., Guze, S. B., Woodruff, R. A. and Winokur, G. (1972). Diagnostic criteria for use in psychiatric research. *Arch. Gen. Psychiat.* **26**, 57–63.

Gardos, G. and Cole, J. O. (1983). Tardive dyskinesia and anticholinergic drugs. *Amer. J. Psychiat.* **140**, 200–2.

Golden, C. J., Moses, J. A., Zelazowski, R., Graber, B., Zatz, L., Horvath, T. B. and Berger, P. A. (1980). Cerebral ventricular size and neuropsychological impairment in young schizophrenics. *Arch. Gen. Psychiat.* **37**, 619–23.

Grimes, J. D, Hassan, M. N. and Preston, D. N. (1982). Adverse neurological effects of metoclopramide. *Can. Med. Assoc. J.* **126**, 23–5.

Gunne, L. M., Haggstrom, J. E. and Sjoquist, B. (1984). Association with persistent neuroleptic-induced dyskinesia of regional changes in brain GABA synthesis. *Nature (Lond.)* **309**, 347–750.

Itil, T. M., Reisberg, B., Hugure, M. and Mehta, D. (1981). Clinical profiles of tardive dyskinesia. *Compr. Psychiat.* **22**, 282–90.

Jeste, D. V. and Wyatt, R. J. (1982). Therapeutic strategies against tardive dyskinesia: two decades of experience. *Arch. Gen. Psychiat.* **39**, 803–16.

Jeste, D. V., Wagner, R. L., Weinberger, D. R., Rieth, K. G. and Wyatt, R. J. (1980). Evaluation of CT scans in tardive dyskinesia. *Amer. J. Psychiat.* **137**, 247–8.

Johnstone, E. C., Crow, T. J., Frith, C. D., Carney, M. W. P. and Price, J. S. (1978). Mechanism of the antipsychotic effect in the treatment of acute schizophrenia. *Lancet* i, 848–51.

Kane, J. M. and Smith, J. M. (1982). Tardive dyskinesia: Prevalence and risk factors, 1959–1979. *Arch. Gen. Psychiat.* **39**, 473–81.

Kane, J. M., Weinhold, P., Kinon, B., Wegner, J. and Leader, M. (1982). Prevalence of abnormal involuntary movements ('spontaneous dyskinesias') in the normal elderly. *Psychopharmacology* **77**, 105–8.

Karson, C. N., Jeste, D. V., Le Witt, P. A. and Wyatt, R. J. (1983). A comparison of two iatrogenic dyskinesias. *Amer. J. Psychiat.* **140**, 1504–6.

Klawans, H. L. and Barr, A. (1982). Prevalence of spontaneous lingual–facial–buccal dyskinesias in the elderly. *Neurology* **32**, 558–9.

Kling, A. S., Kurtz, N., Tachiki, K. and Orzech, A. (1983). CT scans in sub-groups of chronic schizophrenics. *J. Psychiat. Res.* **17**, 375–384.

Lewine, R. R. J., Fogg, L. and Meltzer, H. Y. (1983). Assessment of negative and positive symptoms in schizophrenia. *Schizophrenia Bull.* **9**, 368–76.

Luchins, D. J., Lewine, R. R. J. and Meltzer, H. Y. (1983). Lateral ventricular size in the psychoses: relation to psychopathology and therapeutic and adverse response to medications. *Psychopharmacol. Bull.* **19**, 518–22.

McCreadie, R. G., Barron, E. T. and Winslow, G. S. (1982). The Nithsdale schizophrenia survey: 2. Abnormal movements. *Brit. J. Psychiat.* **140**, 587–90.

Martinelli, P. and Gabellini, A. S. (1982). Essential tremor and bucco–lingo–facial dyskinesias. *Acta Neurol. Scand.* **66**, 705–8.

Maser, J. D. and Keith, S. J. (1983). CT scans and schizophrenia. *Schizophrenia Bull.* **9**, 265–83.

Molloy, A. G. and Waddington, J. L. (1984). Dopaminergic behaviour stereospecifically promoted by the D_1 agonist R-SK&F 38393 and selectively blocked by the D_1 antagonist SCH 23390. *Psychopharmacology* **82**, 409–10.

Mölsa, P. K., Maratila, R. J. and Rinne, U. K. (1982). Extrapyramidal symptoms in dementia. *Acta Neurol. Scand.* **65** (Suppl. 90), 298–9.

O'Boyle, K. M. and Waddington, J. L. (1984). Selective and stereospecific interactions of R-SK&F 38393 with ^{3}H-piflutixol but not ^{3}H-spiperone binding to striatal D_1 and D_2 dopamine receptors: comparisons with SCH 23390. *Eur. J. Pharmacol.* **99**, 433–6.

Owens, D. G. C. and Johnstone, E. C. (1980). The disabilities of chronic schizophrenia: their nature and the factors contributing to their development. *Brit. J. Psychiat.* **136**, 384–95.

Owens, D. G. C., Johnstone, E. C. and Frith, C. D. (1982). Spontaneous involuntary disorders of movement. *Arch. Gen. Psychiat.* **39**, 452–61.

Pandurangi, A. K., Devi, V. and Channabasavanna, S. M. (1980). Caudate atrophy in irreversible tardive dyskinesia. *J. Clin. Psychiat.* **41**, 229–31.

Perenyi, A., Gardos, G., Samu, I., Kallos, M. and Cole, J. O. (1983). Changes in extrapyramidal symptoms following anticholinergic drug withdrawal. *Clin. Neuropharmacol.* **6**, 55–61.

Qureshi, K. N. and Hodkinson, H. M. (1974). Evaluation of a ten-question mental test in the institutionalised elderly. *Age and Ageing* **3**, 152–7.

Rieder, R. O., Donnelly, E. F., Herdt, J. R. and Waldman, I. N. (1979). Sulcal prominence in young chronic schizophrenic patients: CT scan findings associated with impairment on neuropsychological tests. *Psychiat. Res.* **1**, 1–8.

Rieder, R. O., Mann, L. S., Weinberger, D. R., Van Kammen, D. P. and Post, R. M. (1983). Computed tomographic scans in patients with schizophrenia, schizoaffective and bipolar affective disorder. *Arch. Gen. Psychiat.* **40**, 735–9.

Riederer, P., Jellinger, K. and Gabriel, E. (1983). ^{3}H-Spiperone binding to post-mortem human putamen in paranoid and non-paranoid schizophrenics. In *Proceedings of the 7th World Congress of Psychiatry*. Plenum Press, New York.

Rush, A. J., Weissenburger, J., Vinson, D. B. and Giles, D. E. (1983). Neuropsychological dysfunctions in unipolar nonpsychotic major depressions. *J. Affect. Dis.* **5**, 281–7.

Schulz, S. C., Koller, M. M., Kishore, P. R., Hamer, R. M., Gehl, J. J. and Friedel, R. O. (1983*a*). Ventricular enlargement in teenage patients with schizophrenia spectrum disorder. *Amer. J. Psychiat.* **140**, 1592–4.

Schulz, S. C., Sinicrope, P., Kishore, P. and Friedel, R. O. (1983*b*). Treatment response and ventricular brain enlargement in young schizophrenic patients. *Psychopharm. Bull.* **19**, 510–12.

Seidman, L. J. (1983). Schizophrenia and brain dysfunction: an integration of recent neurodiagnostic findings. *Psychol. Bull.* **94**, 195–238.

Smith, J. M. and Baldessarini, R. J. (1980). Changes in the prevalence, severity and recovery in tardive dyskinesia with age. *Arch. Gen. Psychiat.* **37**, 1368–73.

Smith, R. C. and Maser, J. (1983). Morphological and neuropsychological abnormalities as predictors of clinical response to psychotropic drugs. *Psychopharm. Bull.* **19**, 505–22.

Smith, R. C., Strizich, M. and Klass, D. (1978). Drug history and tardive dyskinesia. *Amer. J. Psychiat.* **135**, 1402–3.

Struve, F. A. and Willner, A. E. (1983). Cognitive dysfunction and tardive dyskinesia. *Brit. J. Psychiat.* **143**, 597–600.

Targum, S. D., Rosen, L. N., DeLisi, L. E., Weinberger, D. R. and Citrin, C. M. (1983). Cerebral ventricular size in major depressive disorders: association with delusional symptoms. *Biol. Psychiat.* **18**, 329–36.

Taylor, M. A. and Abrams, R. (1984). Cognitive impairment in schizophrenia. *Amer. J. Psychiat.* **141**, 1196–201.

Toenniessen, L. M., Casey, D. E. and McFarland, B. H. (1985). Tardive dyskinesia in the aged: duration of treatment relationships. *Arch. Gen. Psychiat.* **42**, 278–84.

Waddington, J. L. (1984). Tardive dyskinesia: a critical re-evaluation of the causal role of neuroleptics and of the dopamine receptor supersensitivity hypothesis. In *Recent Advances in Neurology.* N. Callaghan and R. Galvin (eds). Academic Press, London.

Waddington, J. L., Cross, A. J., Gamble, S. J. and Bourne, R. C. (1983a). Spontaneous orofacial dyskinesia and dopaminergic function in rats after 6 months of neuroleptic treatment. *Science* **220**, 530–2.

Waddington, J. L., Cross, A. J., Gamble, S. J. and Bourne, R. C. (1983b). Dopamine receptor function and spontaneous orofacial dyskinesia in rats during 6-month neuroleptic treatments. *Adv. Biochem. Psychopharmacol.* **37**, 299–308.

Waddington, J. L., Youssef, H. A., Molloy, A. G., O'Boyle, K. M. and Pugh, M. T. (1985). Association of intellectual impairment, negative symptoms and ageing with abnormal, involuntary movements ('tardive' dyskinesia) in schizophrenia: clinical and animal studies. *J. Clin. Psychiat.* **46**, (4, Sec. 2), 29–33.

Weinhold, P., Wegner, J. T. and Kane, J. M. (1981). Familial occurrence of tardive dyskinesia. *J. Clin. Psychiat.* **42**, 165–6.

Wiholm, B. -E., Mortimer, O., Boethius, G. and Haggstrom, J. E. (1984).Tardive dyskinesia associated with metoclopramide. *Brit. Med. J.* **288**, 545–7.

Wolf, M. E., Ryan, J. J. and Mosniam, A. D. (1983). Cognitive functions in tardive dyskinesia. *Psychol. Med.* **13**, 671–4.

Woods, B. T. and Wolf, J. (1983). A reconsideration of the relation of ventricular enlargement to duration of illness in schizophrenia. *Am. J. Psychiat.* **140**, 1564–70.

27

The nature of drug delivery can alter the short- and long-term consequences of administering a dopamine agonist

B. Costall, A. M. Domeney and R. J. Naylor

Postgraduate School of Studies in Pharmacology, University of Bradford, Bradford BD7 1DP, UK

27.1. Introduction

The administration of the indirectly acting dopamine agonist amphetamine to the rodent by continous infusion can cause changes in behaviour and brain catecholamine function more marked than observed after a single treatment (Jonsson and Nwanze, 1982; Eison *et al.*, 1983; Nielsen *et al.*, 1983). The present studies investigate the possibility that the actions of the directly acting dopamine agonist, pergolide, can be influenced by the nature of the drug delivery.

27.2. Methods

The studies used male Sprague-Dawley rats selected as 'moderate activity' responders to the dopamine agonist $(-)N$-n-propylnorapomorphine (Costall *et al.*, 1982). Pergolide was administered as a single subcutaneous injection or continuously by infusion (0.3–500 μg kg^{-1} day^{-1}, 0.48 μl h^{-1} for 13 days) from Alzet osmotic minipumps located subcutaneously in the back neck region. Locomotor activity was assessed in individual, screened, perspex cages, each fitted with one photocell unit placed off-centre: interruptions of the photocell beams were recorded electromechanically. Locomotor activity was recorded daily during the infusion of pergolide and at weekly intervals post-infusion for 7 weeks.

27.3. Results

A single acute injection of pergolide significantly reduced locomotor activity (by 65–71%, $P<0.001$) at a dose of 0.025 mg kg^{-1} s.c., whereas 0.05 mg kg^{-1} s.c. pergolide failed to modify locomotion and 0.5 mg kg^{-1} s.c. pergolide caused significant locomotor stimulation (increase of 54–64%, $P<0.001$). There were no changes in the spontaneous locomotion of these groups of animals assessed 7 weeks later (Fig. 27.1).

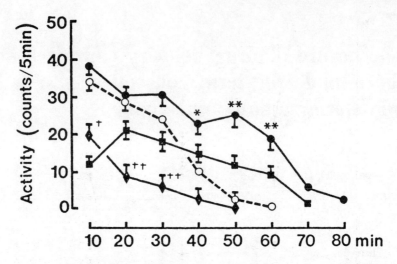

Fig. 27.1. Locomotor activity responses of rats to pergolide administered as an acute subcutaneous injection of 0·5 mg kg^{-1} (●——●), 0·05 mg kg^{-1} (■——■) or 0·025 mg kg^{-1} (◆——◆). (○---○) represents the response of vehicle-treated control rats. Following injection, locomotor activity was measured in photocell cages and is expressed as counts per 5 min. $n = 6$; SEMs are shown. Enhancement of locomotor responding significant to *$P < 0·05$, **$P < 0·01$, and significant reductions indicated as †$P < 0·05$, ††$P < 0·01$ (one-way ANOVA followed by Dunnett's test).

In contrast, pergolide infused continuously at doses as low as 0·6 µg kg^{-1} day^{-1} tended to enhance locomotion, and maximum stimulation (increases of 257–300 %, $P < 0·001$) was recorded when 500 µg kg^{-1} pergolide was infused over a 24 h period (increased levels of hyperactivity were also recorded for 12·5 µg kg^{-1} day^{-1} pergolide). The hyperactivity responses to infused pergolide developed within 2 days of commencing infusion and were maintained throughout the 13-day infusion period. After discontinuing the pergolide infusion, animals exhibited elevated levels of spontaneous locomotion; the intensity depended on the previously infused dose of pergolide, and this persisted for the 7-week post-infusion period (Fig. 27.2).

27.4. Conclusions

The continuous infusion of pergolide can cause locomotor stimulation at doses approximately 40-fold less than those required for locomotor stimulation following acute, single injection. The effect of dopamine receptor stimulation may therefore depend on the continuity of that stimulation. However, the persistent delivery of pergolide, in addition to potently stimulating locomotion during infusion, can also cause long-term

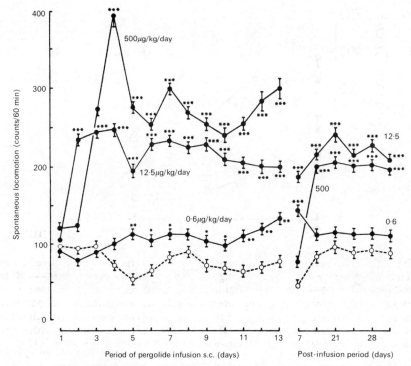

Fig. 27.2. Modification of spontaneous locomotor activity during and following the subcutaneous infusion of pergolide, 500 μg kg^{-1} day^{-1}, 12·5 μg kg^{-1} day^{-1}, 0·6 μg kg^{-1} day^{-1} (●———●) or vehicle (○– – –○). $n = 6$; SEMs are shown. Significant enhancement of spontaneous locomotor activity as compared to vehicle-infused controls is indicated as *$P < 0.05$, **$P < 0.01$, ***$P < 0.001$ (two-way ANOVA followed by Dunnett's test for multiple comparisons).

increases in spontaneous locomotion which last for many weeks after drug withdrawal. It is therefore suggested that, in situations where a continuous stimulation of dopamine receptors is required to cause a continuous elevation in motor performance, a system for the continuous delivery of an agent such as pergolide may be more effective.

Acknowledgement

This work was supported by the S.E.R.C.

References

Costall, B., Domeney, A. M. and Naylor, R. J. (1982). *Neuropharmacology* **21**, 327–35.
Eison, M. S., Eison, A. S. and Iversen, S. D. (1983). *Neurosci. Lett.* **39**, 313–19.
Jonsson, G. and Nwanze, E. (1982). *Brit. J. Pharmacol.* **77**, 335–45.
Nielsen, E. B., Nielsen, M. and Braestrup, C. (1983). *Psychopharmacology* **81**, 81–5.

28

A functional role for dopamine in the habenula nucleus: effects of intracerebral injection of a dopamine agonist

E. W. Thornton and J. A. C. Evans

Department of Psychology, University of Liverpool, Liverpool L69 3BX, UK

28.1. Introduction

A considerable volume of research has demonstrated a critical role for dopamine in schizophrenia and motor disorders. However, little attention has been directed to the neural mechanisms which modulate the activity of cells of origin of ascending dopamine pathways. The habenula nucleus has been propsed to have a pivotal role in a dorsal pathway between the limbic forebrain and midbrain (Sutherland, 1982). As one of few sites of convergence of outputs from the limbic system and the corpus striatum, it is likely to determine the effects of emotions on motor activity. The importance of stress in the development of clinical conditions associated with dopamine dysfunction is well documented. The functional importance of the habenula has been demonstrated by marked behavioural change under stressful conditions following lesion of the nucleus (Thornton and Evans, 1982; Thornton et al., 1983). These changes are most probably effected through the predominant efferent pathways from the habenula to the substantia nigra, ventral tegmental area and the mesencephalic raphe nuclei. Predominant effects on rearing activity suggest a primary influence of the habenula on dopaminergic activity. However, it has not been determined whether lesion-induced changes in behaviour are a consequence of interruption of fibres of passage, or of the destruction of cells within the habenula itself. Certainly, Phillipson and Pycock (1982) have shown that the lateral habenula receives a reciprocal dense dopaminergic innervation from the ventral tegmental area with additional input to the medial nucleus. The present study therefore examined whether direct activation of the dopamine receptors within the habenula led to changes in activity and exploratory behaviour of rats.

28.2. Methods

The animals were 12 male Lister Hooded rats weighing 304 ± 16·7 g adapted to handling and housed singly in standard cages with food and water available *ad libitum.* Stereotaxic surgery under pentobarbitone anaesthesia was used to implant cannulae bilaterally into the habenula nuclei.

Behavioural tests were carried out 6 days after surgery in a standard hole-board apparatus. The enclosure was a square of sides 81 cm with walls 81 cm high. The floor was marked into nine squares with a hole 4 cm in diameter in each corner square, beneath which was placed a different novel object. The enclosure was illuminated with a 60 W lamp 140 cm above the centre of the floor and flooded with background white noise of 70 dB.

Following adaptation to the apparatus in three 30-min exposures on alternate days, animals received bilateral intracerebral injections of saline or one of three doses (1·88 μg, 3·75 μg and 7·5 μg) of apomorphine hydrochloride (Sigma Ltd). Drug doses were prepared freshly each day in isotonic saline containing 2 mg ml^{-1} of ascorbic acid. In a within-groups, repeated-measures design, injections were given in a sequence counterbalanced across animals in a 3-day cycle in which an injection was followed by 2 days in the home cage before testing with a different treatment injection on the third day. Injections of a constant volume of 0·3 μl were infused through an inner cannula over a 3-min period with the cannula left *in situ* for a further minute before removal and replacement with the stylet. Each observation period of 30 min was recorded remotely with video facilities and later scored blind by one experimenter. Behaviours scored included activity, frequency of rearing and exploration of the holes.

28.3. Results

As shown in Table 28.1, significant dose effects were seen for rearing (ANOVA $F = 3·66$, DF $= 3,33$, $P < 0·05$), and number of head dips in hole inspection (ANOVA $F = 4·74$, DF $= 3,33$, $P < 0·01$). Although there was also a tendency to decreased levels of activity at higher drug doses, the ANOVA showed no significant dose effect ($F = 2·08$). The significant dose effects were further analysed by comparing means with the Tukey test, with

Table 28.1. Incidence of behavioural measures in hole-board test (means $\pm$ SD, $n = 12$)

| Behaviour | Saline | Drug dose | | |
		1·88 μg	3·75 μg	7·5 μg
Number of head dips	34·3 $\pm$ 15·0	39·3 $\pm$ 21·5	25·3 $\pm$ 11·2	24·4 $\pm$ 11·8
Number of rears	96·6 $\pm$ 38·7	89·1 $\pm$ 34·5	81·3 $\pm$ 52·5	60·5 $\pm$ 32·9
Incidence of grooming	163·2 $\pm$ 91·1	139·3 $\pm$ 48·4	127·3 $\pm$ 56·7	153·9 $\pm$ 119·9
Number of lines crossed	212·7 $\pm$ 81·2	196·8 $\pm$ 76·8	165·9 $\pm$ 85·9	148·9 $\pm$ 98·3

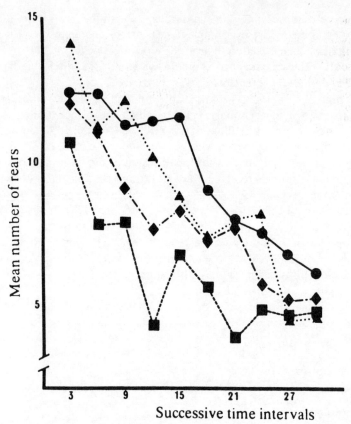

Fig. 28.1.   Changes in rearing activity over successive 3-min periods of the test session following bilateral intra-cerebral injections of saline (●) or one of three doses of apomorphine hydrochloride: 1·88 μg (▲); 3·75 μg (◆), and 7·5 μg (■).

a stringent criterion of α = 0·01 adopted for rejection of the null hypothesis. The analysis showed significant suppression of rearing activity relative to saline controls only at a dose of 7·5 μg apomorphine. (See Fig. 28.1.)

28.4. Conclusion

Changes in behaviour induced by the potent agonist apomorphine confirm that the effects of altered habenular function are dopamine dependent and are not a result of interruption in fibres of passage. The results also provide behavioural data to support evidence which suggests that stress-induced activation of dopaminergic pathways from the ventral tegmental area, through the habenula relay (Lisoprawski *et al.*, 1980), are influenced by a direct dopaminergic feedback loop to the habenula.

References

Lisoprawski, A., Herve, D., Blanc, G., Glowinski, J. and Tassin, J. P. (1980). *Brain Res.* **183**, 229–34.

Phillipson, O. T. and Pycock, C. J. (1982). *Exp. Brain Res.* **45**, 89–94.

Sutherland, R. J. (1982). *Neurosci. Biobehav. Rev.* **6**, 1–13.

Thornton, E. W. and Evans, J. A. C. (1982). *Physiol. Psychol.* **10**, 361–7.

Thornton, E. W., Evans, J. A. C. and Barrow, C. (1982). *IRCS Medical Science* **11**, 779–80.

29

Lack of tolerance of striatal dopamine metabolism to chronic haloperidol in the rat brain

E. F. Marshall, W. N. Kennedy, C. Iniguez* and D. Eccleston

Department of Psychiatry, University of Newcastle upon Tyne, Newcastle upon Tyne NE2 4AL, UK. *Department of Anatomy, University of Valladolid, Valladolid, Spain

29.1. Introduction

The acute effects of haloperidol in the rat include increasing dopamine turnover in the striatum in a dose-dependent fashion, without affecting the concentration of dopamine, and decreasing apomorphine-stimulated stereotyped behaviour (Asper *et al.*, 1973; Matsumoto *et al.*, 1983). Increases in homovanillic acid (HVA) and dihydroxyphenylacetic acid (DOPAC) are considered to be manifestations of the increase in dopamine turnover, although haloperidol most probably acts primarily on the dopamine receptors. Daily administration of haloperidol for anything over 7 days causes tolerance to the acute challenge of the drug to develop in the striatum but not in the frontal cortex (Matsumoto *et al.*, 1983). Increased metabolism of haloperidol is unlikely to account for the development of tolerance as replacement with loxapine does not alter the effect (Asper *et al.*, 1973). However, changes in the sensitivity of the receptors may be responsible (Nicolaou, 1980) and in order to study a system in which receptors are being actively formed, young animals which had received haloperidol from birth have been analysed for changes in striatal dopamine metabolism.

29.2. Methods

29.2.1. *Dose schedule and animal groups*

The dosage schedules of drug administration are detailed below. The *build-up* schedule consisted, initially, of 1 mg kg^{-1} haloperidol (s.c.) at birth, gradually increasing to a final dose of 2 mg kg^{-1} haloperidol (s.c.) after 4 weeks. The build-up was gradual, and animals which were killed at 1, 2 or 3 weeks received on the last day 1·25, 1·5 and 1·75 mg kg^{-1} haloperidol (s.c.) respectively, the daily increment being 0·035 mg kg^{-1}. This increasing dose was believed to allow for the developing brain having increasing receptor numbers and, perhaps, an increasing capacity to metabolise or inactivate the drug. For comparison, three other groups of animals were used. *Control* animals were injected daily with saline. Daily saline injections were also administered to the *single* group, but on the last day they received a dose of haloperidol at the same level as the build-up group of animals of that age group. The third comparison group, referred to as *chronic high*, received daily injections of the dose of haloperidol which the build-up sample of the same age would receive on the last day. This last group was included to determine the difference between gradually increasing the dose or maintaining a constant high dose. All animals received the last injection 3 h prior to sacrifice.

29.2.2. *Assay of metabolites*

Basically, the method adopted for the separation and estimation of DA and HVA in brain samples was a modification of the method detailed by Kilts *et al.* (1981).

Chromatographic conditions. A 25 cm × 4·6 mm i.d. stainless steel column packed with Spherisorb 5 ODS 2 was used. The potential of the glassy carbon TL-5A electrode (Bioanalytical Systems) was set at +0·75 V against an Ag/AgCl reference electrode. The mobile phase was a mixture of Na_2HPO_4 :citrate :methanol :heptanesulphonate (0·075 M :0·1 M :18% :0·75 mM) pH 3·9 and the flow rate was held at 1·0 ml min^{-1}.

Sample preparation. The animals were killed by decapitation and the whole brain was removed, washed with saline and dried with paper towel. The corpus striatum was dissected out and immediately placed in liquid nitrogen. The samples were later stored at -40°C until assayed.

After weighing, 200 μl mobile phase was added; the samples were subject to ultrasonic disintegration for 2 s at number 5 (Kontes Micro-Ultrasonic Cell Disrupter); and the tubes were centrifuged for 5 min at 15 000 g. The supernatant was removed to the smaller tubes (250 μl capacity) required by the autosampler (Magnus 7110). Standards were diluted in mobile phase, subject to the same ultrasonication and assayed once every ten tubes on the autosampler.

29.3. **Results**

The control values for DOPAC and HVA are presented in Table 29.1. Concentrations for DOPAC are higher than the corresponding values for HVA but the increase with age is not so striking as that for HVA.

The percentage increases over control in DOPAC and HVA in the striatum of rats of increasing ages having received drugs appropriate to

Table 29.1.   Control values for DOPAC and HVA. Values are means ± SEM and represent pmol mg^{-1} fresh brain weight. The numbers in brackets are the number of determinations.)

Age (weeks)	1	2	3	4
DOPAC	7·06 ± 2·16 (13)	7·65 ± 3·51 (10)	8·56 ± 3·42 (12)	8·45 ± 3·26 (15)
HVA	1·49 ± 0·37 (13)	5·75 ± 2·43 (10)	5·44 ± 1·1 (12)	6·28 ± 2·43 (15)

their schedule are detailed in Figs 29.1a and b. Although the single doses are increasing in concentration, the effect is most pronounced at 14 days of age. The effects of the build-up schedule in comparison with those of the single doses suggest that there is in fact some tolerance to the drug developing, though the HVA concentrations remain significantly above control even after 4 weeks. For the chronic high schedule, the most pronounced effect on HVA concentrations appears after 1 week. At this age the actual concentration of HVA is very low, and however the haloperidol acts, the amount of enzyme activity in the pathway to HVA or the rate of removal have clearly changed dramatically. However, tolerance does develop to the continued administration of high doses of haloperidol. At 2 weeks, HVA in these animals is still significantly higher than control level although the metabolite concentration is much lower than in the animals which had received either single or build-up doses of the drug. After 4 weeks of chronic high doses, the values for striatal HVA returned to normal.

The effects of single doses of haloperidol on DOPAC concentration, although less pronounced, parallel those on HVA except at the age of 7 days when there is a steep, significant fall in DOPAC levels. There is no effect of either build-up or chronic high doses on the metabolite levels at 1 week, but thereafter the DOPAC concentrations continue to be significantly higher on the build-up dosage scheme than controls, with the peak effect at 2 weeks. High chronic doses of haloperidol have no effect on striatal DOPAC levels until after 4 weeks of administration.

29.4. Discussion

DOPAC is the major metabolite of dopamine in rat brain from an early age, but its constant value with increasing age indicates that the metabolic route from dopamine to DOPAC is already well established at birth. On the other hand the concentrations of HVA increase quite markedly with age, suggesting that the formation of HVA is in some way connected with the development of the dopamine-containing neuronal pathways.

The general picture from all the animal groups is that haloperidol causes a shift of the metabolic pathways from DOPAC to HVA and that the responses to increase HVA are more sensitive than those for DOPAC. This

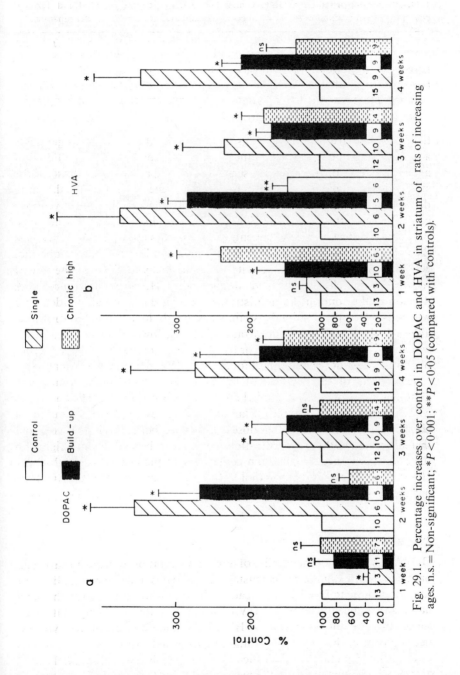

Fig. 29.1. Percentage increases over control in DOPAC and HVA in striatum of rats of increasing ages. n.s. = Non-significant; *$P<0.001$; **$P<0.05$ (compared with controls).

is a different pattern from that observed in the adult animal (Nicolaou, 1980; Matsumoto *et al.*, 1983) but suggests that blocking the receptors in the young brain causes the emphasis to be placed on synaptic release of dopamine with the ultimate formation of HVA. Only in the older animal is there any suggestion that administration of haloperidol could elicit an increase in formation of DOPAC in preference to HVA. Studies in older animals treated from birth would determine whether this apparent breakout of the formation of DOPAC from the restraints of tolerance were a significant phenomenon.

The different effects observed in the animals receiving a build-up schedule compared with those receiving chronic high doses remain a curious phenomenon. The mere occupation of postsynaptic receptors, linked to the feedback loop, by chronic high doses is not effective in terms of maintaining increased responses. Chronic high doses appear to reduce firing of the DA neurone, whereas gradually increasing doses inhibit the postsynaptic receptors but do not reduce normal firing completely, thus keeping the feedback loop open.

These findings may be critical in terms of using drugs such as haloperidol in clinical conditions where the development of tolerance is undesirable.

Acknowledgement

The authors wish to thank the Wellcome Trust for financial support.

References

Asper, H. *et al.* (1973). *Eur. J. Pharmacol.* **22**, 287–94.
Kilts, C. D. *et al.* (1981). *J. Chromat.* **225**, 347–57.
Matsumoto, T. *et al.* (1983). *Eur. J. Pharmacol.* **89**, 27–33.
Nicolaou, N. M. (1980). *Eur. J. Pharmacol.* **64**, 123–32.

Anomalous relationship between serum chlorpromazine and prolactin in tardive dyskinesia

G. Hall, F. J. Rowell and C. G. Rich

Department of Pharmaceutical Chemistry, Sunderland Polytechnic, Sunderland SR2 7EE, UK

and

F. Hassanyeh, A. F. Fairbairn and D. Eccleston

Department of Psychiatry, University of Newcastle upon Tyne, Newcastle upon Tyne NE1 4LP, UK

30.1. Introduction

Neuroleptic drugs are thought to exert their effect by acting as antagonists at dopaminergic receptor sites in the brain. Since the tubero-infundibular dopamine pathway plays a major role in inhibiting prolactin (PRL) release, one effect of neuroleptic dopamine receptor blockade is an increase in PRL secretion. Correlations between plasma neuroleptic and PRL concentrations were demonstrated by Meltzer *et al.* (1979) for both butyrophenone- and phenothiazine-type neuroleptics.

Csernansky *et al.* (1983) reported that in females receiving depot fluphenazine, serum PRL levels were higher in those with severe tardive dyskinesia (TD) than in those with mild symptoms, and that their serum PRL levels were positively correlated with neuroleptic activity. Our study examines serum levels of PRL and chlorpromazine, together with its metabolites in TD and non-TD patients matched for age, weight, sex and dose of orally administered drug and their interrelationship over a 4-week period.

30.2. Methods

Fifteen adult psychiatric in-patients, aged between 29 and 64 years, and diagnosed as schizophrenic (as defined in Research Diagnostic Criteria) were studied. The patients were matched for age and sex, and divided into two groups, one diagnosed as having TD (each patient was seen by two consultant psychiatrists who agreed upon the diagnosis), and the second group TD free, serving as controls. All patients gave their full and informed consent to take part in the study, and Ethical Committee approval was obtained.

The patients were receiving various drugs for the treatment of their schizophrenia and other illnesses. Twelve were receiving oral neuroleptic, and the remaining three received either depot and oral neuroleptics or depot neuroleptic only. The non-essential psychotropic medication was withdrawn 1 month before the neuroleptic changeover took place. Chlorpromazine was substituted for the original neuroleptic (dosage 300 to 600 mg daily), the mechanics of the changeover being dependent on the original neuroleptic regimen.

Blood samples were taken 3 h after chlorpromazine administration at the end of each week throughout the study. After centrifugation, serum was stored at $-20°C$ prior to assay. Serum PRL was measured using an immunoradiometric assay after Hodgkinson *et al.* (1984). Chlorpromazine and its sulphoxide metabolite (CPZ-SO) and N-demethylated metabolites

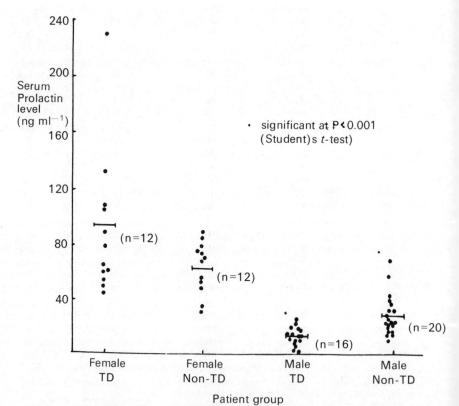

Fig. 30.1.   Serum prolactin levels in groups of patients (for weeks 1 to 4).

Female Tardives

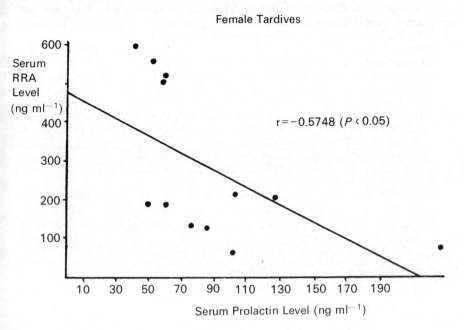

Male Non-Tardives

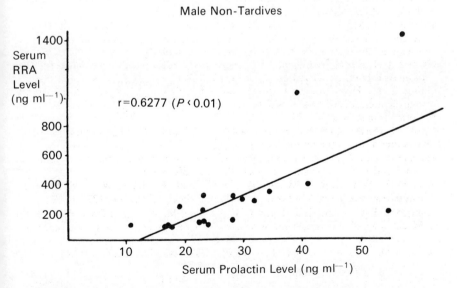

Fig. 30.2. Graphs showing serum radioreceptor activity (RRA) against serum prolactin levels (for weeks 1 to 4).

were measured by a high-performance liquid chromatography assay, after Stevenson and Reid (1981). A radioreceptor assay (RRA) was used to determine total serum neuroleptic activity, after Lader (1980).

30.3. Results

No significant differences in age, weight and dose of chlorpromazine were observed between TD and non-TD males and females. Females were observed to have higher serum PRL levels than males. Although not significantly different, female TD patients were seen to have greater serum levels of PRL than non-TD female patients. In contrast, male non-TD patients had significantly greater levels of PRL than TD males $(P < 0.001)$ (Fig. 30.1).

No significant differences in serum levels of chlorpromazine and its metabolites were observed in female TD and non-TD patients and male TD and non-TD patients, with the exception of the sulphoxide metabolite of chlorpromazine which was significantly higher in non-TD males than TD males $(P < 0.05)$.

Positive correlations between RRA levels and PRL levels in serum were observed for non-TD males $(r = 0.623, P < 0.01)$ and non-TD females $(r = 0.423$, not significant). In contrast, no correlation was observed for TD males $(r = -0.024)$ and a negative correlation was obtained for TD females $(r = -0.575, P < 0.05)$, (Fig. 30.2).

30.4. Discussion

The principal finding of this study is that, in contrast to the reported and observed response to neuroleptic medication for non-TD patients, in TD schizophrenic patients receiving chlorpromazine, the serum PRL levels did not increase in direct proportion to the serum level of neuroleptic. Since no significant differences in serum levels of chlorpromazine and its active metabolites were observed between TD and non-TD patients, this observation may reflect differences in brain response to chlorpromazine and its metabolites in TD and non-TD patients.

Acknowledgement

We would like to thank the N.R.H.A. for partial funding for this study.

References

Csernansky, J. G., Kaplan, J., Holman, C. A. and Hollister, L. E. (1983). *Psychopharmacology* **81**, 115–18.
Hodgkinson, S. C., Landon, J. and Lowry, P. (1984). *Biochem. J.* **217**, 273–9.
Lader, S. (1980). *J. Immunoassay* **1**, 57–75.
Meltzer, H. Y., So, R., Miller, R. J. and Fang, V. S. (1979). *Life Sci.* **24**, 573–84.
Stevenson, D. and Reid, E. (1981). *Anal. Lett.* **14** (B10), 741–61.

31

Topography of action of dihomo-γ-linolenic acid in the guinea-pig striatum to antagonise dyskinesias

B. Costall, M. E. Kelly and R. J. Naylor

Postgraduate School of Studies in Pharmacology, University of Bradford, Bradford BD7 1DP, UK

31.1. Introduction

Striatal dopamine function is considered important for the modulation of normal motor activities, and an excessive stimulation of the striatal mechanisms may be involved with the development of dyskinesias (see review by Costall and Naylor, 1979). A topographical analysis of dopamine agonist action to induce dyskinesias would indicate a differential distribution of striatal sites capable of mediating dyskinesias (Costall *et al.*, 1980). We have recently shown that dihomo-γ-linolenic acid (DHLA) can antagonise dopamine-induced dyskinesias in the rodent (Costall *et al.*, 1984) and in the present study we investigate whether this antagonism is exerted at those striatal sites previously identified for their involvement with dyskinesia induction.

31.2. Methods

Male Dunkin-Hartley guinea-pigs were subject to standard stereotaxic surgery for the implantation of bilateral guide cannulae for the injection of DHLA/vehicle at five different sites within the striatum. The DHLA was injected daily for 1 or 2 days before challenge with the dopamine agonist 2-di-*n*-propylamino-5,6-dihydroxytetralin (tetralin) 0.025 mg kg^{-1} s.c. The peri-oral movements (dyskinesias) induced by tetralin were scored according to the intensity of the response, where score 1 represented a weak response, occasional biting and few abnormal oral movements; score 2 represented a moderate intensity response, periods of biting dominant but clearly broken by brief periods when biting was absent; and score 3 represented an intense response, where biting and oral movements were continuous.

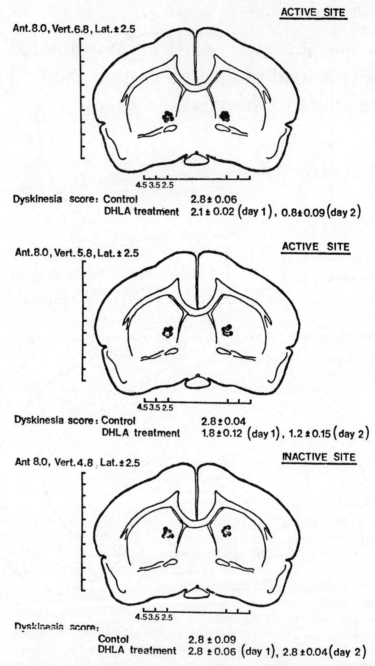

ACTIVE SITE

Ant.8.0, Vert.6.8, Lat.±2.5

4.5 3.5 2.5

Dyskinesia score: Control   2.8±0.06
                 DHLA treatment 2.1±0.02 (day 1), 0.8±0.09 (day 2)

ACTIVE SITE

Ant.8.0, Vert.5.8, Lat.±2.5

4.5 3.5 2.5

Dyskinesia score: Control   2.8±0.04
                 DHLA treatment 1.8±0.12 (day 1), 1.2±0.15 (day 2)

INACTIVE SITE

Ant 8.0, Vert.4.8, Lat.±2.5

4.5 3.5 2.5

Dyskinesia score:
                 Contol   2.8±0.09
                 DHLA treatment 2.8±0.06 (day 1), 2.8±0.04 (day 2)

Fig. 31.1. The active and inactive sites in the striatum at ant. 8·0 for DHLA deposition to antagonise the dyskinesias induced by peripheral tetralin administration (coordinates according to Costall *et al.*, 1980).

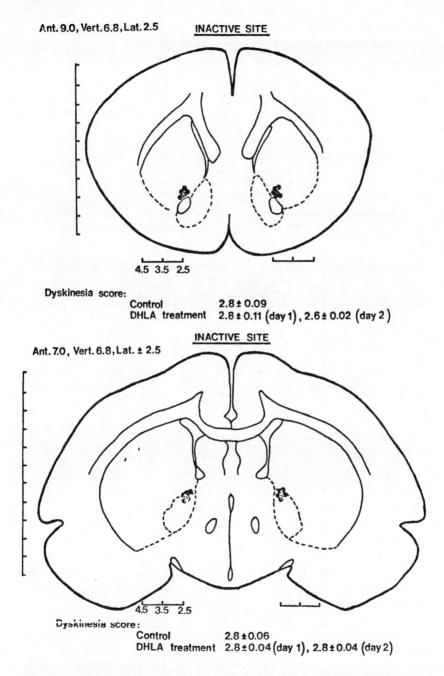

Ant. 9.0, Vert. 6.8, Lat. 2.5   INACTIVE SITE

4.5 3.5 2.5

Dyskinesia score:

Control 2.8 ± 0.09
DHLA treatment 2.8 ± 0.11 (day 1), 2.6 ± 0.02 (day 2)

INACTIVE SITE

Ant. 7.0, Vert. 6.8, Lat. $\pm$ 2.5

4.5 3.5 2.5

Dyskinesia score:

Control 2.8 ± 0.06
DHLA treatment 2.8 ± 0.04 (day 1), 2.8 ± 0.04 (day 2)

Fig. 31.2. Sites in the striatum at ant. 9·0 and 7·0 where DHLA deposition fails to modify the dyskinesias induced by peripheral tetralin administration (coordinates according to Costall *et al.*, 1980).

31.3. Results

A 1-day treatment with DHLA (20–80 μg) was shown to antagonise the tetralin dyskinesias depending on the site of deposition. The injections located at ant. 8·0, lat. ±2·5 and at vert. −6·8 and −5·8 (sites previously identified for their sensitivity to tetralin) were most effective to antagonise the dyskinesias and this was enhanced by repeated treatment (Fig. 31.1). Injections distanced from this active site by 1 mm above (ant. 8·0, lat. ±2·5, vert. −4·8), anterior or posterior (ant. 9·0/ant. 7·0, lat. ±2·5, vert. −6·8), were ineffective (Fig. 31.2). These results are summarised in Table 31.1.

Table 31.1.   Topography of the antidyskinetic action of DHLA in guinea-pig striatum.

Coordinates				Dyskinesia scores			
					DHLA day 1		DHLA day 2
Ant.	Vert.	Lat.	Vehicle	20 μg	80 μg		20 μg
8·0	−6·8	±2·5	2·8 ± 0·06	2·1 ± 0·02	1·4 ± 0·10		0·8 ± 0·09
('Dyskinesia site')							
9·0	−6·8	±2·5	2·8 ± 0·09	2·8 ± 0·11	2·9 ± 0·03		2·6 ± 0·02
7·0	−6·8	±2·5	2·8 ± 0·06	2·8 ± 0·04	2·8 ± 0·09		2·8 ± 0·04
8·0	−5·8	±2·5	2·8 ± 0·04	1·8 ± 0·12	1·6 ± 0·11		1·2 ± 0·15
8·0	−4·8	±2·5	2·8 ± 0·09	2·8 ± 0·06	2·6 ± 0·12		2·8 ± 0·04

31.4. Conclusions

These studies confirm that DHLA may exert an antidyskinetic action in the striatum at a site previously identified as being most sensitive to the actions of tetralin to induce dyskinesias. Thus, DHLA has a highly specific site of antidyskinetic action within the neostriatal complex.

Acknowledgements

This work was supported by the Parkinson's Disease Society. The authors would like to thank Roche Products Ltd for the supply of DHLA and for helpful advice during the preparation of this report.

References

Costall, B. and Naylor, R. J. (1979). In *The Neurobiology of Dopamine*. A. S. Horn, J. Korf and B. H. C. Westerink (eds), pp. 555–76. Academic Press, London.
Costall, B., De Souza, C. X. and Naylor, R. J. (1980). *Neuropharmacology* **19**, 623–31.
Costall, B., Kelly, M. E. and Naylor, R. J. (1984). *Brit. J. Pharmacol.* **83**, 733–40.

Part IV

Basal ganglia disturbances and extra-pyramidal syndrome

32

Huntington's chorea

E. G. S. Spokes

Department of Neurology, Leeds General Infirmary, Great George Street, Leeds LS1 3EX, UK

32.1. Historical and clinical aspects

George Sumner Huntington's classical account of inherited chorea and dementia (Huntington, 1872) caused the eminent physician Sir William Osler (1908) to write: 'In the history of medicine there are few instances in which a disease has been more accurately, more graphically, or more briefly described.' Huntington, an American physician, came from a medical family and first noticed the condition, known locally as 'that disorder', while accompanying his father, a family doctor, on home visits in East Hampton, Long Island. In the 100 years that followed his seminal work, little was added to our understanding of this catastrophic condition, but in the last decade new concepts regarding pathogenesis and pathophysiology have emerged from diverse experimental approaches.

Huntington's chorea is transmitted as an autosomal dominant with complete penetrance and affects about 1 in 10 000 people. First symptoms may arise at any age but most commonly between 30 and 50 years. About 5% of cases start before the age of 20, so-called 'juvenile' chorea, in 80% of whom the disease is transmitted from the father, implying the existence of a sex-related factor in the inheritance (Bird *et al.*, 1974).

The clinical characteristics are a combination of physical and mental symptoms which may appear independently or concurrently. Chorea is the Greek word for dance and describes the jerky uncontrollable movements affecting axial and limb muscles, greatly impeding speech, swallowing, walking, writing and other voluntary motor activities. In contrast, juvenile and some adult cases exhibit the Westphal variant in which the prominent features are muscle rigidity, bradykinesia and tremor (parkinsonism), dystonia, cerebellar ataxia and epilepsy.

Dementia is the other hallmark of this disease and conforms to the subcortical type (Albert, 1978). This is manifest by emotional and personality changes, impairment of recent memory, defective ability to manipulate acquired knowledge and slowing of information processing. In contrast to the cortical dementias, the most common form of which is Alzheimer's disease, aphasia, apraxia and agnosia do not develop. Additional psychiatric features include anxiety, depression, mania and schizophrenia-like psychosis, and suicide is not uncommon in younger patients.

32.2. Neuropathology

The choreic brain shows generalised shrinkage and thinning of the cerebral cortex. The corpus striatum (putamen, caudate and globus pallidus) is grossly atrophied, with a 50% reduction in small cell density (Lange *et al.*, 1976). Similar changes affect the subthalamic nucleus (Lange *et al.*, 1976) and ventrobasal nuclear complex of the thalamus (Dom *et al.*, 1976). These findings are particularly marked in rigid cases. The substantia nigra, especially the pars reticulata, is also shrunken, the changes being most prominent at caudal levels (E. G. S. Spokes, unpublished).

32.3. Neurochemistry

Although the pathogenesis of the disease remains unkown, much has been learned about the pathophysiology of chorea in the last decade from neuropharmacological and neurochemical studies. Evidence suggests that chorea arises from an imbalance of the effects of dopamine, acetylcholine, and GABA released from terminals in the basal ganglia (corpus striatum and substantia nigra). The pertinent neurochemical pathways are shown in Fig. 32.1. Derangements in other nerve cell populations such as those releasing noradrenaline, serotonin, glutamate and neuropeptides may also be involved, but the role of these neurotransmitters in the extrapyramidal control of motor activity has yet to be clarified.

Pharmacological studies suggest that chorea may arise from overactivity in the dopaminergic nigrostriatal tract. Thus drugs which impede dopaminergic neurotransmission such as α-methyl-para-tyrosine (a competitive inhibitor of tyrosine hydroxylase), tetrabenazine (a central monoamine depletor), phenothiazines and butyrophenones (dopamine receptor antagonists) are useful in ameliorating chorea, whereas those which facilitate dopaminergic function such as *L*-dopa, *d*-amphetamine and bromocriptine exacerbate choreic movements. Further support for this concept derives from biochemical analyses of choreic brain tissue obtained at autopsy in which highly significant increases in dopamine concentrations have been found in the basal ganglia (Fig. 32.2) (Spokes, 1980). It is likely that this finding reflects preservation of the dopaminergic nigrostriatal system in the atrophied basal ganglia, producing an increased density of

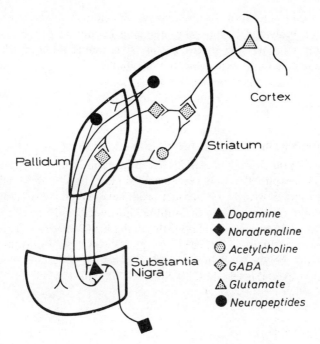

Fig. 32.1. Proposed neurochemical pathways in the basal ganglia. Intrinsic striatal synaptic connections are speculative. (For references, see text.)

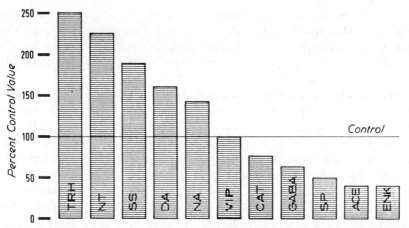

Fig. 32.2. Changes in neurotransmitters and related enzymes in the choreic basal ganglia. *Key:* TRH, thyrotrophin-releasing hormone; NT, neurotensin; SS, somatostatin; DA, dopamine; NA, noradrenaline; VIP, vasoactive intestinal polypeptide; CAT, choline acetyltransferase; GABA, γ-aminobutyric acid; SP, substance P; ACE, angiotensin-converting enzyme; ENK, met-enkephalin. (For references, see text.)

terminals and perikarya in the corpus striatum and substantia nigra respectively. However, despite the increased dopamine concentration, the net content of dopamine in the corpus striatum is reduced when tissue shrinkage is taken into account (Spokes, 1980).

Dopamine levels *per se* give little indication of the functional activity of dopaminergic neurones but this can be inferred from measurements of homovanillic acid (HVA), formed from dopamine released at synapses (Roffler-Tarlov *et al.*, 1971). Since about 75% of the dopamine in mammalian brain is present in the nigrostriatal system, the level of HVA in lumbar cerebrospinal fluid (CSF) is believed largely to mirror the functional activity in this pathway. In choreic patients, HVA concentrations are reduced by 50% (Curzon *et al.*, 1972), suggesting that dopamine turnover is decreased. However, such low levels may relate to the expansion of the CSF compartment consequent upon brain shrinkage. Dopamine levels in basal ganglia tissue from non-rigid and rigid choreic patients are similar (Spokes, 1980) but the net dopamine content in the latter is very low due to gross atrophy. Predictably, such patients have greatly depleted HVA values in the CSF (Curzon *et al.*, 1972).

The extrapyramidal dysfunction in Huntington's chorea cannot, therefore, be attributed to an absolute overactivity of the nigrostriatal dopaminergic system. However, relative overactivity might arise from a deficiency in a neurotransmitter system which is normally antagonistic to dopamine or from hypersensitivity of striatal postsynaptic dopamine receptors. Dopamine receptor hypersensitivity is well documented in experimental animals with nigral lesions, and may underlie the choreiform movements which can complicate the administration of *L*-dopa or dopamine agonists to parkinsonian patients. However, in Huntington's chorea the nigrostriatal system does not degenerate to provide a stimulus for dopamine receptor proliferation. Moreover, studies with radioligands have demonstrated a 50% reduction in D-1 and D-2 receptor densities in choreic striatum (Fig. 32.3) (Reisine *et al.*, 1979; Cross and Rossor, 1983).

The available evidence suggests, therefore, that relative dopamine overactivity results from the loss of an antagonist, in which regard GABA is the obvious candidate. Both GABA and its synthetic enzyme glutamic acid decarboxylase (GAD), localised in striatal interneurones and the striopalli-donigral projections, are markedly depleted in the choreic basal ganglia (Fig. 32.2) (Spokes, 1980; Spokes *et al.*, 1980). A diminished net input of GABA into the substantia nigra might, therefore, decrease the inhibitory influence normally exerted on target nigral dopaminergic neurones. GABA is also lost from the ventrolateral thalamic and subthalamic nuclei, important relay sites for striopallidal fibres (Spokes *et al.*, 1980).

Noradrenaline concentrations are elevated in the choreic basal ganglia (Fig. 32.2), probably through a mechanism similar to that responsible for raised dopamine levels (Spokes, 1980). There is evidence that noradrenergic

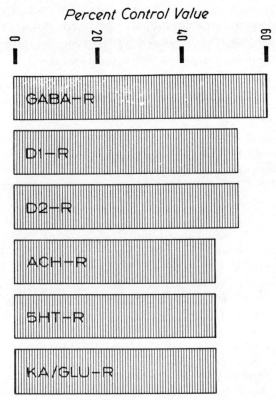

Fig. 32.3. Changes in neurotransmitter receptor densities in the choreic striatum. *Key*: GABA-R, γ-aminobutyric acid receptor; D1-R, dopamine D-1 receptor; D2-R, dopamine D-2 receptor; ACH-R, muscarinic acetylcholine receptor; 5HT-R, 5-hydroxytryptamine receptor; KA/GLU-R, kainic acid/glutamate receptor. (For references, see text.)

fibres derived from the locus coeruleus exert a facilitatory effect on nigral dopaminergic neurones (Donaldson *et al.*, 1977) so that their preservation may contribute to apparent dopaminergic overactivity in chorea.

Some striatal interneurones are cholinergic and are innervated by dopaminergic terminals which exert a tonic inhibitory effect on acetycholine release (Bartholini *et al.*, 1975). Loss of the cholinergic marker enzyme choline acetyltransferase (CAT) from the choreic striatum (Fig. 32.2) (Spokes, 1980) probably reflects the death of such interneurones, disturbing the cholinergic–dopaminergic balance important for normal striatal function. A CAT deficit also occurs in the choreic septohippocampal complex (Spokes, 1980), similar to that reported for Alzheimer's disease (Perry *et al.*, 1977) and may relate to the dysfunction in recent memory observed in these disorders. Cortical cholinergic fibres are believed to originate in the nucleus basalis of Meynert in the substantia innominata (Johnston *et al.*, 1979). This nucleus is spared in Huntington's chorea (Clark *et al.*, 1983) and cortical CAT values are normal (Spokes, 1980). Conversely, in Alzheimer's disease the nucleus degenerates and there is loss of cortical CAT activity (for review, see Rossor, 1982). This poses the interesting speculation that the loss or preservation of cortical cholinergic fibres may, in part, underlie the clinical differences between cortical and subcortical dementias.

At the time of writing, some thirty neuropeptides have been recognised as candidates for neurotransmitters or neuromodulators in the CNS. Those investigated in Huntington's chorea are shown in Fig. 32.2. Levels of substance P (Gale *et al.*, 1978), enkephalin and angiotensin-converting enzyme (Arregui *et al.*, 1979) the latter being a marker for angiotensin-containing neurones, are depleted by 50 % or more in the choreic pallidum and substantia nigra, indicating degeneration of striopallidonigral peptidergic pathways. In contrast, striatal concentrations of thyrotrophin-releasing hormone, neurotensin and somatostatin are considerably increased (Spindel *et al.*, 1980; Nemeroff *et al.*, 1983), suggesting a condensation either of surviving intrinsic neurones or afferent fibres of extrinsic origin. It is too early to speculate on the role of these various peptides in basal ganglia function.

32.4. Clinical pharmacology

Unlike Parkinson's disease, which selectively affects dopaminergic and noradrenergic fibres, permitting the efficacious use of replacement therapy in the form of L-dopa, Huntington's chorea involves diverse cell populations throughout the corpus striatum, rendering replacement treatment unfeasible. Attempts to ameliorate chorea, other than with drugs impeding dopaminergic neurotransmission, have concentrated on enhancement of GABA- and acetylcholine-mediated activities. GABA will not cross the blood-brain barrier and the functional capacity of GABA systems can be facilitated only by indirect methods. Inhibitors of GABA-transaminase, the main degradative enzyme for GABA, such as sodium valproate, isoniazid and γ-vinyl GABA are ineffective (Shoulson *et al.*, 1976; Paulson *et al.*, 1979), likewise lipid-soluble GABA agonists such as muscimol and tetrahydroisoxazolopyridinol (THIP) (Foster *et al.*, 1983).

Pharmacological manipulation of central cholinergic systems has also proved disappointing. No consistent beneficial response has been obtained with the use of oral choline, dimethylaminoethanol or cholinomimetics such as arecoline or pilocarpine (Chase, 1976; Nutt *et al.*, 1978; Growdon and Wurtman, 1979).

A factor which must be considered in design of therapy is the integrity of postsynaptic receptor sites. Both GABA and muscarinic cholinergic receptors are reduced in density in the choreic striatum (Fig. 32.3) (Hiley and Bird, 1974; Lloyd *et al.*, 1977), suggesting loss of cells normally responsive to these neurotransmitters. However, GABA receptor density is increased in the choreic substantia nigra (Enna *et al.*, 1976), in parallel with raised dopamine concentrations (Spokes, 1980), probably arising from closer packing of dopaminergic cells which carry GABA receptors. Thus an effective GABA-mimetic might be expected to inhibit activity in the nigrostriatal tract. On the other hand, striatal cholinergic neurones exert a positive feedback influence on dopaminergic nigral cells so that potentiated cholinergic activity may be offset by increased dopamine release (Bunney and Aghajanian, 1976).

32.5. Kainic acid model

The gene for Huntington's chorea has now been localised to chromosome 4 by the finding of a closely linked cloned DNA sequence (Gusella *et al.*, 1983). In the long term this raises the possibility of isolation of the abnormal gene itself and elucidation of the basic biochemical defect. However, a predictive test for preclinical and prenatal diagnosis is now a reality and poses serious ethical questions implicit in detection without prevention or cure. Consequently, there is even greater impetus to develop treatments which will prevent or slow the progression of the disease. One faint ray of hope has been the discovery of the kainic acid animal model for Huntington's chorea.

Intrastriatal injections of kainic acid, a rigid cyclic analogue of glutamate with potent neuroexcitatory effects, evokes structural and biochemical sequelae reminiscent of those seen in the choreic striatum (Coyle *et al.*, 1977). The salient features are striatal shrinkage and loss of intrinsic neurones containing GABA, CAT, substance P and angiotensin-converting enzyme, accompanied by reductions in GABA-, substance P- and angiotensin-converting-enzyme-containing terminals in the substantia nigra. Alterations in receptor densities are also found in parallel with those observed in Huntington's chorea. However, in contrast to Huntington's chorea, the kainate-lesioned striatum shows normal dopamine concentrations so that some loss of dopaminergic afferents is likely. Moreover, the kainate-lesioned rat does not display true chorea and, clearly, the time course of striatal damage is quite different from Huntington's chorea where pathological changes evolve over many years. These criticisms aside, qualitatively at least, there would appear to be striking biochemical similarities between Huntington's chorea and the kainate 'model'.

The cytotoxic effect of kainate is indirect and depends on glutamate release from corticostriatal terminals. Kainate may overstimulate the corticostriatal pathway or inhibit glutamate reuptake, or both, cell death arising from an 'excitotoxic' effect of glutamate producing a sustained increase in cell permeability from overstimulation of glutamate receptors

(McGeer *et al.*, 1978). Extrapolating from these findings, it is possible that striatal cell death in Huntington's chorea stems from overactivity in the corticostriatal pathway, a defect in glutamate reuptake from the synaptic cleft or supersensitivity of postsynaptic glutamate receptors. Readily compatible with these proposals has been the claim by several research groups that the disease may be associated with a diffuse generalised membrane abnormality (for review, see Butterfield and Markesbery, 1981). However, other workers have failed to replicate these findings (Beverstock and Pearson, 1981; Dubbelman *et al.*, 1981) and further evaluation is needed for the membrane hypothesis.

If the inborn error of metabolism in Huntington's chorea does indeed prove to involve a cell membrane constituent, it is conceivable that the glutamate receptor might also be affected. Although it would by myopic to suggest that the similar morphological and biochemical profiles seen in the choreic and kainate-lesioned striata result from the same cause, it is an important area for further investigation with practical implications for the design of treatment.

References

Albert, M. L. (1978). Subcortical dementia. In *Alzheimer's Disease: Senile Dementia and Related Disorders*. R. Katzman, R. D. Terry and K. L. Bick (eds), pp. 173–9. Raven Press, New York.

Arregui, A., Iversen, L. L., Spokes, E. G. S. and Emson, P. C. (1979). Alterations in postmortem brain angiotensin-converting enzyme activity and some neuropeptides in Huntington's disease. In *Advances in Neurology*, Vol. 23. T. N. Chase, N. S. Wexler and A. Barbeau (eds), pp. 517–25. Raven Press, New York.

Bartholini, G., Stadler, H. and Lloyd, K. G. (1975). Cholinergic–dopaminergic interregulations within the extrapyramidal system. In *Cholinergic Mechanisms*. P. G. Waser (ed.), pp. 411–18, Raven Press, New York.

Beverstock, G. C. and Pearson, P. L. (1981). Membrane fluidity measurements in peripheral cells from Huntington's disease patients. *J. Neurol. Neurosurg. Psychiat.* **44**, 684–9.

Bird, E. D., Caro, A. and Pilling, J. (1974). A sex-related factor in the inheritance of Huntington's chorea. *Ann. Hum. Genet.* **37**, 255–60.

Bunney, B. S. and Aghajanian, G. K. (1976). Dopamine influence in the basal ganglia: evidence for striatonigral feedback regulation. In *The Basal Ganglia*. M. D. Yahr (ed.), pp. 249–67. Raven Press, New York.

Butterfield, D. A. and Markesbery, W. R. (1981). Huntington's disease: a generalised membrane defect. *Life Sci.* **28**, 1117–31.

Chase, T. N. (1976). Rational approaches to the pharmacotherapy of chorea. In *The Basal Ganglia*. M. D. Yahr (ed.), pp. 337–49. Raven Press, New York.

Clark, A. W., Parhad, I. M., Folstein, S. E., Whitehouse, P. J., Hedreen, J. C., Price, D. L. and Chase, G. A. (1983). The nucleus basalis in Huntington's disease. *Neurology* **33**, 1262–7.

Coyle, J. T., Schwarcz, R., Bennett, J. P. and Campochiaro, P. (1977). Clinical, neuropathologic and pharmacologic aspects of Huntington's disease: correlates with a new animal model. *Progr. Neuro-psychopharmacol.* **1**, 13–30.

Cross, A. and Rossor, M. (1983). Dopamine D-1 and D-2 receptors in Huntington's disease. *Eur. J. Pharmacol.* **88**, 223–9.

Curzon, G., Gumpert, J. and Sharpe, D. (1972). Amine metabolites in the cerebrospinal fluid in Huntington's chorea. *J. Neurol. Neurosurg. Psychiat.* **35**, 514–19.

Dom, R., Malfroid, M. and Baro, F. (1976). Neuropathology of Huntington's chorea. *Neurology* **26**, 64–8.

Donaldson, I. M., Dolphin, A. O., Jenner, P., Pycock, C. and Marsden, C. D. (1977). Rotational behaviour produced in rats by unilateral electrolytic lesions of the ascending noradrenergic bundles. *Brain Res.* **138**, 487–509.

Dubbelman, T. M., de Bruijne, A. W., van Steveninck, J. and Bruyn, G. W. (1981). Studies on erythrocyte membranes of patients with Huntington's disease. *J. Neurol. Neurosurg. Psychiat.* **44**, 570–3.

Enna, S. J., Bennett, J. P., Bylund, D. B., Snyder, S. H., Bird, E. D. and Iversen, L. L. (1976). Alterations of brain neurotransmitter receptor binding in Huntington's chorea. *Brain Res.* **116**, 531–7.

Foster, N. L., Chase, T. N., Denaro, A., Hare, T. A. and Tamminga, C. A. (1983). THIP treatment of Huntington's disease. *Neurology* **33**, 637–9.

Gale, J., Bird, E. D., Spokes, E. G. S., Iversen, L. L. and Jessell, T. M. (1978). Human brain substance P: distribution in controls and in Huntington's chorea. *J. Neurochem.* **30**, 633–4.

Growdon, J. H. and Wurtman, R. J. (1979). Oral choline administration to patients with Huntington's disease. In *Advances in Neurology*, Vol. 23. T. N. Chase, N. S. Wexler and A. Barbeau (eds), pp. 765–76, Raven Press, New York.

Gusella, J. F., Wexler, N. S., Conneally, P. M., Naylor, S. L., Anderson, M. A., Tanzi, R. E., Watkins, P. C., Ottina, K., Wallace, M. R., Sakaguchi, A. Y., Young, A. B., Shoulson, I., Bonilla, E. and Martin, J. B. (1983). A polymorphic DNA marker genetically linked to Huntington's disease. *Nature (Lond.)* **306**, 234–8.

Hiley, C. R. and Bird, E. D. (1974). Decreased muscarinic receptor concentration in post-mortem brain in Huntington's chorea. *Brain Res.* **80**, 355–8.

Huntington, G. S. (1872). On chorea. *Med. Surg. Rep. Philadelph.* **26**, 320–1.

Johnston, M. V., McKinney, M. and Coyle, J. T. (1979). Evidence for a cholinergic projection to neocortex from neurons in basal forebrain. *Proc. nat. Acad. Sci. USA* **76**, 5392–6.

Lange, H., Thorner, G., Hopf, A. and Schroder, K. F. (1976). Morphometric studies of the neuropathological changes in choreatic diseases. *J. Neurol. Sci.* **28**, 401–25.

Lloyd, K. G., Dreksler, S. and Bird, E. D. (1977). Alterations in ^{3}H-GABA binding in Huntington's chorea. *Life Sci.* **21**, 747–54.

McGeer, E. G., McGeer, P. L. and Singh, K. (1978). Kainate-induced degeneration of neostriatal neurons: dependency upon corticostriatal tract. *Brain Res.* **139**, 381–3.

Nemeroff, C. B., Youngblood, W. W., Manberg, P. J., Prange, A. J. and Kizer, J. S. (1983). Regional brain concentrations of neuropeptides in Huntington's chorea and schizophrenia. *Science* **221**, 972–5.

Nutt, J. G., Rosin, A. and Chase, T. N. (1978). Treatment of Huntington's disease with a cholinergic agonist. *Neurology* **28**, 1061–4.

Osler, W. (1908). Historical note on hereditary chorea. *Neurographs* **1**, 113–16.

Paulson, G. W., Malarkey, W. B. and Shaw, G. (1979). Huntington's disease, INH, and prolactin levels. In *Advances in Neurology*, vol. 23. T. N. Chase, N. S. Wexler and A. Barbeau (eds), pp. 797–801. Raven Press, New York.

Perry, E. K., Gibson, P. H., Blessed, G., Perry, R. H. and Tomlinson, B. E. (1977). Neurotransmitter enzyme abnormalities in senile dementia. *J. Neurol. Sci.* **34**, 247–65.

318 Neurobiology of dopamine systems

Reisine, T. D., Beaumont, K., Bird, E. D., Spokes, E. and Yamamura, H. I. (1979). Huntington's disease: alterations in neurotransmitter receptor binding in the human brain. In *Advances in Neurology*, Vol. 23. T. N. Chase, N. S. Wexler and A. Barbeau (eds), pp. 717–26. Raven Press, New York.

Roffler-Tarlov, S., Sharman, D. F. and Tegerdine, P. (1971). 3,4-Dihydroxyphenylacetic acid and 4-hydroxy-3-methoxyphenylacetic acid in the mouse striatum: a reflection of intra- and extra-neuronal dopamine? *Brit. J. Pharmacol.* **42**, 343–51.

Rossor, M. N. (1982). Dementia. *Lancet* **ii**, 1200–4.

Shoulson, I., Kartzinel, R. and Chase, T. N. (1976). Huntington's disease: treatment with dipropylactic acid and gamma-aminobutyric acid. *Neurology* **26**, 61–3.

Spindel, E. R., Wurtman, R. J. and Bird, E. D. (1980). Increased TRH content of the basal ganglia in Huntington's disease. *New Engl. J. Med.* **303**, 1235–6.

Spokes, E. G. S. (1980). Neurochemical alterations in Huntington's chorea: a study of post-mortem brain tissue. *Brain* **103**, 179–210.

Spokes, E. G. S., Garrett, N. J., Rossor, M. N. and Iversen, L. L. (1980). Distribution of GABA in post-mortem brain tissue from control, psychotic and Huntington's chorea subjects. *J. Neurol. Sci.* **48**, 303–13.

33

Dopamine deficiency and dopamine substitution in Parkinson's disease

O. Hornykiewicz

Institute of Biochemical Pharmacology, University of Vienna, Borschkegasse 8A, A-1090 Vienna, Austria

33.1. Parkinson's disease and the meso-telencephalic dopamine loss

33.1.1. *Striatal dopamine*

The most consistent neurochemical finding in Parkinson's disease (PD) is the profound loss of dopamine in the nuclei of the basal ganglia, especially the substantia nigra, the caudate nucleus and the putamen (together called the striatum), and the globus pallidus (Table 33.1). This dopamine loss is characteristic of PD regardless of its aetiology; it has been found in cases with idiopathic PD, postencephalic parkinsonism, and parkinsonism associated with senile-arteriosclerotic brain changes or chronic manganese poisoning (Bernheimer *et al.*, 1973). The only determinant of the dopamine loss in the basal ganglia is the degree of cell loss in the dopamine- (and melanin-) containing neurones in the compact zone of the substantia nigra (Fig. 33.1). Normally, the nigral zona compacta neurones give rise to a prominent dopamine pathway innervating both the caudate nucleus and the putamen; the terminal branches of this nigrostriatal dopamine pathway contain about 80 % of the total brain dopamine. In PD, the nigral cells of origin of this dopaminergic fibre system regularly degenerate (Hassler, 1938; Greenfield and Bosanquet, 1953; Bernheimer *et al.*, 1973). The crucial role in PD of the nigral cell loss is borne out by the recent observation of the so-called 1-methyl-4-phenyl-1,2,3,6-tetrahydropyridine- (MPTP-) induced extrapyramidal syndrome in man, which not only imitates PD clinically, but also is accompanied by morphological changes in the compact zone of the substantia nigra quite analogous to those seen in genuine PD (Davis *et al.*, 1979; Langston *et al.*, 1983).

Since in PD all components of the nigrostriatal dopamine pathway degenerate, the dopamine loss is accompanied by a marked reduction in all other neurochemical markers of presynaptic dopamine neurones, including the dopamine metabolite homovanillic acid as well as the synthesising

Table 33.1. Presynaptic dopamine neurone indices in the nigrostriato-pallidal complex in Parkinson's disease—comparison with controls.

Brain region	Dopamine (μg g^{-1} wet tissue)	Homovanillic acid (μg g^{-1} wet tissue)	L-tyrosine hydroxylase (nmol CO$_2$ 0·5 h^{-1} 100 mg^{-1} protein)	L-dopa decarboxylase (nmol CO$_2$ 2 h^{-1} 100 mg^{-1} protein)
Substantia nigra				
Controls	0·46 (13)	2·32 (7)	17·4 (1)	549·3±293·7 (15)
Parkinson	0·07 (10)	0·41 (9)	6·1±1·5 (3)	21·4±6·0 (10)
Caudate nucleus				
Controls	4·06±0·47 (18)	2·92±0·37 (19)	18·7±2·0 (3)	364·0±95·4 (19)
Parkinson	0·20±0·19 (3)	1·19±0·10 (3)	3·2±0·5 (3)	54·3±14·0 (13)
Putamen				
Controls	5·06±0·39 (17)	4·92±0·32 (16)	17·4±2·4 (3)	431·7±109·1 (13)
Parkinson	0·14±0·13 (3)	0·54±0·13 (3)	3·1±1·2 (3)	32·3±7·1 (13)
Globus pallidus				
Controls	0·5 (6)	2·25 (8)	3·5 (1)	22·1±2·7 (9)
Parkinson	0·2 (4)	0·72 (9)	1·7±0·2 (3)	18·0±2·6 (12)

The values are means±SEM (where applicable); number of cases in parentheses. The data are taken from the following references: substantia nigra, Hornykiewicz (1963); Bernheimer and Hornykiewicz (1965); caudate nucleus and putamen, Lloyd et al. (1975); globus pallidus, Ehringer and Hornykiewicz (1960); Bernheimer and Hornykiewicz (1965).

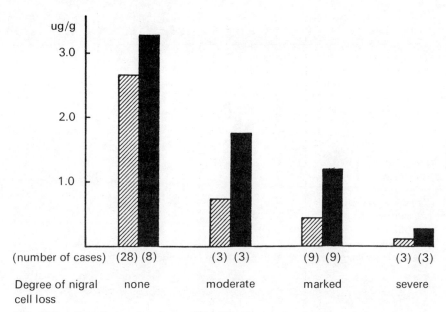

Fig. 33.1. Correlation between cell loss in the zona compacta of the substantia nigra and dopamine (cross-hatched) and homovanillic acid (shaded) concentrations in the caudate nucleus in patients with Parkinson's disease. (Data from Bernheimer *et al.*, 1973.)

enzymes L-tyrosine hydroxylase and L-dopa decarboxylase (cf. Hornykiewicz, 1982) (Table 33.1).

33.1.2. *Limbic dopamine*

In addition to the severe loss of dopamine in the striatum, in PD dopamine is also reduced in several extra-striatal forebrain regions, especially within the subcortical and cortical components of the limbic system; they include the nucleus accumbens, the olfactory regions and the parolfactory gyrus (Brodmann area 25) (Farley *et al.*, 1977; Price *et al.*, 1978) (Table 33.2). These limbic dopamine changes are probably due to neuronal loss in the paranigral midbrain region (area A10 of Dahlström and Fuxe) (cf. Hassler, 1938), from where the limbic forebrain receives most of its dopamine innervation (cf. Moore and Bloom, 1978).

33.1.3. *Correlation between symptoms and dopamine loss*

Of crucial importance for the understanding of the role of dopamine in PD are the observations showing that the degree of striatal dopamine loss is correlated with the severity of the cardinal (motor) symptoms of PD, especially akinesia (Bernheimer *et al.*, 1973) (Table 33.3). In this respect,

Table 33.2. Limbic dopamine and homovanillic acid in Parkinson's disease—comparison with controls.

Brain region	Dopamine (μg g^{-1} wet tissue)	Homovanillic acid (μg g^{-1} wet tissue)
Nucleus accumbens		
Controls	$3\cdot79 \pm 0\cdot82$ (8)	$4\cdot38 \pm 0\cdot64$ (8)
Parkinson	$1\cdot61 \pm 0\cdot28$ (4)	$3\cdot13 \pm 0\cdot13$ (3)
Lateral hypothalamus		
Controls	$0\cdot51 \pm 0\cdot08$ (4)	$1\cdot96 \pm 0\cdot28$ (3)
Parkinson	$<0\cdot03$ (2)	$1\cdot03 \pm 0\cdot23$ (3)
Parolfactory gyrus		
Controls	$0\cdot35 \pm 0\cdot09$ (4)	$0\cdot98$ (2)
Parkinson	$0\cdot03$ (2)	—

The values are means $\pm$ S,EM; number of cases in parentheses. Data from Price *et al.* (1978).

Table 33.3. Relationship between the degree of dopamine loss in the putamen and the severity of Parkinsonian symptoms.

Severity of symptoms	Dopamine (μg g^{-1} wet tissue)	Per cent of control
Mild akinesia	$0\cdot44$ (13)	13
Marked akinesia	$0\cdot05$ (9)	$1\cdot5$
Hemiparkinson (right side)		
Right putamen	$0\cdot93$ (1)	20
Left putamen	$0\cdot13$ (1)	$3\cdot5$

The values are means (where applicable), with the number of cases in parentheses. Data from Barolin *et al.* (1964); Bernheimer *et al.* (1973).

there seems to exist a regional selectivity for each of the dopamine-dependent symptoms: loss of dopamine in caudate nucleus correlated best with akinesia, whereas homovanillic acid loss in globus pallidus was best correlated with tremor. The direct relationship between the severity of PD symptoms and striatal dopamine deficiency is also borne out by the results obtained in a case of hemiparkinsonism (Barolin *et al.*, 1964). In this case, the striatum contralateral to the symptoms was more severely depleted of its dopamine than the ipsilateral striatum (Table 33.3). There is a wealth of experimental evidence obtained in laboratory animals demonstrating the relationship between parkinsonian symptoms, especially hypokinesia, and striatal dopamine (cf. Schultz, 1984).

Recent pharmacological observations made in laboratory animals have indicated that the accumbens dopamine system is involved with control of locomotor behaviour (Pijnenburg *et al.*, 1976; Iversen and Koob, 1977). This makes it very likely that dopamine deficiency in this region, if sufficient in degree, may substantially contribute to the motor deficits, especially akinesia, in patients with PD. With regard to the dopamine loss in the limbic cortex in PD, it is not unlikely that this change plays a part in the pathophysiology of the cognitive, affective and psychotic disturbances often seen in PD patients. An area in the rhesus monkey brain in part identical with the human parolfactory gyrus has recently been suggested to be involved in memory processes (Mishkin and Bachevalier, 1983).

33.2. Pathophysiology of dopamine substitution in Parkinson's disease

The most important result of the discovery of the brain dopamine deficiency in PD was the development of dopamine substitution with L-dopa and the direct acting dopamine agonists (e.g. ergolines such as bromocriptine or lisuride) as the most rational and efficacious drug treatment of PD. In the concept of dopamine substitution, the following three aspects are of special importance: adaptive changes in the dopamine neurones; the synergistic role of brain noradrenaline; and the functional heterogeneity of the striatum.

33.2.1. *Adaptive changes in the partly damaged dopamine system*

Small dopamine losses do not produce overt clinical symptoms. Biochemical clinical correlations have shown that clinical symptoms do not become apparent unless the striatal dopamine loss reaches the critical value of about 80 % (this is, only 20 % of dopamine remaining: Bernheimer *et al.*, 1973; cf. Table 33.3). This stage of PD represents the 'decompensated stage' of striatal dopamine deficiency; milder than 80 % dopamine losses can very effectively be compensated by the remaining dopamine neurones, thus providing the basis for the 'compensated stage' of PD (Fig. 33.2). An analogous situation has been observed in rats with experimental lesions of the nigrostriatal dopamine path (Zigmond and Stricker, 1972). Obviously, the therapeutic goal of dopamine substitution is to revert the decompensated stage of PD to the stage of functional compensation.

Molecular mechanisms. With regard to the biochemical changes that are important for the compensatory activity in the partly damaged nigrostriatal doperminergic system, these events take place in presynaptic as well as postsynaptic dopaminergic elements; they include: (a) overactivity in the remaining dopamine neurones as indicated by the marked shifting, in the PD striatum, of the ratio 'homovanillic acid: dopamine' in favour of the

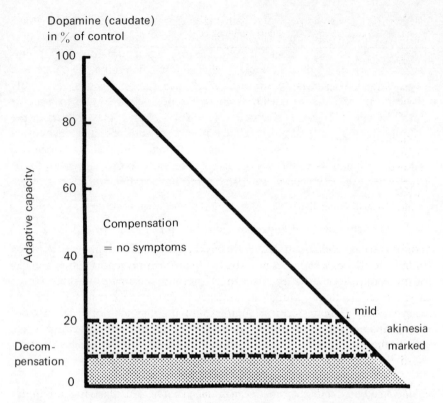

Fig. 33.2.   Adaptive capacity of the dopamine system: compensation and decompensation in Parkinson's disease. (Diagram based on data obtained by Bernheimer *et al.*, 1973.)

metabolite (Bernheimer and Hornykiewicz, 1965; Bernheimer *et al.*, 1973) (Table 33.4); and (b) postsynaptic dopamine receptor supersensitivity, as judged from the increased number of specific [^{3}H]neuroleptic binding sites measured in the striatum (especially the putamen) of PD patients not treated with L-dopa (Lee *et al.*, 1978; Rinne, 1982; Bokobza *et al.*, 1984); see Table 33.4. Both these changes are aimed at maintaining striatal dopamine function in the face of a marked reduction of nigrostriatal dopamine neurone number. Of these two adaptive changes, the presynaptic increase in dopamine turnover seems to be the much more sensitive mechanism, being activated at considerably smaller degrees of striatal dopamine loss (about −25%) (Agid *et al.*, 1973) than the postsynaptic dopamine receptor increase, which is only observed at over 90% dopamine loss (Creese and Snyder, 1979). Recently, Zigmond *et al.* (1984) have carefully analysed several quantitative aspects of the presynaptic dopamine compensation in an

Table 33.4. Pre- and postsynaptic compensatory changes in the remaining striatal dopamine elements in Parkinson's disease—comparison with controls.

Biochemical index	Control	Parkinson
Presynaptic change (molar ratio HVA/DA)[a]		
Caudate	1·16	1·75
Putamen	1·09	3·58
Postsynaptic change (D-2 receptor density)[b]		
Caudate	42 ± 4	53 ± 12
Putamen	44 ± 5	68 ± 3

[a]From Berheimer and Hornykiewicz (1965).
[b]Specific [^{3}H]haloperidol binding (fmol mg^{-1} protein); from Lee *et al.* (1978).

animal (rat) model of preclinical parkinsonism. Taken together, the available evidence clearly demonstrates that the presynaptic increase in dopamine turnover represents the main mechanism operating during the early, clinically silent, compensated stage of PD; in contrast, increase in postsynaptic dopamine receptor number represents a 'last ditch' attempt at functional compensation during stages of PD when the dopamine loss exceeds the 80% mark, i.e. during the late, decompensated stages of this disorder (cf. Schultz, 1982).

Significance for effectiveness of dopamine substitution. The plasticity and high adaptive capacity of the nigrostriatal dopamine neurones form the basis for the special sensitivity of the PD patient to dopamine substitution, especially by virtue of the increased number of postsynaptic dopamine receptors. Furthermore, since these compensatory measures can be assumed to be especially prominent in the most severely affected brain regions, they are bound to confer on the dopamine replacement therapy a high degree of regional specificity. However, from the pharmacological point of view, it is evident that long-term dopamine substitution, especially with high drug doses, will inevitably down-regulate the compensatory changes in the remaining dopamine elements, especially by reducing the striatal dopamine receptor numbers; this will result in reduction of clinical effectiveness as well as increased incidence of side-effect due to the loss of regional specificity of the treatment. It has indeed been observed that in patients receiving L-dopa, the number of striatal dopamine receptors (measured as [^{3}H]neuroleptic binding) was either in the normal range or below normal (Reisine *et al.*, 1977; Lee *et al.*, 1978).

33.2.2. *Brain noradrenaline and dopamine substitution*

It is well established that in PD brain, there are, in additin to the severe dopamine loss, other neurotransmitter changes, including acetylcholine, serotonin, noradrenaline, γ-aminobutyric acid, glycine, neurotensin, the enkephalins, substance P, and cholecystokinin (Table 33.5). Many of these non-dopamine changes are quantitatively small (cf. Table 33.5) and therefore may lack any functional significance. Nevertheless, the loss of noradrenaline in the PD basal ganglia (Farley and Hornykiewicz, 1976) deserves special consideration because from animal studies we know that there exists a strong brain noradrenaline–dopamine synergy in respect of locomotor activity (cf. Donaldson *et al.*, 1978; Ögren *et al.*, 1983; Plaznik and Kostowski, 1983). Thus, amphetamine (which greatly increases locomotor activity in mice and releases synaptic dopamine as well as noradrenaline) loses a significant part of its locomotor stimulant effect under conditions of selective brain noradrenaline depletion; the full amphetamine effect can, however, be restored by threo-3,4-dihydroxy-phenylserine (threo-dops) (Derkach, 1975), a synthetic aromatic amino acid which in tissues is directly converted to noradrenaline. Since in normal animals selective increase in brain noradrenaline levels has little effect on motor activity, noradrenaline seems to exert physiologically a 'tuning control' on the dopaminergic motor activity, which is augmenting the sensitivity of the locomotor brain mechanisms to dopamine. The decrease in brain noradrenaline in PD is therefore likely to contribute significantly to the overall severity of the motor deficits which, primarily, are due to the striatal dopamine loss. The exceptionally strong antiparkinsonian activity

Table 33.5. Parkinson's disease: changes in some basal ganglia neurotransmitters other than dopamine.

Neurotransmitter markers	Site(s) of change	References
Noradrenaline (-50%)	SN, ACC, STR	Farley and Horneykiewicz, 1976
Serotonin (-50%)	SN, STR, GP	Bernheimer *et al.*, 1961
Met^5-enkephalin (-65%)	SN, PUT, GP	Taquet *et al.*, 1983
Leu^5-enkephalin (-35%)	PUT, GP	Taquet *et al.*, 1983
Cholecystokinin-8 (-35%)	SN	Studler *et al.*, 1982
Substance P (-35%)	SN, GP, PUT	Mauborgne *et al.*, 1983; Tenovuo *et al.*, 1984
Glutamate decarboxylase (-50%)	STR, GP, SN	Lloyd and Hornykiewicz, 1973
Specific 'receptor' density		
γ-Aminobutyric acid (-69%)	SN	Lloyd *et al.*, 1977
Neurotensin (-65%)	SN	Uhl *et al.*, 1984
Met^5-enkephalin ($+42\%$)	PUT	Rinne, 1982
Leu^5-enkephalin ($+43\%$)	PUT	Rinne, 1982
Glycine (-33%)	SN	DeMontis *et al.*, 1982

Abbreviations: ACC = nucleus accumbens; GP = globus pallidus; PUT = putamen; SN = substantia nigra; STR = caudate and putamen.

of L-dopa is most likely due to the fact that—unlike the direct-acting dopamine agonists such as the apomorphine or ergoline derivatives with their more selective dopaminergic activity— L-dopa is converted in the body to both dopamine and noradrenaline, and thus in principle stimulates both catecholamine receptors. The successful use of threo-dops as an adjunct to L-dopa treatment (Narabayashi *et al.*, 1981) supports the role of brain noradrenaline in the motor symptomatology of PD. From this it seems logical to conclude that development of new direct-acting antiparkinsonian drugs should aim at imitating the dual property of L-dopa as a dopaminergic as well as noradrenergic drug.

33.2.3. *Novel approaches to dopamine substitution: importance of striatal heterogeneity*

In the context of the relationship between striatal dopamine deficiency and dopamine substitution, the early observation (cf. Bernheimer *et al.*, 1973) that in idiopathic PD the dopamine loss in putamen is considerably greater than in the caudate nucleus (Fig. 33.3) is of considerable significance. This differential striatal dopamine loss assumes special significance in view of the recent neuroanatomical as well as neurophysiological and behavioural studies indicating that these striatal nuclei possess a high degree of anatomical and functional heterogeneity (cf. Divac and Öberg, 1979). Thus, of the two nuclei, only the putamen receives projections from the somatic

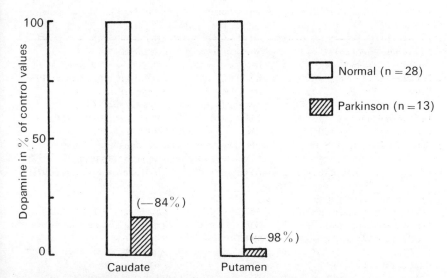

Fig. 33.3. Different degree of dopamine loss in caudate and putamen in idiopathic Parkinson's disease. (Data from Bernheimer *et al.*, 1973.)

motor cortex (Künzle, 1975), whereas the caudate nucleus preferentially receives projections from the prefrontal (limbic) cortex (Goldman and Nauta, 1977). This striking difference suggests that on the functional level the caudate nucleus is to a large extent involved with the diverse psychomotor aspects of motor behaviour, whereas the putamen primarily participates in movement control and motor programming.

Since in recent years intrastriatal grafting of nigral (or other dopamine-producing) cells in animal models of PD has become a feasible way of functionally restoring lost dopamine function (Dunnett *et al.*, 1983 *a*, *b*; see also Chapter 34), it is obvious that the success of these novel dopamine substituting techniques will crucially depend on the appropriate localisation of such implants. Work with the recently introduced MPTP model of PD in primates is likely to contribute substantially to the solution of these and related problems (see chapters 35 and 36, this volume).

References

Agid, Y., Javoy, F. and Glowinski J. (1973). Hyperactivity of remaining dopaminergic neurones after partial destruction of the nigro-striatal dopaminergic system in the rat. *Nature (Lond.)* **245**, 150–1.

Barolin, G. S., Bernheimer, H. and Hornykiewicz, O. (1964). Seitenverschiedenes Verhalten des Dopamins (3-Hydroxytyramin) im Gehirn eines Falles von Hemiparkinsonismus. *Schweiz. Arch. Neurol. Psychiat.* **94**, 241–8.

Bernheimer, H. and Hornykiewicz, O. (1965). Herabgesetzte Konzentration der Homovanillinsäure im Gehirn von Parkinsonkranken Menschen als Ausdruck der Störung des Zentralen Dopaminstoffwechsels. *Klin. Wschr.* **43**, 711–5.

Bernheimer, H., Birkmayer, W. and Hornykiewicz, O. (1961). Verteilung des 5-Hydroxytryptamins (Serotonin) im Gehirn des Menschen und sein Verhalten bei Patienten mit Parkinson-Syndrom. *Klin. Wschr.* **39**, 1056–9.

Bernheimer, H., Birkmayer, W., Hornykiewicz, O., Jellinger, K. and Seitelberger, F. (1973). Brain dopamine and the syndromes of Parkinson and Huntington. *J. Neurol. Sci.* **20**, 415–55.

Bokobza, B., Ruberg, M., Scatton, B., Javoy-Agid, F. and Agid, Y. (1984). [³H]Spiperone binding, dopamine and HVA concentrations in Parkinson's disease and supranuclear palsy. *Eur. J. Pharmacol.* **99**, 167–75.

Creese, I. and Snyder, S. H. (1979). Nigrostriatal lesions enhance striatal [³H]apomorphine and [³H]spiroperidol binding. *Eur. J. Pharmacol.* **56**, 277–81.

Davis, G. C., Williams, A. C., Markey, S. P., Ebert, M. H., Caine, E. D., Reichert, C. M. and Kopin, I. J. (1979). Chronic parkinsonism secondary to intravenous injection of meperidine analogues. *Psychiat. Res.* **1**, 249–54.

DeMontis, G., Beaumont, K., Javoy-Agid, F., Agid, Y., Constandinidis, J., Lowenthal, A. and Lloyd, K. G. (1982). Glycine receptors in the human substantia nigra as defined by [³H]strychnine binding. *J. Neurochem.* **38**, 718–24.

Derkach, P. (1975). The pharmacology of locomotor activity: relative roles of brain dopamine and noradrenaline. *M.Sc. Thesis, University of Toronto*, 1–180.

Divac, I. and Öberg, R. G. E. (1979). Current conceptions of neostriatal functions: history and evaluation. In *The Neostriatum*. I. Divac and R. G. E. Öberg (eds), pp. 215–30. Pergamon Press, Oxford.

Donaldson, I. MacG., Dolphin, A. C., Jenner, P., Pycock, C. and Marsden, C. D. (1978). Rotational behaviour produced in rats by unilateral electrolytic lesions of the ascending noradrenergic bundles. *Brain Res.* **138**, 487–509.

Dunnett, S. B., Björklund, A., Schmidt, R. H., Stenevi, U. and Iversen, S. D. (1983*a*). Intracerebral grafting of neuronal cell suspensions IV. Behavioural recovery in rats with unilateral 6–OHDA lesions following implantation of nigral cell suspensions in different forebrain sites. *Acta Physiol. Scand.* Suppl. 522, 29–37.

Dunnett, S. B., Björklund, A., Schmidt, R. H., Stenevi, U. and Iversen, S. D. (1983*b*). Intracerebral grafting of neuronal cell suspensions V. Behavioural recovery in rats with bilateral 6–OHDA lesions following implantation of nigral cell suspensions. *Acta Physiol. Scand.* Suppl. 522, 39–47.

Ehringer, H. and Hornykiewicz, O. (1960). Verteilung von Noradrenalin und Dopamine (3-Hydroxytyramin) im Gehirn des Menschen und ihr Verhalten bei Erkrankungen des extrapyramidalen Systems. *Klin. Wschr.* **38**, 1238–9.

Farley, I. J. and Hornykiewicz, O. (1976). Noradrenaline in subcortical brain regions of patients with Parkinson's disease and control subjects. In *Advances in Parkinsonism*. W. Birkmayer and O. Hornykiewicz (eds), pp. 178–85. Editiones Roche, Basel.

Farley, I. J., Price, K. S. and Hornykiewicz, O. (1977). Dopamine in the limbic regions of the human brain: normal and abnormal. *Adv. Biochem. Psychopharmacol.* **16**, 57–64.

Goldman, P. S. and Nauta, W. J. H. (1977). An intricately pattern prefronto-caudate projection in the rhesus monkey. *J. Comp. Neurol.* **171**, 369–86.

Greenfield, S. G. and Bosanquet, F. D. (1953). The brain-stem lesions in parkinsonism. *J. Neurol. Neurosurg. Psychiat.* **16**, 213–26.

Hassler, R. (1938). Zur Pathologie der Paralysis agitans und des postenzephalitischen Parkinsonismus. *J. Psychol. Neurol. (Lpz.)* **48**, 387–476.

Hornykiewicz, O. (1963). Die topische Lokalisation und das Verhalten von Noradrenalin und Dopamin (3–Hydroxytyramin) in der Substantia nigra des normalen und Parkinsonkranken Menschen. *Wien. Klin. Wschr.* **75**, 309–12.

Hornykiewicz, O. (1982). Brain neurotransmitter changes in Parkinson's disease. In *Movement Disorders*. C. D. Marsden and S. Fahn (eds), pp. 41–58. Butterworth Scientific, London.

Iversen, S. D. and Koob, G. F. (1977). Behavioral implications of dopaminergic neurones in the mesolimbic system. *Adv. Biochem. Psychopharmacol.* **16**, 209—14.

Künzle, H. (1975). Bilateral projections from precentral motor cortex to the putamen and other parts of the basal ganglia. An autoradiographic study in *Macaca fascicularis*. *Brain Res.* **88**, 195–209.

Langston, J. W., Ballard, P., Tetrud, J. W. and Irwin, I. (1983). Chronic parkinsonism in humans due to a product of meperidine-analog synthesis. *Science.* **219**, 979–80.

Lee, T., Seeman, P., Rajput, A., Farley, I. J. and Hornykiewicz, O. (1978). Receptor basis for dopaminergic supersensitivity in Parkinson's disease. *Nature (Lond,)* **273**, 59–61.

Lloyd, K. G. and Hornykiewicz, O. (1973). L-glutamic acid decarboxylase in Parkinson's disease: effect of L-dopa therapy. *Nature (Lond.)* **243**, 521–3.

Lloyd, K. G., Davidson, L. and Hornykiewicz, O. (1975). The neurochemistry of Parkinson's disease: effect of L-dopa therapy. *J. Pharmacol. Exp. Ther.* **195**, 453–64.

Lloyd, K. G., Sheman, L. and Hornykiewicz, O. (1977). Distribution of high affinity sodium-independent [^{3}H]gamma-aminobutyric acid ([^{3}H]-GABA) binding in the human brain: alterations in Parkinson's disease. *Brain Res.* **127**, 269–78.

Mauborgne, A., Javoy-Agid, F., Legrand, J. C., Agid, Y. and Cesselin, F. (1983).

Decrease of substance P-like immunoreactivity in the substantia nigra and pallidum of parkinsonian brains. *Brain Res.* **268**, 167–70.

Mishkin, M. and Bachevalier, J. (1983). Object recognition impaired by ventromedial but not dorsolateral prefrontal cortical lesions in monkeys. *Soc. Neurosci. Abstr.* **9**, 29.

Moore, R. Y. and Bloom, F. E. (1978). Central catecholamine neurone systems: anatomy and physiology of the norepinephrine and epinephrine systems. *Ann. Rev. Neurosci.* **1**, 129–69.

Narabayashi, H., Kondo, T., Hayashi, A., Suzuki, T. and Nagatsu, T. (1981). L-threo-3,4-dihydroxyphenylserine treatment for akinesia and freezing of Parkinsonism. *Proc. Jap. Acad.* **57**, 351–4.

Ögren, S. O., Archer, T. and Johansson, C. (1983). Evidence for selective brain noradrenergic involvement in the locomotor stimulant effect of amphetamine in the rat. *Neurosci. Lett.* **43**, 327–31.

Pijnenburg, A. J. J., Honig, W. M. M., Van der Heyden, J. A. M. and Van Rossum, J. M. (1976). Effects of chemical stimulation of the mesolimbic dopamine system upon locomotor activity. *Eur. J. Pharmacol.* **35**, 45–58.

Plaznik, A. and Kostowski, W. (1983). The interrelationship between brain noradrenergic and dopaminergic neuronal systems in regulating animal behavior: possible clinical implications. *Psychopharmacol. Bull.* **19**, 5–11.

Price, K. S., Farley, I. J. and Hornykiewicz, O. (1978). Neurochemistry of Parkinson's disease: Relation between striatal and limbic dopamine. *Adv. Biochem. Psychopharmacol.* **19**, 293–300.

Reisine, T. D., Fields, J. Z., Yamamura, H. I., Bird, E. D., Spokes, E., Schreiner, P. S. and Enna, S. J. (1977). Neurotransmitter receptor alterations in Parkinson's disease. *Life Sci.* **21**, 335–44.

Rinne, U. K. (1982). Brain neurotransmitter receptors in Parkinson's disease. In *Movement Disorders*. C. D. Marsden and S. Fahn (eds), pp. 59–74. Butterworth Scientific, London.

Schultz, W. (1982). Depletion of dopamine in the striatum and an experimental model of parkinsonism: direct effects and adaptive mechanisms. *Progr. Neurobiol.* **18**, 121–66.

Schultz, W. (1984). Recent physiological and pathophysiological aspects of parkinsonian movement disorders. *Life Sci.* **34**, 2213–23.

Studler, J. M., Javoy-Agid, F., Cesselin, F., Legrand, J. C. and Agid, Y. (1982). CCK-8 immunoreactivity distribution in human brain: selective decrease in the substantia nigra from parkinsonian patients. *Brain Res.* **243**, 176–9.

Taquet, H., Javoy-Agid, F., Hamon, M., Legrand, J. C., Agid, Y. and Cesselin, F. (1983). Parkinson's disease affects differently Met5 and Leu5-enkephalin in the human brain. *Brain Res.* **280**, 379–82.

Tenovuo, O., Rinne, U. K. and Viljanen, M. K. (1984). Substance P immunoreactivity in the post-mortem parkinsonian brain. *Brain Res.* **303**, 113–16.

Uhl, G. R., Whitehouse, P. J., Price, D. L., Tourtelotte, W. W. and Kuhar, M. J. (1984). Parkinson's disease: depletion of substantia nigra neurotensin receptors. *Brain Res.* **308**, 186–90.

Zigmond, M. J. and Stricker, E. M. (1972). Deficits in feeding behavior after inraventricular injection of 6-hydroxydopamine in rats. *Science* **177**, 1211–14.

Zigmond, M. J., Acheson, A. L., Stachowiak, M. K. and Stricker, E. M. (1984). Neurochemical compensation after nigrostriatal bundle injury in an animal model of preclinical parkinsonism. *Arch. Neurol.* **41**, 856–61.

34

Dopaminergic transplantation in an animal model of Parkinson's disease

S. B. Dunnett[1] and A. Björklund[2]

[1]Department of Experimental Psychology, University of Cambridge, Downing Street, Cambridge CB2 3EB, UK and [2]Department of Histology, University of Lund, Biskopsgatan 5, S-223 62 Lund, Sweden

34.1. Introduction

The movement disorder first described by James Parkinson in 1817 in his classic paper '*An essay on the shaking palsy*' is the commonest single disease of middle age, and is characterised by akinesia, rigidity, postural deformity and tremor. The description a century later of pigmented cell loss from the substantia nigra of parkinsonian patients (Tretiakoff, 1919), and the akinesia seen in experimental animals treated with dopamine-depleting drugs (Carlsson, 1959), led to the identification by Ehringer and Hornykiewicz (1960) of a decline in the dopamine content of the neostriatum of parkinsonian patients, which, moreover, correlated with their degree of motor impairment. This led naturally to the introduction of dopamine replacement therapy, for example using the precursor drug L-dopa (Birkmeyer and Hornykiewicz, 1961; Cotzias *et al.*, 1967), which has revolutionised the treatment of Parkinson's disease over the last two decades. However, the new pharmacological regimes are not without problems. In particular, during advanced stages of the disease, drug response failure and fluctuations and the incidence of 'on-off' symptoms become increasingly problematic. Moreover, the drug treatments do not appear to slow the progress of underlying neuronal degeneration.

In parallel with the clinical studies, our understanding of the neuro-anatomy, pharmacology and functions of forebrain dopamine systems has progressed rapidly through animal experimentation, in particular in studies on rodents. Two major ascending forebrain dopamine systems have been described, the first originating in the substantia nigra and innervating the neostriatum (caudate nucleus and putamen of the basal ganglia), and the second originating in the more medially placed cell group of the ventral tegmental area and innervating several cortical and limbic areas including the nucleus accumbens, olfactory tubercle, lateral septum, amygdala and the frontal and entorhinal cortex (Ungerstedt, 1971; Lindvall and Björklund, 1978). Although often considered as two systems, there exists a degree of overlap both at the dopamine cell body level and in their terminal projections, and the axons ascend together in the medial forebrain bundle (Lindvall and Björklund, 1978).

Forebrain dopamine depletions, either by pharmacological treatments such as reserpine or α-methyltyrosine, or by lesions using the specific catecholamine neurotoxin 6-hydroxydopamine (6-OHDA), produce a profound akinesia, postural rigidity, sensorimotor neglect and motivational impairments such as aphagia and adipsia in experimental animals, which have frequently been regarded as an animal model of Parkinson's disease (Schultz, 1982). Moreover, many of these impairments can be reversed by treatment with L-dopa or the dopamine receptor agonist apomorphine (Carlsson *et al.*, 1957; Ljungberg and Ungerstedt, 1976; Marshall and Gotthelf, 1979), although such recovery may be dependent on the presence of residual dopamine terminals spared by the lesion in mediating the effects.

34.2. Transplantation of dopamine neurones

Although attempts had been made to transplant nervous tissue to the brain as long ago as the last century (Thompson, 1890; Dunn, 1917), the principles and procedures for achieving reliable graft survival have only been characterised over the last decade (Stenevi *et al.*, 1976; Das *et al.*, 1979). In particular it was found necessary to utilise donor tissue taken from specific periods of embryonic development and to provide adequately vascularised sites in the host brain to nourish the newly transplanted tissue. With the provision of these requirements, immunological rejection of the graft tissue is not a problem due to the 'immunological privilege' of the brain (Barker and Billingham, 1977), although the detailed nature of this privilege remains poorly understood. We were therefore interested to ask whether intracerebral grafts could have any influence on host behaviour and in particular whether grafts could reverse any functional impairments induced by brain damage. Because of the availability of the selective toxin, 6-OHDA, the well characterised nature of the 6-OHDA lesion syndrome, and the possible implications for human parkinsonism, the transplantation of dopaminerich tissue to dopamine-depleted rats was chosen as the first model system for study.

Three alternative transplantation procedures have been adopted by ourselves and others for grafting dopamine neurones to the forebrain of 6-OHDA lesioned rats (see Fig. 34.1). The donor tissue is dissected from the dopamine cell-rich ventral mesencephalon of embryonic day 15–16 rat fetuses and is then either injected into the lateral ventricle via a modified lumbar puncture needle adjacent to the medial surface of the neostriatum (Fig. 34.1B; Freed *et al.*, 1980), placed into a previously prepared cortical

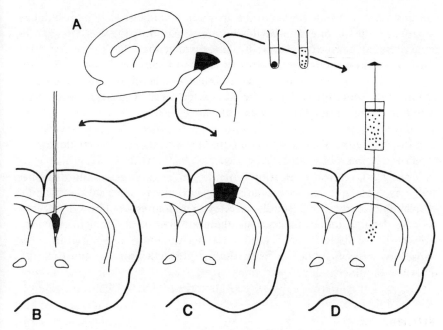

Fig. 34.1. Three experimental procedures for the transplantation of dopamine-rich tissue taken from the embryonic ventral mesencephalon (A) to the dopamine-denervated neostriatum of the rat. B: Insertion of graft tissue via a cannula into the lateral ventricle, medial to the host neostriatum. C: Placement of graft tissue into a previously prepared cortical cavity, dorsal to the host neostriatum. D: Stereotaxic injection of dissociated cell suspension of graft tissue directly into the host neostriatum.

cavity overlying the dorsal or lateral surface of the neostriatum (Fig. 34.1C; Björklund *et al.*, 1980), or stereotaxically injected as a dissociated cell suspension directly into any site within the host parenchyma (Fig. 34.1D; Björklund *et al.*, 1983a). The alternative procedures have different advantages and disadvantages. For example, the placement of a solid piece of graft tissue permits subsequent access to the graft for removal (Björklund *et al.*, 1980) or electrode implantation (Fray *et al.*, 1983). The dissociated cell suspension enables graft placement at deep sites not reachable by other procedures or the placement of multiple deposits at different sites in the host brain (Dunnett *et al.*, 1983a).

34.3. Restructuring host circuitry by transplantation

The majority of studies of graft anatomy, biochemistry, pharmacology and behavioural function have been conducted in rats with unilateral 6-OHDA lesions. This not only leaves the contralateral side to serve as a within-animal control in all post-mortem analysis, but additionally provides an intact system on one side of the brain to maintain the basic regulatory functions such as eating and drinking, which are life-threatening when bilateral lesions are made. Graft survival is generally seen to be good, with 80–95% survival rates, and once established the transplanted tissue will continue to develop and become integrated with host tissue for the full lifetime of the animal (up to 18 months have been followed in rats, without decline).

Histofluorescence microscopy has permitted visualisation of dopamine-containing neurones which are seen to cluster within the grafts, often adopting a layered organisation with perpendicularly extending dendrites, reminiscent of the normal organisation within the intact substantia nigra (Björklund et al., 1980). Bundles of dopamine-fluorescent fibres are seen to course through the grafts, to cross the graft-host border and to provide a moderate to dense catecholamine-fluorescent terminal plexus within the host neostriatum, which is patterned at the light-microscope level in a manner similar to intact terminal areas, sparing the white matter fibre bundles of the internal capsule (Björklund et al., 1980, 1983b; Freed et al., 1980). Whereas extensive graft-derived dopamine innervation is seen in host dopamine-denervated terminal areas, a sparser outgrowth is seen if the intrinsic system is not lesioned prior to transplantation. Moreover, graft dopamine fibres do not grow into areas of the host brain which normally do not receive any intrinsic dopaminergic inputs (Björklund et al., 1980, 1983b), indicating that quite specific trophic and guidance mechanisms are regulating axonal outgrowth from the transplants.

Biochemical measurements of the reinnervated host neostriatum indicate that as much as 10–15% of normal striatal dopamine content can be reinstated by a single graft (Freed et al., 1980; Schmidt et al., 1982; see Fig. 34.2), and multiple graft placements using the suspension procedure have been seen to replace up to 50% of total unilateral forebrain dopamine content (Schmidt et al., 1983). Dopamine synthesis (DOPA : DA ratio following DOPA decarboxylase inhibition), dopamine turnover (DOPAC : DA ratio) and dopamine receptor density (spiroperidol binding) have all been seen to become partially or completely normalised in grafted animals (Schmidt et al., 1982, 1983; Freed et al., 1983). Thus, in certain respects at least, a new striatal dopamine innervation after dopamine-depleting lesions appears to have been formed by the grafts, which may be able to subserve some recovery of lesion-induced behavioural deficits.

34.4. Functional recovery in transplanted animals

As in clinical Parkinson's disease, animals with dopamine-depleting lesions manifest a syndrome of behavioural impairments with many different

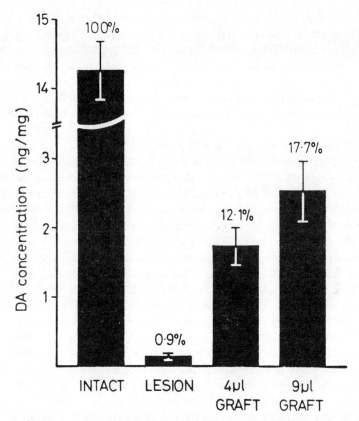

Fig. 34.2. Dopamine concentrations (nanograms per milligram) and percentage of intact levels, in the normal intact neostriatum, in the neostriatum following unilateral forebrain dopamine denervation by 6-OHDA lesion, and in the neostriatum following lesion and subsequent intrastriatal grafts or dopamine-rich cell suspensions. The grafts comprised either 2×2 μlitre (4 μlitre) suspension injection into the centre of the head of the neostriatum, or 3×3 μlitre (9 μlitre) suspensin injections into the dorsal and lateral neostriatum and nucleus accumbens. (Data from Schmidt *et al.*, 1983.)

components. Thus a unilateral lesion in rats produces not only a postural bias, analogous to the scoliosis of hemi-parkinsonism, but also an ipsilateral motor turning response ('rotation') when the animal is activated either by external stimulation or by stimulant drugs such as amphetamine, a sensorimotor neglect of contralateral space and body side, and a locomotor activation by low doses of the dopamine receptor agonist apomorphine. Lesion studies indicate that these various functional components are mediated by different terminal areas within forebrain dopamine projections. For example, the sensorimotor impairments result from lesions within ventrolateral neostriatum but not in other striatal sites (Dunnett and Iversen, 1982), whereas the hyperactivity to apomorphine at doses that have no effect on intact animals is believed to be attributable to the drug's action on supersensitive receptors, in particular within the nucleus accumbens (Kelly *et al.*, 1975). In parallel with these observations, graft-derived reinnervation of the neostriatum has been seen to provide a partial or complete recovery of many components of the full lesion syndrome, but only when that reinnervation invades regions of the total forebrain terminal fields critical for a particular component (see Fig. 34.3). Thus, single solid grafts reinnervating dorsal or medial parts of the head of the neostriatum compensate the rats' postural and turning bias but have no effect on the sensorimotor neglect, whereas, conversely, reinnervation of the ventrolateral neostriatum can ameliorate the contralateral sensorimotor impairments but has no effect on the turning asymmetries (Björklund *et al.*, 1980; Freed *et al.*, 1980; Dunnett *et al.*, 1981*a*). In a recent study comparing different single and multiple suspension graft placements (Dunnett *et al.*, 1983*a*) not only were these effects replicated (see Fig. 34.3), but additionally three grafts placed into the dorsal and lateral neostriatum and nucleus accumbens had an additive effect, such that the host rats were compensated to near normal levels on rotational bias, sensorimotor neglect and hyperactivity to apomorphine.

In all cases where functional recovery has been achieved, the grafts have been placed in ectopic locations in denervated terminal areas of the host brain. Graft placement back into a more normal location in the substantia nigra, or into the lateral hypothalamus where the ascending axons are at their densest, has not been seen to have any beneficial effect on any behavioural measure so far investigated (Dunnett *et al.*, 1983*a*). Although these grafts survive well, and extend fibres that cross the graft-host border, no axonal elongation or extensive outgrowth is seen within host tissue at these locations. Thus, a graft placed into the striatum can reinnervate deafferented host terminal areas, but is in a location unlikely to receive inputs from host systems that normally innervate intact substantia nigra neurones. Conversely, a graft placed in the denervated nigra has a chance of attracting appropriate host innervation, but then the grafted dopamine neurones do not establish contact with the denervated host terminals. This failure to restructure host circuitry fully may account for the absence of any recovery in several components of the 6-OHDA syndrome, in particular the aphagia and adipsia following bilateral lesions (Dunnett *et al.*, 1981*b*, 1983*b*). The functional recovery that is nevertheless seen in many components of the syndrome, as described above, may therefore suggest that intrinsic dopamine systems function in a relatively 'permissive' manner, providing a level of activation that enables an otherwise intact striatal circuitry to function normally, rather than by providing specific organised information to striatal motor control (Björklund *et al.*, 1981).

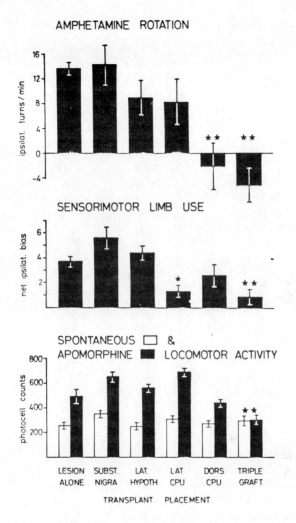

Fig. 34.3. Behavioural asymmetry and functional recovery in six groups of rats on three behavioural tests. One group received unilateral 6-OHDA lesion of forebrain dopamine systems alone. The other groups received unilateral 6-OHDA lesions plus additional grafts in different forebrain or midbrain placements ipsilateral to the lesion: the substantia nigra; the lateral hypothalamus; the lateral caudate-putamen (neostriatum), the dorsal caudate-putamen, or a triple graft placed in the two caudate-putamen sites and the nucleus accumbens. The three behavioural tests involved: rotation in a hemispheric bowl over 90 min following injection i.p. of 5 mg kg^{-1} metamphetamine; ratings of coordinated limb use on the two sides of the body in a battery of neurological tests involving limb placing, grasping and withdrawal reflexes; apomorphine-induced stimulation of locomotor activity compared with spontaneous baseline levels, recorded over 10 min in an Animex activity meter. **$P < 0.01$; *$P < 0.05$ in comparison with the lesion alone group. (Data from Dunnett *et al.*, 1983*a*.)

34.5. Potential clinical application?

The current interest in the studies of intracerebral transplantation in animals originates in large part from the possibilities for a radical new therapy for neurodegenerative diseases such as parkinsonism. Indeed, at the time of writing, two parkinsonian patients at the Karolinska hospital in Sweden have received autografts of their own adrenal medulla, containing catecholamine- (including some dopamine-) secreting cells, into the caudate nucleus unilaterally (Kolata, 1982). In these first clinical attempts the beneficial effects of the graft have been short-lived, providing some alleviation of symptoms for only a few days, and several key issues await resolution before the long-term feasibility of such techniques can be assessed.

First, the aetiology of Parkinson's disease, at least in its idiopathic form, remains completely unknown, although a viral agent is suspected by some. Does the causative agent become established within the patient's system, rendering any grafted neurones as vulnerable as the host's own dopamine neurones? Furthermore, is the more widespread degeneration, which is characteristic of advanced stages of the disease, secondary to the primary dopaminergic cell loss, and, if not, what implications does this have for the timing and likely efficacy even if the damaged dopamine system can be reformed by transplantation?

Secondly, in rats, the most effective graft tissue has been obtained from fetuses of the same strain, whereas cross-species grafting, or use of mature peripheral catecholamine-rich tissue such as the adrenal medulla, has been generally less effective in promoting regeneration or functional recovery (Björklund *et al.*, 1982; Freed, 1983). Consequently the identification of optimal sources of donor tissue for human transplantation is likely to prove ethically as well as practically controversial.

Finally, many practical questions related to immunological factors, graft placement, and the number of grafts necessary to provide adequate reinnervation of a structure as large as the human caudate nucleus and putamen, remain largely unsolved.

Nevertheless, dopamine-rich transplants have been seen to produce a dramatic functional recovery in dopamine-lesioned rats. Although the answers to the above issues are as yet unknown, each appears to be soluble. It may be expected that the next few years will see a rapid increase in research in functional brain transplantation, including in primate species, that will soon indicate whether such procedures indeed may have clinical application.

References

Barker, C. F. and Billingham, R. E. (1977). Immunologically privileged sites. *Adv. Immunol.* **25**, 1–54.

Birkmayer, W. and Hornykiewicz, O. (1961). Der L-dioxyphenylalanin (= L-dopa) effect bei der Parkinson-Akinese. *Wien. Klin. Wschr.* **73**, 787.

Björklund, A., Dunnett, S. B., Stenevi, U., Lewis, M. E. and Iversen, S. D. (1980). Reinnervation of the denervated striatum by substantia nigra transplants: functional consequences as revealed by pharmacological and sensorimotor testing. *Brain Res.* **199**, 307–33.

Björklund, A., Stenevi, U., Dunnett, S. B. and Iversen, S. D. (1981). Functional reactivation of the deafferented neostriatum by nigral transplants. *Nature (Lond.)* **289**, 497–9.

Björklund, A., Stenevi, U., Dunnett, S. B. and Gage, F. H. (1982). Cross-species neural grafting in a rat model of Parkinson's disease. *Nature (Lond.)* **298**, 652–4.

Björklund, A., Stenevi, U., Schmidt, R. H., Dunnett, S. B. and Gage, F. H. (1983a). Intracerebral grafting of neuronal cell suspensions. I. Introduction and general methods of preparation. *Acta Physiol. Scand.* Suppl. **522**, 1–7.

Björklund, A., Stenevi, U., Schimdt, R. H., Dunnett, S. B. and Gage, F. H. (1983b). Intracerebral grafting of neuronal cell suspensions. II. Survival and growth of nigral cell suspensions implanted in different brain sites. *Acta. Physiol. Scand.* Suppl., **522**, 9–18.

Carlsson, A. (1959). The occurrence, distribution and physiological role of catecholamines in the nervous system. *Pharm. Rev.* **11**, 490–3.

Carlsson, A., Lindquist, M. and Magnusson, T. (1957). 3, 4–dihydroxyphenylalanine and 5–hydroxytryptophan as reserpine antagonists. *Nature (Lond.)* **180**, 1200.

Cotzias, G. C., Van Woert, M. H. and Schiffer, L. M. (1967). Aromatic amino acids and modification of Parkinsonism. *New Engl. J. Med.* **276**, 374–9.

Das, G. D., Hallas, B. H. and Das, K. G. (1979). Transplantation of neural tissue in the brains of laboratory mammals: technical details and comments. *Experientia.* **35**, 143–53.

Dunn, E. H. (1917). Primary and secondary findings in a series of attempts to transplant cerebral cortex in the albino rat. *J. Comp. Neurol.* **27**, 565–82.

Dunnett, S. B. and Iversen, S. D. (1982). Sensorimotor impairments following localised kainic acid and 6–hydroxydopamine lesions of the neostriatum. *Brain Res.* **248**, 121–7.

Dunnett, S. B., Björklund, A., Stenevi, U. and Iversen, S. D. (1981a). Grafts of embryonic substantia nigra reinnervating the ventro-lateral striatum ameliorate sensorimotor impairments and akinesia in rats with 6–OHDA lesions of the nigrostriatal pathway. *Brain Res.* **229**, 209–17.

Dunnett, S. B., Björklund, A., Stenevi, U. and Iversen, S. D. (1981b). Behavioural recovery following transplantation of substantia nigra in rats subjected to 6–OHDA lesions of the nigrostriatal pathway. II. Bilateral lesions. *Brain Res.* **229**, 457–70.

Dunnett, S. B., Björklund, A., Schmidt, R. H., Stenevi, U. and Iversen, S. D. (1983a). Intracerebral grafting of neuronal cell suspensions. IV. Behavioural recovery in rats with unilateral implants of nigral cell suspensions in different forebrain sites. *Acta Physiol. Scand.* Suppl. **522**, 29–37.

Dunnett, S. B., Björklund, A., Schmidt, R. H., Stenevi, U. and Iversen, S. D. (1983b). Intracerebral grafting of neuronal cell suspensions. V. Behavioural recovery in rats with bilateral 6–OHDA lesions following implantation of nigral cell suspensions. *Acta Physiol Scand.* Suppl. **522**, 39–47.

Ehringer, H. and Hornykiewicz, O. (1960). Verteilung von Noradrenalin und Dopamin (3-Hydroxytyramin) im Gehirn des Menschen und ihr Verhalten bei Erkrankungen. *Wien Klin. Wschr.* **38**, 1236–9.

Fray, P. J., Dunnett, S. B., Iversen, S. D., Björklund, A. and Stenevi, U. (1983). Nigral transplants reinnervating the dopamine-depleted neostriatum can sustain intracranial self-stimulation. *Science* **219**, 416–19.

Freed, W. J. (1983). Functional brain tissue transplantation: reversal of lesion-induced rotation by intraventricular substantia nigra and adrenal medulla grafts, with a note on intracranial retinal grafts. *Biol. Psychiat.* **18**, 1205–67.

Freed, W. J., Perlow, H. J., Karoum, F., Seiger, Å., Olson, L., Hoffer, B. J. and Wyatt, R. J. (1980). Restoration of dopaminergic function by grafting of fetal rat substantia nigra to the caudate nucleus: long-term behavioral, biochemical, and histochemical studies. *Ann. Neurol.* **8**, 510–19.

Freed, W. J., Ko, G. N., Niehoff, D. L., Kuhar, M. J., Hoffer, B. J., Olson, L., Cannon-Spoor, H. E., Morihisa, J. M. and Wyatt, R. J. (1983). Normalization of spiroperidol binding in the denervated rat striatum by homologous grafts of substantia nigra. *Science* **222**, 937–9.

Kelly, P. H., Seviour, P. W. and Iversen, S. D. (1975). Amphetamine and apomorphine responses in the rat following 6–OHDA lesions of the nucleus accumbens septi and corpus striatum. *Brain Res.* **94**, 507–22.

Kolata, G. (1982). Grafts correct brain damage. *Science* **217**, 342–4.

Lindvall, O. and Björklund, A. (1978). Organization of catecholamine neurons in the rat central nervous system. In *Handbook of Psychopharmacology*, Vol. 9, L. L. Iversen, S. D. Iversen and S. H. Snyder (eds), pp. 139–231. Plenum Press, New York.

Ljungberg, T. and Ungerstedt, U. (1976). Reinstatement of eating by dopamine agonists in aphagic dopamine denervated rats. *Physiol. Behav.* **16**, 277–83.

Marshall, J. F. and Gotthelf, T. (1979). Sensory inattention in rats with 6–hydroxydopamine-induced lesions of ascending dopaminergic neurons: apomorphine-induced reversal of deficits. *Exp. Neurol.* **65**, 389–411.

Schmidt, R. H., Ingvar, M., Lindvall, O., Stenevi, U. and Björklund, A. (1982). Functional activity of substantia nigra reinnervating the striatum: neurotransmitter metabolism and $[^{14}C]$–2–deoxy–D–glucose autoradiography. *J. Neurochem.* **38**, 737–48.

Schmidt, R. H., Björklund, A., Stenevi, U., Dunnett, S. B. and Gage, F. H. (1983). Intracerebral grafting of neuronal cell suspensions. III. Activity of intrastriatal nigral suspension implants as assessed by measurements of dopamine synthesis and metabolism. *Acta. Physiol. Scand.* Suppl. 522, 19–28.

Schultz, W. (1982). Depletion of dopamine in the striatum as an experimental model of Parkinsonism: direct effects and adaptive mechanisms. *Prog. Neurobiol.* **18**, 121–66.

Stenevi, U., Björklund, A. and Svendgaard, N. -Aa. (1976). Transplantation of central and peripheral monoamine neurons to the adult rat brain: techniques and conditions for survival. *Brain Res.* **114**, 1–20.

Thompson, W. G. (1890). Successful brain grafting. *N.Y. Med. J.* **51**, 701–2.

Tretiakoff, C. (1919). Contribution a l'étude de l'anatomie pathologique du locus nigra de Soemmering avec quelques déductions rélatives à la pathogenie des troubles du tonus musculaire de la maladie de Parkinson. Thèse, Paris.

Ungerstedt, U. (1971). Stereotaxic mapping of the monoamine pathways in the rat brain. *Acta. Physiol. Scand.* Suppl. 367, 1–48.

35

Long-term complications of levodopa therapy

D. B. Calne

Division of Neurology, University of British Columbia,
Health Sciences Centre Hospital—ACU,
2211 Wesbrook Mall, Vancouver, British Columbia, Canada V6T 1W5

35.1. Introduction

Conventional treatment of patients with Parkinson's disease comprises prolonged administration of L-dopa, which is usually given in combination with a peripheral decarboxylase inhibitor (carbidopa or benserazide) to minimise nausea. Therapy is generally continued until the death of the patient; because of the usual onset of Parkinson's in the sixth or seventh decade, treatment commonly comprises a 10 to 20 year period, over which time the underlying pathology of the disease continues to progress. There is a natural tendency, shared for most medications, for dosage to be reduced in senescence because of adverse reactions becoming more prominent in late life. We are therefore dealing with a complex array of variables: the age of the patient, the profile of neurological deficits, the rate of progress of the disease, the pattern of side-effects and the nature of concomitant medications. In this setting, the identification of 'long-term complications of therapy' is not a straightforward task. In particular, the question arises as to which problems derive from the inexorable advance of the pathology, which from the ultimate senescence of the patient, and which from treatment other than with L-dopa.

We do not have answers to all of these questions, so caveats, qualifications and speculations are frequent in the literature addressing these topics. Recently the subject has been well reviewed by Rinne (1983). Possible long-term complications of L-dopa therapy include declining efficacy, increasing dyskinesia, fluctuations in response (wearing-off reactions and on–off reactions), and dementia. I shall discuss each of these in turn, though it should be stressed that none has been proved to be consequent upon prolonged exposure to L-dopa. It will emerge that although some are probably related to the cumulative intake of L-dopa, others are most unlikely to be caused by such treatment.

35.2. Declining efficacy

The question 'Does L-dopa therapy induce a decreased therapeutic response to itself? can be reformulated into four components:

(1) Does L-dopa modify its own pharmacokinetics to result in a decreased response (e.g. by a mechanism such as enzyme induction)?

(2) Does L-dopa alter its own pharmacodynamics, leading to a reduced response (e.g. by blocking dopamine receptors)?

(3) Does L-dopa cause side-effects which limit treatment and might have been avoided by decreasing the cumulative dose?

(4) Does L-dopa hasten the advance of the underlying pathology of Parkinson's disease?

The practical impacts of these questions, for the management of Parkinson's disease, are widely disparate. There is no firm evidence for a positive answer to questions (1) and (2), and even if a mechanism was shown to exist for pharmacokinetic or pharmacodynamic down-regulation, it would be unlikely to have a catastrophic impact on treatment, because such a mechanism should be at least partly correctable by simply readjusting the dose of L-dopa (increasing or decreasing, according to the nature of the dose–response relationship).

Of greater importance are questions (3) and (4). Question (3) is addressed in the discussion of individual adverse effects (see below). Question (4) is currently controversial, and it is appropriate to review the issues here.

The first point to be made in the context of question (4) is that data on progress of the pathology of Parkinson's disease are limited. Riederer and Wuketich (1976) speculated on the rate of loss of dopamine from the basal ganglia, but more direct information should become available now that 6–fluorodopa has become available for positron emission tomography.

The clinical features of Parkinson's disease advance at a rate that varies considerably from patient to patient. Prior to the advent of L-dopa, the prognosis ranged from 5 to 33 years (Hoehn and Yahr, 1967). This has led to such concepts as the subcategorisation of Parkinson's disease into 'benign' and 'malignant' forms (Birkmayer *et al.*, 1979).

Following the availability of L-dopa, there have been further observations on the progress of the disease. These have led to conflicting conclusions. On the one hand, Markham and Diamond (1981) and Muenter (1984) have argued that progress of the disease is not related to L-dopa exposure. Fahn and Bressman (1984) and Rajput *et al.* (1984) have drawn the opposite conclusion. Fahn and Bressman attributed the discrepancy, at least in part, to bias introduced by non-linear clinical scoring protocols. (Fig. 35.1).

Studies of the clinical rate of progress of Parkinson's disease have relevance since they represent the net result of numerous variables acting upon the patient. However, they do not really provide a satisfactory

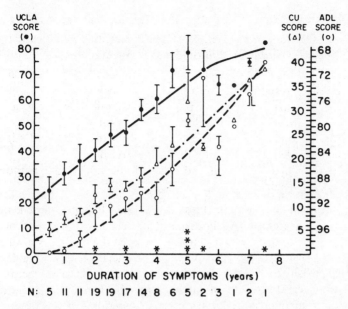

Fig. 35.1. Severity of parkinsonism in 26 patients prior to starting Sinemet therapy, as assessed by the UCLA Scoring Scale (solid circles), the Columbia University Parkinson's Disease Rating Scale (triangles), and the modified Schwab and England ADL Scoring Scale (open circles) as a function of duration of the illness. Each point represents the mean score obtained for the number of patients at each assessment point, which is presented along the bottom of the figure. Each asterisk represents individual patients who started on Sinemet immediately after the assessment at the time point the asterisk is located. Once Sinemet was started, subsequent scores obtained on that patient are no longer entered into the analysis. The vertical lines represent the standard errors of the means. From Fahn and Bressman (1984).

measure of the advancing pathology, because over the period of observation the dose of L-dopa will have been changed and this will have a profound effect on evaluation by any clinical scoring protocol. Over the first months of treatment, dosage is gradually increased, and after a 5- to 15-year period of stable dosage, the dose is usually decreased because of side-effects. Furthermore, other powerful anti-parkinsonian medications, such as bromocriptine, may have been added, often with a substantial simultaneous readjustment of L-dopa intake. These therapeutic manipulations will be reflected by alteration of the clinical scores, irrespective of the status of the underlying pathology.

Another issue for discussion is the mechanism by which L-dopa might contribute towards accelerating the pathological progress of Parkinson's disease. It has been suggested that catecholamine cells in the brain are particularly susceptible to damage by free radicals such as superoxide

(Heikkila and Cohen, 1971), and that free radical trapping agents are deficient (Ambani *et al.*, 1975). This concept has been employed to explain the neurotoxic action of substances such as 6–hydroxydopamine. Free radicals are produced in the course of metabolising catecholamines; administration of L-dopa can be expected to increase the turnover of catecholamines and hence augment the formation of free radicals. Furthermore, as pointed out by Barbeau (1984), there is a natural trend for the production of free radicals to increase while the pathology of Parkinson's disease advances—as dopaminergic neurones in the zona compacta of the substantia nigra die, the remaining dopamine cells will be more active to compensate. Thus each surviving nigral dopaminergic cell will produce more dopamine than previously, and this, in turn, will lead to the output of more free radicals. In a sense, a vicious circle is established such that the decay will 'feed upon itself'.

35.3. Increasing dyskinesia

As soon as administration of large doses of L-dopa was recognised to induce major amelioration of parkinsonian symptoms, it was observed that the drug caused involuntary movements manifested as one or more of the following: chorea, athetosis, dystonia, ballism, myoclonus. Subsequently, L-dopa was given to normal subjects (Ansel and Markham, 1970) and depressed patients (Goodwin *et al.*, 1970); none of these developed dyskinesia. In consequence, it seemed possible that the dyskinesia induced by L-dopa was the product of increased dopamine at sites of abnormally enhanced receptors (formed by the process of denervation supersensitivity consequent upon the underlying pathology).

Although this hypothesis is unproven, relevant observations should accrue from the new technique of positron emission tomography with ligands that bind to dopamine receptors (Wagner *et al.*, 1983). It is of interest that artificial dopamine agonists, such as bromocriptine, are less likely to produce dyskinesia than L-dopa (Calne *et al.*, 1984). This observation raises certain questions. The general query posed is what actions of L-dopa are distinct from the ergot agonists? The possibility exists that dyskinesia might result from activation of D1-1 receptors because most of the dopaminomimetic ergots are D1-1 antagonists (Calne, 1982). More observations are required with the few drugs that stimulate D1-1 receptors, such as pergolide (Calne *et al.*, 1984), or the abeorphine CJ201–678 (Markstein *et al.*, 1984).

It is irrefutable that L-dopa causes dyskinesia in the majority of parkinsonian patients. Dyskinesia is more likely after prolonged treatment with L-dopa (Fig. 35.2). In some patients the dyskinesia gradually becomes more severe in amplitude, more widespread in distribution, and more prevalent through waking hours. In these patients, there is no doubt that

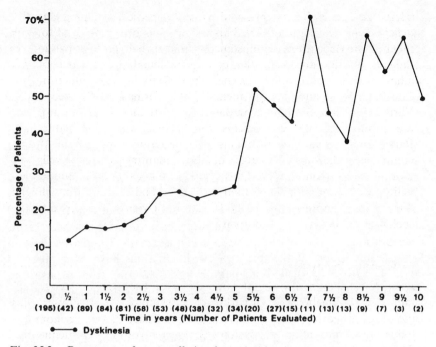

Fig. 35.2. Percentage of cases suffering from dyskinesia, among those evaluated at a particular 6-month interval. Not every patient had strict 6-month evaluations. The evaluation nearest to the assigned interval was taken into consideration. There is progressive increase in frequency of dyskinesia ($P = 0.00005$) with duration of treatment. From Rajput *et al.* (1984).

dyskinesia constitutes a late complication of L-dopa therapy, but it is not possible to determine whether the late deterioration is due to more advanced pathology irrespective of treatment, or damage accelerated by administration of L-dopa, with the production of free radicals as outlined in the discussion of declining efficiency.

35.4. Fluctuations in response

Prior to the use of L-dopa, sudden brief exacerbations of immobility—so-called 'freezing attacks'— were well recognised in Parkinson's disease. They last anything from 5 to 100 seconds, and are most characteristically seen when the patient is starting to walk, or to change direction. These attacks have nothing to do with L-dopa.

Some 3 to 4 years after the introduction of L-dopa, other sudden transient deteriorations in mobility were described. These differed from 'freezing attacks', because they lasted anything from 10 min to 6 h, and commonly developed while the patient was sitting or lying. Various categories of these changes were described. At present, there is general acceptance of at least two types, one tending to occur at the end of the interdose period, generally termed a 'wearing-off' reaction, and the other appearing less predictably, called an 'on-off' reaction. Some of these fluctuations may derive from alterations in the availability of L-dopa to the brain, but others occur in spite of steady intravenous infusion of L-dopa without administration of amino acids which could compete for active transport across the blood–brain barrier (Nutt *et al.*, 1984). It has been suggested that reduced intraneuronal storage sites for dopamine, consequent upon the death of dopaminergic neurones, may account for some of these attacks. Positron emission tomography with fluorodopa should help to confirm or refute this hypothesis.

Fluctuations become more frequent with the passage of time (McDowell and Sweet, 1976)–see Fig. 35.3. However, it is not clear whether this deterioration correlates with advancing disease, or the cumulative dose of L-dopa. In support of the former view, fluctuations have been observed relatively early in patients with parkinsonism resulting from exposure to methylphenyltetrahydropyridine (MPTP); these patients are characterised by the sudden onset of severe parkinsonian deficits (Burns *et al.*, 1984; Langston and Ballard, 1984).

Here again, the central question therefore becomes 'Does L-dopa hasten dopaminergic cell loss?', the same question that emerge from the discussion on efficacy and dyskinesia.

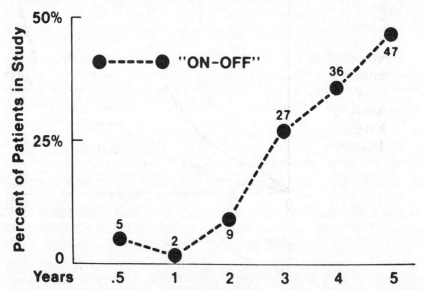

Fig. 35.3.   The percentage of patients with clinical fluctuations at various periods of follow-up. From McDowell and Sweet (1976).

35.5. Dementia

Dementia occurs with an abnormally high frequency in Parkinson's disease. By the end of 5 years from the time of diagnosis, dementia was found in 18·4% of parkinsonian patients (Fig. 35.4), compared with 5·5% in controls (Rajput *et al.*, 1984).

The arguments concerning mechanism are identical to those advanced for the other long-term problems of L-dopa therapy. Dementia certainly becomes more common with the passage of time. Is this because it is causally related to the severity of disease, or to exposure to L-dopa? Again there are insufficient data to answer this question. Documentation of the prevalence of dementia in Parkinson's disease prior to L-dopa therapy is difficult to interpret. Many patients of this era were chronically confused because of high levels of intake of anticholinergic drugs. The criteria for diagnosing dementia were less clearly defined. Prospective studies are necessary, but difficult to design because of the multiplicity of variables. Even if correlation is established with severity of disease, the question of whether this, in turn, is related to L-dopa treatment will remain unanswered; in general, patients with more prominent deficits will have received a larger intake of L-dopa.

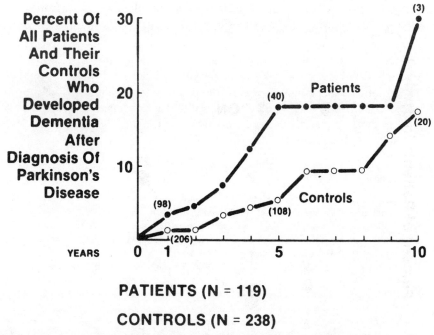

PATIENTS (N = 119)

CONTROLS (N = 238)

Fig. 35.4. New cases of dementia after the index date among cases and two matched controls. (From Rajput, 1984.)

35.6. Conclusions

In conclusion, it is not easy to define the late complications of L-dopa therapy because the cumulative dose of L-dopa is usually higher in patients with more advanced disease; this makes if difficult to distinguish whether problems of management stem from treatment or pathology. Comparisons with pre-L-dopa experience are obscured by toxicity from anticholinergic drugs, often imprecise quantification of clinical features, possible complications deriving from stereotactic surgery, and the shorter life expectation from the time of diagnosis. Prospective studies are limited, because it is obviously not possible to withhold L-dopa for a prolonged period of time to obtain control observations. As a compromise, it should be possible to compare the outcome in patients who are treated with L-dopa alone, and those given L-dopa together with an artificial dopamine agonist, such as bromocriptine, from relatively early in the natural history of the disease. The cumulative dose of L-dopa will be lower in the latter group, but of course the agonists may generate late management problems of their own. However, this comparison should at least shed light on the issue of whether L-dopa accelerates the underlying pathology by contributing to increased production of free radicals. For this study to succeed, a carefully controlled series of observations are required, from which it should be possible to resolve a number of practical questions such as whether to give L-dopa early or late, and whether L-dopa should be given by itself or in combination with a dopamine agonist. The design of such a study should also focus attention on important methodological issues such as how to record the severity of an underlying disease when powerful palliative treatment is continuously necessary, and how to obtain an index of clinical status when many patients display frequent, profound and uncontrollable fluctuations in response to therapy. We should not expect quick answers; it will be no mean task to develop a suitable design, execute the study, analyse the results, and formulate a conclusion.

References

Ambani, L. M., Van Woert, M. H. and Murphy, S. (1975). Brain peroxidase and catalase in Parkinson's disease. *Arch Neurol.* **32**, 114–18.

Ansel, R. D. and Markham, C. H. (1970). Effects of L-Dopa in normal humans. In *L-Dopa and Parkinsonism*. A. Barbeau and F. H. McDowell (eds), pp. 69–72, F. A. Davis Company, Philadelphia, Pa.

Barbeau, A. (1984). Etiology of Parkinson's disease: a research strategy. *Can. J. Neurol. Sci.* **11**, 24–8.

Birkmayer, W., Riederer, P. and Youdim, M. B. H. (1979). Distinction between benign and malignant type of Parkinson's disease. *Clin. Neurol. Neurosurg.* **81**, 158–64.

Burns, R. S., Markey, S. P., Phillips, J. M. and Chiueh, C. C. (1984). The neurotoxicity of 1-Methyl-4-phenyl-1, 2, 3, 6-tetrahydropyridine in the monkey and man. *Can. J. Neurol. Sci.* **11** (Suppl. 1), 166–8.

Calne, D. B. (1982). Dopamine receptors in movement disorders. In *Movement Disorders*. C. D. Marsden and S. Fahn (eds), pp. 348–55, Butterworth, London.

Calne, D. B., Burton, K., Beckman, J. and Wayne Martin, W. R. (1984). Dopamine agonists in Parkinson's disease. *Can. J. Neurol. Sci.* **11** (Suppl. 1), 221–4.

Fahn, S. and Bressman, S. B. (1984). Should levodopa therapy for parkinsonism be started early or late? Evidence against early treatment. *Can. J. Neurol. Sci.* **11** (Suppl. 1), 200–6.

Goodwin, F. K., Brodie, H. K., Murphy, D. L. and Bunney, W. E. (1970). Administration of a peripheral decarboxylase inhibitor with L-Dopa to depressed patients. *Lancet* **1**, 908–11.

Heikkila, R. E. and Cohen, G. (1971). Inhibition of biogenic amine uptake by hydrogen peroxide: a mechanism for toxic effects of 6–hydroxydopamine. *Science* **172**, 1257–8.

Hoehn, M. M. and Yahr M. D. (1967). Parkinsonism: onset, progression, and mortality. *Neurology (Minneap.)* **17**, 427–42.

Langston, J. W. and Ballard, P. (1984). Parkinsonism induced by 1-methyl-4-phenyl-1,2,3,6-tetrahydropyridine (MPTP): implications for treatment and the pathogenesis of Parkinson's disease. *Can. J. Neurol. Sci.* **11** (Suppl. 1), 160–5.

McDowell, F. H. and Sweet, R. D. (1976). The 'on–off' phenomenon. In *Advances in Parkinsonism*. W. Birkmayer and O. Hornykiewicz (eds), pp. 603–12, Editiones Roche, Basle.

Markham, C. H. and Diamond, S. G. (1981). Evidence to support early levodopa therapy in Parkinson's disease. *Neurology (N.Y.)* **31**, 125–31.

Markstein, R., Jaton, A. L., Vigouret, J. M., Giger, R., Closse, A., Briner, U. and Enz, A. (1984). Pharmacological properties of 201–678, a new dopamine agonist. *Clin. Neuropharmacol.* **7**, (Suppl. 1), 800–1.

Muenter, M. D. (1984). Should levodopa therapy be started early or late? *Can. J. Neurol. Sci.* **11** (Suppl. 1), 195–9.

Nutt, J. G., Woodward, W. R., Hammerstad, J. P., Carter, J. and Anderson, J. L. (1984). The 'on–off' phenomenon in Parkinson's disease: relation to levodopa absorption and transport. *New Engl. J. Med.* **310**, 483–8.

Rajput, A. H. (1984). Epidemiology of Parkinson's disease. *Can. J. Neurol. Sci.* **11** (Suppl. 1), 156–9.

Rajput, A. H., Stern, W. and Laverty, W. H. (1984). Chronic low-dose levodopa therapy in Parkinson's disease. *Neurology (Cleveland)* **34**, 991–6.

Riederer, P. and Wuketich, St. (1976). Time course of nigrostriatal degeneration in Parkinson's disease. A detailed study of influential factors in human brain amine analysis. *J. Neural Trans.* **38**, 277–301.

Rinne, V. K. (1983). Problems associated with long-term levodopa treatment in Parkinson's disease. *Acta Neurol. Scand.* Suppl. 95, 19–26.

Wagner, H. N., Burn, H. D., Dannals, R. F., Wong, D. F. *et al.* (1983). Imaging dopamine receptors in the human brain by positron tomography. *Science* **221**, 1264–6.

36

Parkinsonism induced by 1-methyl-4-phenyl-1,2,3,6-tetrahydropyridine in man and animals

P. Jenner and C. D. Marsden*

MRC Movement Disorders Research Group, University Department of Neurology and Parkinson's Disease Society Research Centre, Institute of Psychiatry and King's College Hospital Medical School, Denmark Hill, London SE5 8AF, UK

36.1. Introduction

Idiopathic Parkinson's disease in man is characterised by akinesia, rigidity, postural abnormalities and tremor. The pathology of this disorder consists of a loss of pigmented brainstem nuclei, with Lewy body inclusions in remaining neurones (see Forno, 1982). In particular, there is a gross degeneration of the dopamine-containing cells of zona compacta of substantia nigra. Biochemical studies in post-mortem tissue from parkinsonian patients show a general deficiency of dopamine content throughout the brain (Hornykiewicz, 1982). The loss of dopamine-containing neurones is thought to be the primary change responsible for the motor deficits of the disease. However, a whole range of other changes also occur in related neurotransmitter systems, with alterations in brain noradrenaline, 5-HT, GABA, acetylcholine and neuropeptides (Hornykiewicz, 1982; Rinne, 1982; Javoy-Agid et al., 1984). Although the pathology and biochemistry of Parkinson's disease are well documented, the cause of the disease remains unknown. Much attention has been paid to the involvement of melanin pathways and the possibility of the formation of toxic radicals which might cause cell death (see, for example, Graham, 1979). However, the pathology of Parkinson's disease extends to non-pigmented areas, such as the substantia innominata. Wherever cellular destruction occurs in Parkinson's disease, Lewy bodies are evident, and these inclusions may hold the key to the cause of the illness (Marsden, 1983).

The primary involvement of dopamine neuronal loss in Parkinson's disease is confirmed by the effectiveness of treatment with L-dopa or dopamine agonist drugs. Drugs affecting other neuronal systems in brain

---

*To whom correspondence should be addressed.

are generally without effect. This raises the question as to whether changes in other transmitter pathways contribute to the symptoms of the illness, or whether they are merely secondary to changes in dopamine.

Although L-dopa therapy is initially effective in controlling the symptoms of Parkinson's disease, benefit wanes after some years of treatment and a variety of long-term complications of L-dopa therapy, including 'on–off' effects and the induction of dyskinesias, become evident (Marsden *et al.*, 1982). The reasons for the emergence of these problems are unclear; arguments have been put forward to suggest pharmacokinetic and pharmacodynamic changes induced by chronic L-dopa therapy or the progression of the underlying disease (Fahn, 1982).

Despite major advances in the understanding of Parkinson's disease, a number of critical questions remain unanswered. One of the reasons for this situation lies in the lack of an effective animal model of the illness. The use of amine-depleting drugs, of 6-hydroxydopamine, or of other specific lesions in basal ganglia has not created an animal model that mimics the symptomatology, pathology and biochemistry of Parkinson's disease as it occurs in man. Recently a chance occurrence resulted in the discovery that 1-methyl-4-phenyl-1,2,3,6-tetrahydropyridine (MPTP), a dopaminergic neurotoxin, may cause typical parkinsonism in man and other primates. The object of this chapter is to review the effects of MPTP in man and to examine its potential for producing parkinsonism in animals.

36.2. MPTP-induced parkinsonism in man

In 1979 Davis and his colleagues published a case report of a man who developed persistent parkinsonism following drug abuse (Davis *et al.*, 1979). The patient was a student, with access to a synthetic chemistry laboratory, who had a history of addiction to pethidine. In his search for more potent derivatives he attempted to synthesise 1-methyl-4-phenyl-4-propionoxypiperdine (MPPP) which he successfully achieved on a number of occasions. However, as a result of an attempt to decrease the time required for synthesis, he altered the reaction conditions. Within a few days of self-administering the product of this new procedure he was admitted to hospital mute and with akinesia, rigidity and tremor. Due to a prior history of behavioural disturbance he was initially diagnosed as a catatonic schizophrenic and treated with a course of ECT and haloperidol. However, his condition persisted and on examination by a neurologist he was diagnosed as suffering from a parkinsonian syndrome, and was treated with L-dopa (plus a peripheral decarboxylase inhibitor and an anticholinergic) which considerably improved his condition. On subsequent examination he was found to show some blepharospasm, and to be bradykinetic and depressed, but there were no signs of generalised psychiatric illness or other neurological condition which might be mistaken for Parkinson's disease.

The diagnosis of parkinsonism was confirmed when cessation of drug treatment led to the emergence of a classical parkinsonism. In addition, examination of CSF revealed low HVA concentrations but normal levels of 3-methoxy-4-hydroxyphenylethylene glycol (HMPG) and 5-hydroxyindole-acetic acid (5-HIAA). Subsequently the patient was maintained using bromocriptine. Although it was suspected that MPPP or some derivative was responsible for the parkinsonism, the causative agent was not identified.

The original report by Davis *et al.* (1979) appears to have been almost totally overlooked. However, in 1983 Langston and colleagues reported four further cases of young drug addicts developing parkinsonism, which responded to the administration of L-dopa plus carbidopa, and, in two cases, the addition of bromocriptine or lisuride. Again, two patients showed low CSF HVA concentrations but relatively normal 5-HIAA and HMPG levels. These subjects had also been self-administering MPPP. However, since pure MPPP had been successfully taken for some months without ill effect by the patient studied by Davis and colleagues, it seemed unlikely that MPPP itself was the toxic agent. Consequently, Langston and colleagues examined the nature of the material injected by the addicts. They found that it was contaminated with varying percentages of another synthetic product, 1-methyl-4-phenyl-1,2,3,6-tetrahydropyridine (MPTP), and it appeared that this substance was responsible for the induction of parkinsonism. Since this report, approximately 150 individuals who were exposed to MPTP have been uncovered (Langston *et al.*, 1984b). These include an organic chemist who synthesised large batches of MPTP as part of a pharmaceutical research programme and who may have been exposed to vapour inhalation or cutaneous absorption of MPTP (Langston and Ballard, 1983).

The patient initially described by Davis and his colleagues subsequently died of drug overdosage. On examination of the brain, gross neuronal loss and gliosis were discovered, limited to the zona compacta of substantia nigra. In contrast, the locus coeruleus appeared intact. A single darkly staining eosinophilic body, somewhat similar to a Lewy body, was seen in substantia nigra.

Patients developing parkinsonism as a result of MPTP administration exhibited a relatively advanced form of the disease (Hoehn and Yahr stage IV and V). Initially, all responded well to the administration of L-dopa or other dopamine agonists. Langston and Ballard (1984) used this as an opportunity to follow the effects of newly initiated L-dopa therapy in patients with advanced parkinsonism. Surprisingly, within a few weeks or months of starting drug treatment, all patients started to exhibit the classical long-term side-effects (wearing off, 'on–off' and dyskinesias) of L-dopa therapy. Investigation of the treatment of patients exhibiting advanced parkinsonism caused by MPTP adminstration may throw light on the cause of many of the side-effects experienced by patients treated for long periods with L-dopa but whose underlying Parkinson's disease is progressing. Clearly MPTP induces a parkinsonian state in man which is accompanied by loss of dopamine-containing neurones and so mimics the idiopathic disease. An understanding of the mechanism of action of MPTP might therefore throw light on the cause of Parkinson's disease in man.

36.3. The effect of MPTP administration to rodents

In general, attempts to reproduce the parkinsonian syndrome induced by MPTP in man using rats have not been successful. A number of investigations have failed to show neurotoxicity as a result of the systemic administration of MPTP to this species.

In our study, MPTP (10 mg kg^{-1} i.p.) was administered to male Wistar rats for 16 days (Boyce *et al.*, 1984). The dose of MPTP used was based on calculation of the highest daily intake of MPTP by drug addicts (*c.* 1 mg kg^{-1}; Langston *et al.*, 1983) multiplied by a factor of 10 to compensate for the generally higher drug metabolism of rodents. Behavioural tests were performed on the animals both during drug administration and following drug withdrawal. Biochemical examinations were performed 9–11 days after drug withdrawal.

Animals receiving MPTP (10 mg kg^{-1} day^{-1} i.p.) for 16 days did not gain weight during the treatment period, but on drug withdrawal they gained weight such that after 1 week their body weight was not different from that of control animals. Adminstration of MPTP (10 mg kg^{-1} i.p.) for 16 days did not alter spontaneous locomotor activity compared with that in control rats when assessed using photocell cages or a hole-board apparatus (Fig. 36.1). Nine days following cessation of MPTP treatment, spontaneous activity again was unchanged when compared with that of control animals (Fig. 36.1).

Amphetamine (2·5 mg kg^{-1} i.p.) administration caused an identical marked increase in locomotor activity both in rats treated with MPTP (10 mg kg^{-1} i.p.) for 16 days and then withdrawn for 9 days, and in control animals. Administration of amphetamine (2·5 and 5·0 mg kg^{-1} i.p.; 45 min previously) or apomorphine (0·25 mg kg^{-1} s.c.; 20 min previously) also produced similar stereotyped behaviour in both control and MPTP-treated animals, which did not differ between the groups when examined 13–16 days following MPTP treatment (Table 36.1).

Measurement of striatal dopamine, HVA and DOPAC concentrations 13 days following withdrawal from a 16-day period of administration of MPTP revealed no difference between control and MPTP-treated animals (Table 36.1). The uptake of [^{3}H]dopamine and [^{3}H]5-HT into striatal synaptosomes from rats treated with MPTP for 16 days and then withdrawn for 19 days was not different from values found in tissue from control animals (values obtained from a single experiment: [^{3}H] dopamine, control 36·1 pmol h^{-1} mg^{-1} protein; MPTP-treated 35·5 pmol h^{-1} mg^{-1}

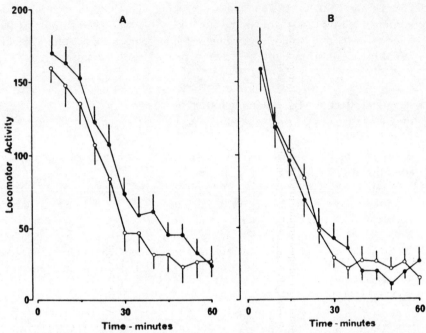

Fig. 36.1.   The effect on spontaneous locomotor activity assessed using photocell cages of (A) the intraperitoneal (i.p.) administration to rats of MPTP (10 mg kg^{-1} day^{-1} i.p.) for 16 days, and (B) 9 days following withdrawal of MPTP treatment, in comparison with vehicle-injected control animals; (●) control. (○) MPTP (10 mg kg^{-1} day^{-1} i.p.). No significant difference between groups using one-way analysis of variance and Student's t-test.

protein. [^{3}H]5-HT: control 40·0 pmol h^{-1} mg^{-1} protein; MPTP-treated 37·7 pmol h^{-1} mg^{-1} protein).

   These findings are in general agreement with those of others. Chiueh *et al.* (1984) found that immediately following the administration of 2·5 or 5·0 mg kg^{-1} s.c. MPTP for 28 days, nigral dopamine, HVA and DOPAC concentrations were reduced by approximately 50%. However, the dopamine content of the striatum was unaffected although the levels of HVA and DOPAC were greatly reduced. Interestingly the dopamine, HVA and DOPAC content of the nucleus accumbens was unaffected by MPTP treatment. Saghal *et al.* (1984) were unable to show any persistent change in amphetamine- or apomorphine-induced locomotor activity following administration of 2·5–10 mg kg^{-1} i.p. MPTP for 2 weeks. Striatal dopamine, HVA and DOPAC concentrations were unaffected. In contrast, Steranka *et al.* (1983), using a bolus injection of 35 mg kg^{-1} i.p. MPTP followed by a subcutaneous infusion of 160 mg kg^{-1} over the following 24 h, showed a persistent 40% decrease in striatal dopamine content 7 days later.

Table 36.1. The effect of intraperitoneal (i.p.) administration of MPTP (10 mg kg^{-1} day^{-1} i.p.) for 16 days followed by 13–16 days' drug withdrawal on stereotyped behaviour induced by subcutaneous (s.c.) administration of apomorphine (0·25 mg kg^{-1} s.c.) and intraperitoneal (i.p.) administration of amphetamine (2·5 and 5 mg kg^{-1} i.p.) and on striatal DA, HVA and DOPAC concentrations compared with control animals.

Treatment group	Stereotypy			Striatal concentration			
	Amphetamine (mg kg^{-1})		Apomorphine (0·25 mg kg^{-1})	DA (μg g^{-1})	HVA (ng g^{-1})	DOPAC (ng g^{-1})	HVA/DOPAC ratio
	2·5	5·0					
Control	1·8±0·2	2·7±0·4	2·7±0·2	8·34±0·48	569±70	746±37	0·76±0·07
MPTP	1·7±0·2	2·0±0	2·5±0·3	8·47±0·61	458±23	778±72	0·61±0·03

Values are expressed as mean ±1 SEM for 6 to 8 animals in each group. No significant difference was found between control and MPTP groups. Student's t-test for parametric data; Mann–Whitney U test for non-parametric data.

The dopamine-containing cells of zona compacta of substantia nigra in the rat do not appear to be as susceptible to the neurotoxic actions of MPTP, when administered systemically, as those of man. However, when given in sufficiently high dosage, some neurotoxicity is evident. Bradbury *et al.* (1984) have recently shown that the direct unilateral or bilateral intranigral infusion of MPTP (1–10 μg per 24 h for up to 13 days) caused persistent motor deficits in rats. The behavioural changes were accompanied by a fall in nigral dopamine and DOPAC concentrations but similar changes were not observed in the striatum. Similar to the toxic actions of MPTP on direct infusion into rat brain, the exposure of explants of embryonic rat mesencephalon to MPTP results in the loss of dopamine cell bodies and fibre outgrowths (Mytilineou and Cohen, 1984).

In mice, more success has been obtained in attempting to mimic the neurotoxic actions of MPTP observed in man. Administration of MPTP in high doses (30 mg kg^{-1} i.p. for up to 20 days) produced up to an 80% decrease in striatal dopamine content, and 70 and 50% losses in DOPAC and HVA content, respectively, when measured up to 18 days after treatment (Heikkila *et al.*, 1984*a*). These changes were accompanied by a marked reduction of [^{3}H] dopamine uptake into striatal synaptosomes and a profound loss of dopamine-containing neurones within zona compacta of substantia nigra. This effect, however, may not be specific since in another study the administration of MPTP (2 × 10 mg kg^{-1} i.v.) caused a persistent decrease in brain noradrenaline content (Hallman *et al.*, 1984). In none of the mouse studies has any evidence for altered motor behaviour been reported.

So far the mouse appears to be the only rodent species which shows any degree of susceptibility to MPTP. The rat, guinea pig and gerbil appear resistant (Heikkila *et al.*, 1984*b*). Even in the mouse the doses of MPTP required to produce neurotoxic effects are grossly in excess of those needed to produce a profound parkinsonian state in man. Possible reasons for this will be discussed later. However, although the mouse might prove suitable for biochemical studies of MPTP action, there is at present no evidence to suggest that the systemic administration of MPTP in this species mimics the features of Parkinson's disease as it occurs in man.

36.4. Administration of MPTP to rhesus and squirrel monkeys

The difficulties encountered in attempting to reproduce the neurotoxic actions of MPTP in rodents led to its use in non-human primate species. Administration of MPTP to rhesus or squirrel monkeys has produced dramatic effects.

The studies to date have been carried out by two groups, namely those of Langston and Burns. Although different dosages schedules and routes of administration of MPTP have been employed, the results obtained have been in good agreement (Burns *et al.*, 1983, 1984; Langston *et al.*, 1984*c*).

Thus, after two or more doses of MPTP, the animals develop an acute behavioural response to the drug. This takes the form of eyelid closure and 'nodding off', a paucity of movement with salivation, rotation of the head, extension of the limbs and shaking. As administration of MPTP continues, a persistent impairment of motor function becomes apparent such that within a few days the animals exhibit a profound parkinsonian state characterised by akinesia, rigidity, postural tremor, flexed posture, impaired righting reflexes, loss of vocalisation, eyelid closure, drooling and freezing episodes. These persistent effects of MPTP are reversed by the administration of L-dopa (plus a peripheral decarboxylase inhibitor).

Biochemical analysis has shown that the parkinsonism produced by MPTP in rhesus and squirrel monkeys is associated with low CSF HVA concentrations and low striatal dopamine and HVA levels. In general, no persistent effects on CSF or brain HMPG or 5-HIAA concentrations were observed. Histological examination has shown a loss of dopamine-containing cells in substantia nigra zona compacta but, interestingly, not in the adjacent ventral tegmental area or in the nucleus locus coeruleus.

36.5. Studies with MPTP in the common marmoset

Our own studies with MPTP in primates have centred on the use of the common marmoset, a species bred in captivity and more amenable to biochemical analysis. These are now reported in detail.

36.5.1. *Initial dose ranging studies*

MPTP (1, 2, 3, or 4 mg kg^{-1} i.p.) was administered for 4 days to four adult female marmosets (295–420 g). A single animal was used as a vehicle-injected control. Behavioural examination was carried out on a daily basis and the animals were killed on day 10 for biochemical examination (Jenner *et al.*, 1984).

For the measurement of dopamine, HVA and DOPAC, samples of caudate were homogenised in 1·4 M perchloric acid containing 0·03% EDTA. To the resulting supernatant 3,4–dihydroxybenzylamine (DHBA) was added as internal standard. Aliquots were injected on to a Waters HPLC connected with a Bio-Analytical System electrochemical detector. The HPLC was fitted with a Spherisorb ODS 5μ column. The mobile phase was methanol/0·5 M phosphate buffer containing octanesulphonic acid and 0·03% sodium EDTA, pH 2·7.

For [^{3}H]dopamine uptake experiments, crude P2 synaptosomal preparations of putamen (in 20 volumes original tissue weight) were prepared as described previously. Aliquots were added to oxygenated Krebs bicarbonate buffer (pH 7·4) containing 5×10^{-7} M pargyline and 0·17 μM ascorbic acid. A final concentration of 5×10^{-8} M [^{3}H]dopamine (13·5 Ci $mmol^{-1}$; Amersham International) was added to each sample. After 5 min incubation at 37°C, samples were rapidly filtered under vacuum. Blank values were obtained by incubation of tissue on ice at 0°C. The results were expressed as pmoles [^{3}H]dopamine taken up per milligram protein per hour.

For binding experiments samples of caudate were homogenised in Tris-HCl buffer containing 120 mM NaCl, pH 7·4. Washed membrane preparations were diluted to 600 volumes and aliquots were incubated in triplicate with [^{3}H]spiperone (0·05–1·0 nM; 19 Ci mmole^{-1}; Amersham International). Specific binding was defined by the incorporation of 10^{-5} M (±)-sulpiride. Samples were separated by rapid vacuum filtration. Data were analysed using Eadie–Hofstee plots to give B_{max} and K_D values. Regression coefficients ranged between 0·97 and 0·99.

The behavioural effects induced by administration of MPTP were evident by day 2 and could be divided into those appearing acutely after each dose of MPTP and those resulting from more prolonged effects of the drug. Animals receiving 2, 3 or 4 mg kg^{-1} i.p. MPTP showed akinesia, prostration, abnormal limb movements and a fixed gaze starting some 5 min following MPTP administration and lasting for approximately 15 min. If the animals were disturbed by either tactile or auditory stimuli, they appeared normal transiently, but then reverted to the abnormal state.

The animal receiving 1 mg kg^{-1} MPTP appeared unaffected. By 48 h after starting, MPTP animals receiving 2, 3 or 4 mg kg^{-1} of MPTP developed persistent behavioural abnormalities; they showed little spontaneous movement, an abnormal posture, and a lack of vocalisation.

On day 8, 3 days following the cessation of MPTP administration, a more detailed behavioural observation was made. Compared with the control animal, the animal receiving 1 mg kg^{-1} MPTP appeared normal. The animal receiving 2 mg kg^{-1} MPTP showed reduced spontaneous movement but appeared normal when disturbed. There was loss of vocalisation but no loss of grasp or blink reflexes. The animal treated with 3 mg kg^{-1} MPTP was profoundly akinetic, although it could make slow but clumsy movements when disturbed; its limbs were rigid, there was some postural tremor of the hindquarters and it exhibited a poor grasp reflex with the front limbs. The animal treated with 4 mg kg^{-1} MPTP was even more parkinsonian; it was totally akinetic, stooped and rigid with postural tremor of the hindquarters, and was unable to make anything but minimal movements when disturbed; there was no vocalisation and it had a staring appearance although the blink reflex appeared normal. If the animal was touched, clonus developed, particularly in the hindquarters. There was a loss of the grasp reflex in the front paws, which impaired the animal's ability to eat. Animals receiving 2, 3 and 4 mg kg^{-1} MPTP showed a considerable decrease in body weight over the 10-day course of the experiment.

Biochemical examination was carried out on day 10 from the start of MPTP administration (Table 36.2). MPTP caused a dose-dependent decrease in the concentration of dopamine, HVA and DOPAC in the

Table 36.2. Alterations in dopamine parameters in the caudate-putamen of four individual marmosets receiving MPTP (1–4 mg kg⁻¹ i.p.) for 4 days compared with a control animal.

Treatment with MPTP (mg kg⁻¹)	Dopamine (μg g⁻¹)	HVA (μg g⁻¹)	DOPAC (μg g⁻¹)	[³H]Dopamine uptake (pmol mg⁻¹ protein h⁻¹)	[³H]Spiperone binding B_{max} (pmol g⁻¹)	K_D (nM)
None	11·8	5·6	2·9	591	19·3	0·051
1	4·4	0·45	0·42	282	29·4	0·060
2	0·02	ND*	ND	546	23·3	0·039
3	ND	ND	ND	189	22·8	0·039
4	ND	ND	ND	54	20·7	0·043

*ND = None detected

Table 36.3. Alterations in dopaminergic parameters in the caudate-putamen induced by MPTP treatment for up to 6 weeks.

Treatment group (n=6–10)	Dopamine (μg g⁻¹)	HVA (μg g⁻¹)	DOPAC (μg g⁻¹)	[³H]Dopamine uptake (pmol mg⁻¹ protein h⁻¹)	[³H]Spiperone B_{max} (pmol g⁻¹)
Controls	8·9±0·6	5·6±0·6	5·2±0·5	228±47	17·2±1·1
10 days	0·35±0·20	0·26±0·03*	0·28±0·08*	77±10*	19·2±1·5
4–6 weeks	0·89±0·32*	1·4±0·4*†	0·87±0·17*†	34±12*	18·9±0·9

*Mean ±1 SEM shown; *$P < 0.05$ Student's t-test compared with controls; †$P < 0.05$ Student's t-test compared with animals at 10 days.

caudate nucleus. In tissue from the animal treated with 1 mg kg^{-1} MPTP there was a 63 % loss of dopamine content, but greater losses of HVA and DOPAC. With MPTP 2 mg kg^{-1} there was a 99 % loss of dopamine content and the levels of HVA and DOPAC were below the sensitivity of the analytical technique. In animals receiving 3 and 4 mg kg^{-1} of MPTP, no dopamine, HVA or DOPAC was detectable. MPTP caused a dose-related decrease in the uptake of [^{3}H]dopamine into synaptosomal preparations from the putamen. In contrast, the number of specific binding sites (B_{max}) of [^{3}H]spiperone was, if anything, slightly increased by MPTP treatment, and there was no obvious change in the dissociation constant (K_D).

From this initial experiment it appears that administration of MPTP to the marmoset, as in man and other primate species, can induce a parkinsonian state accompanied by evidence of dopamine neuronal loss. In animals receiving higher doses or MPTP there was virtually no dopamine or its metabolites detectable in the samples examined. However, these animals varied considerably in the extent of their parkinsonism. There appears to be a high degree of individual susceptibility to the effects of MPTP on motor behaviour. As a consequence, in subsequent experiments a fixed dosage regime was abandoned and animals were treated with individualised doses of MPTP so as to render them obviously parkinsonian.

36.5.2. *Behavioural and biochemical consequences of MPTP administration for up to 6 weeks*

Marmosets (290–440 g) of either sex received MPTP (1–4 mg kg^{-1} i.p.) daily for up to 5 days. Within 2–3 days of starting drug treatment, the animals exhibited varying degrees of bradykinesia or akinesia, rigidity, postural abnormalities, loss of vocalisation and some postural tremor. This state was maintained at 10 days following the start of MPTP treatment, but by 4–6 weeks some reversal of the motor deficits had occurred. By then animals usually were not obviously rigid, some spontaneous vocalisation occurred and rapid movements were possible when attempts were made to remove the animals from their home cage. However, they still exhibited fewer and less co-ordinated spontaneous movements compared with control animals.

At 10 days and at 4–6 weeks, animals were killed for biochemical examination. At 10 days following the start of MPTP treatment, concentrations of dopamine, HVA and DOPAC were reduced by more than 90 % compared with values from control animals (Table 36.3). The uptake of [^{3}H]dopamine into putamen synaptosomal preparations was also markedly reduced, but to a lesser extent than the caudate dopamine loss. In the same preparations [^{3}H]glutamate uptake was not different from values obtained in control animals. The number of specific [^{3}H]spiperone binding sites (B_{max}) and the dissociation constant (K_D) in caudate were unchanged.

At 4–6 weeks following the start of MPTP treatment, the caudate concentrations of dopamine, HVA and DOPAC remained reduced. Similarly, the uptake of [^{3}H]dopamine in putamen was decreased but this loss now exceeded that for dopamine concentrations. Again, uptake of [^{3}H]glutamate into putamen preparations did not differ from that observed in control animals. At this time more variation between individual animals was observed, perhaps reflecting differing degrees of behavioural recovery. Again, the B_{max} and K_D values for specific [^{3}H]spiperone binding in caudate were unchanged following MPTP treatment. The concentrations of noradrenaline, 5-HT and 5-HIAA in cortical regions and in the caudate-putamen in MPTP-treated animals did not differ from those found in control animals at any time. In addition, measurement of glutamic acid decarboxylase activity, choline acetyl transferase activity or [^{3}H]quinuclidinylbenzilate binding to muscarinic receptors did not show any change with MPTP treatment.

Histological analysis is being undertaken (by Drs A. Lees, W. Gibb, S. Hunt, B. Berger and P. Gaspa). Although this is not complete, some general observations can be made. The administration of MPTP caused a profound loss of dopamine-containing cells in the zona compacta of substantia nigra as judged by light microscopy and by immunocytochemical techniques visualising tyrosine hydroxylase. Some cell loss may also have occurred in the adjacent ventral tegmental area although this was not nearly so dramatic as in the zona compacta. The cells of the locus coeruleus appeared intact.

The administration of MPTP to the marmoset induces a syndrome which contains the major motor features, biochemical changes and cellular destruction characterising Parkinson's disease in man. MPTP appears to selectively destroy dopamine-containing neurones since markers for other neuronal systems were unchanged. It even appears to be selective for those dopamine-containing cells of the zona compacta of substantia nigra since only slight damage to the ventral tegmental area was apparent, as previously reported in other primates. Changes in other neuronal systems were not apparent, unlike Parkinson's disease. This may be due to the rapid destruction of dopamine neurones induced by MPTP treatment and the short interval between administration of MPTP and sacrifice, as compared with the slow rate of cellular loss occurring in Parkinson's disease.

36.5.3. The effect of dopamine agonists on the MPTP-treated marmosets
Marmosets (280–430 g) of either sex were treated with MPTP (1–4 mg kg^{-1} i.p.) for up to 5 days so as to render them parkinsonian. Following treatment, animals were allowed to recover from the acute effects of MPTP treatment for up to 6 weeks. They were then examined for their behavioural response to L-dopa (20 mg kg^{-1} i.p.) plus carbidopa (5 mg kg^{-1} i.p.), to the selective D-2 agonist LY 141865 (0·5 mg kg^{-1} i.p.) or to the selective D-1

agonist SK&F 38393 (1–20 mg kg⁻¹ i.p.). Behavioural testing was carried out by removing each animal from its home cage and placing it in a novel cage environment. Animals were allowed 30 min to acclimatise to this novel environment after which a video sequence of their behaviour was filmed. Following drug administration the behaviour was then observed over the following 120 min by videorecording. The video tapes were assessed blindly and the number of movements occurring in given time periods was assessed.

Within 4 min of the administration of L-dopa plus carbidopa the parkinsonian syndrome began to disappear. Initial improvement consisted of restoration of checking movements (and an increase in limb tremor). A gradual increase in locomotor activity then occurred, so by 6 min the animal explored its environment and could vocalise. Within 25 min there was almost total reversal of the parkinsonism; animals could make a full repertoire of movements although many still appeared clumsy. The effects of L-dopa lasted for approximately 8 h after which the animals reverted to the pre-drug state.

Prior to the administration of LY 141865 or SK&F 38393, all animals were immobile on either the perch or the floor of the cage and showed little or no spontaneous checking movements 30 min following introduction into the novel cage environment. The administration of SK&F 38393 (1–20 mg kg⁻¹ i.p.) did not alter the behaviour of any animal over the next 30–120 min (Fig. 36.2). The animals remained immobile in a fixed position, showing little or no checking movements.

In contrast, the administration of LY 141865 (0·5 mg kg⁻¹ i.p.) caused profound behavioural activation and reversal of MPTP-induced akinesia which lasted for up to 2 h (Fig. 36.2). Within a few minutes of drug administration, two or the four animals either retched or vomited for up to 5 min. After approximately 10 min all animals started to become alert, with an increase in the number of checking movements and a gradual increase in general movement from place to place within the test cage. Movements were co-ordinated and included normal running on the floor of the cage, jumping up to high perches or down to the floor, grasping of perches and cage walls, feeding and grooming, and checking behaviour. Most of this repertoire of movement appeared to be normal behaviour; occasionally there were brief periods of explosive motor behaviour and sometimes inappropriate chewing occurred. The response in individuals varied, some animals showing only a moderate increase in activity whereas others showed a dramatic reversal of their initial akinesia. The durations of effect of LY 141865 also varied between animals from approximately 30 min to 2 h. Drug treatment also increased aggression, which was noticed when the experimenter attempted to remove the animal from the test cage back to its home environment.

The ability of L-dopa to reverse the parkinsonian state induced by the administraton of MPTP emphasises the primary role played by the loss of

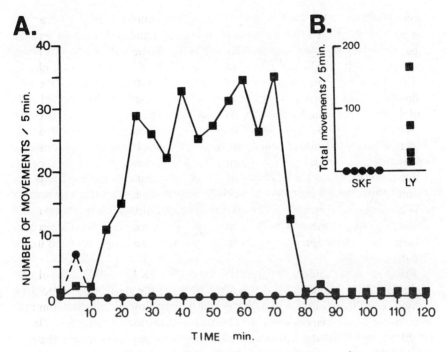

Fig. 36.2. The effect of administration of LY 141865 (0·5 mg kg^{-1} i.p.; ■) or SKF 38393 (20 mg kg^{-1} i.p.; ●) on the number of movements made by a single animal previously treated with MPTP (1–4 mg kg^{-1} i.p.). (B) The total number of movements made by MPTP-treated animals receiving either LY 141865 (0·5 mg kg^{-1} i.p.; ■) or SKF 38393 (1·0–20 mg kg i.p.; ●) over a 5-min period starting 30 min following drug administration.

dopamine-containing neurones and the resemblance to idiopathic Parkinson's disease in man. However, L-dopa administration results in the stimulation of both D-1 and D-2 receptors in brain. Using the selective dopamine agonists LY 141865 and SK&F 38393, it appears that it is the activation of D-2 receptors but not D-1 receptors that reverses the parkinsonian akinesia induced by the prior treatment of marmosets with the nigrostriatal dopamine neurotoxin MPTP.

36.6. Mechanism of action of MPTP

In addition to providing a model of parkinsonism in animals, an understanding of the mechanism by which MPTP induces its neurotoxic effects might shed light on how Parkinson's disease is produced in man. Initially it is necessary to differentiate between the acute and chronic effects of MPTP, for the drug does produce acute behavioural syndromes in animals as well as the more persistent neurotoxic actions.

Comparisons have been made between the acute effects of MPTP and 5-HT syndromes which can be induced in animals by the administration of 5-HT agonists and hallucinogens (Chiueh *et al.*, 1984). Thus, the acute administration of MPTP has been associated with the occurrence of prostration, limb extension, head weaving and straub tail phenomenon. No detailed investigation of the acute MPTP syndrome has been made. However, in *in vitro* experiments, MPTP appears in micromolar concentrations to alter the release and uptake of [^{3}H]dopamine, [^{3}H]5-HT and [^{3}H]noradrenaline (Denton and Howard, 1984; Kula *et al.*, 1984; Schmidt *et al.*, 1984).

Of more interest to the present discussion, however, is the mechanism of the chronic neurotoxic action of MPTP. From the outset it must be pointed out that this is not known, so much of what follows is necessarily conjecture. Whatever the mechanism, it must involve a high degree of specificity since MPTP neurotoxicity is not only selective for dopamine neurones but also appears to be relatively selective for those neurones located in the zona compacta of substantia nigra. One possible explanation is that MPTP itself is not the active moiety but rather it is some product of the metabolism of MPTP which is neurotoxic. The need for the production of an active metabolite might explain the relative lack of neurotoxicity of MPTP in rodent species compared with human and non-human primates.

Studies of the metabolism of MPTP are at an early stage but both *in vitro* investigations using brain mitochondrial preparations and *in vivo* examination have revealed the production of large quantities of the corresponding pyridinium ion, MPP$^+$ (Chiba *et al.*, 1984; Markey *et al.*, 1984). This compound cannot be formed by a single-stage metabolic process but requires at least two intermediate steps. Potentially the metabolic route for the formation of the pyridinium ion might involve a number of possible toxic reactive intermediates. Certainly toxicity appears to involve a high degree of structural specificity, since the saturated derivative of MPTP, 1-methyl-4-phenylpiperidine (MPP), does not induce parkinsonism in monkeys (Langston *et al.*, 1984*d*).

The general systemic production of some toxic derivative of MPTP would not, however, explain the highly selective nature of the neuronal destruction produced. A number of other approaches have been tried to explain the neurotoxicity of MPTP, but so far unsuccessfully. For example, *in vitro* in rat brain there is no specific localisation of binding sites for [^{3}H]MPTP in brain (Wieczorek *et al.*, 1984). These studies employed rat tissue where the compound is not strongly neurotoxic. Recently, autoradiographic studies in human brain have shown very high receptor densities for [^{3}H]MPTP in the caudate, substantia nigra and locus coeruleus (Javitch *et al.*, 1984). This localisation was used to explain the neurotoxic effect of MPTP in man compared with rodents. However, MPTP does not appear to destroy the locus coeruleus in man so this cannot be the sole explanation.

The administration of MPTP to rats and guinea pigs did cause a dramatic increase in $(2\text{-}^{14}C)$-deoxyglucose uptake in zona compacta of substantia nigra, the ventral tegmental area and the locus coeruleus, but not in other brain regions (Palacios and Wiederhold, 1984). Such effects might be related to the acute effects of MPTP but again occur in species where MPTP is not neurotoxic and in areas where cellular destruction is not observed in primates or man. Similarly, *in vitro* MPTP binds with micromolar affinity to melanin, but this pigment is not restricted to the dopamine-containing cells of substantia nigra (Lyden *et al.*, 1983).

Some clue as to the mechanism underlying the neurotoxic effects of MPTP may come from the nature of the enzyme systems involved in its metabolism. Unexpectedly, monooxygenases do not appear to be primarily involved in the metabolism of MPTP; its neurotoxicity is not prevented in primates by the administration of SKF 525A, an inhibitor of monooxygenase action (unpublished data). However, and perhaps surprisingly, MPTP toxicity and the metabolism of the compound to the pyridinium ion are prevented by monoamine oxidase inhibitors (Heikkila *et al.*, 1984b; Markey *et al.*, 1984). This suggests that the first stage of MPTP metabolism to give rise to a toxic intermediate can be mediated by monoamine oxidase enzymes. It was also suggested that MPTP metabolism is carried out selectively by monoamine oxidase B; toxicity and metabolism were prevented by the selective monoamine oxidase B inhibitor deprenyl. Although this remains to be demonstrated conclusively, it gives rise to an intriguing possibility. In primates monoamine oxidase B is responsible for the metabolism of dopamine. It might be that the toxicity of MPTP in primates is due to its metabolism to a toxic substance in close proximity to the vulnerable dopaminergic system. In contrast in rodents, it is monoamine oxidase A which is responsible for the metabolism of dopamine so conversion of MPTP to the active moiety might not occur so readily. Indeed it has recently been demonstrated that the specific binding sites for [^{3}H]MPTP in rodent brain correspond exactly to those identified by [^{3}H]pargyline (Parsons and Rainbow, 1984), so strengthening the link between the actions of MPTP and monoamine oxidase activity. However, such a hypothesis, although tempting, would not explain why the toxicity of MPTP would appear to be limited to the nigral dopamine neurones and why the adjacent cells of the ventral tegmental area are spared. Indeed, monoamine oxidase B is distributed widely throughout the body so a more general neurotoxic action might be expected.

36.7. Conclusions

The chance occurrence of MPTP-induced toxicity to nigral dopamine neurones in man may have provided one of the major advances to be made in the study of Parkinson's disease in the past decade. This substance can be used to produce animal models of Parkinson's disease in which drug effects may be studied alongside the key biochemical parameters responsible for causing the symptoms of the disorder. Furthermore, MPTP may provide the clue as to the mechanism of cellular death in Parkinson's disease.

However, attention must also be paid to the differences that exist between MPTP-induced parkinsonism in non-human primates and the idiopathic disease in man. Of major importance is the fact that the MPTP-induced syndrome in primates is of rapid onset, so does not mimic the slow progression of the disease as it occurs in man. It is, however, not known if animals given a subthreshold lesion using MPTP would later develop parkinsonian symptoms as they aged. The rapid onset of experimental MPTP parkinsonism might also explain the pathological difference between this and the idiopathic disease. MPTP-induced experimental parkinsonism does not appear to be associated with the characteristic presence of Lewy bodies. This might be interpreted as the presence of Lewy bodies being due to slow progressive cellular death. Also, in the same vein, the limited pathology of MPTP-induced experimental parkinsonism does not resemble the wide disturbance of neurotransmitter systems that is observed in man. MPTP appears to lesion dopamine cells in zona compacta of substantia nigra relatively selectively, whereas in Parkinson's disease there is generalised loss of dopamine-containing neurones throughout the brain, and considerable changes in other neurotransmitter systems. The differences in the time course of the two syndromes might explain these differences. However, in the MPTP-induced syndrome, it is clear that most of the classical symptoms of Parkinson's disease are to be produced through loss of nigrostriatal dopamine neurones without involvement of other dopamine systems or neuronal pathways. This in itself is an important observation with respect to interpretation of the human disease. No-one has yet investigated the biochemical changes that occur in the brain of the MPTP-treated primates months or years following lesioning with MPTP.

Clearly, there is now going to be an era of research into Parkinson's disease based on the actions of MPTP. If the promise of the initial studies is maintained, then we can look forward to some exciting advances in our understanding of the aetiology and treatment of Parkinson's disease.

36.8. Summary

Self-administration of 1-methyl-4-phenyl-1,2,3,6-tetrahydropyridine (MPTP) by drug addicts induced a persistent parkinsonian syndrome which was reversed by administration of L-dopa. MPTP-induced parkinsonism was accompanied by low CSF HVA concentrations and in one patient, who subsequently died, by a loss of dopamine neurones in zona compacta of substantia nigra.

In rodent species, administration of MPTP is toxic to dopamine-containing neurones only when administered in high doses. However, non-human primate species appear susceptible to the neurotoxic actions of this agent.

The administration of MPTP to the common marmoset produces a partially reversible parkinsonian syndrome characterised by akinesia or bradykinesia, rigidity, loss of vocalisation, postural abnormalities and some postural tremor. The onset of the syndrome is accompanied by a profound loss of caudate-putamen dopamine, HVA and DOPAC concentrations and [^{3}H]dopamine uptake but no change in specific [^{3}H]spiperone binding. Histological analysis revealed a loss of dopamine-containing cells in the zona compacta of substantia nigra. The MPTP-induced parkinsonian syndrome was reversed by the administration of L-dopa and the selective D-2 agonist LY 141865 but not by the D-1 agonist SK&F 38393.

Although the mechanism of action of MPTP is unknown, this substance may provide a viable animal model of Parkinson's disease in which to study the problems associated with its aetiology and drug treatment.

Acknowledgements

This study was supported by the Medical Research Council, the Parkinson's Disease Society and the Research Funds of the Bethlem Royal and Maudsley Hospitals and King's College Hospital.

Addendum

Since this paper was written much more has been discussed about MPTP. Of particular importance has been the observation that MPP$^+$ is accumulated in nigrostriatal neurones via the dopamine uptake system (Javitch, J. A. and Snyder, S. H. (1985). Uptake of MPP$^+$ by dopamine neurons explains selectivity of parkinsonism-producing neurotoxin, MPTP. *Euro. J. Pharmac.* **106**, 455–456.

References

Boyce, S., Kelly, E., Reavill, C., Jenner, P. and Marsden, C. D. (1984). Repeated administration of *N*-methyl-4-phenyl-1,2,5,6-tetrahydropyridine to rats is not toxic to striatal dopamine neurones. *Biochem. Pharmacol.* **23**, 1747–52.

Bradbury, A. J., Costall, B., Domeney, A. M., Jenner, P., Marsden, C. D. and Naylor, R. J. (1984). MPTP infusion into rat substantia nigra causes hypokinesia and bradykinesia, which are reversed by L-dopa treatment. Presented at *Dopamine 84*, Southampton.

Burns, R. S., Chiueh, C. C., Markey, S. P., Ebert, M. H., Jacobowitz, D. M. and Kopin, I. J. (1983). A primate model of parkinsonism: selective destruction of dopaminergic neurones in the pars compacta of the substantia nigra by *N*-methyl-4-phenyl-1,2,3,6-tetrahydropyridine. *Proc. nat. Acad. Sci. USA* **80**, 4546–50.

Burns, R. S., Markey, S. P., Phillips, J. M. and Chiueh, C. C. (1984). The neurotoxicity of 1-methyl-4-phenyl-1,2,3,6-tetrahydropyridine in the monkey and man. *Can. J. Neurol. Sci.* **11**, Suppl. 1. 166–8.

Chiba, K., Trevor, A. and Castagnoli, N. Jr (1984). Metabolism of the neurotoxic tertiary amine, MPTP, by brain monoamineoxidase. *Biochem. Biophys. Res. Commun.* **120**, 574–8.

Chiueh, C. C., Markey, S. P., Burns, R. S., Johanessen, J. N., Pert, A. and Kopin, I. J. (1984). Neurochemical and behavioural effects of systemic and intranigral administration of *N*-methyl-4-phenyl-1,2,3,6-tetrahydropyridine in the rat. *Eur. J. Pharmacol.* **100**, 189–94.

Davis, G. C., Williams, A. C., Markey, S. P., Ebert, M. H., Caine, E. D., Reichert, C. M. and Kopin, I. J. (1979). Chronic Parkinsonism secondary to intravenous injection of meperidine analogues. *Psychiat. Res.* **1**, 249–54.

Denton, T. and Howard, B. D. (1984). Inhibition of dopamine uptake by *N*-methyl-4-phenyl-1,2,3,6-tetrahydropyridine, a cause of Parkinsonism. *Biochem. Biophys. Res. Commun.* **119**, 1186–90.

Fahn, S. (1982). Fluctuation of disability in Parkinson's disease: pathophysiology. In *Movement Disorders.* C. D. Marsden and Fahn S. (eds), pp. 123–45. Butterworth, London.

Forno, L. S. (1982). Pathology of Parkinson's disease. In *Movement Disorders.* C. D. Marsden and S. Fahn (eds), pp. 25–40. Butterworth, London.

Graham, D. G. (1979). On the origin and significance of neuromelanin. *Arch. Path. Lab. Med.* **103**, 359–62.

Hallman, H., Olson, L. and Jonsson, G. (1984). Neurotoxicity of the meperidine analogue *N*-methyl-4-phenyl-1,2,3,6-tetrahydropyridine on brain catecholamine neurones in the mouse. *Eur. J. Pharmacol.* **97**, 133–6.

Heikkila, R. E., Cabbat, F. S., Manzino, L. and Duvoisin, R. C. (1984*a*). Effects of 1-methyl-4-phenyl-1,2,3,6-tetrahydropyridine on neostriatal dopamine in mice. *Neuropharmacology* **3**, 711–13.

Hekkila, R. E., Manzino, L., Cabbat, F. S. and Duvoisin, R. C. (1984*b*). Protection against the dopaminergic neurotoxicity of 1-methyl-4-phenyl-1,2,3,6-tetrahydropyridine by monoamineoxidase inhibitors. *Nature (Lond.)* **311**, 467–9.

Hornykiewicz, O. (1982). Brain neurotransmitter changes in Parkinson's disease. In *Movement Disorders.* C. D. Marsden and S. Fahn (eds), pp. 41–58. Butterworth, London.

Javitch, J. A., Uhl, G. R. and Snyder, S. H. (1984). Parkinsonism-inducing neurotoxin *N*-methyl-4-phenyl-1,2,3,6-tetrahydropyridine: characterisation and localisation of receptor binding sites in rat and human brain. *Proc. nat. Acad. Sci. USA* **81**, 4591–5.

Javoy-Agid, F., Taquet, H., Cesselin, F., Epelbaum, J., Grouselle, D., Mauborgne, A., Studler, J. M. and Agid, Y. (1984). Neuropeptides in Parkinson's disease. In *Catecholamines: Part C Neuropharmacology and Central Nervous Systems— Therapeutic Aspects*, pp. 35–42. Alan R. Liss, New York.

Jenner, P., Rupniak, N. M. J., Rose, S., Kelly, E., Kilpatrick, G., Lees, A. and Marsden, C. D. (1984). 1-Methyl-4-phenyl-1,2,3,6-tetrahydropyridine-induced Parkinsonism in the common marmoset. *Neurosci. Lett.* **50**, 85–90.

Kula, N. S., Baldessarini, R. J., Campbell, A., Finklestein, S., Ram, V. J. and Neumeyer, J. L. (1984). Effects of *N*-substituted phenyltetrahydropyridines on cerebral high-affinity synaptosomal uptake of dopamine and other monoamines in several mammalian species. *Life Sci.* **34**, 2567–75.

Langston, J. W., Ballard, P. A. Jr (1983). Parkinson's disease in a chemist working with 1-methyl-4-phenyl-1,2,5,6-tetrahydropyridine. *New Engl. J. Med.* **309**, 310.

Langston, J. W., Ballard, P., Tetrud, J. W. and Irwin, I. (1983). Chronic Parkinsonism in humans due to a product of meperidine-analog synthesis. *Science* **219**, 979–80.

Langston, J. W. and Ballard, P. A. Jr. (1984). Parkinsonism induced by 1-methyl-4-phenyl-1,2,3,6-tetrahydropyridine (MPTP): implications for treatment and the pathogenesis of Parkinson's disease. *Canad. J. Neurol. Sci.* **11**, Suppl. 1, 160–5.

Langston, J. W., Irwin, I., Langston, E. B. and Forno, L. S. (1984a). Pargyline prevents MPTP-induced Parkinsonism in primates. *Science* **225**, 1480–2.

Langston, J. W., Langston, E. B. and Irwin, I. (1984b). MPTP-induced Parkinsonism in human and non-human primates—clinical and experimental aspects. *Acta Neurol. Scand.* **70**, Suppl. 100, 49–54.

Langston, J. W., Forno, L. S., Rebert, C. S. and Irwin, I. (1984c). Selective nigral toxicity after systemic administration of 1-methyl-4-phenyl-1,2,5,6-tetrahydropyridine (MPTP) in the squirrel monkey. *Brain Res.* **292**, 390–4.

Langston, J. W., Irwin, I., Langston, E. B. and Forno, L. S. (1984d). The importance of the '4–5' double bond for neurotoxicity in primates of the pyridine derivative MPTP. *Neurosci. Lett.* **50**, 289–4.

Lyden, A., Bondesson, U., Larsson, B. S. and Lindquist, N. G. (1983). Melanin affinity of 1-methyl-4-phenyl-1,2,5,6-tetrahydropyridine, an inducer of chronic Parkinsonism in humans. *Acta Pharmacol. Toxicol.* **53**, 429–32.

Marsden, C. D. (1983). Neuromelanin and Parkinson's disease. *J. Neural. Trans. Suppl.* **19**, 121–41.

Marsden, C. D., Parkes, J. D. and Quinn, N. (1982). Fluctuations of disability in Parkinson's disease—clinical aspects. In *Movement Disorders*. C. D. Marsden and S. Fahn (eds), pp. 96–122, Butterworth, London.

Markey, S. P., Johannessen, J. N., Chiueh, C. C., Burns, R. S. and Herkinham, M. A. (1984). Intraneuronal generation of a pyridinium metabolite may cause drug-induced Parkinsonism. *Nature (Lond.)* **311**, 464–7.

Mytilineou, C. and Cohen, G. (1984). 1-Methyl-4-phenyl-1,2,3,6-tetrahydropyridine destroys dopamine neurons in explants of rat embryo mesencephalon. *Science* **225**, 529–31.

Palacios, J. M. and Wiederhold, K. H. (1984). Acute administration of 1-N-methyl-4-phenyl-1,2,3,6-tetrahydropyridine (MPTP), a compound producing Parkinsonism in humans, stimulates [2-^{14}C]deoxyglucose uptake in the regions of the catecholaminergic cell bodies in the rat and guinea pig briains. *Brain Res.* **301**, 187–91.

Parsons, B. and Rainbow, T. C. (1984). High-affinity binding sites for ^{3}H-MPTP may correspond to monoamine oxidase. *Eur. J. Pharmacol.* **102**, 375–7.

Rinne, U. K. (1982). Brain neurotransmitter receptors in Parkinson's disease. In *Movement Disorders*. C. D. Marsden and S. Fahn (eds), pp. 59–74, Butterworth, London.

Saghal, A., Andrews, J. S., Biggins, J. A., Candy, J. M., Edwardson, J. A., Keith, A. B., Turner, J. D. and Wright, C. (1984). N-methyl-4-phenyl-1,2,3,6-tetrahydropyridine (MPTP) affects locomotor activity without producing a nigro-striatal lesion in the rat. *Neurosci. Lett.* **48**, 179–84.

Schmidt, C. J., Matsuda, L. A. and Gibb, J. W. (1984). *In vitro* release of tritiated monoamines from rat CNS tissue by the neurotoxic compound 1-methyl-phenyl-tetrahydropyridine. *Eur. J. Pharmacol.* **103**, 255–60.

Steranka, L. R., Polite, L. N., Perry, K. W. and Fuller, R. W. (1983). Dopamine depletion in rat brain by MPTP (1-methyl-4-phenyl-1,2,3,6-tetrahydropyridine). *Res. Comm. Substances Abuse* **4**, 315–22.

Wieczorek, C. M., Parsons, B. and Rainbow, T. C. (1984). Quantitative autoradiography of ^{3}H-MPTP binding sites in rat brain. *Eur. J. Pharmacol.* **98**, 453–4.

37

Delineation of a particular subregion within the neostriatum involved in the expression of muscular rigidity

B. Ellenbroek*, M. Schwarz and K. H. Sontag

Max Planck Institute for Experimental Medicine,
Hermann Reinstr. 3, 3400 Göttingen, Federal Republic of Germany
and
R. Jaspers and A. Cools

Psychoneuropharmacology Research Unit, University of Nijmegen,
The Netherlands

37.1 Introduction

It has long since been recognised from clinical studies that neuroleptics can induce 'extrapyramidal' (or Parkinson-like) side-effects, including muscular rigidity. These side-effects have been classically attributed to the striatal dopamine receptor blocking properties of neuroleptic drugs. Nevertheless, conflicting data exist as to whether or not a neuroleptic drug like haloperidol can induce muscular rigidity in animals. In order to try to clarify this controversy, we studied the effects of different doses of intrastriatally applied haloperidol on the spontaneous muscle tone of the gastrocnemius soleus (GS) muscle in awake rats.

37.2. Methods

Adult male Wistar rats (TNO/W 70, F. Winkelmann, Borchen, F.R.G.) weighing 200–220 g were stereotaxically implanted with bilateral guide cannulae aimed at various parts of the neostriatum. After a recovery period of at least 5 days, saline, haloperidol (Janssen Pharmaceutica) and/or apomorphine hydrochloride (ACF Chemiefarma) was bilaterally injected by means of a Hamilton syringe in a volume of 0·5 μl per side. All animals were only used once.

*To whom all correspondence should be addressed; present address: Psychoneuropharmacology Research Unit, University of Nijmegen, P.O. Box 9101, 6500 HB Nijmegen, The Netherlands.

After this injection rats were individually placed in ventilated polyacryl boxes for EMG recording. The left hindleg was gently fixed in a position avoiding extreme flexion and extreme extension and two teflon-insulated stainless steel wire electrodes (Cooner Wire AS 632 SS) were percutaneously inserted into the GS muscle. The recorded signal was amplified, filtered, rectified, and fed into an integrator which automatically reset after having reached a predetermined voltage. The reciprocal of this reset time was the measure of the EMG activity. The EMG was recorded continuously from 10 to 60 min after the injection and the mean activity of 5-minute periods was calculated. Short bursts lasting less than 1 min due to movements of the animal were discarded to ensure that only tonic EMG activity was recorded. After completion of the experiments the injection sites were histologically verified using cresyl violet stained serial sections.

37.3. Results

Bilateral injections of haloperidol into the most rostral part of the neostriatum, viz. an area delineated by the coordinates A 8620–9650

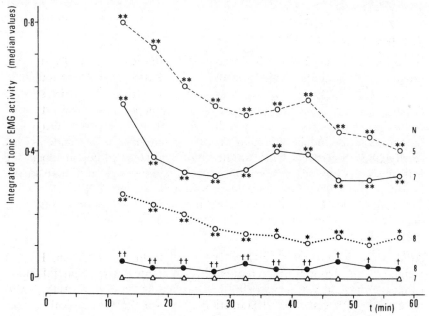

Fig. 37.1. Integrated tonic EMG activity after bilateral injections of haloperidol and/or apomorphine into the most rostral part of the neostriatum. N = number of animals used. △——△: Saline ($N=7$); ○· · · · · · ·○: haloperidol 250 ng ($N=8$); ○——○: haloperidol 500 ng ($N=7$); ○- - - - -○: haloperidol 750 ng ($N=5$); ●——●: haloperidol 500 ng + apomorphine 500 ng ($N=8$). Significances: $*P<0.02$; $**P<0.01$, when compared with saline; $†P<0.01$, $††P<0.002$ when compared with haloperidol 500 ng.

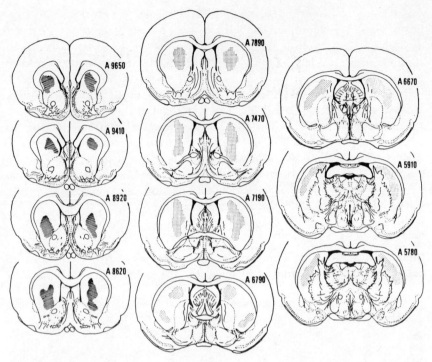

Fig. 37.2.   Schematic representation of the area in which haloperidol injections were effective (hatched area, between A 8620 and 9650) and in which haloperidol injections were ineffective (dotted area, between A 7890 and 5780) in inducing tonic EMG activity. (According to König and Klippel, 1963.)

according to König and Klippel (1963) induced a tonic EMG activity in the GS muscle which was dose dependent (250–750 ng) and statistically different from the effect seen after saline injections (Fig. 37.1). When haloperidol (500 ng) was injected together with apomorphine (500 ng), a significant reduction of the tonic EMG activity was recorded (Fig. 37.1), whereas injections of apomorphine (500 ng) alone did not induce any tonic EMG activity ($N = 7$, data not shown).

In contrast to injections into the most rostral part of the neostriatum, haloperidol (500 ng) failed to induce any tonic EMG activity when it was injected into the neostriatum at sites caudal to A 8620 (Fig. 37.2).

37.4. Discussion

Controversy exists as to whether or not neuroleptic drugs can induce muscular rigidity in rats. In anaesthetised preparations it has been shown that haloperidol can increase the reflex activity of α-motoneurones (Arvidsson *et al.*, 1966). In contrast, a decrease in the reflex activity of the α-motoneurones has also been reported (Langer *et al.*, 1979). Recently haloperidol has been reported to fail to induce muscular rigidity when it was injected into the neostriatum of awake rats in relatively high doses (12 μg; Havemann and Kuschinsky, 1981). However, due to the use of different doses and experimental techniques, these data are hard to compare with each other as well as with data obtained in the present study. Nevertheless the present data provide direct evidence that haloperidol can induce muscular rigidity by blocking neostriatal dopamine receptors.

A second important result is the finding that the neostriatum has been found to consist of two distinct parts with regard to haloperidol-induced tonic EMG activity. The most rostral part between A 8620 and 9650 constitutes an area where haloperidol is very effective in inducing tonic EMG activity, whereas in the more caudal part haloperidol is completely ineffective in this respect. This also provides a possible explanation for the fact that, in the earlier mentioned study, haloperidol has been reported to fail to induce muscular rigidity (Havemann and Kuschinsky, 1981), since in the latter study the injections were placed caudally from the target area for tonic EMG activity delineated in the present study.

Acknowledgements

This study was supported by a grant from the Deutsche Forschungsgemeinschaft (SFB 33 and So 136/2–1). The fellowship of the Max Planck Gesellschaft for B. E., and the skilful technical assistance of Christine Bode, Robert Meseke and Harald Ropte are gratefully acknowledged.

References

Arvidsson, J., Roos, B. -E. and Steg, G. (1966). *Acta Physiol. Scand.* **67**, 398–404.
Havemann, U. and Kuschinsky, K. (1981). *Naunyn-Schmiedebergs Arch. Pharmacol.* **317**, 321–5.
König, J. and Klippel, R. (1963). *The Rat Brain: A Stereotaxic Atlas of the Forebrain and Lower Parts of the Brain Stem.* Williams and Wilkins, Baltimore.
Langer, J., Seeber, U., Kuschinsky, K. and Sontag, K. -H. (1979). *Naunyn-Schmiedeberg's Arch. Pharmacol.* **308**, 149–54.

Use of the 2–deoxyglucose technique to study the neural mechanisms that mediate basal ganglia dyskinesias

I. J. Mitchell, A. Jackson, M. A. Sambrook and A. R. Crossman*

Department of Anatomy, Medical School, University of Manchester, Oxford Road, Manchester M13 9PT, UK

38.1. Introduction

Disturbances of basal ganglia function result in disorders of movement as in Parkinson's disease, chorea and ballism. Despite this well established link between the basal ganglia and the control of movement, relatively little is known about the neural mechanisms responsible for mediating dyskinesias. To study experimentally the activity of abnormal function of the basal ganglia *in vivo* is particularly problematic. Such a study would require the use of an animal model of a basal ganglia disorder and the simultaneous monitoring of the activity of all the basal ganglia and associated nuclei. The development of primate models of ballism (Crossman *et al.*, 1984) and chorea (Jackson and Crossman, 1984) has provided suitable models with which to study these mechanisms, though the problem of recording from large numbers of cells in different nuclei of a behaving and unrestrained primate makes traditional electrophysiological recording methods unsuitable for this task. An alternative approach has recently been made possible by the development of the 2–deoxyglucose (2–DG) technique by Sokoloff (Sokoloff *et al.*, 1977) in which the regional distribution of radiolabelled 2–DG, a glucose analogue, is visualised autoradiographically. The differential labelling between neuronal regions gives an indication of local cerebral glucose utilisation and by implication of neuronal electrical activity. This chapter describes an experiment in which the 2–DG technique was used to investigate local cerebral metabolic activity in primate models of chorea and ballism.

38.2. Methods

Hemiballismus was induced in two adult macaque monkeys by the injection of a γ-aminobutyric acid antagonist into the subthalamic nucleus. The

*To whom correspondence should be addressed.

procedure involved injecting 30 μg of bicuculline methiodide (dissolved in 2 μl of sterile saline) unilaterally into the subthalamic nucleus of an awake unrestrained animal using a permanently embedded cannula and a remotely operated pump. Hemichorea was induced in two other monkeys by similar injections of bicuculline into the lateral portion of the lateral segment of the globus pallidus. A normal unoperated animal served as a control. [^{3}H]–2–DG (2–3 mCi kg^{-1}, Amersham Radiochemicals, specific activity 30–60 Ci mMol^{-1}) was administered intravenously and the animals were killed by barbiturate overdose 45 min later. The brains were frozen *in situ* in isopentane at −50°C, the tissue was blocked stereotaxically and the brain was exposed by careful drilling. Cryostat-cut sections (20 μm) were thaw-mounted on to gelatin-coated slides, freeze-dried and exposed to tritium-sensitive film (LKB Ultrofilm) for approximately 23 days at −25°C. The films were developed using Kodak D19 developer and the autoradiographs were examined densitometrically for side-to-side asymmetries in 2–DG uptake. The cryostat sections were stained with cresyl violet for histological examination.

38.3. Results

The STN microinjections which were located in the caudal third of the nucleus resulted in a dyskinesia of the contralateral limbs. The nature of the dyskinesia was typical of that seen in human hemiballismus in which the axial and proximal appendicular musculature are affected such that the limbs are involuntarily thrown in a continuous and violent manner. The microinjections made in the corpus striatum were located on the ventrolateral border of the lateral pallidal segment at the level of the massa intermedia. These injections resulted in a dyskinesia of the contralateral limbs, predominantly affecting the musculature of the wrist, ankle and digits. The dyskinesia possessed complex choreic and athetoid components.

No side-to-side differences in optical density were seen in the autoradiographs from the control animal. In contrast, the autoradiographs from the hemiballistic animals showed a decrease in 2–DG uptake in the injected STN and in both pallidal segments on the injected side (see Fig. 38.1). The decrease in glucose metabolism in the pallidum, which receives the major output of the subthalamic nucleus (Carpenter and Sutin, 1983), is interpreted as showing a decrease in the synaptic activity in the dendritic–terminal complex of the subthalamo-pallidal pathway. The autoradiographs from the choreic animals showed an increase in optical density in the dorsolateral portion of the subthalamic nucleus ipsilateral to the injection (which receives an input from the injected region of the lateral pallidal segment), but a decrease in optical density in the ipsilateral medial pallidal segment.

38.4. Conclusions

Thus it would appear that in experimental hemiballismus both the

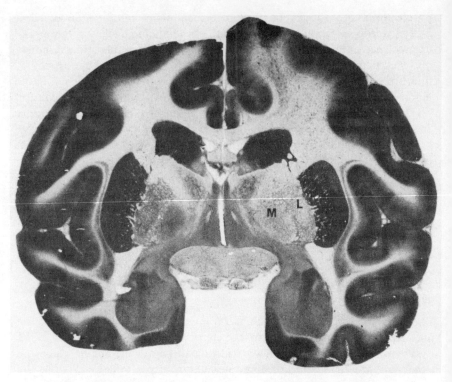

38.1.   A direct reversal print of a [³H]-2-DG autoradiograph from a hemiballistic animal. The medial pallidal segment (M) and the lateral pallidal segment (L) ipsilateral to the injection are paler than the contralateral nuclei. This decrease in labelling is interpreted as being due to decreased activity in the subthalamo-pallidal pathway.

subthalamic nucleus and the subthalamo-pallidal pathway are underactive and in hemichorea the pallido-subthalamic pathway is overactive whereas the subthalamo-pallidal pathway is underactive. These results demonstrate the power of the 2–DG technique to map the activity of the neural pathways which mediate movement disorders due to basal ganglia dysfunction. It is also interesting to note that hemiballismus and chorea, two disease states which in man have sometimes been considered to have separate and distinct neuropathologies and clinical characteristics, may share a common neural mechanism.

Acknowledgements

This work was supported by the Medical Research Council. We thank Mr D. Peggs and Mr D. Brandon for their invaluable technical assistance.

References

Carpenter, M. B. and Sutin, J. (1983). *Human Neuroanatomy*, 8th Edn, Williams and Wilkins, Baltimore.

Crossman, A. R., Sambrook, M. A. and Jackson, A. (1984). *Brain* **107**, 579–96.

Jackson, A. and Crossman, A. R. (1984). *Neurosci. Lett.* **46**, 41–5.

Sokoloff, L., Reivich, H., Kennedy, L., Des Rosiers, M. B. and Patlak, G. B. (1977). *J. Neurochem.* **28**, 897–916.

39

Unilateral infusion of 1-methyl-4-phenyl-1,2,3,6-tetrahydropyridine into the rat substantia nigra: behavioural and biochemical consequences

A. J. Bradbury[1], B. Costall[1], A. M. Domeney[1], P. Jenner[2], C. D. Marsden[2], and R. J. Naylor[1]

[1] Postgraduate School of Studies in Pharmacology, University of Bradford, Bradford BD7 1DP
[2] University Department of Neurology and Parkinson's Disease Society Research Centre, Institute of Psychiatry and King's College Hospital Medical School, Denmark Hill, London SE5 8AF, UK

39.1. Introduction

1-Methyl-4-phenyl-1,2,3,6-tetrahydropyridine (MPTP) administered peripherally to primates can cause Parkinson-like symptoms which have been associated with a neurotoxic action on the nigrostriatal projection (see Langston et al., 1984). It has been difficult to reproduce such changes in the rat (Chiueh et al., 1983; Boyce et al., 1984). The present studies utilise a unilateral infusion of MPTP into the rat substantia nigra (SN) to show that this species is sensitive to its neurotoxic action.

39.2. Methods

Male Sprague–Dawley rats with stereotaxically implanted cannulae were subjected to unilateral infusion of MPTP or vehicle into the SN effected by Alzet osmotic minipumps ($0.48 \ \mu l \ h^{-1}$ for 13 days). Spontaneous and apomorphine-induced ($0.25 \ mg \ kg^{-1}$ s.c.) asymmetries and circling behaviour were assessed daily during infusion and for 60 days post-infusion.

Spontaneous asymmetry was assessed on handling and was scored: 1, consistent bending of the body in one direction only when released from handling, with movement in wide circles; 2, twisting of body detectable both on handling and when released, with movement in wide circles; 3, marked twisting of the body both on handling and when released, with immediate circling within a small radius.

Asymmetry following apomorphine challenge was assessed in an open field and scored: 1, a distinct tendency to move in one direction only, with some inflexion of ipsilateral front limb, movements always in wide circles; 2, movements in one direction only with clear bending of the body, inflexion of both ipsilateral limbs, extension of contralateral limbs, twisting of body not sufficient to impede movement; 3, intense twisting of the body in one direction, marked inflexion of ipsilateral limbs and extension of contralateral limbs, movements nose to tail, frequently so tight as to cause animal to 'trip' over its tail.

Active circling behaviour following apomorphine challenge was measured in revolutions per minute. An animal was categorised as circling when its movements were in one direction only.

Groups of rats were killed at critical stages of behavioural change for determination of levels of dopamine (DA), noradrenaline (NA), serotonin (5-HT), 3, 4-dihydroxyphenylacetic acid (DOPAC), homovanillic acid (HVA), and 5-hydroxyindoleacetic acid (5-HIAA) in the frontal cortex, suprarhinal cortex, tuberculum olfactorium/nucleus accumbens, striatum and SN, using HPLC with electrochemical detection.

39.3. Results

The unilateral infusion of MPTP (10 μg per 24 h) into the SN caused spontaneous asymmetries, and both asymmetric and circling behaviours were recorded following apomorphine challenge. These effects were apparent within 12 h of starting the infusion and followed a defined time

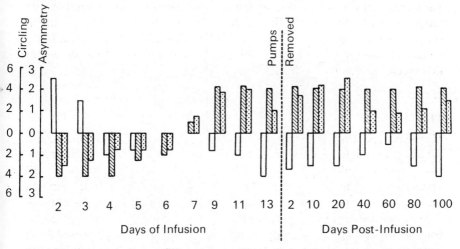

Fig. 39.1. Asymetric body posturing and circling behaviour recorded during unilateral infusion of MPTP, 10 μg per 24 h, into the SN of rat. Unshaded: spontaneous asymmetries; cross-hatched: apomorphine-induced asymmetries; stippled: apomorphine-induced circling. Mean values are given for six rats. SEMs < 12%. All symmetries significant to $P < 0.01 - P < 0.001$ as compared with vehicle-infused controls (Mann–Whitney U-test). All circling significant to $P < 0.01 - P < 0.001$ as compared with vehicle-infused controls (Student's *t*-test).

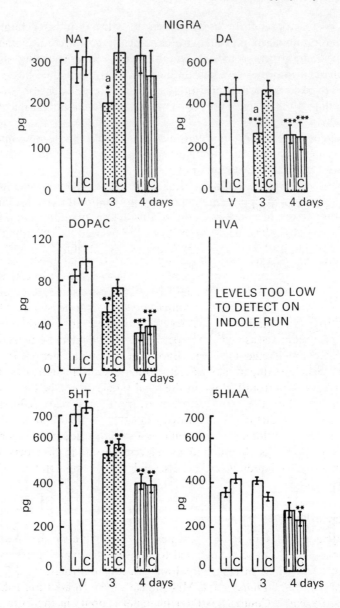

Fig. 39.2.   Biochemical consequences of MPTP infusions into the SN (10 μg per 24 h for 3 or 4 days). Levels of noradrenaline (NA), dopamine (DA), dihydroxyphenylacetic acid (DOPAC), serotonin (5-HT), and 5-hydroxyindolacetic acid (5-HIAA) are shown in the ipsilateral (I) and contralateral (C) SN (in pg mg^{-1} wet weight). v, tissue from vehicle-treated animals. $n = 5$–6. SEMs given. *$P < 0.05$, **$P < 0.01$, ***$P < 0.001$ (Mann–Whitney U-test) for differences between levels in tissue from vehicle- and MPTP-treated animals ('a' represents a significant difference between the two hemispheres).

course. Thus, up to day 3 of infusion, rats exhibited spontaneous ipsilateral asymmetry which was converted to a contralateral response, associated with circling, by apomorphine. On day 4 of infusion the direction of spontaneous asymmetry changed to contralateral, and the apomorphine response changed to ipsilateral over a 3-day period such that by day 5 of infusion spontaneous contralateral asymmetries were associated with ipsilateral responding to apomorphine. This situation was maintained for the remainder of the infusion period and for 60 days post-infusion (Fig. 39.1).

The most important biochemical changes detectable on day 3 of infusion were within the substantia nigra: on the side of infusion the levels of dopamine and DOPAC were reduced from 444 ± 42 to 254 ± 61 pg mg^{-1} tissue ($P < 0.001$) and 84 ± 5 to 52 ± 12 pg mg^{-1} tissue ($P < 0.01$) respectively, with no change in the contralateral SN. On day 4 the reductions in dopamine and DOPAC levels (decreased further to 34 ± 11 pg mg^{-1} tissue) on the side of infusion were matched by reductions in the contralateral SN. Significant changes in striatal dopamine and DOPAC content were not detectable until day 4 when modest (27 %) reductions were apparent in both hemispheres (Fig. 39.2).

39.4. Conclusions

The body asymmetry and circling responses following unilateral infusion of MPTP into the SN are complex, probably involving a disruption of nigral dopamine function and, via compensatory inter-hemispheric processes, a disturbance between the two hemispheres.

Acknowledgements

This work was supported by the Wellcome Trust and the SERC.

References

Boyce, S., Kelly, E., Reavill, C., Jenner, P. and Marsden, C. D. (1984). *Biochem. Pharmacol.* **33**, 1747–52.
Chiueh, C. C., Markey, S. P., Burns, R. S., Johannessen, J., Jacobowitz, D. M. and Kopin, I. J. (1983). *Pharmacologist* **25**, 131.
Langston, J. W., Forno, L. S., Rebert, C. S. and Irwin, I. (1984). *Brain Res.* **292**, 390.

Part V

Further roles for dopamine

40

Dopamine in the kidney

B. C. Williams

*University Department of Medicine, Western General Hospital,
Edinburgh EH4 2XU, UK*

40.1. Introduction

Dopamine plays a fundamental and therefore crucial role as a neurotransmitter in the brain. In recent years it has become apparent that dopamine could also have important physiological functions outside the central nervous system in a number of different tissues. In the kidney, dopamine has been shown to exert three actions:

(a) It causes vasodilatation of the renal vasculature resulting in an increase in renal blood flow and a decrease in renal vascular resistance.
(b) It acts on tubular sodium transport as a natriuretic agent to increase sodium excretion.
(c) It acts on the juxtaglomerular cells of the renal cortex to stimulate the release of renin.

Although the mechanisms responsible for these effects have not been fully characterised, much of the evidence suggests that they could all involve the activation of specific dopaminergic receptors. The concentrations of dopamine required to elicit these responses are higher than those normally found in the circulation. However, since it has now been established that there are intrarenal sources of dopamine, it is possible that all three effects could be of physiological as well as pharmacological significance.

The purpose of this chapter is to review our current understanding of the role played by dopamine in the kidney. Particular attention is given to its effect on renin release, the functional significance of which is yet to be fully realised.

40.2. Sources of intrarenal dopamine

40.2.1. *Renal dopaminergic nerves*

The first indirect evidence for a dopaminergic innervation in the kidney was reported by Bell and Lang (1973) who showed that stimulation of the midbrain and hypothalamus in guanethidine-treated anaesthetised dogs caused dilatation of the renal vasculature which was attenuated by the dopamine antagonist haloperidol. The same workers (Bell *et al.*, 1978) later demonstrated, by fluorescence histochemistry, the presence of numerous catecholamine-containing axons in association with renal cortical arteries and arterioles. The fluorescence was abolished by treatment of the dogs with reserpine but was restored to some axons if reserpine treatment was followed by administration of L-dopa. They concluded that there was a pharmacologically distinct dopamine pool within the kidney consistent with the existence of dopamine-containing nerves. Direct microspectrofluorometric measurements of Falck–Hillarp formaldehyde-induced fluorescence from individual nerve fibres in the canine kidney parenchyma were carried out by Dinerstein and co-workers (Dinerstein *et al.*, 1979) who distinguished between noradrenaline and dopamine-derived fluorescence using a procedure developed by Bjorklund *et al.* (1972). Their results indicated that the neuronal catecholamine at the glomerular vascular pole in any cortical location is predominantly dopamine whereas in the periadventitial layer of the arcuate arteries it is predominantly noradrenaline. Consistent with these findings, several biochemical studies (Bell *et al.*, 1978; Bell and Gillespie, 1981; Dinerstein *et al.*, 1983) have indicated a higher dopamine/noradrenaline ratio in the renal cortex compared with the medulla. Evidence against significant dopaminergic innervation in the rat kidney was recently reported by McGrath *et al.* (1983) who showed a correlation between changes in nerve fluorescence and tissue noradrenaline but not dopamine levels following treatment with 6-hydroxydopamine. In contrast to these observations, other workers (Morgunov and Baines, 1981; Stephenson *et al.*, 1982; Bell, 1983) demonstrated that surgical renal denervation in Sprague–Dawley and Wistar strains of rat reduced the renal content of both noradrenaline and dopamine, suggesting that both neurotransmitters are released from renal nerves. Further support for a dopaminergic innervation in the rat kidney has been obtained from renal nerve stimulation experiments. Chapman *et al.* (1979) demonstrated that renal nerve stimulation caused vasodilatation effects which were reversed by the specific dopamine antagonist sulpiride. More detailed investigations in dogs (Kopp *et al.*, 1983) showed that low-level renal nerve stimulation caused small changes in renal venous outflow of dopamine and increased noradrenaline outflow whereas high-level renal nerve stimulation reduced renal blood flow by 50% and increased renal venous outflow of both noradrenaline and dopamine. Urinary excretion of noradrenaline but not dopamine was increased by low- and high-level renal nerve stimulation. These results suggest that renal nerves release dopamine as well as noradrenaline and also emphasise that catecholamine excretion may not be a good indicator of catecholamine release in the kidney.

40.2.2. *Intrarenal synthesis of dopamine from circulating L-dopa*

Several convincing studies have been reported to indicate that circulating L-dopa is actively decarboxylated within the kidney to free dopamine. The enzyme aromatic L-amino acid decarboxylase, which is involved in the conversion of L-dopa to dopamine appears to be located in both the proximal and distal tubules (Goldstein *et al.*, 1972). Moreover, L-dopa is reabsorbed from the proximal tubule in an active process (Chan, 1976) which is dependent on the sodium concentration (Fox *et al.*, 1964; Ullrich *et al.*, 1974). Micropuncture studies (Baines and Chan, 1980) showed that when tritiated L-dopa was injected into the proximal tubules or peritubular space in both innervated and denervated kidneys, L-dopa was actively converted into urine dopamine. Approximately 30 % of urine-free dopamine appeared to be derived from circulating L-dopa in the rat. Investigations in human subjects (Brown and Allison, 1981) focused on the measurement of arteriovenous differences of L-dopa and dopamine across the kidney, simultaneously with their excretion in urine following oral L-dopa administration. L-dopa was found to be extracted by the kidney to a much higher degree than could be accounted for by its urinary excretion. The results were consistent with the conversion of L-dopa to urine-free dopamine. The decarboxylation of L-dopa could not, however, account for all the dopamine measured in the urine, suggesting that L-dopa was not the only source of urine dopamine. More recent experiments carried out by Ball (Ball *et al.*, 1982) in anaesthetised dogs showed by analysis of arteriovenous differences across the kidney that a fall in the concentration of L-dopa was accompanied by a rise in venous plasma dopamine concentration. The L-dopa extracted by the kidney did not appear in the urine. These observations lend further support to the idea that a significant proportion of intrarenal dopamine originates from circulating L-dopa which is extracted by the kidney.

40.2.3. *Intrarenal synthesis of dopamine from tyrosine*

Although relatively high concentrations of tyrosine can occur in plasma (10 μg ml^{-1}), since denervated kidneys lacked the enzyme tyrosine hydroxylase (Nagatsu *et al.*, 1969) and innervated and denervated rat kidneys excreted similar levels of dopamine, it was originally thought that circulating tyrosine did not make a significant contribution to dopamine production in the kidney (Baines and Chan, 1980). However, recent work seriously questions this conclusion. Administration of tyrosine (200 mg kg^{-1} i.p.) to rats that had been pretreated with 6-hydroxy-dopamine produced an increase in urinary dopamine excretion (Agharanya and Wurtman, 1982). More specific studies (Baines and Drangova, 1984) demonstrated that isolated perfused chemically denervated kidneys increased their excretion of dopamine when the tyrosine concentration in the perfusate was increased from 0·03 to 0·2 mM. These observations suggest that there is a constant conversion of tyrosine to L-dopa and urinary dopamine in the kidney.

40.2.4. *Dopamine production from dopamine conjugates*

The idea that renal dopamine may originate from the adrenal medulla was developed by Kuchel and co-workers (Unger *et al.*, 1980). They suggested that dopamine released from the adrenal medulla is rapidly converted to the sulphate and glucuronide conjugates which circulate to the kidney and are deconjugated by the renal enzymes arylsulphatase and β-D-glucuronidase, respectively, to free dopamine. More recent work (Akpaffiong *et al.*, 1983), however, showed that sodium chloride administration to adrenalectomised rats still caused a marked increase in urinary dopamine excretion. Further arguments which suggest that the adrenal pool of dopamine is of minor importance in determining the intrarenal production of dopamine have been discussed in a comprehensive review by Lee (1982).

40.2.5. *Dopamine production from unknown sources*

In very recent experiments, Baines and Drangova (1984) demonstrated that isolated perfused denervated rat kidneys produced dopamine at the same rate as denervated kidneys *in vivo* (0·7 ng min^{-1} g^{-1}; Stephenson *et al.*, 1982). However, they established that less than 33% of the dopamine produced could have come from intrarenal stores of L-dopa, free dopamine and conjugated dopamine. The rest appeared to be synthesised from precursors other than L-dopa and tyrosine. These rather puzzling observations imply that sources for intrarenal dopamine production may exist which have not yet been identified.

40.3. **Action of dopamine on the renal vasculature**

40.3.1. *Pharmacological characterisation of renal vascular dopamine receptors*

In vivo experiments on anaesthetised dogs established that peripheral dopamine receptors are preferentially located in the intrarenal vascular bed and that activation of these receptors induces vasodilatation which is selectively antagonised by dopaminolytic neuroleptics (Goldberg, 1972; Goldberg *et al.*, 1978). The most convenient model for the pharmacological study of renal vascular receptors would appear to be the isolated perfused kidney. Using this model Woodman and Lang (1980) showed that dopamine (5–100 μg) produced a dose-dependent renal vasodilatation which was inhibited by ergometrine (50 μg) but not by the β-adrenoreceptor antagonist propranolol (0·5 mg), suggesting that the vasodilatation involved the activation of specific dopamine receptors. Other workers (Schmidt and Imbs, 1980) demonstrated that the vasodilatation produced by dopamine (ED_{50} $2 \cdot 5 \times 10^{-6}$ M) was competitively inhibited in order of potency by (+)-butaclamol > haloperidol > sulpiride. On the basis of these results they tentatively classified the renal vascular dopamine receptors as belonging to the D-1 type as originally defined by Kebabian and Calne (1979). Woodman *et al.* (1980) on the other hand demonstrated that the dopamine agonists bromocriptine, SK&F 38393 and apomorphine caused vasodilatation of the renal vasculature and all three agonists were antagonised by metoclopramide. Since bromocriptine and metoclopramide act specifically on D-2 receptors, they argued that the dopamine receptor which mediates renal vasodilatation could not be a D-1 receptor. It should perhaps be emphasised that the dopamine receptor classification introduced by Kebabian and Calne in 1979 was originally based on experiments relating to the central nervous system. Extrapolation of these findings to peripheral vascular dopamine receptors requires a degree of caution (Goldberg *et al.*, 1978). Nevertheless, some workers (Schmidt *et al.*, 1982) still claim, on the basis of the relative potency of dopamine agonists, that the renal vascular receptor is analogous to the dopamine receptor associated with the activation of adenylate cyclase (Kebabian and Calne, 1979), despite the fact that there is as yet no evidence to suggest that dopamine-induced vasodilatation is mediated by cyclic AMP (Brodde, 1982).

40.4. Natriuretic effects of dopamine

40.4.1. *Dopamine and salt intake*

Dopamine infusion in man has been shown to augment sodium excretion, glomerular filtration rate and renal plasma flow (McDonald *et al.*, 1964). The mean plasma dopamine concentration in man has been estimated as $0 \cdot 74$ nmol litre^{-1} (Da Prada and Zürcher, 1976) whereas urinary dopamine levels range from 650 to 2285 nmol per 24 h (Crout, 1963). An increase in dietary sodium chloride intake from 9 mmol to over 200 mmol day^{-1} increased urinary dopamine from 130 μg to 195 μg per 24 h in man (Alexander *et al.*, 1974). Other workers have also reported a positive correlation between increases in salt intake and urinary dopamine excretion (Oates *et al.*, 1979) whereas plasma dopamine concentrations either remained unchanged (Oates *et al.*, 1979) or were decreased (Romoff *et al.*, 1979). One very interesting study in rats (Ball *et al.*, 1978) indicated that dopamine excretion was correlated with chloride intake rather than sodium, which raises the possibility that the natriuretic effect of dopamine could be secondary to a dopamine-induced chloruresis.

40.4.2. *Mechanism of dopamine-induced natriuresis*

Infusion of dopamine in man is associated with an increase in urinary cyclic

AMP which parallels the increase in sodium excretion (Vlachoyannis *et al.*, 1976). Since both β-adrenoreceptor stimulation and cyclic AMP can cause a decrease in sodium reabsorption from the proximal tubules, it could be argued that the natriuretic effect of dopamine may be mediated by β-adrenoreceptor activation. Other studies have shown that dopamine-induced natriuresis can be blocked by specific dopamine receptor antagonists (Wassermann *et al.*, 1980; McGrath and Jablonski, 1981), which points to the involvement of a specific dopaminergic receptor. Evidence for a direct stimulatory effect of dopamine on tubular sodium transport was obtained from micropuncture studies (Greven and Klein, 1977). Several studies have indicated that the dopamine responsible for the natriuretic effect is derived from L-dopa which is decarboxylated in the kidney tubules. Pretreatment of rats with L-dopa leads to a natriuresis whereas carbidopa, a drug that inhibits L-dopa decarboxylase, has an antinatriuretic effect (Ball and Lee, 1977). There are also strong indications, however, to suggest that dopaminergic renal nerves could be involved in dopamine-induced natriuresis (Chapman *et al.*, 1980, 1982). Urinary excretion of dopamine derived from renal nerves was found to be decreased by salt restriction (Morgunov and Baines, 1981).

40.5. Dopamine effects on renin release

40.5.1. *In vivo studies*

Ayers *et al.* (1969) observed that intrarenal infusion of dopamine (12 μg kg^{-1} min^{-1}) in concious dogs with chronic renovascular hypertension caused an increase in plasma renin activity (PRA) and mean arterial pressure (MAP). The increase in PRA was attributed to the vasodilatory effect of dopamine. Similar conclusions were reached in a more recent study (Holdaas *et al.*, 1982) using anaesthetised dogs with renal arterial constriction or ureteral occlusion. In contrast, Chokshi *et al.* (1972) reported that dopamine (3 μg kg^{-1} min^{-1}) infusion in anaesthetised dogs caused an increase in renal blood flow (RBF) and a decrease in renal vascular resistance (RVR) with no change in PRA, suggesting that changes in renin secretion could be independent of the changes in RBF or RVR. Other workers (Imbs *et al.*, 1975) demonstrated that intrarenal infusion of dopamine (6 μg kg^{-1} min^{-1}) in anaesthetised dogs produced increases in renin secretion and RBF. Both these effects were inhibited by haloperidol but neither was affected by propranolol. This suggested that specific dopaminergic receptors were involved in these actions of dopamine. A more detailed study using conscious dogs (Mizoguchi *et al.*, 1983) showed that dopamine in the dose range 0·28–3 μg kg^{-1} min^{-1} caused an increase in renin secretion rate (Fig. 40.1) without altering the MAP. Increases in RBF and decreases in RVR were also observed (Fig. 40.2). Over the same dose range, dopamine induced a natriuresis. The effect of dopamine on PRA

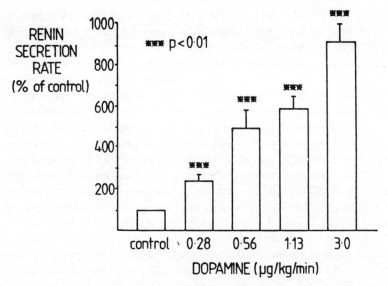

Fig. 40.1.   Effect of dopamine infusion on renin release in dogs. ***$P < 0.01$ (paired *t*-test).

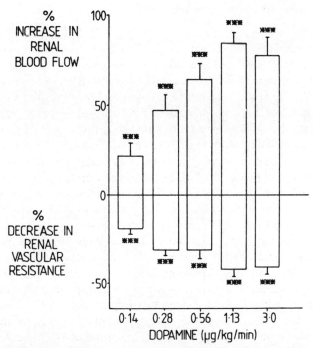

Fig. 40.2.   Effect of dopamine infusion on renal blood flow and renal vascular resistance in dogs. ***$P < 0.01$ (paired *t*-test).

could not be attributed to its vasodilatory action since two other vasodilators, papaverine and acetylcholine, produced a rise in RBF similar to that seen by dopamine with no change in PRA. Intrarenal infusion of sulpiride significantly inhibited the dopamine-induced rise in renin release (Fig. 40.3) and the dopamine-induced vasodilatation. These results suggested that specific receptors for dopamine exist which, when activated, could induce an increase in renin release from the juxtaglomerular cells. In addition to demonstrating that acute intrarenal infusion of dopamine caused renal vasodilatation, natriuresis and increased renin release, Brown et al. (1984) studied the chronic effect of dopamine ($1 \cdot 13 \, \mu g \, kg^{-1} \, min^{-1}$) infusion over 7 days in conscious dogs. In this situation RBF decreased and hypertension developed. It appeared from these studies that the renin-stimulatory actions of dopamine in the chronic situation impair its ability to produce renal vasodilatation, natriuresis and hypotension. It is possible that the continuous elevated levels of intrarenal dopamine lead to high levels of intrarenal renin and hence high levels of angiotensin II which is known to reduce RBF and stimulate sodium reabsorption from the

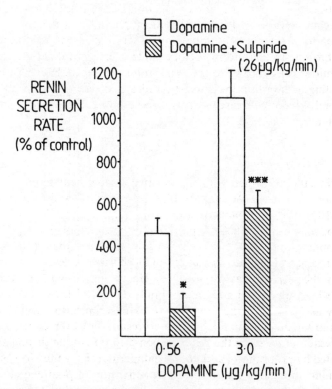

Fig. 40.3.   Effect of sulpiride on dopamine-induced renin release in dogs. ($n = 7$).
*$P < 0 \cdot 05$; ***$P < 0 \cdot 01$ (paired t-test).

proximal tubule (Carey *et al.*, 1980). This effect would then override the direct effect of dopamine in promoting sodium excretion and inducing renal vasodilatation. Relatively few studies on renin release in response to dopamine have been carried out in man. In one study Wilcox *et al.* (1974) compared the renin responses to intravenous infusion of dopamine and noradrenaline in three normal subjects, three patients with autonomic insufficiency, and three patients with Parkinson's disease. PRA increased in all three groups with dopamine but was unchanged or decreased with noradrenaline. The patients with autonomic insufficiency were more sensitive to dopamine than the normal subjects. Wilcox suggested that specific dopamine receptors may be involved since the renin response occurred in association with an increase in both MAP and RBF, and previous studies had suggested that these haemodynamic changes may lead to a decrease in renin release (Hollenberg *et al.*, 1969, 1973). Studies carried out by Sowers *et al.* (1982) showed that oral administration of the dopamine agonist bromocriptine to nine patients with essential hypertension caused a decrease in both PRA and plasma aldosterone. Similarly, in rats (Sowers *et al.*, 1980, 1981) an inhibitory effect of dopamine on PRA was indicated by the observation that the dopamine antagonist metoclopramide caused a significant increase in PRA. It is difficult to interpret these experiments in relation to direct effects of dopamine on renin release from the kidney since both bromocriptine and metoclopramide exert central actions and metoclopramide is also known to act as a serotonin agonist. However, one should be willing to entertain the interesting idea that central actions of dopamine could inhibit renin release whereas direct actions of dopamine in the kidney stimulate renin release.

40.5.2 *In vitro studies*

Experiments performed with the isolated perfused rat kidney preparation (Quesada, *et al.* 1979) showed that relatively high doses of dopamine ($4 \cdot 7 \times 10^{-5}$ M) are capable of stimulating renin release. The renin response to dopamine was significantly inhibited by haloperidol (5×10^{-5} M) but not by propranolol (2×10^{-4} M), which again suggested that specific dopaminergic receptors mediate dopamine-induced renin release. An *in vitro* rat kidney slice system was employed by Henry and co-workers (Henry *et al.*, 1977) who showed that dopamine stimulated renin release at concentrations as low as 10^{-8} M if the monoamine oxidase inhibitor pheniprazine was added to the incubation medium. Their results, however, contradicted those of Quesada *et al.* since the renin response to dopamine in this system was inhibited by propranolol and not by haloperidol. They therefore argued that dopamine stimulated renin release by activation of β-adrenergic receptors. Studies with kidney slices suffer the disadvantage that the macula densa cells may still be capable of influencing the function of the juxtaglomerular cells (Khayat *et al.*, 1981). In order to overcome this problem so that direct effects of catecholamines on renin secretion could be studied, Williams *et al.* (1983) have developed an isolated renal cortical cell superfusion system based on a technique originally used for isolated adrenal cells (McDougall and Williams, 1978). This system allows dynamic studies to be carried out on isolated cell suspensions at controlled flow rates and avoids the problem of metabolism which occurs in static incubation studies. In superfused isolated rat renal cortical cell suspensions, a dose- dependent increase in renin release in response to dopamine was observed (Williams *et al.*, 1983) which was not affected by propranolol but was significantly inhibited by sulpiride (Fig. 40.4). The dopamine agonist bromocriptine was also observed to stimulate renin release at concentrations above 10^{-5} M. These results demonstrated a direct effect of dopamine on renin release at the cellular level by activation of specific dopamine receptors, and support the conclusions derived from

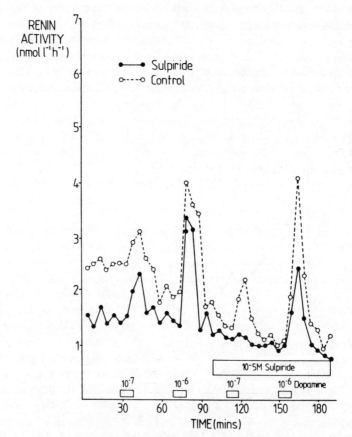

Fig. 40.4.   Effect of sulpiride on dopamine-induced renin, release in superfused rat renal cortical cells. Mean data from two experiments.

the *in vitro* and *in vivo* studies of other workers (Imbs *et al.*, 1975; Quesada *et al.*, 1979; Mizoguchi *et al.*, 1983). Since dopamine is known to exert a natriuretic effect in the kidney, and dietary salt intake can alter the intrarenal levels of dopamine, we have recently studied the effect of dopamine on renin release from isolated renal cortical cells obtained from rats on low, normal and high salt intake (Williams *et al.*, unpublished results). In one series of experiments (Fig. 40.5) both the basal and dopamine-induced renin responses were enhanced in cells obtained from rats on low salt intake compared with rats on normal salt intake. A second series of experiments (Fig. 40.6) demonstrated that the renin response to dopamine was almost abolished in cells prepared from rats on high salt intake. This indicates that dietary salt intake can modify the renin response to dopamine in isolated rat renal cortical cells. The physiological meaning of these results is not yet clear but at this stage it is tempting to put forward the hypothesis that in the low salt state, where the kidney is attempting to retain sodium, the increased sensitivity of the renin response to dopamine could lead to an increase in intrarenal angiotensin II which then increases sodium reabsorption from the proximal tubule. In the high salt state the decreased sensitivity of the renin response to dopamine would lead to a decrease in intrarenal angiotensin II, which would then tend to reinforce the natriuretic effect of dopamine. This would lead me to postulate that the

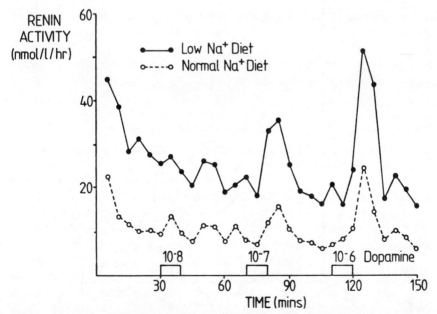

Fig. 40.5. Effect of low salt diet on the renin response to dopamine in superfused rat renal cortical cells. Mean data from two experiments.

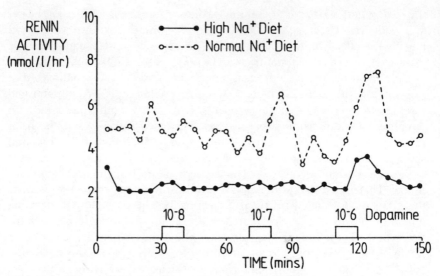

Fig. 40.6.   Effect of high salt diet on the renin response to dopamine in superfused rat renal cortical cells. Mean data from two experiments.

intrarenal effect of dopamine on renin release may be important in the maintenance of sodium homeostasis and normal renal function. The source of intrarenal dopamine responsible for the effects on renin release has not been identified. It is possible that it could originate from dopaminergic nerve fibres which lie in close proximity to the juxtaglomerular apparatus (Dinerstein *et al.*, 1979). It is also conceivable that dopamine generated locally within the tubules could circulate to the juxtaglomerular cells via the macula densa to alter renin release.

40.6.   Clinical applications of dopamine action in the kidney

The effects of intrarenal dopamine on vasodilatation and natriuresis suggest that drugs based on dopamine would be beneficial in the treatment of disorders such as essential hypertension (Perkins *et al.*, 1980), acute renal failure and chronic renal failure (Casson *et al.*, 1983*b*). The problem of synthesising a specific dopamine agonist which is devoid of both central effects and interaction with α- or β- adrenoreceptors has tended to inhibit progress in this area. Nevertheless, recent advances have led to the synthesis of a compound SK&F 82526 which has been shown to increase RBF and to normalise blood pressure when given orally to patients with essential hypertension (Carey *et al.*, 1983). Rather than attempting to synthesise highly selective and specific dopamine agonists, a more promising approach may be the development of dopaminergic prodrugs (Lee, 1982) which are converted within the kidney and act as a local source of dopamine. Work by

Casson *et al.* (1983*a*) demonstrated that the dopamine prodrug γ-glutamyl-dopa reduced significantly the damage caused by glycerol in a rat model for acute renal failure. Another dopamine prodrug, ibopamine, was successful in the treatment of chronic renal failure in man (Stefoni *et al.*, 1981). The same drug, however, did not significantly affect heart rate or blood pressure when given to normal volunteers (Harvey *et al.*, 1984). It is possible that the effect of dopamine on renin release could oppose the effects of DA as a natriuretic and vasodilator agent in some situations. This is an important point to consider when designing drugs based on dopamine for the treatment of hypertension.

40.7. Summary and conclusions

Dopamine occurs in the kidney in association with dopaminergic renal nerves. It can also be synthesised intrarenally from circulating L-dopa and from tyrosine. Other as yet unidentified sources of intrarenal dopamine may exist. Within the kidney, dopamine acts as a vasodilator and natriuretic agent and also stimulates renin release. These three effects are all mediated by specific dopamine receptors. The sources of intrarenal dopamine which are responsible for stimulating the receptors leading to vasodilatation, natriuresis and renin release have not been precisely located. Some studies have indicated involvement of the dopaminergic renal nerves in the natriuretic effect of dopamine whereas other research indicates that dopamine synthesised from L-dopa in the renal tubules could be the source. It is postulated that the vasodilator, natriuretic and renin effects of dopamine may form part of an integrated mechanism for sodium handling in the kidney. The development of specific and selective dopamine agonists and dopamine prodrugs should find useful application in disorders that relate to renal sodium handling such as essential hypertension and acute and chronic renal failure.

Acknowledgements

I thank my colleagues in the Department of Medicine at the Western General Hospital who have contributed to some of the experimental work discussed in this review. In particular I am grateful to Professor C. R. W. Edwards, Mrs Frances Duncan and Mrs Angela Eglen. In addition I thank the Medical Research Council for their financial support in the form of a project grant given for the study of control mechanisms related to renin release.

References

Agharanya, J. C. and Wurtman, R. J. (1982). Studies on the mechanism by which tyrosine raises urinary catecholamines. *Biochem. Pharmacol.* **31**, 3577–80.

Akpaffiong, M. J., Redfern, P. H. and Woodward, B. (1983). An investigation of the importance of the adrenal gland to the action of dopamine in the rat kidney. *Brit. J. Pharmacol.* **79**, 103–8.

Alexander, R. W., Gill, J. R. Jr, Yamabe, H., Lovenberg, W. and Keiser, H. R. (1974). Effect of dietary sodium and of acute saline infusion on the interrelationship between dopamine excretion and adrenergic activity in man. *J. Clin. Invest.* **54**, 194–200.

Ayers, C. R., Harris, R. H. Jr and Lefer, L. G. (1969). Control of renin release in experimental hypertension. *Circ. Res.* **24** and **25** (Suppl. I), 1–103.

Baines, A. D. and Chan, W. (1980). Production of urine-free dopamine from DOPA: a micropuncture study. *Life Sci.* **26**, 253–9.

Baines, A. D. and Drangova, R. (1984). Dopamine production by the isolated perfused rat kidney. *Can. J. Physiol. Pharmacol.* **62**, 272–6.

Ball, S. G. and Lee, M. R. (1977). The effect of carbidopa administration on urinary sodium excretion in man. Is dopamine an intrarenal natriuretic hormone? *Brit. J. Clin. Pharmacol.* **4**, 115–19.

Ball, S. G., Oates, N. S. and Lee, M. R. (1978). Urinary dopamine in man and rat: effects of inorganic salts on dopamine excretion. *Clin. Sci. Mol. Med.* **55**, 167–73.

Ball, S. G., Gunn, I. G. and Douglas, I. H. S. (1982). Renal handling of dopa; dopamine, norepinephrine and epinephrine in the dog. *Am. J. Physiol.* **242**, F56–F62.

Bell, C. (1983). Location of dopamine stores in rat kidney. *Experimentia* **39**, 733–4.

Bell, C. and Gillespie, J. S. (1981). Dopamine and noradrenaline levels in peripheral tissues of several mammalian species. *J. Neurochem.* **36**, 703–6.

Bell, C. and Lang, W. J. (1973). Neural dopaminergic vasodilator control in the kidney. *Nature (Lond.)* **246**, 27–9.

Bell, C., Lang, W. J. and Laska, J. (1978). Dopamine-containing vasomotor nerves in the dog kidney. *J. Neurochem.* **31**, 77–83.

Bjorklund, A., Ehinger, B. and Falck, B. (1972). Analysis of fluorescence excitation peak ratios for the cellular identification of noradrenaline, dopamine, or their mixtures. *J. Histochem. Cytochem.* **20**, 56–64.

Brodde, O. E. (1982). Vascular dopamine receptors: demonstration and characterization by *in vitro* studies. *Life Sci.* **31**, 289–306.

Brown, A. J., Lohmeier, T. E. and Carroll, R. G. (1984). Chronic intrarenal dopamine infusion produces high renin hypertension. *Fed. Proc.* **43**, 1000.

Brown, M. J. and Allison, D. J. (1981). Renal conversion of plasma DOPA to urine dopamine. *Brit. J. Clin. Pharmacol.* **12**, 251–3.

Carey, R. M., Levens, N. R. and Peach, M. J. (1980). Studies of the functional role of the intrarenal renin–angiotensin system. *Prog. Biochem. Pharmacol.* **17**, 6–13.

Carey, R. M., Townsend, L. H., Rose, C. E. Jr, Kaiser, D. L., Linday, C. C. and Ragsdale, N. V. (1983). The specific dopamine agonist, SKF 82526-J increases renal blood flow and lowers blood pressure in essential hypertension. *Clin. Res.* **31**, 487A.

Casson, I. F., Clayden, D. A., Cope, G. F. and Lee, M. R. (1983a). The protective effect of γ-glutamyl L-dopa on the glycerol-treated rat model of acute renal failure. *Clin. Sci.* **65**, 159–64.

Casson, I. F., Lee, M. R., Brownjohn, A. M., Parsons, F. M., Davison, A. M., Will, E. J. and Clayden, A. D. (1983b). Failure of renal dopamine response to salt loading in chronic renal disease. *Brit. Med. J.* **286**, 503–6.

Chan, Y. L. (1976). Cellular mechanisms of renal tubular transport of L-dopa and its derivatives in the rat: microperfusion studies. *J. Pharmacol. Exp. Ther.* **199**, 17–24.

Chapman, B. J., Horn, N. M., Munday, K. A. and Robertson, M. J. (1979). Changes in renal blood flow in the rat during renal nerve stimulation: the effects of α-adrenergic blockers and a dopaminergic blocker. *J. Physiol. (Lond.)* **291**, 64P–65P.

Chapman, B. J., Horn, N. M., Munday, K. A. and Robertson, M. J. (1980). The actions of dopamine and of sulpiride on regional blood flows in the rat kidney. *J. Physiol. (Lond.)* **298**, 437–52.

Chapman, B. J., Horn, N. M. and Robertson, M. J. (1982). Renal blood flow changes during renal nerve stimulation in rats treated with α-adrenergic and dopaminergic blockers. *J. Physiol (Lond.)* **325**, 67–77.

Chokshi, D. S., Yeh, B. K. and Samet, P. (1972). Effect of dopamine and isoproterenol on renin secretion in the dog. *Proc. Soc. Exp. Biol. Med.* **140**, 54–7.

Crout, J. R. (1963). Sampling and analysis of catecholamines and metabolites. *Anesthesiology* **29**, 661–9.

Da Prada, M. and Zürcher, G. (1976). Simultaneous radioenzymatic determination of plasma and tissue adrenaline, noradrenaline and dopamine within the femtomole range. *Life Sci.* **19**, 1161–74.

Dinerstein, R. J., Vannice, J., Henderson, R. C., Roth, L. J., Goldberg, L. I. and Hoffmann, P. C. (1979). Histofluorescence techniques provide evidence for dopamine-containing neuronal elements in canine kidney. *Science* **205**, 497–9.

Dinerstein, R. J., Jones, R. T. and Goldberg, L. I. (1983). Evidence for dopamine-containing renal nerves. *Fed. Proc.* **42**, 3005–8.

Fox, M., Thier, S., Rosenberg, L. and Segal, S. (1964). Ionic requirements for amino acid transport in the rat kidney cortex slice. I. Influence of extracellular ions. *Biochim. Biophys. Acta* **79**, 167–76.

Goldberg, L. I. (1972). Cardiovascular and renal actions of dopamine: potential clinical applications. *Pharmacol. Rev.* **24**, 1–29.

Goldberg, L. I., Volkman, P. H. and Kohli, J. D. (1978). A comparison of the vascular dopamine receptor with other dopamine receptors. *Ann. Rev. Pharmacol. Toxicol.* **18**, 57–79.

Goldstein, M., Fuxe, K. and Hökfelt, T. (1972). Characterization and tissue localization of catecholamine-synthesising enzymes. *Pharmacol. Rev.* **24**, 293–308.

Greven, J. and Klein, H. (1977). Effects of dopamine on whole kidney function and proximal transtubular volume fluxes in the rat. *Naunyn-Schmiedeberg's Arch. Pharmacol.* **296**, 289–92.

Harvey, J. N., Worth, D. P., Brown, J. and Lee, M. R. (1984). Lack of effect of ibopamine, a dopamine prodrug, on renal function in normal subjects. *Brit. J. Clin. Pharmacol.* **17**, 671–7.

Henry, D. P., Aoi, W. and Weinberger, M. H. (1977). The effects of dopamine on renin release *in vitro*. *Endocrinology* **101**, 279–83.

Holdaas, H., Langaard, O., Eide, I. and Kiil, F. (1982). Conditions for enhancement of renin release by isoproterenol, dopamine, and glucagon. *Am. J. Physiol.* **242**, F267–F273.

Hollenberg, N. K., Epstein, M., Basch, R. I., Merrill, J. P. and Hickler, R. B. (1969). Renin secretion in the patient with hypertension. *Circ. Res.* **24** and **25** (Suppl. 1) 113–22.

Hollenberg, N. K., Adams, D. F., Mendell, P., Abrams, H. L. and Merrill, J. P. (1973). Renal vascular responses to dopamine: haemodynamic and angiographic observations in normal man. *Clin. Sci. Mol. Med.* **45**, 733–42.

Imbs, J. L., Schmidt, M. and Schwartz, J. (1975). Effect of dopamine on renin secretion in the anaesthetized dog. *Eur. J. Pharmacol.* **33**, 151–7.

Kebabian, J. W. and Calne, D. B. (1979). Multiple receptors for dopamine. *Nature (Lond.)* **277**, 93–6.

Khayat, A., Gonda, S., Sen, S. and Smeby, R. R. (1981). Responses of juxtaglomerular cell suspensions to various stimuli. *Hypertension* **3**, 157–67.

Kopp, U., Bradley, T. and Hjemdahl, P. (1983). Renal venous outflow and urinary excretion of norepinephrine, epinephrine, and dopamine during graded renal nerve stimulation. *Am. J. Physiol.* **244**, E52–E60.

Lee, M. R. (1982). Dopamine and the kidney. *Clin. Sci.* **62**, 439–48.

McDonald, R. H. Jr, Goldberg, L. I., McNay, J. L. and Tuttle, E. P. Jr (1964). Effects of dopamine in man: augmentation of sodium excretion, glomerular filtration rate and renal plasma flow. *J. Clin. Invest.* **43**, 1116–24.

McDougall, J. G. and Williams, B. C. (1978). An improved method for the superfusion of dispersed rat adrenal cells. *J. Endocrinol.* **78**, 157–8.

McGrath, B. P. and Jablonski, P. (1981). Effects of dopamine on renal function in the isolated perfused rat kidney. *Proc. 8th Int. Congress on Nephrology*, p. 174.

McGrath, B. P., Lim, A. E., Bode, K., Willis, G. L. and Smith, G. C. (1983). Differentiation of noradrenergic and dopaminergic nerves in the rat kidney: evidence against significant dopaminergic innervation. *Clin. Exp. Pharmacol. Physiol.* **10**, 543–53.

Mizoguchi, H., Dzau, V. J., Siwek, L. G. and Barger, A. C. (1983). Effect of intrarenal administration of dopamine on renin release in conscious dogs. *Am. J. Physiol.* **244**, H39–H45.

Morgunov, N. and Baines, A. D. (1981). Renal nerves and catecholamine excretion. *Am. J. Physiol.* **240**, F78–F81.

Nagatsu, T., Rust, L. A. and DeQuattro, V. (1969). The activity of tyrosine hydroxylase and related enzymes of catecholamine biosynthesis and metabolism in dog kidney. *Biochem. Pharmacol.* **18**, 1441–6.

Oates, N. S., Ball, S. G., Perkins, C. M. and Lee, M. R. (1979). Plasma and urine dopamine in man given sodium chloride in the diet. *Clin. Sci. Mol. Med.* **56**, 261–4.

Perkins, C. M., Casson, I. F., Cope, G. F. and Lee, M. R. (1980). Failure of salt to mobilize renal dopamine in essential hypertension. *Lancet* **2**, 1370.

Quesada, T., Garcia Torres, L., Alba, F. and Garcia Del Rio, C. (1979). The effects of dopamine on renin release in the isolated perfused rat kidney. *Experimentia* **35**, 1205.

Romoff, M. S., Keusch, G., Campese, V. M., Wang, M. S., Friedler, R. M., Weidmann, P. and Massry, S. G. (1979). Effect of sodium intake on plasma catecholamines in normal subjects. *J. Clin. Endocrinol. Metab.* **48**, 26–31.

Schmidt, M and Imbs, J. L. (1980). Pharmacological characterization of renal vascular dopamine receptors. *J. Cardiovasc. Pharmacol.* **2**, 595–605.

Schmidt, M., Imbs, J. L., Giesen, E. M. and Schwartz, J. (1982). Vasodilator effects of dopaminomimetics in the perfused rat kidney. *Eur. J. Pharmacol.* **84**, 61–70.

Sowers, J. R., Sollars, E. G., Tuck, M. L. and Asp, N. D. (1980). Dopaminergic modulation of renin activity and aldosterone and prolactin secretion in the spontaneously hypertensive rat. *Proc. Soc. Exp. Biol. Med.* **164**, 598–603.

Sowers, J. R., Barrett, J. D. and Sambhi, M. P. (1981). Dopaminergic modulation of renin release. *Clin. Exp. Hypertension* **3**, 15–25.

Sowers, J. R., Stern, N., Nyby, M. D. and Jasberg, K. A. (1982). Dopaminergic regulation of circadian rhythms of blood pressure, renin and aldosterone in essential hypertension. *Cardiovasc. Res.* **16**, 317–23.

Stefoni, S., Coli, L., Mosconi, G. and Prandini, R. (1981). Ibopamine (SB 7505) in normal subjects and in chronic renal failure: a preliminary report. *Brit. J. Clin. Pharmacol.* **11**, 69–72.

Stephenson, R. K., Sole, M. J. and Baines, A. D. (1982). Neural and extraneural catecholamine production by rat kidneys. *Am. J. Physiol.* **242**, F261–F266.

Ullrich, K. J., Rumrich, G. and Klöss, S. (1974). Sodium dependence of the amino acid transport in the proximal convolution of the rat kidney. *Pflügers Arch.* **351**, 49–60.

Unger, T., Buu, N. T. and Kuchel, O. (1980). Conjugated dopamine: peripheral origin, distribution and response to acute stress in the dog. *Can. J. Physiol. Pharmacol.* **58**, 22–7.

Vlachoyannis, J., Weismuller, G. and Schoeppe, W. (1976). Effects of dopamine on kidney function and on the adenyl cyclase phosphodiesterase system in man. *Eur. J. Clin. Invest.* **6**, 131–7.

Wassermann, K., Huss, R. and Kullmann, R. (1980). Dopamine-induced diuresis in the cat without changes in renal haemodynamics. *Naunyn–Schmiedeberg's Arch. Pharmacol.* **312**, 77–83.

Wilcox, C. S., Aminoff, M. J., Kurtz, A. B. and Slater, J. D. H. (1974). Comparison of the renin response to dopamine and noradrenaline in normal subjects and patients with autonomic insufficiency. *Clin. Sci. Mol. Med.* **46**, 481–8.

Williams, B. C., Duncan, F. M., Drury, P. L., Train, L. M. C. and Edwards, C. R. W. (1983). Dopamine stimulates renin release in isolated rat renal cortical cells by activation of specific dopaminergic receptors. *J. Hypertension* **1** (Suppl. 2), 177–9.

Woodman, O. L. and Lang, W. J. (1980). Dopamine-induced vasodilatation in the isolated perfused rat kidney. *Arch. Int. Pharmacodyn.* **243**, 228–35.

Woodman, O. L., Rechtman, M. P. and Lang, W. J. (1980). A comparison of the responses to some dopamine-receptor agonists and antagonists in the isolated perfused rat kidney. *Arch. Int. Pharmacodyn.* **248**, 203–11.

41

Dopamine super-reactivity in the 'quasi-abstinence' syndrome

F. Sicuteri, M. Fanciullacci, E. Baldi and M. G. Spillantini

Institute of Internal Medicine and Clinical Pharmacology, University of Florence, Viale Morgagni 85, Florence, Italy

41.1. Introduction

As with all other transmitters, endogenous opioids (EOs) act by interacting with their specific receptors. Even though each opioid ligand can bind with more than one subtype of opioid receptor, opioids exhibit a peculiar preference for only one, according to their molecular conformation. By means of binding techniques, neuronal receptor localisation can be determined in either central (brain) or peripheral structures. Clinical, biochemical and electrophysiological changes resulting from a pharmacological stimulation of specific receptors can be used to further indicate the localisation of the endogenous opiate neuronal pathway. This approach is artificial and corresponds only in part to what happens in nature. Opioid receptors are not tonically stimulated by their natural agonists, as large doses of specific antagonists, naloxone and naltrexone do not provoke appreciable changes in either animal or man.

Since the main task of EOs is that of inhibiting various other aminergic or peptidergic neurones, their activity is expected to emerge when these opioid-dependent neurones are persistently stimulated. Thus, in order to specify the role of EOs, it is necessary to consider experimental conditions where the opiate-dependent neurone (the executive neurone) is repetitively stimulated. If opioid-dependent neurones are maintained under a strict inhibition (opiate addiction), an impairment of their function will occur. This impaired function, which is latent during addiction, manifests itself during acute abstinence.

The best information concerning the functional and topographical location of opiate receptors, in particular mu type, is therefore obtained by analysing the phenomena arising during abstinence following opiate withdrawal. The dimension of the clinical phenomena of acute abstinence is the most precise index of a hierarchy of the opioid-dependent structures and their relative functions. The phenomenological succession of symptomatology seen in the abstinence syndrome is helpful in understanding which structures are eventually liberated from opiate inhibition.

If a spontaneous condition clinically comparable to acute abstinence exists, then a pathophysiological analogy with the opiate withdrawal syndrome could be proposed. Migraine is apparently unique in this respect (Sicuteri, 1979). In fact, by comparing a crisis of acute opiate abstinence with that of migraine, it is impossible to specify substantial differences in either quality or succession of the clinical manifestations. One difference is in the distribution and intensity of the pain, which is usually moderate and localised in muscle and bone during opiate abstinence, and which occurs preferentially but not exclusively in the head region in migraine. Extra-pain phenomena (yawning, sleepiness, cold–warm sensations, gooseflesh, nasal and conjunctival congestion, lacrimation, irritability, depression, nausea, vomiting, and at times diarrhoea and sexual excitation) occur in both migraine and abstinence. Therefore, as a working hypothesis, one can assume migraine to be a 'quasi-self endorphin abstinence syndrome'. The clinical phenomena are consistent with a failure of the physiological opioid tonus. The possible consequence of a smaller inhibitory opioid tonus is an exhaustion, for instance, of noradrenergic neurones, resulting in an 'empty neurone' with insufficient secretion of noradrenaline (NA), and compensatory super-reactivity of the effector cell. The theory of the 'empty neurone' may explain more than one analogous or paradoxical responsiveness of migraine patients to drugs which in some manner affect the opioid-dependent neurones (Sicuteri, 1980).

41.2. The 'empty neurone' theory

Circulating EOs have only a slight influence on other neurones. It is generally believed that only neuronally released EOs exert an inhibitory action on subserved neurones. EOs coexist in neurones with other peptidergic or dopaminergic transmitters (i.e. they are 'mixed' neurones) (Hökfelt *et al.*, 1980; Lundberg and Hökfelt, 1983) or exist in a separate neurone which impinges on the executive or host neurone. Neuropeptide transmission exhibits two main characteristics: first, peptides are not subject to neuronal re-uptake, and, secondly, the neurone must be in a condition of continual recharge (Hökfelt *et al.*, 1980). Buffering the effect of repetitive stimuli to the subserved neurones appears to be the main physiological task of EOs. A silent surveyor in resting conditions, the local EOs apparatus is progressively enhanced when the host neurone is persistently stimulated.

What can be expected from a deficient and exhaustible EOs homeostasis? A succession of consequences is theoretically forseeable.

(a) Chronic leakage of the subserved neurone exhausts the reduced stores of transmitter (e.g. NA).

(b) The affected neurone (e.g. a sympathetic neurone) is highly vulnerable to drugs that inhibit transmitter release by acting on presynaptic receptors. Among different presynaptic receptors with inhibitory actions, the D-2 receptor is suitable for clinical investigations.

(c) The neurone with deficient and exhaustible EOs inhibition does not fatigue, and does not benefit from the progressive buffering action of EOs. The effector cell (e.g. smooth muscle) will respond with the same reaction to the repetitive stimuli.

(d) A post-junctional super-reactivity that compensates for the poor transmitter secretion of the effector cell is also possible. It is not clear if this super-reactivity depends on a supersensitivity of the post-synaptic receptor and/or on an increase of the intrinsic energy of the cell.

Here we report and discuss examples of the first two points, as they are pertinent to the subject matter of this book.

41.3. Decreased noradrenaline release from sympathetic neurones

41.3.1. *Postural reflex*

An increased secretion of NA at the level of the neuromuscular junction of the heart (postural tachycardia) and at the level of the arterioles (vasoconstriction) are the final key events for maintaining arterial blood pressure in the standing position. This hypersecretion is detectable in plasma, where NA levels double on standing (Mefford *et al.*, 1981). Syncopal migraine is a common subtype of migraine, characterised by a sudden loss of consciousness (Fig. 41.1). This fainting is due to a precipitous fall of

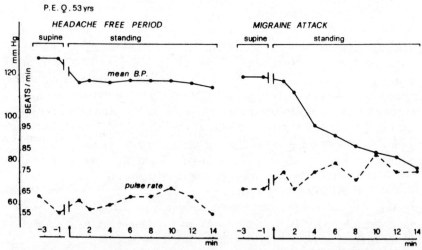

Fig. 41.1. An example of postural hypotension during migraine attack in a syncopal migraine patient. During a migraine attack a marked decrease of mean arterial blood pressure (BP) occurs without a tachycardia (relative bradycardia).

blood pressure without the usual compensatory tachycardia. In studied cases the physiological increase of the plasma NA following assumption of the upright position is either reduced or absent (DeCarolis *et al.*, 1983).

41.3.2. *Iris*

In cluster headache (CH), a spontaneous miosis in the eye ipsilateral to the pain is observable during the attack (Kudrow, 1980). In the majority of patients in free period this miosis is absent. On the other hand, administration bilaterally of conjunctival tyramine (a NA releaser) provokes an enlargement of the pain-side pupil. The pupil ipsilateral to the pain dilates only slightly or not at all. The result is an anisocoria, lasting from 1 to 2 h. This is of diagnostic importance (Fanciullacci *et al.*, 1982).

Evidently, when entering into the iris sympathetic terminal, tyramine cannot release further NA. The lack of mydriasis does not depend on a post-synaptic defect of α-adrenoceptors, as the iris enlarges more than normal when phenylephrine (a pure α-adrenoceptor agonist) is locally administered (Fanciullacci *et al.*, 1982). The empty adrenoneurone in CH is a stable condition, and can be demonstrated several months following the disappearance of clinical manifestations.

Even in common migraine a bilateral reactivity (mydriasis) to conjunctival tyramine, apparently to the same extent as normal, is noted. But NA release must be considered as substantially reduced when one accounts for the exaggerated mydriasis resulting from locally applied phenylephrine, which is capable of unmasking an adrenoreceptor supersensitivity (Fanciullacci, 1979).

41.4. **Vulnerability to agents which inhibit noradrenaline release**

It is generally assumed that migraine sufferers rarely become heroin addicts. In reality the temptation to interrupt a migraine attack with an injection of a narcotic is not unusual. The use of morphine is promptly abandoned following one or two attempts as the opiate offers little or no benefit and produces side-effects such as nausea and vomiting and a feeling of depersonalisation. This observation suggests the distinct possibility of a defect of mu opiate receptors of the antinociceptive system, or alternatively a deficiency of correlated neurones (serotonergic and dopaminergic), which cooperate physiologically with the endorphin system in regulating the autoanalgesia. Morphine, therefore, does not appear to be a suitable drug to explore the vulnerability of the sympathetic neurones in migraine (M) sufferers.

Bromocriptine binds tightly to the presynaptic receptors of sympathetic vascular neurones (Clark *et al.*, 1978). Clinical doses in healthy individuals do not provoke appreciable side-effects, but rather a decrease of plasma NA

(Ziegler *et al.*, 1979). In headache sufferers small oral doses of bromocriptine (2·5 mg) can induce a partial or total paralysis of the postural reflex. This phenomenon is more frequent and dramatic in patients who experience spontaneous fainting during some migraine attacks (Boccuni *et al.*, 1981).

The most impressive phenomenon of the postural reflex is the precipitous fall of the arterial pressure together with the partial or total absence of compensatory tachycardia. This relative bradycardia plays an important role in the mechanism of the fall of the orthostatic arterial pressure. Bromocriptine-induced inhibition of NA secretion is of pathological importance at the level of the myocardium and the arteries in these patients. This event can explain the 'ergotamine paradox' of migraine sufferers, particularly of the syncopal type. It can be defined as a serious collapse following parenteral administration of ergotamine given to interrupt a migraine attack. This effect is paradoxical because ergotamine is a powerful vasoconstrictor. In healthy individuals ergotamine provokes a moderate hypertension when acutely administered (Wolff, 1963). In these individuals ergotamine, as with the other ergot derivatives evidently acts like the dopaminomimetic, bromocriptine, via inhibitory presynaptic dopamine (DA) receptors (Horowski, 1982).

41.5. Apomorphine headache

Apomorphine (APO), a classical DA agonist, when given parenterally in doses of 10–30 μg kg, i.m. to normal individuals provokes a gamut of clinical phenomena which are identical to the non-pain symptoms of migraine. In these individuals APO does not, in fact, induce headache but rather the minimal effects of yawning, somnolence and a vague nausea (Corsini *et al.*, 1979). In M sufferers, lower sublingual doses (100 μg) provoke the entire range of the above-described phenomena in addition to a spontaneous headache (Cangi *et al.*, 1985) (Table 41.1). The acute pre-administration of domperidone prevents either totally or partially the extra-pain phenomenon but not the headache.

Another powerful dopaminomimetic agent, lisuride, in doses of 25 μg i.v., evokes the same phenomenon as APO in M sufferers. Those same doses in healthy individuals have only slight and inconsistent effects without pain (Table 41.1). Daily administration of the same doses of these two dopaminomimetics brings about a gradual tolerance to the side-effects.

An important aspect of this desensitisation, which roughly parallels the clinical improvement of migraine, is the cross-tolerance between the two dopaminometics. The following major inferences can be made:

(a) APO and other dopaminomimetics, when administered in sufficient doses, mimic the clinical phenomenon of both acute opiate abstinence and migraine attack.

(b) M sufferers exhibit a net super-reactivity to dopaminomimetics.

Table 41.1. Apomorphine (100 μg sublingually) and lisuride (25 μg i.v.) induced symptoms in migraine sufferers.

Symptom	Placebo n = 12	Apomorphine n = 24	Placebo n = 10	Lisuride n = 18
Headache	1	19***	1	9*
Drowsiness	0	12**	0	6*
Nausea	1	10*	1	9*
Yawning	0	7*	0	10*
Warmth	1	9	0	3
Coldness	1	6	0	1
Shivering	1	4	0	0
Dizziness	0	4	0	2
Photophobia	0	3	1	2
Hypersalivation	1	4	1	3
Tremor	0	3	0	1

Significance vs placebo (χ^2 test): $*P < 0.05$, $**P < 0.01$, $***P < 0.001$, $n =$ number of patients tested.

(c) An unusual APO-induced phenomenon is headache (exclusive to headache sufferers).

(d) The APO-headache is central in nature, since it escapes the inhibition of domperidone, a drug not capable of crossing the blood–brain barrier, but not that of haloperidol, a drug capable of crossing the blood–brain barrier.

(e) The clinical benefit of chronic treatment with APO or lisuride apparently correlates with the phenomenon of DA desensitisation.

In view of the exact reproduction of spontaneous-pain and extra-pain phenomena, the APO headache undoubtedly represents an important clue to the understanding of the mechanism of headache (Table 41.2). Clearly, the headache itself is evoked by a reduced secretion of DA, which, together with EOs and serotonin (5-HT), is one of the main transmitters of the

Table 41.2. Dopamine super-reactivity in migraine.

Nociception	Apomorphine is a unique drug capable of mimicking (preventable by haloperidol, but not domperidone) the pain and extra-pain phenomena of the spontaneous attack
Postural pressure reflex	Haloperidol and domperidone prevent bromocriptine-induced paralysis (dramatic in syncopal migraine)
Vomiting	Haloperidol and domperidone prevent hyperemesis provoked by all dopaminomimetics
Sexual excitation	Haloperidol and domperidone prevent apomorphine penile hypererection
Smooth muscle	Haloperidol and domperidone prevent DA hypervenospasm

antinociceptive system. Recent investigations (Paalzow and Paalzow, 1983), stress the double effect of APO on pain perception. This dopaminomimetic acts as a hyperalgesic agent in the rat when given in small doses, binding only to presynaptic DA receptors, resulting in the hyposecretion of DA intended as an analgesic transmitter. When APO is administered in high doses, it acts on the postsynaptic receptors, evoking analgesia, after having saturated the presynaptic receptors. Unfortunately, with large doses of APO, humans react by vomiting, a side-effect not seen in rats. This limits the anti-headache therapeutic effect of APO in man.

41.6. Chemical analogies between migraine and opiate abstinence-dependence

41.6.1. Lowering of plasma and cerebrospinal fluid EOs

Decreased levels of EOs have been detected in cerebrospinal fluid (CSF) (Sicuteri et al., 1978; Terenius, 1979; Anselmi et al., 1980; Baldi et al., 1982a; Sicuteri et al., 1982; Genazzani et al., 1984), and in plasma (Facchinetti et al., 1981; Baldi et al., 1982b) of migraine and other pain patients.

Plasma β-endorphin levels have been found to be decreased during both migraine attack and status hemicranicus (Baldi et al., 1982b), where an increase due to the stress from the pain is anticipated (Fig. 41.2). Furthermore, acupuncture, which generally enhances plasma β-endorphin levels in control subjects, is unable to provoke such an increase in M

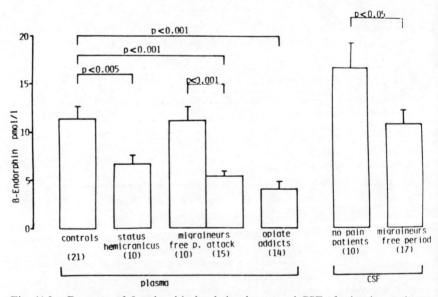

Fig. 41.2.   Decrease of β-endorphin levels in plasma and CSF of migraine patients and in plasma of opiate addicts. The number of patients is given in parentheses. Mean ± SEM.)

sufferers (Facchinetti *et al.*, 1981). In heroin addicts, decreased levels of opioid peptides have been detected in CSF (Clement-Jones *et al.*, 1979) and plasma (Clement-Jones *et al.*, 1979; Ho *et al.*, 1980; Baldi *et al.*, 1984) (Fig. 41.2).

The chronic use of opiates seems to depress the endogenous opioid system. We can therefore infer that, in the first phase of the abstinence, the endogenous opioid system cannot be promptly mobilised after withdrawal of chronic opiate treatment. In this context the biochemical analogies between migraine and abstinence might be the common failure of the EOs.

41.6.2. Enhancement of plasma enkephalinase

The enkephalin-degrading enzyme, enkephalinase, is an indirect index of the turnover of EOs. It has been well characterised in mouse brain (Schwartz *et al.*, 1981). In human plasma and CSF it is detectable and measurable (Spillantini *et al.*, 1985b). In the brain of the mouse rendered opioid dependent, enkephalinase is augmented (Malfroy *et al.*, 1978). In mice, pain-induced stress does not modify brain enkephalinase levels (Spillantini *et al.*, unpublished). In man, various enkephalin inhibitors, such as thiorphan, d-phenylalanine and captopril, reduce the enkephalinase activity in plasma and CSF (Spillantini *et al.*, 1984a).

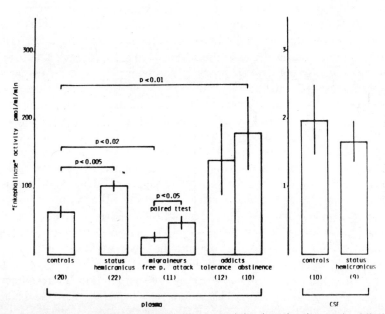

Fig. 41.3. Increase of plasma enkephalinase activity in migraine and addiction. (The number of patients is given in parentheses. Mean ± SEM.)

An increase of enkephalinase activity was found in status hemicranicus when compared with control subjects and in migraine patients during attack periods when compared with levels in pain-free periods (Spillantini *et al.*, 1985*e*). Patients suffering from status hemicranicus do not show any significant modifications of CSF enkephalinase activity when compared with pain-free patients who were considered as controls (Fig. 41.3). In addition, a significant increase of enkephalinase activity was found in plasma of heroin addicts during opiate abstinence syndrome, whereas during the tolerance–abstinence period the increase of plasma enkephalinase activity is not significant (Fig. 41.3). The increase of plasma enkephalinase activity found in both opiate addicts and migraine sufferers may offer an explanation for the decrease of EOs present in these conditions, and furthermore suggests an impairment of the endogenous opioid system.

41.7. Loss of dynamic opioid homeostasis in abstinence and migraine

When the effect of either pharmacological or physiological stimulation in a system is modified by naloxone, this implies that this stimulation is either partially or totally, or directly or indirectly, mediated by EOs, e.g. the human vein spasms *in vivo* when a pharmacological stimulus such as 5-HT is locally applied. This 5-HT spasm depends to a large extent on the release of NA into the synaptic cleft. When the 5-HT stimulus is repeated at regular intervals (10 min), the spasmogenic effect of 5-HT declines. This phenomenon is due to the gradual reduction of NA secretion. It is our opinion that this reduction of NA secretion is not due to the gradual exhaustion of intraneuronal NA, but rather to the progressive enhancement of the local opioid apparatus, which is silent in resting conditions and is subsequently activated by the 5-HT stimulus on the sympathetic neurone resulting in a progressive inhibition of NA release (Sicuteri *et al.*, 1983*a*). Naloxone in the first hypothetical mechanism of the exhaustion could be inactive and in the second case could be operating. Following the final administration of 5-HT, naloxone (which is normally inactive when applied to veins) revives the effect of 5-HT on the fatigued vein. This suggests a dynamic homeostasic role for local EOs on the neuromuscular junctions of the vein.

If one assumes a partial failure of the opioid apparatus in both abstinence and migraine, dynamic homeostasis by EOs can be partially or totally lost. In fact, in both conditions the vein does not show tiredness in response to repetitive inoculation of 5-HT. The loss of 5-HT tachyphylaxis is detectable even during intercritical periods, that is between migraine attacks (Sicuteri *et al.*, 1983*b*). During attacks, there is frequently an inversion of tachyphylaxis that is a progressive increase of the entity of the spasm.

Abstainers also do not exhibit physiological tachyphylaxis. During the

Table 41.3. Morphine abstinence–Migraine correlates.

Clinical phenomena	
psychic	anhedonia, irritation
sensorial	photophonophobia
sensitive	hyperalgesia, 'spontaneous' pains
vegetative	yawning, somnolence, cold–warm sensations, gooseflesh, nausea, vomiting, fainting, occasional sexual excitation
Chemical phenomena	
β-endorphin	decreased in plasma
enkephalinase	increased in plasma

Pharmacological phenomena
non-deviation (post-junctional) multiple (5-HT, NA, DA) *supersensitivity* (iris and vein tests)
loss of opioid homeostasis (absence or inversion of the naloxone reversible 5-HT tachyphylaxis: venotest)

initial phase of abstinence, an inversion, even of impressive entity, is observable. The inversion of tachyphylaxis suggests that opioids not only have an inhibitory effect on NA secretion, but also regulate the intraneuronal mechanism rendering neuronal NA more available. The inversion of tachyphylaxis is compatible with a modulating effect of EOs on the availability of promptly releasable NA (Sicuteri *et al.*, 1983*c*).

In conclusion, we have seen how either a spontaneous (migraine) or induced (addiction) impairment of the EOs can be involved in these diseases (Table 41.3). We have also seen that the unique drug capable of reproducing the entire symptomatology (pain and extra-pain phenomena of the migraine attack) is the pure dopaminomimetic agent, APO. A convincing correlation seems to exist between the tandem EOs–DA in physiological and pathological (migraine-abstinence) conditions. This correlation is substantiated by histoimmunofluorescence data which demonstrate a similarity in the distribution of DA and EOs (particularly enkephalins) in discrete areas of the brain (Hökfelt *et al.*, 1977).

The decrease of β-endorphin and the increase of enkephalinase activity further links the two clinical entities, which seems to be the basis of the impairment of the endogenous opioid apparatus. Thus data are emerging that correlate the basic mechanisms of the two seemingly distinct pathological entities, headache and opiate addiction.

41.8. Summary

The clinical picture of both migraine attack and acute opiate abstinence is remarkably similar, suggesting that in both conditions there may be a functional disorder of neurones subserved by receptors of the endogenous opioid ligands. The aetiology is obviously distinct, but a common mechanism of altered endogenous opioid homeostasis may be proposed as a link between these two diseases. This supposition is supported by similarly altered levels of plasma β-endorphin and plasma enkephalinase.

In addition, in migraine there is also dopaminergic involvement, probably as a consequence of the endogenous opioid impairment. In fact, the only drugs capable of evoking the entire picture of a migraine attack (painful and non-painful), are two dopaminomimetics, apomorphine and lisuride. Low doses of apomorphine, when administered to migraine sufferers, appear to bind only with presynaptic dopamine receptors, resulting in a migraine-like picture (including pain). In migraine sufferers this may be the partial or total consequence of an inhibition of the dopamine pain-inhibiting neurones.

The symptomatology evoked by apomorphine in migraine sufferers is identical to that of acute opiate abstinence. The chronic administration of these two dopaminomimetics induces two parallel results, i.e. a desensitisation of receptors inhibiting dopamine release and clinical benefit.

Acknowledgements

The authors wish to thank Mr G. O'Donnell for revision of the text. This work has been supported by grants from the 'Consiglio Nazionale delle Ricerche' (CNR, Rome), finalised projects: 'Chimica Fine e Secondaria' and 'Medicina Preventiva e Riabilitativa' (subproject 8: 'Controllo del Dolore').

References

Anselmi, B., Baldi, E., Casacci, F. and Salmon, S. (1980). Endogenous opioids in cerebrospinal fluid and blood in idiopathic headache sufferers. Headache 20, 294–9.

Baldi, E., Conti, P., Conti, R., Michelacci, S., Salmon, S., Spillantini, M. G. and Sicuteri, F. (1982a). Livelli liquorali di β-endorfina in pazienti cefalaegici. Giornale di Neuropsicofarmacologia 2, 58.

Baldi, E., Salmon, S., Anselmi, B., Cappelli, G., Brocchi, A., Spillantini, M. G. and Sicuteri, F. (1982b). Intermittent hypoendorphinemia in migraine attack. Cephalalgia 2, 77–81.

Baldi, E., Ferretti, C., Geppetti, P., Pini, L. A. and Spillantini, M. G. (1984). Reduced plasma β-endorphin levels in heroin addicts: lack of correlation between indexes of opiate abuse. Abstract Book: International Congress 'Opioid peptides in periphery', Rome, 23–25 May, p. 65.

Boccuni, M., Fanciullacci, M., Michelacci, S. and Sicuteri, F. (1981). Impairment of postural reflex in migraine: possible role of dopamine receptors. In Apomorphine and Other Dopaminomimetics, vol. 2. G. U. Corsini and G. L. Gessa (eds), pp. 267–73. Raven Press, New York.

Cangi, F., Fanciullacci, M., Pietrini, U., Boccuni, M. and Sicuteri, F. (1985). Emergence of pain and extra pain phenomena from dopaminomimetics in migraine. In Updating Headache, V. Pfaffenrath, P.O. Lundberg and O. Sjaastad (eds), pp. 181–5, Springer, Berlin.

Clark, B. J., Scholtysik, G. and Flückiger, E. (1978). Cardiovascular actions of bromocriptine. *Acta Endocrinol.* **88** (Suppl. 216), 75–81.

Clement-Jones, V., McLoughlin, L., Lowry, P. J., Besser, G. M., Rees, L. H. and Wen, H. L. (1979). Acupuncture in heroin addicts: changes in met-enkephalin and β-endorphin in blood and CSF. *Lancet* **ii** 380–2.

Corsini, G. U., Riccardi, M. P., Bacchetta, A., Bernardi, F. and Del Zompo, M. (1979). Behavioral effects of apomorphine in man: dopamine receptors implications. In *Apomorphine and Other Dopaminomimetics*, vol. 2, G. U. Corsini and G. L. Gessa (eds), pp. 13–24. Raven Press, New York.

De Carolis, P., Merlo Pich, E., Costelli, P., Ligabue, A., Ferrara, R., D'Alessandro, R. and Sacquegna, T. (1983). Risposta cardiocircolatoria e variazioni catecolaminiche al tilt test in pazienti emicranici fuori e dentro attacco. Abstract Book, VII Congr. Nazionale Soc. It. per lo Studio delle Cefalee, Capo Boi, 23–25 Sept., p. 20.

Facchinetti, F., Nappi, G., Savoldi, F. and Genazzani, A. R. (1981). Primary headaches: reduced circulating β-lipotropin and β-endorphin levels with impaired reactivity to acupuncture. *Cephalalgia* **1**, 195–201.

Fanciullacci, M. (1979). Iris adrenergic impairment in idiopathic headache. *Headache* **19**, 8–13.

Fanciullacci, M., Pietrini, U., Gatto, G., Boccuni, M. and Sicuteri, F. (1982). Latent dysautonomic pupillary lateralization in cluster headache. A pupillometric study. *Cephalalgia* **2**, 135–44.

Genazzani, A. R., Nappi, G., Facchinetti, F., Micieli, G., Petraglia, F., Bono, G., Monittola, C., Savoldi, F., (1984). Progressive impairment of CSF β-EP levels in migraine sufferers. *Pain* **18**, 127–33.

Ho, W. K. K., Wen, H. L. and Ling, N. (1980). β-endorphin-like immunoreactivity in the plasma of heroin addicts and normal subjects. *Neuropharmacology* **19**: 117–20.

Hökfelt, T., Elde, R., Johansson, O., Terenius, L. and Stein, L. (1977). The distribution of enkephalin-immunoreactive cell bodies in the rat central nervous system. *Neurosci. Lett.* **5**, 25–31.

Hökfelt, T., Johansson, O., Ljungdal, J. M. and Schultzberg, M. (1980). Peptidergic neurons. *Nature (Lond.)* **284**, 515–21.

Horowski, R. (1982). Some aspects of the dopaminergic action of ergot derivatives and their role in the treatment of migraine. In *Advances in Neurology: Headache*, vol. 33. M. Critchley, A. P. Friedman, S. Gorini and F. Sicuteri (eds), pp. 325–34. Raven Press, New York.

Kudrow, L. (1980). *Cluster Headache, Mechanisms and Management.* Oxford University Press, Oxford.

Lundberg, J. M. and Hökfelt, T. (1983). Coexistence of peptides and classical neurotransmitters. *TINS* August 1983, 325–33.

Malfroy, B., Swerts, J. P., Guyon, A., Roques, B. P. and Schwartz, J. C. (1978). High affinity enkephalin-degrading peptidase in brain is increased after morphine. *Nature (Lond.)* **276**, 523–6.

Mefford, I. M., Ward, M. M., Miles, L., Taylor, B., Chesuey, M. A., Keegan, D. L. and Barchas, J. D. (1981). Determination of plasma catecholamines and free 3, 4-dihydroxyphenylacetic acid in continuously collected human plasma by high performance liquid chromatography with electrochemical detection. *Life Sci.* **28**, 477–83.

Paalzow, G. H. M. and Paalzow, L. K. (1983). Opposing effects of apomorphine on pain in rats. Evaluation of the dose–response curve. *Eur. J. Pharmacol.* **88**, 27–35.

Schwartz, J. C., Malfroy, B. and De La Baume, S. (1981). Biological inactivation of enkephalin and the role of enkephalin-degrading peptidase (enkephalinase) as neuropeptidase. *Life Sci.* **29**, 1718–40.

Sicuteri, F. (1979) Headache as the most common disease of the antinociceptive system: analogies with morphine abstinence. In *Advances in Pain Research and Therapy*, vol. 3. J. J. Bonica, J. C. Liebeskind, and D. Albe-Fessard (eds), pp. 358–65. Raven Press, New York.

Sicuteri, F. (1980). Empty neuron theory as a research perspective of diseases of the nociceptive system. *J. Drug Res.* **5**, 73–8.

Sicuteri, F., Anselmi, B., Curradi, C., Michelacci, S. and Sassi, A. (1978). Morphine-like factors in CSF of headache patients. In *Advances in Biochemistry and Psychopharmacology*, Vol. 18. E. Costa and M. Trabucchi (eds), pp. 363–6. Raven Press, New York.

Sicuteri, F., Fanciullacci, M., Michelacci, S., Boccuni, M., Conti, P., Baldi, E., Spillantini, M. G. and Gatto, G. (1982). Diminution des β-endorphines et avantage de la naloxone dans les migraines opthalmiques: un paradoxe? In *Migraine et Céphalées*, Colloque du Marseille, 15 May 1982, Sandoz, Basel, pp. 87–97.

Sicuteri, F., Del Bianco, P. L., Anselmi, B., Franchi, G. C. and Panconesi, A. (1983a). Naloxone reversal of 5-hydroxytryptamine tachyphylaxis in humans. *Clin. Neuropharmacol.* **6**(4), 305–9.

Sicuteri, F., Panconesi, A., Franchi, G. C., Del Bianco, P. L. and Anselmi, B. (1983b). Impaired 5-hydroxytryptamine tachyphylaxis in migraine. *Cephalalgia* **3**, 139–42.

Sicuteri, F., Panconesi, A., Pietrini, U. and Geppetti, P. (1983c). The inversion of serotonin tachyphylaxis and monoamine hyperreactivity in heroin addicts. In *Pharmacological Basis of Anaesthesiology: Clinical Pharmacology of new Analgesics and Anaesthetics*. M. Tiengo and M. J. Cousins (eds), pp. 237–47, Raven Press, New York.

Spillantini, M G., Fanciullacci, M., Michelacci, S. and Sicuteri, F. (1985a). Effects of inhibitors on enkephalinase activity and ACE activity in CSF and plasma in man. Abstract Book, Collegium Intern. Neuro-Psychopharmacologicum, 14th CINP Congress, Florence, 19–23 June 1984, p. 212.

Spillantini, M. G., Malfroy, B. and Schwartz, J. C. (1984b). Identification of "enkephalinase" activity in cerebrospinal fluid by a two-step fluorimetric method. *Eur. J. Biochem.* Submitted.

Spillantini, M. G., Fanciullacci, M., Michelacci, S. and Sicuteri, F. (1985b). Enkephalinase activity in plasma and CSF in idiopathic headache patients. In *Updating Headache*, V. Pfaffenrath, P. O. Lundberg and O. Sjaastad (eds), pp. 276–80. Springer, Berlin.

Terenius, L. (1979). Endorphins in chronic pain. In *Advances in Pain Research and Therapy*, vol. 3. J. J. Bonica, J. C. Liebeskind and D. Albe-Fessard (eds), pp. 459–71. Raven Press, New York.

Wolff, H. G. (1963). *Headache and Other Pain* (2nd edn), Oxford University Press, New York.

Ziegler, M. G., Lake, C. K., Williams, A. C., Teychenne, P. F. Shoulson, I. and Stainsland, O. (1979). Bromocriptine inhibits norepinephrine release. *Clin. Pharmacol. Ther.* **25**, 137–42.

Dopamine as a deuteragonist in migraine: implications for clinical pharmacology of dopamine agonists

R. Horowski, H. Wachtel and R. Dorow

Research Laboratories, Schering AG,
Berlin (West)/Bergkamen, Federal Republic of Germany

42.1. Introduction

Headache is one of the most common complaints of patients who go to see a doctor. It is estimated that 5–10% of the population apply for medical aid for this condition. Since a headache attack not only disables those affected directly, but also interferes often dramatically with their work and their social environment, its treatment should be of great importance. However, headache attacks are not linked to some organ pathology and they do not cause irreversible damage; therefore most people, including doctors, still feel that the causes of headache are of psychogenic origin. This view is very old: indeed, Plato quotes Socrates as saying that a headache remedy (leaves obtained from Thracian doctors) should only be used in those who previously had opened their soul to 'good talks', which can be considered as a form of psychological intervention (Plato, *Charmides*).

That this type of intervention, at least in some cases, is not sufficient is demonstrated quite clearly by the case of Sigmund Freud, who opened important new areas to psychotherapy of several very disabling conditions but could not obtain relief for his own severe and often disabling headache. Thus, a consensus should be reached among psychologists and doctors that (a) psychological factors can trigger a migraine attack and therefore need to be identified and modified; (b) important steps in the pathogenesis of a migraine attack have more or less well defined biological substrates. Because the real cause of migraine remains unknown, a growing number of hypotheses have appeared and can be expected to appear in the future. Some of these at least are not mutually exclusive and might well be combined into a sequence of pathophysiological events associated with a migraine attack (see Table 42.1).

As can be seen, a central origin of an attack associated with visual symptoms, aura and nausea is quite compatible with a subsequent change in

Table 42.1. Pathophysiological events in migraine: a summary of several hypotheses.

	Contributing or triggering factor	Mechanism involved	Hypothesis by:	Symptoms caused
I. Aetiology	Genetic factors	Unknown	—	—
II. Central between attacks	Stress	Enhanced sensitivity of monoaminergic systems? Altered monoamine-endorphin balance.	Sicuteri et al., 1979	Enhanced sensitivity towards physiological and pharmacological stimuli.
During an attack	Stress, climatic factors, reserpine, tyramine	Sudden endorphin withdrawal? Hyperactivity of N. raphe dorsalis 5-HT neurones.	Sicuteri et al., 1979 Raskin and Appenzeller, 1980	Aura, visual symptomatology (scotoma, fortification spectra), nausea, after attack sedation
		Spreading, depressing	Richards, 1971	
III. Vascular events	Phenylethylamine-containing food	Vasoconstriction followed by vasodilatation.	Wolff, 1948	Paleness, unilateral pain, pulsating blood vessels, tissue hypoxia.
		Opening of arteriovenous anastomoses	Heyck, 1975	
IV. Humoral events	Tyramine? Hormonal factors?	5-HT release from platelets; low 5-HT blood levels; enhanced blood levels; enhanced capillary permeability and extravasation of pain-producing compounds (kinins and prostaglandins). Thrombocyte dysfunction	Lance, 1973	Oedema; dull, throbbing headache.

the tone of carotid vessels and other vascular changes which again precede humoral events; and a variety of trigger factors operating at different levels may contribute to the genesis of an attack.

Although the extreme sensitivity of migraine sufferers to external stimuli such as light, noise or drugs during an attack is well known, interest now focuses on the altered reactivity of these patients that also occurs in the intervals between them (see Table 42.2). Avoidance of these stimuli may be sufficient to achieve a kind of rehabilitation of migraine patients who are often just described as too sensitive and too weak to withstand the stresses and strains of life. Whether a 'migraine personality' really exists is still a matter of debate. However, clinicians are well aware of migraine patients with a rather typical so-called neurotic, anarchistic, pedantic and oversensitive psychic structure, and it remains to be seen whether this is not just another consequence of higher biological reactivity. It is quite understandable that people who after drinking a glass of red wine or after eating chocolate (both tyramine- and phenylethylamine-containing foods), have to pay with a subsequent day of severe pain and disability, develop personality changes which are considered as social rigidity and loss of spontaneity, and that they try to organise their lives so that any exposure to strong stimuli is avoided.

Before any kind of treatment of migraine is considered, a careful exploration of the patient and a thorough differential diagnosis is necessary. Other organic or functional causes of headache need to be identified and treated appropriately, and even if the diagnosis of migraine is confirmed, an analysis of precipitating factors may be extremely helpful for the patient.

Migraine in women is more frequent and often associated with rapidly changing hormonal levels, and this condition is rare during pregnancy and after the menopause (when steroid hormone levels are more stable). There is a considerable overlap between cycle disorders, e.g. so-called premenstrual

Table 42.2. Enhanced sensitivity of migraine patients

Stimulus	Symptom induced	Reference
Pharmacological		
Bromocriptine	Hypotension	Sicuteri, 1986
Apomorphine	Emesis, headache	Sicuteri, 1986
Lisuride	Headache	Sicuteri, 1986
Sulpiride	Hyperprolactinaemia	Horowski, 1983
	Enhanced cerebral blood flow and side-effects of piribedil	Bès *et al.*, 1982
Physiological	Perception of higher stimulus	Klein, 1983
	Longer latency of P_2; higher amplitudes in visual evoked potentials	Kennard *et al.*, 1978

Table 42.3. Clinical results obtained with lisuride

Authors	Number of patients	Diagnosis	Dosage (mg) and duration of treatment	Design	Results	Remarks
Podvalová and Dlabač, 1972	170	Migraneous cephalalgia	2–3 × 0·025; 1 or 2–several months	open	77 % efficacy (total disappearance or reduction of frequency and intensity), no severe side-effects	Summary report of studies involving 8 centres in Czechoslovakia
Ripka, 1972	229	Headache from various causes, among them migraine	2–3 × 0·025 (3 × 0·050); 4–23 months	open	Best effects in patients with migraine and hypertension headache; laboratory data from the hypertensive group ($n = 97$) reveal no disturbance	
Herrmann et al., 1977	130 (+123 on methysergide)	Severe migraine ≥ 2 attacks per per month	3 × 0·025 (vs 3 × 2 mg methysergide); 3 months	double-blind randomised vs methysergide	Same efficacy, but significantly less drop-outs initially and over 3 months than with methysergide (2 vs 20 % and 17 vs 39 %)	Multicentre study
Herrmann et al., 1978	117 (+123 on placebo)	Severe migraine	3 × 0·025; 6 months	double-blind randomised vs placebo	Significantly more effective than placebo, well tolerated (2 drop-outs vs 0 on placebo)	Multicentre study in Czechoslovakia

Somerville and Hermann, 1978	66 (+66 on placebo)	Migraine	3 × 0·025; 3 months	double-blind randomised vs placebo	Superior to placebo, significantly less drop-out due to lack of efficacy	Study in Australia high placebo response rate
Tessitore and Battaglia, 1982	20	Migraine	3 × 0·025; 12 months	open	Long-term improvement in 15 patients	
Del Bene et al., 1983	23 (+22 on pizotifene)	Childhood migraine (age 8–14 years)	1st week 0·0125, 2nd week 0·0250, 3rd week 0·0500; 42 days	open controlled	Good improvement in 12 cases (vs 10 on pizotifen), fair improvement in 5 vs 10, one drop-out due to side effects on lisuride, no significant differences between groups	
Franchi and Mallucci, 1983	100	Essential headache (12 chronic headache)	3 × 0·025; 3–6 months	open	72% of patients with improvement, 5 drop-outs due to side-effects	
Rafaelli et al., 1983	10	Cluster headache	2 × 0·025– 16 × 0·025	open	Good improvement at higher doses, no drop-outs	Basic medication with ergotamine, methysergide or lithium

(Cuvalit®) in the prevention of headache.

syndrome, and premenstrual migraine attacks, and even the type of oral contraceptive taken may warrant special consideration. For other patients, dietary factors or stress situations are of importance, and the climate also has a significant but still ill defined role in precipitating an attack.

42.2. Therapeutic approaches in migraine

If medical therapy becomes necessary, weak analgesics, mostly acetyl salicylic acid (ASA), should first be tried because many patients get relief from this single medication. If gastric absorption is slow, or if there is nausea associated with the attack, metoclopramide or domperidone may be a helpful addition. Because of the risk of blood or kidney disease, pyrazol- or phenacetin-containing drugs need to be avoided, as well as barbiturates with their risk of dependence. If this type of treatment fails, ergotamine combined with caffeine becomes the treatment of choice. It is most effective when given during the prodromal stage, but due to its variable absorption and bioavailability and its small therapeutic range, each patient must be titrated very carefully (Ala-Hurula, 1982; Eadie, 1983). If this treatment alone is not sufficient, it can be combined with sedatives and the patients will stay in bed until the attack is over. Ergotamine-containing drugs should not be given frequently because, with overdosage, not only nausea, emesis, collapse and headache itself, but also ergotism, i.e. peripheral vaso-constriction, may appear. On chronic ergotamine use, withdrawal headache may complicate the therapy further; therefore, in cases of frequent and severe attacks, prophylactic treatment is useful.

For migraine prophylaxis, β-blocking agents such as propranolol as well as calcium antagonists have shown to be effective (Peroutka, 1983). More efficient, however, and maybe more specific, are drugs interfering with serotonergic function such as pizotifen, methergoline, methysergide and lisuride (*Lancet* Editorial, 1982; Horowski, 1982).

Thus, as in the treatment of acute attacks, ergot derivatives are still the most effective and specific treatment. Even in the first paper on the use of standardised extracts of ergot in the symptomatic treatment of migraine by Eulenburg 100 years ago, it was reported that these compounds were more potent on parenteral administration and that side-effects such as nausea, emesis and headache occurred on overdose. The poor bioavailability of ergot derivatives is still a major problem with ergotamine, and side-effects prevented methysergide, the most effective prophylactic agent (introduced in therapy by Sicuteri in 1959), from being used on a large scale.

Assuming serotonin to be a protagonist in the pathophysiology of migraine, other drugs that interfere with this transmitter have been developed for the prevention of migraine attacks, among them pizotifen— which is structurally related to tricyclic antidepressants—and lisuride, another ergot derivative with outstanding affinity for serotonin and dopamine receptors (McDonald and Horowski, 1983). Lisuride has been shown in double-blind controlled studies to be more effective than placebo, and at least as effective but much better tolerated than methysergide in the prevention of attacks at extremely low dosage (3×0.025 mg day^{-1} per oral). A review of clinical studies is given in Table 42.3. At higher dosages, it is also a very potent dopamine agonist with prolactin-lowering and antiparkinsonian properties. In spite of many possible effects of lisuride on central monoamine function (see Table 42.4), its mechanisms of action in migraine are still unknown.

Classically, the antimigraine effect of lisuride is attributed to its high-affinity interaction with 5-HT receptors both in the CNS and in peripheral tissues. This interpretation is in line with the assumed role of 5-HT as the protagonist in the pathophysiology of migraine. Lisuride as a serotonin partial agonist can inhibit or substitute for 5-HT effects, depending on the neuronal situation, or even potentiate it. An 'antihallucinogenic' effect of lisuride has recently been proposed to prevent the visual aura of migraine attacks which could also be related to 5-HT receptors (Kranda and Kulikowski, 1984). Since, however, other 'protagonists' such as endorphins, substance P or other peptides emerge, we want to discuss in the following paragraphs the role of dopamine as the first 'deuteragonist' in migraine and other possible actions of lisuride related to its dopaminergic activity.

In our clinical pharmacology studies with single oral or intravenous doses

Table 42.4. Possible effects of lisuride related to the pathophysiological events of a migraine attack (for discussion see Horowski, 1982).

Effect	Reference
1. Central inhibition of N. raphe dorsalis firing rate by a stimulation of presynaptic 5-HT receptors	Raskin and Appenzeller, 1980
2. Change in monoamine receptor sensitivity of neurones interacting with endorphinergic systems modulating pain	Sicuteri *et al.*, 1979
3. Peripheral vascular effects mediated by 5-HT or DA-receptors e.g. prevention of opening of arteriovenous anastomoses or carotid vasodilation by substituting for 5-HT, or by effects on 5-HT or DA vascular receptors including 5-HT potentiation	Del Bianco *et al.*, 1982
4. Humoral effects such as inhibition of 5-HT-induced enhanced permeability of the capillaries, extravasation, edema and pain, and inhibition of 5-HT or adrenaline-induced thrombocyte aggregation and deformation	Podvalovà and Dlabač, 1972, Glusa *et al.*, 1979
5. Hormonal effects such as an inhibition of prolactin secretion, e.g. in premenstrual migraine	

of lisuride and many other dopamine agonists, including the classical agonist apomorphine, we have observed a typical spectrum of side-effects after overdosage, which includes nausea, emesis, orthostatic hypotension, drowsiness and sedation, and, in our healthy volunteers, sometimes long-lasting episodes of headache. With the exception of this last symptom, administration of peripheral dopamine such as sulpiride, metoclopramide and domperidone was able to prevent or treat these symptoms quite effectively, and we therefore can conclude that they are mediated by dopamine receptors outside the blood–brain barrier, such as the chemoreceptor trigger zone or other parts of the medulla oblongata and the hypothalamus. Another rare side-effect we observed was constipation, which could be related to dopamine systems involved in the control of gastrointestinal motility. These symptoms represent parts of the migraine symptomatology. Indeed, years before our observations, Sicuteri and his co-workers have reported that, in patients suffering from migraine, dopamine agonists were able to produce consistently parts or the whole spectrum of migraine symptomatology at dose levels at which side-effects are quite rare in healthy subjects (see Chapter 41, this volume). Since these observations were most pronounced after sublingual or parenteral application of the pure dopamine agonist apomorphine, they cannot be due to an altered gastrointestinal milieu, leading to increased absorption, but seem to relate to more central dopaminergic systems. Recently, the same group has reported that all symptoms elicited by sublingual apomorphine could be antagonised by the peripheral dopamine antagonists tiapride or domperidone with the exception of the headache itself. This symptom, which in migraine sufferers is reported to resemble sometimes very closely an individual attack, but which in our observations in non-migrainous volunteers consists of a uniform, bilateral, throbbing head pain, sometimes associated with lightheadedness, was antagonised only by the central dopamine antagonist haloperidol (Cangi *et al.*, 1984). Thus, it is likely that most symptoms or side-effects caused by dopamine agonists are caused by stimulation of peripheral dopamine receptors (including receptors in the CNS but outside the blood–brain barrier, such as the chemoreceptor trigger zone or other parts of the medulla oblongata and the hypothalamus), but that the headache itself is of central origin.

42.3. **Pathological changes in migraine**

If one compares symptoms of migraine with effects caused by dopamine agonists, (see Table 42.5), one can conclude that during an attack of migraine, an activation of several dopaminergic systems occurs—probably together with other systems and possibly caused by disturbance of the protagonist 5-HT system. This does not mean that dopamine has a causal role in migraine, but simply that it contributes to the polymorphic and often

Table 42.5. Some similarities between acute side-effects of dopamine agonists (apomorphine, bromocriptine and lisuride) in healthy volunteers, and the symptomatology of an acute migraine attack.

Dopamine-agonist side-effects[a]	System involved	Migraine attack side-effects
Nausea, emesis	Chemoreceptor trigger zone	Nausea, sometimes emesis
Reduced gastrointestinal motility and constipation	Gastrointestinal dopamine receptors	Reduced gastrointestinal motility
Headache	?	Headache
Drowsiness	?	Sedation

[a]On repeated intake, effects undergo tolerance development—in contrast to prolactin-lowering and antiparkinson properties.

changing symptomatology of this affliction. This is confirmed by the often quite effective relief migraine sufferers can have during an attack by using dopamine antagonists such as metoclopramide or domperidone. The old observation that reserpine and the related compound tetrabenazine can also trigger acute migraine attacks must not invalidate this concept, because both compounds are functional dopamine antagonists but initially release dopamine from its stores.

If we accept the similarities between dopaminergic effects and migraine symptomatology, we have also to ask why these systems have an enhanced sensitivity in migraine. This could be due to a more activated receptor state and a higher affinity of these receptors to dopamine and dopamine agonists. Comparable changes in receptor activity are known to be caused by peptides in the case of most of the classical neurotransmitters; therefore, a sudden change in endorphins or other circulating peptide levels—which may either enhance or reduce the effects of dopamine on its receptors—could explain the fact that normally silent dopaminergic systems become activated. This activation of dopaminergic receptors by an unknown cause then results in nausea, emesis, drowsiness and gastrointestinal stasis, as observed during a migraine attack.

It is well known that peptides exert some of their effects not by a direct mechanism, but by an indirect effect on 'classical' neurotransmitters such as noradrenaline, dopamine or GABA; they can enhance the receptor binding affinity of these compounds or in other ways locally change the biological effects of constant neuronal activity. Thus a sudden withdrawal of an endogenous compound which prevents neurotransmitter release presynaptically—a suggested mechanism of action of endorphins—or a rapid release of a similar compound that enhances (e.g. by an allosteric mechanism) dopamine receptor affinity could cause (at unchanged dopaminergic neurone activity and dopamine levels) nausea, emesis and other symptoms associated with a migraine attack. Similar events could involve central or peripheral 5-HT receptors and therefore cause the variable symptomatology of attacks. Candidates for such factors could be the endorphins, substance P or some as yet unknown central peptide ('migrainine' or 'antimigrainine'). Furthermore, it is well known that all acute side-effects of dopamine agonists as shown in Table 42.5 undergo tolerance development after repeated application of these compounds. Therefore, it is not too unlikely that chronic treatment with lisuride not only stabilises 5-HT function, but also desensitises supersensitive dopaminergic systems in migraine sufferers and, therefore, prevents symptoms associated with an attack of migraine.

42.4. Summary

In conclusion, we propose:

(a) that dopaminergic systems are activated during a migraine attack and thereby cause nausea and other symptoms;
(b) that these systems in migraine sufferers are more sensitive than in healthy subjects;
(c) that dopamine agonists given acutely can trigger an attack or at least mimic it;
(d) that dopamine antagonists given acutely can alleviate some of the migraine symptomatology;
(e) that chronic treatment with compounds that also have dopamine agonist properties can desensitise dopaminergic receptors and thus prevent migraine attacks in part or completely.

We know that these observations cannot explain the origin of migraine and that some protagonist of this disease must exist; we feel, however, that dopamine could be a good candidate as a 'deuteragonist' in migraine, which contributes to migraine symptomatology and mediates in part the therapeutic efficacy of some drugs in this severe and disabling condition.

This hypothesis of a role of dopamine in the pathophysiology of migraine attacks could be tested further by using lisuride and lisuride derivatives as pharmacological tools. Indeed, our pharmacological and clinical studies show that dopamine receptors should not be considered as a kind of 'on–off' system, but rather as a continuum: at this level, apomorphine acts as a strong agonist, lisuride and proterguride act as high-affinity, low-activity agonists (which have higher activity mostly on 'empty' neurones), terguride as a partial agonist (which at higher doses acts as an antagonist) and bromerguride (2-bromlisuride) as an antagonist (Wachtel *et al.*, 1984). In agreement with our hypothesis, antimigraine activity of these compounds should switch from a prophylactic effect to symptomatic efficacy during an acute attack. Future studies will tell us whether such a pharmacological approach indeed leads to improvement in knowledge and therapy.

References

Bès Ala-Hurala, V. (1982). Correlation between pharmacokinetics and clinical effects of ergotamine in patients suffering from migraine. *Eur. J. Clin. Pharmacol.* **21**, 397–402.

Cangi, F., Boccuni, M., Fanciullacci, M. Pietrini, U. and Sicuteri, F. (1984). Algesic effect of apomorphine in dysnociceptive man. 14th CINP Congress, Florence, Abst. P867.

Del Bene, E., Poggioni, M. and Michelacci, S. (1983). Lisuride as a migraine prophylactic in children: an open clinical trial. *Int. J. Clin. Pharm. Res. III*, 137–41.

Del Bianco, O. I., Franchi, G., Anselmi, B. and Sicuteri, F. (1982). Monoamine sensitivity of smooth muscle *in vivo* in nociception disorders. In *Advances in Neurology*, vol. 33. M. Critchley *et al.*, (eds), pp.391–8. Raven Press, New York.

Eadie, M. J. (1983). Ergotamine pharmacokinetics in man: an editorial. *Cephalalgia* **3**, 135–8.

Eulenberg, A. (1883). Uber Hemikranie. *Dtsch. Med. Wschr.* **1883**, 637–9.

Franchi, G. and Mallucci, C. (1983). Il lisuride maleato acido nella profilassi della cefalea emicranica. *Minerva Med.* **74**, 1749–53.

Glusa, E., Markwardt, F. and Barthel, W. (1979). Studies on the inhibition of adrenaline-induced aggregation of blood platelets. *Pharmacology* **19**, 196–201.

Herrmann, W. M., Horowski, R., Dannehl, K., Kramer, U. and Lurati, K. (1977). Clinical effectiveness of lisuride hydrogen maleate: a double blind trial versus methysergide. *Headache* **17**, 54–60.

Herrmann, W. M., Kristof, M. and Sastre y Hernandez, M. (1978). Preventive treatment of migraine headache with a new isoergolenyl derivative. *J. Int. Med. Res.* **6**, 476–82.

Heyck, H. (1975). *Der Kopfschmerz.* Georg Thieme, Stuttgart.

Horowski, R. (1982). Some aspects of the dopaminergic action of ergot derivatives and their role in the treatment of migraine. In *Advances in Neurology: Headache*, vol. 33. M. Critchley, A. P. Friedman, S. Gorini and F. Sicuteri (eds), pp. 325–34. Raven Press, New York.

Horowski, R. (1983). Pharmacological effects of lisuride and their potential role in further research. In *Lisuride and Other Dopamine Agonists.* D. B. Calne, R. Horowski, R. McDonald and W. Wuttke (eds), pp. 127–41. Raven Press, New York.

Kennard, C., Gawel, M., de Rudolph, M. and Clifford, F. (1978). Visual evoked potentials in migraine subjects. *Res. Clin. Stud. Headache.* **6**, 73–80.

Klein, S. H. (1983). Perception of stimulus intensity by migraine and non-migraine subjects. *Headache* **23**, 158–61.

Kranda, K. and Kulikowski, J. J. (1984). Visual aura in classical migraine. In *The Neurobiology of Pain.* A. V. Holden and W. Winlow (eds), pp. 243–66. Manchester University Press, Manchester.

Lance, J. W. (1973). *The Mechanism and Management of Headache.* Butterworth, London.

Lancet Editorial (1982). Treatment of migraine. *Lancet* **i**, 1338–9.

McDonald, R. J. and Horowski, R. (1983). Lisuride in the treatment of parkinsonism. *Eur. Neurol.* **22**, 240–528.

Peroutka, St. J. (1983). The pharmacology of calcium channel antagonists: a novel class of antimigraine agents? *Headache* 278–83.

Podvalovà, I. and Dlabač, A. (1972). Lysenyl, a new antiserotonin agent. *Res. clin. Stud. Headache* **3**, 325–34.

Raffaelli, E., Martins, O. J., dos Santos, P. and Dágua Filho, A. (1983). Lisuride in cluster headache. *Headache*, **23**, 117–21.

Raskin, N. H. and Appenzeller, O. (1980). Headache. In *Major Problems in Internal Medicine*, vol. 19. L. H. Smith (ed.).

Richards, W. (1971). The fortification illusions of migraine. *Sci. Amer.* **224**, 88–96.

Ripka, O. (1972). Effeti della somministrazione a lungo termine dell'anti-serotoninico lysenyl nelle cefalee di varia eziologia. *Minerva Med.* **63**, 3266–71.

Sicuteri, F. (1959). Prophylactic and therapeutic properties of 1-methyl-lysergic acid butanolamide in migraine: preliminary report. *Int. Arch. Allergy Appl. Immunol.* **15**, 300–7.

Sicuteri F., Fanciullacci, M. and Michelacci, S., (1979). Decentralisation supersensitivity in headache and central analgesia. In *Research and Clinical Studies in Headache*, vol. 6, A. P. Friedmann *et al.* (eds), pp. 19–33, Karger, Basel.

Somerville, B. M. and Herrmann, W. M. (1978). Migraine prophylaxis with lisuride hydrogen maleate—a double blind study of lisuride versus placebo. *Headache* **18**, 75–9.

Tessitore, A. and Battaglia, A. (1982). L'utilità della lisuride nel trattamento delle cefalee idiopatiche. *Rassegna Internazionale di Clinica e Terapia* **LXII**, 541–4.

Wachtel, H., Dorow, R. and Sauer, G. (1984). Novel 8α-erolines with inhibitory and stimulatory effects on prolactin secretion in rats. *Life Sci.* **35**, 1859–67.

Wolff, H. G. (1948). *Headache and Other Head Pain*. Oxford University Press, New York.

43

Use of dopaminergic agents in the treatment of senile dementia

R. J. McDonald

Sandoz Research Institute, Sandoz Inc.
East Hanover, New Jersey, USA

43.1. Introduction

Senile dementia is a neurological disorder estimated to affect between 5 and 20% of persons aged 65 and over (Kay and Bergmann, 1980; Mortimer *et al.*, 1981). It is a progressive disorder characterised initially by a loss of recent memory and followed by a worsening of other cognitive functions, such as attention, orientation, judgement and language (NIA Task Force, 1980; Beck *et al.*, 1982). Even though the cause of the disease is unknown, present medical consensus is that senile dementia is a pathological disorder and not an inevitable consequence of ageing.

Currently, great emphasis is being placed on identifying the neurotransmitter systems involved in the pathogenesis and symptomatic manifestations of the disease. Compelling evidence is available relating disturbances in central cholinergic neurotransmitter systems to memory loss and cognitive dysfunctions in patients with senile dementia (for reviews, see Crook and Gershon, 1981; Bartus *et al.*, 1982 Corkin *et al.*, 1982; Coyle *et al.*, 1983). Cholinergic systems are receiving the most interest at this time, but the acknowledged role of other neuronal systems, such as noradrenergic, serotoninergic and dopaminergic, in arousal, attention and learning, suggest that these systems may also be involved in the cognitive impairment associated with senile dementia (Carlsson *et al.*, 1980; Mann *et al.*, 1982; Rossor, 1982; Gottfries *et al.*, 1983). One of these neurotransmitter systems, the dopaminergic, is the focus of this chapter.

In 1969, Gottfries *et al.* suggested the possibility that the monoamine systems were involved in the aetiology of senile dementia or at least in the symptom formation. They reported reduced levels of homovanillic acid (HVA) in the basal ganglia in patients with senile dementia. These patients were identified at autopsy and the degree of their dementia was measured post mortem with a rating scale. A negative correlation was found between brain HVA levels and the degree of mental impairment. Gottfries *et al.*

(1970) also reported a lower concentration of HVA in the cerebrospinal fluid in patients with senile dementia compared with age-matched controls. Subsequent postmortem studies (Adolfsson *et al.*, 1979; Carlsson *et al.*, 1980) demonstrated a lower concentration of dopamine and HVA in different parts of the brain of patients with senile dementia. In summarising the above findings, Gottfries (1981) concluded that there is diminished dopaminergic activity in the brain in senile dementia. Furthermore, since there were negative correlations between the lower levels of HVA and the degree of dementia, it may be assumed that the reduced activity in the dopaminergic system has pathophysiological significance for this disease.

Such findings helped to form a rationale for the testing of dopaminergic agents in senile dementia. As in Parkinson's disease, the demonstration of a possible dopamine deficiency in senile dementia led to speculation that replacement therapy with the dopamine precursor levodopa or with dopamine agonists bromoriptine and lisuride might be beneficial in the treatment of this disease. The purpose of this chapter, therefore, is to review the clinical work conducted with these dopaminergic agents. The review is limited to double-blind studies (except for those by Phuapradit *et al.* and Borenstein *et al.*) which involved either patients diagnosed as having presenile/senile dementia or elderly volunteers.

43.2. Clinical studies with levodopa

Levodopa is mainly used in the treatment of symptoms of idiopathic Parkinson's disease. Current evidence indicates that symptoms of Parkinson's disease are related to depletion of dopamine in the corpus striatum. Administration of dopamine is ineffective in the treatment of Parkinson's disease because it does not readily cross the blood–brain barrier. However, its precursor, L-dopa, does cross the blood–brain barrier, and presumably is converted to dopamine in the basal ganglia, the area of the brain thought to control motor activities (Marsden, 1982a). The most obvious mechanism for L-dopa's therapeutic action is replacement of deficient striatal dopamine. Evidence in support of this mechanism includes a positive correlation between the symptoms of Parkinson's disease and loss of nigrostriatal neurones (Marsden, 1982b). The usual optimal therapeutic dosage of L-dopa in Parkinson's disease is between 3 and 4 g per day but should not exceed 8 g. Therapeutic response is usually achieved within 2 months but may not be obtained until after 6 months of treatment. A summary of the L-dopa studies in senile dementia now follows.

Kristensen *et al.* (1977) studied the effects of L-dopa in a sample of 18 out-patients with presenile dementia (mean age 51 years, range 30–63 years). Patients were given placebo or L-dopa in a randomised double-blind design. The initial dose of L-dopa was 500 mg day^{-1} and was increased by 500 mg day^{-1} until 4 g day^{-1} were reached. The duration of treatment was

6 months. Testing was conducted before the start of treatment and after 6 months of treatment. Cognitive functioning was assessed by the Wechsler Adult Intelligence Scale, WAIS (Wechsler, 1955), a non-verbal memory test (visual gestalts), paired associates, finger maze, and finger tapping. A behavioural rating scale (12 items), assessing a broad spectrum of symptoms (such as impairment of memory, concentration and of the ability for abstract thinking), was also used.

The results indicated that there were no significant effects due to treatment either on the psychiatric items or on the cognitive functioning. The overall impression demonstrated that there was no convincing proof of a positive change after 6 months of treatment with 4 g of L-dopa in comparison with placebo. The differences between the groups were small and, according to the authors, perhaps due more to personality variables than to specific treatment. The authors concluded that the results of the psychological investigation did not provide any evidence that L-dopa treatment has any beneficial effect on patients diagnosed as having presenile dementia.

Lewis *et al.* (1978) investigated the effects of L-dopa in 14 in-patient females diagnosed as suffering from senile dementia (mean age 79 years). The study was a double-blind crossover trial; patients received 4 weeks of treatment with placebo and 4 weeks of treatment with the active drug. The starting dose was 125 mg daily which was increased over 2 weeks to 875 mg daily. Behavioural and intellectual functioning were assessed by the Crichton Geriatric Behavior Rating Scale and Crichton Intellectual Rating Scale (Robinson, 1961). All ratings were made blind at the end of weeks 0, 2, 4, 6 and 8. The results indicated a significant improvement in intellectual function during the drug phase; there were no significant changes in any other areas of behaviour with the exception of continence.

Johnson *et al.*(1978), in a follow-up study, followed five of the 14 patients with senile dementia who had shown a favourable response in the original trial. After a break of 1 month, 750 mg of L-dopa was given for a further 6 months. Five control patients with senile dementia, matched for age and sex, were selected and both groups were rated monthly on a Crichton Intellectual and Behavioral Rating Scale. During the 1-month drug-free period between the original and the more extended trial, there was a slight decline in the mean intellectual scores. During the drug phase, the authors reported slight improvements in intellectual functioning in three out of five patients during the 6-month treatment period. Even though there was an indication of improvement, the authors concluded that the clinical relevance of these findings may be questionable, and did not endorse the findings or recommend L-dopa for the routine treatment of senile dementia.

Jellinger *et al.* (1980) examined the effects of long-term administration of L-dopa in 12 outpatients in the earlier stages of organic brain syndrome (mean age 57 years, age range 54–60 years). The investigation was a controlled double-blind study. Half of the patients were given L-dopa orally as Madopar (200 mg of L-dopa plus 50 mg of benserazide) and the other half placebo for 24 weeks. The drug was given once daily at 08.00. Clinical evaluations included neurologic and psychiatric examinations and a mental status examination consisting of a 12-item behavioural rating scale. Psychometric assessment included three subtests of the WAIS: digit span, coding B test, and block design. Evaluations were performed before onset of treatment, and at weeks 4, 12 and 24.

Clinical findings indicated that there was a significant improvement in the L-dopa group on the total score of the behavioural rating scale in comparison with the placebo group at week 24. The psychometric evaluation showed a slight increase in the scores of all three subtests in the L-dopa group which only differed slightly after 4, 12 and 24 weeks, respectively. The scores in all subtests differed from those in the control group, which showed only slight variation. The increases were more pronounced in the digit span and coding B test in the L-dopa group and were significant ($P < 0.05$) at each post-evaluation. The block design test was significant only at the last two measurement points, 12 and 24 weeks. The authors interpreted these results as indicating that L-dopa improved visuo-spatial function, visual memory, and visual motor co-ordination in this sample. They concluded, in spite of the limited number of test persons, that L-dopa treatment has some beneficial effects in the early stages of presenile dementia and that this improvement increases with the duration of the treatment.

Adolfsson *et al.* (1982) investigated the therapeutic effect of L-dopa (Madopar) in 37 outpatients with dementia of the Alzheimer's type (mean age 64 years, age range 50–77 years, mean duration of disease 3·4 years). Patients were randomly assigned to one of three groups: a low-dose group (375 mg day^{-1}), a high-dose group (750 mg day^{-1}), and a placebo group. Ten patients were in the low-dose group, 12 patients were in the high-dose group, and 15 patients were in the placebo group. The treatment period was 10 weeks. Patients were evaluated at the hospital for 1 week before the start of the study, and another 1-week investigation was done after the 10-week treatment period. Psychometric testing included subtests of the WAIS (digit-span and similarities) (Wechsler, 1955), the Fuld Objective Memory Test (Fuld, 1980: immediate recall, storage, retrieval, recall failure, and long-term memory), the Bender–Gestalt Test and psychomotor functions (reaction time and tapping speed). A psychiatric behavioural rating scale was used, which focused on items measuring depression and dementia. The scale consisted of 69 items which were assessed either in an interview or by observation.

Following 10 weeks of treatment, there were no significant differences between the two L-dopa groups and placebo on any of the psychometric tests except for a negative effect observed for the low-dose group on the

Bender–Gestalt Test. Likewise, there were no significant changes observed within the groups or between the three groups on any of the items on the behavioural rating scale with the exception of a significant improvement in long-term memory and dementia score in the high-dose group. The results clearly demonstrated that no objective clinical improvement was registered with either a low or high dose of L-dopa (Madopar) in comparison with placebo for this sample of Alzheimer's patients. Despite the discouraging results, the authors offered suggestions for future trials with L-dopa:

(1) test patients in an earlier stage of the disease;
(2) test for a longer duration;
(3) use a higher dose.

Ferris *et al.* (1982) examined the effects of L-dopa on 55 out-patients who showed objective evidence of very mild to moderately severe cognitive impairment secondary to Alzheimer's disease (age range 60–85 years). A crossover design was employed. Patients received either active medication or placebo in randomised order, each for a 6-week treatment period. A 2-week placebo phase preceded each treatment period. Thus, the total study duration was 16 weeks (2–6–2–6). The L-dopa dose was 2 g day^{-1} (given in four equal doses per day). The assessment measures were obtained at baseline and after 2 and 6 weeks of treatment. They included an extensive cognitive test battery and behavioural rating scales. The cognitive battery included measures of short-term, long-term and remote memory, as well as measures of attention and perceptual-motor function.

The findings did not reveal any significant differences between the two groups on either the cognitive tests or behavioural ratings. L-dopa had neither a positive nor negative effect on any of the cognitive parameters. Changes from baseline at 2 weeks and 6 weeks for both L-dopa and placebo were identical. Based on these findings, it was concluded that L-dopa had no beneficial effect in patients with very mild to moderately severe cognitive impairment secondary to Alzheimer's disease.

Newman *et al.* (1984) studied the effects of Sinemet (carbidopa/L-dopa) on effortful and automatic episodic memory and semantic memory in 10 healthy normal volunteers (mean age 64 years, range 61–67 years). Volunteers were given placebo or Sinemet for 7 weeks in a randomised double-blind crossover design. Active medication was administered in doses of 25/100 mg (carbidopa/L-dopa), increased by one tablet every day for a week to a total of eight tablets daily (800 mg of L-dopa).

In both study phases, volunteers were given equivalent forms of a test procedure that measures effortful and automatic episodic memory (Hasher and Zacks, 1979). In brief, volunteers were read, at 3-s intervals, a list of 40 common imaginable English words. Each word was read one to seven times and in a random order. After hearing all the words, followed by a 5-min distractor-filled delay, volunteers were asked to remember as many words as possible—an effort-demanding task. This was followed by an automatic memory-processing task during which each of the words was read, and volunteers were asked how often a word was read. A semantic memory task was also performed. Voluteers were asked to generate items in response to letters as well as superordinate category names, such as vegetables, flowers and animals.

L-dopa treatment significantly facilitated effortful/episodic memory process (free recall) for both infrequently and frequently processed words but did not affect access to semantic or automatic memory processing tasks. Based on these findings, the authors suggested L-dopa may be useful clinically in attenuating impairments that involve disruptions in tasks demanding effortful cognitive capacity, as in hyperactivity syndromes, depression, or Parkinson's disease. They also speculated that L-dopa influences effort-demanding cognitive operations in a manner similar to that of other drugs with noradrenergic effects.

43.3. Clinical studies with bromocriptine treatment

Bromocriptine, a dopamine receptor agonist, is indicated for the short-term treatment of amenorrhoea/galactorrhoea with hyperprolactinaemia due to varied aetiologies, the treatment of female infertility associated with hyperprolactinaemia, in the prevention of physiological lactation, and the treatment of the signs and symptoms of idiopathic or postencephalitic Parkinson's disease. Bromocriptine produces its therapeutic effect in Parkinson's disease by directly stimulating the striatal dopamine receptors in the corpus striatum. The optimal therapeutic dosage of bromocriptine in Parkinson's disease is between 15 and 40 g per day provided it is gradually built up over a period of 4 to 8 weeks. A summary of the bromocriptine studies in senile dementia follows.

Phuapradit et al. (1978) investigated, in a single-blind trial, the effects of bromocriptine in nine patients with early presenile dementia, presumed to be Alzheimer's disease (mean age 64, mean duration of disease 3 years). Higher cortical functions were assessed by behavioural rating scales and a variety of cognitive performance tests: visual retention test, digit symbol, block design, reaction time, and colour matrix test. Treatment was begun with bromocriptine 2·5 mg daily and increased by 2·5 mg increments at 4-day intervals to a maximum dose of 20 mg for 4 weeks. Patients were assessed at 14-day intervals. Placebo was then substituted for bromocriptine unknown to the patient and a further assessment was made after 2 weeks.

Bromocriptine failed to improve cognitive disabilities. None of the behavioural or cognitive parameters assessed indicated any improvement following the bromocriptine treatment. Side-effects, including confusion and nausea, were common and only six patients attained a dose of 20 mg. Relatives of three patients, however, reported improvement in the patients' ability to cope with their disease as a result of a mild sedative effect. The authors concluded that bromocriptine as well as presently available dopamine agonists are unlikely to benefit the intellectual and memory deficits associated with presenile dementia.

Smith *et al.* (1979) examined the effects of bromocriptine in patients with senile dementia. Seventeen patients were randomly assigned to either bromocriptine or placebo in a double-blind placebo-controlled crossover trial. Patients received 7·5 mg day^{-1} for a period of 4 weeks. Cognitive changes were assessed by two geriatric behavioural rating scales. The authors reported that bromocriptine failed to demonstrate any beneficial effect on any of the parameters assessed. They did report, however, a significant decrease in prolactin levels.

Borenstein *et al.* (1981) studied (in an open trial) the effects of bromocriptine in 22 patients with senile dementia (age range 60–80 years). Patients were treated for 12 weeks with a daily dose of 5 to 10 mg bromocriptine which was added to a low dose of chlorpromazine. Effects of treatment were examined by behavioural rating scales and EEG. The clinical results indicated improvement in motor, cognitive and emotional behaviour. EEG data demonstrated in 70 % of the cases a return of alpha rhythm, improvement in arousal reaction and the disappearance of delta rhythm. The authors reported a concordance between the clinical and the EEG results.

43.4. Clinical studies with lisuride

Lisuride, a dopamine receptor agonist, is indicated in the treatment of hyperprolactinaemia and acromegaly, prevention of post-partum lactation, and parkinsonism (Horowski, 1983). Like bromocriptine, lisuride produces its therapeutic effect in Parkinson's disease by directly stimulating the dopamine receptor in the corpus striatum. The optimal therapeutic dosage of lisuride in Parkinson's disease is between 1·5 to 4·5 mg per day provided the daily dose is built up gradually over a period of 4 to 8 weeks (McDonald and Horowski, 1983). A summary of the lisuride studies follows.

Suchy and McDonald (1978) investigated the effects of lisuride in a sample of nursing-home patients with mild to moderate cognitive impairment ($N = 160$, mean age 78 years). Patients were randomly assigned to one of four groups: a low-dose group (75 μg day^{-1}), a middle-dose group (150 μg day^{-1}), a high-dose (300 μg day^{-1}), or placebo. The design was double-blind and the length of treatment was 16 weeks. Patients were evaluated at 4-week intervals. Psychometric testing included simple and disjunctive reaction time, continuous performance, and short-term, long-term and recognition memory. Behavioural rating scales and self-rating scales were also used.

Following 16 weeks of treatment, there were no significant differences found among the four treatment groups on any of the parameters assessed. There was a slight indication that mood was improved with the two higher doses of lisuride. Results were also analysed according to the severity of the patients' cognitive impairment, mild vs moderate. The analyses yielded no significant differences among the three treatment groups.

Suchy and McDonald (1980) studied the effects of lisuride in a sample of 90 out-patients with symptoms of senile dementia (mean age 65 years, range 56–75). The study was a multicentre double-blind trial. Patients were randomly assigned to one of three treatment groups: a low-dose group (75 μg day^{-1}); a high-dose group (300 μg day^{-1}), or placebo. Thirty-one patients were in the low-dose group, 30 patients were in the high-dose group, and 29 patients were in the placebo group. The length of treatment was 24 weeks. Evaluations were conducted at the beginning of the study and at weeks 8, 16 and 24. The psychological test battery consisted of short-term, long-term and recoginition memory, concentration and crystalised intelligence. Two self-rating scales were also given out. All testing was performed at the home of the patient.

The results of the psychometric tests indicated a trend in favour of the lisuride groups; however, none of the findings were significant. The most pronounced differences between the lisuride and placebo groups were found on the self-rating scales for sense of well-being and anxiety. Taken together, the results suggested that lisuride did not have a beneficial effect on any cognitive parameters but did have a slight antidepressant effect in a sample of out-patients with complaints of memory impairment.

McDonald and Rohloff (1983) investigated the effects of lisuride on psychomotor functions and short-term memory recall in a sample of 24 healthy elderly volunteers (mean age 70, range 61–75). The trial was double-blind and subjects were randomly assigned either to the lisuride (150 μg day^{-1}) or the placebo group, 12 per group. The duration of the study was 7 weeks; the first week was an adaptation period followed by 6 weeks of active treatment. There were two evaluations per study week. The indices of evaluation were simple and disjunctive reaction times, a continuous performance test involving short-term memory recall, and a self-rating adjective checklist.

Findings indicated that lisuride slightly enhanced psychomotor functions but did not hinder or facilitate memory processing. Significant differences were not obtained on any of the parameters tested. The authors questioned the practicality of using relatively healthy elderly persons to demonstrate the clinical merits of a compound considered as a possible treatment for behavioural and mental disturbances in older patients.

Suchy *et al.* (1984, unpublished), in a follow-up study, examined the effects of lisuride on a variety of psychomotor tests in healthy elderly volunteers. The sample consisted of 20 healthy elderly persons over the age of 60. The duration of the study was 12 weeks. The trial was double-blind and subjects were randomly assigned to either the lisuride (150 μg day^{-1}) or placebo group. Assessments consisted of rating scales and psychometric tests. They were conducted at the beginning of the trial and once again at the end of the study.

Following 12 weeks of treatment, there was no indication that lisuride had either a positive or negative effect on any of the test parameters. Changes from baseline to post-evaluation on the psychomotor tests indicated no change in performance level in either group. On the continuous performance test there was a fairly pronounced learning effect from pre- to post-evaluation. This learning pattern was noted in both treatment groups. These findings were in agreement with those reported by McDonald and Rohloff (1983), who also noted a pronounced learning effect on the same continuous performance test for both the lisuride and placebo groups.

43.5. **Discussion**

The goal of the present review was to evaluate the clinical merits of dopaminergic agents used for the treatment of senile dementia. The studies reviewed were limited to three drugs, the dopamine precursor, L-dopa, and two dopamine receptor agonists, bromocriptine and lisuride.

Collectively, the studies failed to provide any consistent evidence that dopaminergic agents are beneficial in treating this disease. Among the possible explanations for this lack of efficacy may be methodological deficiencies. In general, many of the studies suffered from one or more deficiencies, such as poorly defined patient samples, inappropriate testing instruments, too small a sample, too low a dose, or too short a study period. Three of these points were also cited by Adolfsson *et al.* (1982) as possible reasons for the discouraging results obtained in their study. They recommended that future trials with dopaminergic agents should test (a) patients who are in an earlier stage of the disease, (b) for a longer duration, and (c) with a higher dose. As noted, none of the studies cited employed these criteria.

The only study which indicated a positive effect on cognitive parameters was done with L-dopa in elderly normal volunteers (Newman *et al.*, 1984). The authors interpreted their findings as indicating that L-dopa could be clinically useful in attenuating cognitive impairments that involve disruptions in tasks demanding effortful cognitive capacity, which require attention, as in hyperactivity syndromes, depression or Parkinson's disease. Furthermore, they stated that the cognitive impairments of Alzheimer's disease, which include both effortful and automatic cognition, are instead due to disruption of semantic memory (Weingartner *et al.*, 1983). They concluded that stimulating dopamine activity, therefore, would not be expected to affect the type of cognitive impairment observed in Alzheimer's disease.

Their concluding point, i.e. stimulating dopamine activity, may be another reason why these agents were not beneficial in this disorder. Dopamine is associated with several functions of the brain, each mediated through its own neuronal pathways. The first pathway, the dopaminergic nigrostriatal tract, subserves motor movements. A reduction of dopamine in this pathway results in Parkinson's disease. A second pathway is the tuberinfundibular one, which is associated with endocrine functions, namely the production and release of prolactin. A third pathway is the mesolimbic. The system is likely to undertake functions different from those of the nigrostriatal system although all influence motor activity (Marsden, 1982*b*) and may be associated with affect and cognitive functioning. This suggests that the stimulation of the dopaminergic pathways may not be helpful in dementia since the cognitive impairment observed in these patients does not appear to be mediated by dopamine neuronal pathways. Furthermore, it may be that the dopaminergic deficiency reported in this disease is not related to cognitive dysfunctions as is the cholinergic neurotransmitter system. On the other hand, it may be that the dosages given in these studies were too small and not enough drug crossed the blood–brain barrier and had access to the striatum. This explanation is similar to that observed more than a decade earlier when low doses of L-dopa and later when low doses of bromocriptine were given to Parkinson patients. In both instances it was necessary to increase the dosages in order to obtain better therapeutic responses. Therefore, it may be the case that greater success would be achieved in dementia with dopaminergic agents by giving larger doses for a longer duration of time.

43.6. Conclusion

The findings do not suggest that dopaminergic agents are effective in the treatment of senile dementia. However, due to numerous methodological deficiencies in the studies reviewed, it may be short-sightedness to conclude that dopaminergic compounds will have no role in the treatment of senile dementia. Systematic studies involving a better defined, homogeneous sample of patients who are treated for a longer period of time with higher doses could provide more insight into the relation of the dopaminergic system and senile dementia.

References

Adolfsson, R., Gottfries, C. G., Roos, B. E. *et al.* (1979). Changes in the brain catecholamines in patients with dementia of Alzheimer type. *Brit. J. Psychiat.* **135**, 216–23.
Adolfsson, R., Brane, G., Bucht, G., Karlsson, I., Gottfries, C. -G., Persson, S. and Winblad, B. (1982). A double-blind study with levodopa in dementia of Alzheimer

type. In *Alzheimer's Disease: a Report of Progress in Research*, S. Corkin, K. L. Davis, J. H. Growdon, E. Usdin and R. J. Wurtman (eds), pp. 469–74. Raven Press, New York.

Bartus, R. T., Dean III, R. L., Beer, B. and Lippa, A. S. (1982). The cholinergic hypothesis of geriatric memory dysfunction. *Science* **217**, 408–17.

Beck, J. C., Benson, F. D., Scheibel, A. B., Spar, J. E. and Rubenstein, L. Z. (1982). Dementia in the elderly: the silent epidemic. *Ann. Int. Med.* **97**, 231–41.

Borenstein, P., Soret, C. and Graille, D. (1981). Interet de la bromocriptine dans les demences de la vieillesse. *Ann. Med. Psychol.* **139**, 326–34.

Carlsson, A., Adolfsson, R., Aquilonius, S. M., Gottfries, C. G., Oreland, L., Svennerholm, L. and Winblad, B. (1980). Biogenic amines in human brain in normal aging, senile dementia and chronic alcoholism. In *Ergot Compounds and Brain Function: Neuroendocrine and Neuropsychiatric Aspects*, M. Goldstein, D. B. Calne, A. Lieberman and M. O. Thorner (eds), pp. 295–304. Raven Press, New York.

Corkin, S., Davis, K. L., Growdon, J. H., Usdin, E. and Wurtman, R. J. (eds) (1982). *Alzheimer's Disease: a Report of Progress in Research* vol. 19. Raven Press, New York.

Coyle, J. T., Price, D. L. and DeLong, M. R. (1983). Alzheimer's disease: a disorder of cortical cholinergic innervation. *Science* **219**, 1184–190.

Crook, T. and Gershon, S. (eds) (1981). *Strategies for the Development of an Effective Treatment for Senile Dementia*. Mark Powley Associates, Inc., Connecticut.

Ferris, S. H., Reisberg, B., Crook, T., Friedman, E., Schneck, M. K., Mir, P., Sharman, K. A., Corwin, J., Gershon, S. and Bartus, R. T. (1982). Pharmacologic treatment of senile dementia: choline, L-DOPA, piracetam, and choline plus piracetam. In *Alzheimer's Disease: a Report of Progress in Research*. S. Corkin, K. L. Davis, J. H. Growdon, E. Usdin and R. J. Wurtman (eds), pp. 475–82. Raven Press, New York.

Fuld, P. A. (1980). Guaranteed stimulus-processing in the evaluation of memory and learning. *Cortex* **16**, 255–71.

Gottfries, C. G. (1981). Etiological and treatment considerations in SDAT. In *Strategies for the Development of an Effective Treatment for Senile Dementia*. T. Crook and S. Gershon (eds), pp. 107–20. Mark Powley Associates, Inc. Connecticut.

Gottfries, C. G., Gottfries, I. and Roos, B. E. (1969). The investigation of homovanillic acid in the human brain and its correlation to senile dementia. *Brit. J. Psychiat.* **115**, 563–74.

Gottfries, C. G., Gottfries, I. and Roos, B. E. (1970). Homovanillic acid and 5-hydroxyindoleacetic acid in cerebrospinal fluid related to rated mental and motor impairment in senile and presenile dementia. *Acta Psychiat. Scand.* **46**, 99–105.

Gottfries, C. G., Adolfsson, R., Aquilonius, S. M., Carlsson, A., Eckernas, S. A., Nordberg, A., Oreland, L., Svennerholm, L., Weiberg, A. and Winblad, B. (1983). Biochemical changes in dementia disorders of Alzheimer type. *Neurobiology of Aging* **4**, 261–71.

Hasher, L. and Zacks, R. T. (1979). Automatic and effortful processes in memory. *Exp. Psychol. Gen.* **108**, 356–88.

Horowski, R. (1983). Pharmacological effects of lisuride and their potential role in further research. In *Lisuride and Other Dopamine Agonists*. D. B. Calne, R. Horowski, R. McDonald and W. Wutke (eds), pp. 127–41. Raven Press, New York.

Jellinger, K., Flament, P., Riderer, H. and Ambrozi, L. (1980). Levodopa in the treatment of (pre)senile dementia. *Mechanisms of Aging and Development* **14**, 253–64.

Johnson, K., Presly, A. S. and Ballinger, B. R. (1978). Levodopa in senile dementia. *Brit. Med. J.* **1**, 1625.

Kay, D. W. K. and Bergmann, K. (1980). Epidemiology of mental disorders among the aged in the community. In *Handbook of Mental Health and Aging*. J. E. Birren and R. B. Sloane (eds), pp. 32–56.

Kristensen, V., Olsen, M. and Theilgaard, A. (1977). Levodopa treatment of presenile dementia. *Acta Psychiat. Scand.* **55**, 41–51.

Lewis, C., Ballinger, B. R. and Presley, A. S. (1978). Trial of levodopa in senile dementia. *Brit. Med. J.* **1**, 550.

Mann, D. A., Yates, P. O. and Hawkes, J. (1982). The noradrenergic system in Alzheimer's and multi-infarct dementias. *J. Neurol. Neurosurg. Psychiat.* **45**, 113.

Marsden, C. D. (1982a). The mysterious motor function of the basal ganglia: the Robert Wartenberg lecture. *Neurology* **32**, 514–39.

Marsden, C. D. (1982b). Basal ganglia disease. *Lancet* **20**, 1141–6.

McDonald, R. J. and Rohloff, A. (1983). Effects of lisuride on psychomotor functions and recall memory in elderly healthy volunteers. In *Lisuride and Other Dopamine Agonists*. D. B. Calne, R. Horowski, R. McDonald and W. Wutke (eds), pp. 515–28. Raven Press, New York.

McDonald, R. J. and Horowski, R. (1983). Lisuride in the treatment of parkinsonism. *Eur. Neurol.* **22**, 240–55.

Mortimer, J. A., Schulman, L. M. and French, L. R. (1981). Epidemiology of dementing illness. In *The Epidemiology of Dementia*. J. A. Mortimer and L. M. Schulman (eds), pp. 3–23. Oxford University Press, New York.

Newman, R. P., Weingartner, H., Smallberg, S. A. and Calne, D. B. (1984). Effortful and automatic memory: effects of dopamine. *Neurology* **34**, 805–7.

NIA Task Force (1980). Senility reconsidered: treatment possibilities for mental impairment in the elderly. *JAMA* **244**, 259–63.

Phuapradit, P., Philips, M., Lees, A. J. and Stern, G. M. (1978). Bromocriptine in presentile dementia. *Brit. Med. J.* **1**, 1052–3.

Robinson, R. A. (1961). Some problems of clinical trials in elderly people. *Geront. Clin.* **3**, 247.

Rossor, M. N. (1982). Dementia. *Lancet*, 1200–4.

Smith, H. W., Kay, D. S. G., Johnson, K. and Ballinger, B. K. (1979). Bromocriptine in senile dementia: a placebo-controlled double-blind trial. In *IRCS Medical Sciences: Clinical Medicine, Clinical Pharmacology and Therapeutics; Endocrine System; Nervous System; Psychiatry; Reproduction, Obstetrics and Gynecology*, **7**, 463.

Suchy, I. and McDonald, R. J. (1978). Die Wirkung von LHM auf gestorte mental Funktionen bei Patienten eines Hauses Für Chronisch-Kranke. Data on file, Schering AG, West Berlin, Germany.

Suchy, I. and McDonald, R. J. (1980). Relationship between subjective complaints, emotion and performance in the elderly: results from a survey and its impact on a clinical drug trial. Communication presented at the *12th CINP Congress*, Gothenburg.

Suchy, I., Rohloff, A. and Runge, I. (1984). Messung der Bioverfugbarkeit von 3×50 mg Lisuride on 20 Probanden über 60 Jahren. Data on file, Schering AG, West Berlin, Germany.

Wechsler, D. (1955). *Manual for the Wechsler Adult Intelligence Scale*. The Psychology Corporation.

Weingartner, H., Kaye, W., Gold, P. *et al.* (1981). Vasopressin treatment of cognitive function on progressive dementia. *Life Sci.* **29**, 2721–6.

Weingartner, H., Grafman, J., Boutelle, W., Kaye, W. and Martin, P. R. (1983). Forms of memory failure. *Science* **221**, 380–2.

44

Can dopaminergic substances be useful in patients with symptoms of senile dementia?

I. H. Suchy

Research Laboratories, Schering AG,
Berlin (West)/Bergkamen, Federal Republic of Germany

44.1. Introduction

One in twenty people over the age of 65 and one in five of those over 80 suffer from some form of dementia. Since, in Western society, life expectancy is 73–76 years, one in eight people can expect to suffer from senile dementia. Efforts are increasing to find the reasons for this apparently inevitable fate and to establish a cure.

A possible therapeutic approach is the use of dopamine agonists. It was shown by McGeer *et al.* (1971) that tyrosine hydroxylase in the basal ganglia of healthy people declines with age. In 1977, McGeer *et al.* showed that the biochemical changes in the nigrostriatal dopaminergic system are due to cell loss in the substantia nigra (see Fig. 44.1). Their findings, as well as those of other authors, confirmed the strong correlation of this decline with age. Marshall and Berrios (1979) showed that aged rats lost some of their ability to swim, an ability that could be restored by means of apomorphine. Experience with L-dopa in Parkinson's disease (Barbeau *et al.*, 1961; Birkmayer and Hornykiewicz, 1961) has shown that substituting a missing neurotransmitter can abolish pathological symptoms that had been caused by the loss of this neurotransmitter. Although the role of the dopaminergic nigrostriatal system in cognitive processes is unclear, it seemed worth while to investigate whether dopamine agonists can reverse some of the pathological symptoms seen in elderly persons.

44.2. Clinical trials with substances that enhance central dopaminergic transmission

The following substances have been used to test the dopamine hypothesis on pathological symptoms in elderly persons:

L-dopa is converted to dopamine, which stimulates D-1 and D-2 receptors and is also to some extent metabolised to noradrenaline.

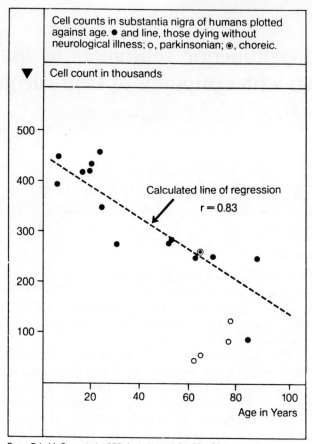

From: P. L. McGeer et al., 1977, Arch. Neurol. 34, 33 – 35

Fig. 44.1.   Cell counts in substantia nigra of humans plotted against age. ● and line, those dying without neurological illness; O, parkinsonian; ⊙, choreic. (From McGeer *et al.*, 1977.)

Bromocriptine is a semisynthetic ergot derivative which stimulates D-2 receptors. It also interferes with serotonergic and adrenergic transmission.

Lisuride is another semisynthetic ergot derivative with strong dopamine-agonist effects on D-2 receptors. Like bromocriptine, it interferes with serotonergic and adrenergic neurotransmission. However, the effective dose to bring about the various effects seems to be different from that found with bromocriptine. Binding data comparing both substances have been described by Suchy *et al.*, (1983).

Dihydroergotoxine mesilate is composed of four peptide ergot alkaloids. Besides its α-adrenolytic effect, which has prompted its therapeutic use in cerebrovascular disorders (CVI) long before interest focused on the symptoms of senile dementia, dihydroergotoxine exerts agonistic effects on dopamine and serotonin receptors.

Nicergoline interferes with α-adrenergic transmission and stimulates metabolism. It does not reduce prolactin levels and therefore has been considered to be inactive in the dopaminergic system. Recently, Carfagna and Moretti (1984) published results showing an increase of homovanillic acid (HVA) in the mesolimbic system, which they interpreted as a sign of enhanced dopaminergic transmission after treatment with nicergoline in animal experiments.

The following tables, together with additional new information, are derived from Suchy (1983).

L-dopa (Table 44.1) has no detectable effect on presenile dementia of Alzheimer's disease. In patients with senile dementia of the Alzheimer type (SDAT), the benefit of L-dopa on cognitive functions may become apparent only in the presence of parkinsonian symptoms. The improvement of effortful memory in elderly volunteers will certainly stimulate continuing research.

Bromocriptine (Table 44.2) has not been studied extensively in elderly patients. The results in SDAT are not conclusive, especially if one takes into account the possible effects of concomitant medication with neuroleptics. Bromocriptine seems to be helpful in alcohol withdrawal: a selective loss of noradrenergic transmitters has been described in chronic alcoholism. It has been shown to be of therapeutic benefit in hepatic encephalopathy, a state that shows widespread changes in transmitter states which are similar to those observed in SDAT. Bromocriptine has also been found to have antidepressive properties (Waehrens and Gerlach, 1981).

Lisuride (Table 44.3) has been studied double-blind in patients with organic psychosyndrome, in patients with symptoms of SDAT, and in psychogeriatric patients who suffered partly from SDAT but to a greater extent from cerebrovascular disorders. Various effects have been observed; those influencing the emotional level are the most consistent ones.

Dihydroergotoxine (Table 44.4) has been used for many years in numerous trials. McDonald in 1982 reviewed 38 double-blind trials using dihydroergotoxine (McDonald, 1982), and Fanchamps (1983) in a review summarised all trials that had used the SCAG scale (Sandoz Clinical Assessment of Geriatric Populations; Shader *et al.*, 1974) for evaluating the clinical effects. In the SCAG scale, 19 items, together with a global impression, are rated from 1 (absent) to 7 (severe). The 19 items represent the factors: cognitive dysfunction, interpersonal relationships, affective, apathy, and somatic functioning. Trials were performed on patients with symptoms of senile dementia, SDAT or CVI. They showed an improvement in most factors, with emphasis on the emotional level. Recently a dose-dependent effect on cognition has been reported by Yesavage (1981), Yoshikawa *et al.* (1983), and Van Loveren-Huyben *et al.* (1984).

Table 44.1. Trials with L-dopa.

Diagnosis and no. of patients	L-dopa dose per day	Duration of study	Study design	Relevant positive effects on:	References
Out-patients with presenile dementia (9)	4 g	6 months	Double-blind, placebo controlled	None	Kristensen et al., 1977
Senile dementia, non-parkinsonian (4)	4–6 g	2 months	Open, uncontrolled	None	Van Woert et al., 1970
Senile dementia without 'serious sign of Parkinson's disease' (14)	0·875 g (+PDI)*?	4 weeks	Double-blind	Intellectual rating	Lewis et al., 1978
Senile dementia, drug responders of former trial (5)	0·750 g (+PDI)?	6 months	Open	Continued slight intellectual improvements in 3 of 5	Johnson et al., 1978
Senile dementia, signs of Parkinson's disease (9)	1–3·75 g	3–24 months	Open, uncontrolled	Intellectual and motor impairment	Drachman and Stahl, 1975
Senile dementia, partly with symptoms of Parkinson's disease (25)	0·375 g (+PDI) or 0·750 (+PDI)	10 weeks	Double-blind, placebo controlled	None	Adolfsson, pers. comm.
Parkinson's disease	4–8 g	5–13 months	Open, uncontrolled	WAIS verbal and performance IQ	Loranger, 1972
Parkinson's disease	800 mg (+PDI)	4 weeks	Open, healthy control group matched for age and education	Only partial remission of memory deficits	Jacobi et al., 1976
Parkinson's disease without signs of dementia (40)	?	22–40 months	Cross-sectional study, healthy control group matched for age	Short-term patients superior to long-term due to progression of disease	Halgin 1977
Parkinson's disease (39)	?	?	Open, health control group matched for age and education	Perception organisation, memory, verbal comprehension	Meier and Martin, 1970
Old volunteers (10)	0·800 g (+PDI)	7 weeks	Double-blind, placebo controlled	Effortful memory	Newman et al., 1984

Table 44.2. Trials with bromocriptine.

Diagnosis and no. of patients	Bromocriptine dose (mg day^{-1})	Duration of study	Study design	Relevant positive effects on:	References
Hepatic encephalopathy (6)	15	?	Double-blind crossover, placebo controlled	Clinical picture: psychometric testing, EEG, CBF O$_2$ + glucose consumption	Morgan et al., 1984
Senile dementia (10)	7·5	4 weeks	Double-blind, placebo controlled	No effect	Smith et al., 1979
Alcohol withdrawal (12)	7·5	10 days?	Double-blind, controlled against apomorphine	Anxiety, depression, tremor	Borg and Weinhold, 1980
Senile dementia (25)	5-10	?	Open	Behaviour, personality, mood, EEG	Borentein et al., 1982

Table 44.3. Trials with lisuride.

Diagnosis and no. of patients	Lisuride dose (μg day^{-1})	Duration of therapy	Study design	Relevant positive effects on:	References
In-patients with organic psycho-syndrome (15)	75	8 weeks	Double-blind, matched pairs, placebo controlled	Orientation, restlessness, suicidal thoughts, irritability	Misurec et al., 1978
Out-patients with symptoms of senile dementia	30 × 75 (in 30 patients) 30 × 300 (in 30 patients)	24 weeks	Double-blind, placebo controlled	Depression, anxiety, no difference between two dosages	Suchy et al., 1983
In-patients, mainly cerebro-vascular disorders (182)	75	12 weeks	Double-blind, controlled against dihydroergotoxine	No obvious statistical difference between two drugs	Aizawa et al., 1980

Table 44.4. Trials with dihydroergotoxine (only very few examples; for others see cited reviews).

Diagnosis and no. of patients	Dihydroergotoxine dose (mg day^{-1})	Duration of therapy	Study design	Relevant positive effects on:	References
Cerebrovascular disorders (10)	4·5	12 weeks	Double-blind, controlled vs placebo	Subjective physical symptoms, EEG	Arrigo et al., 1973
Healthy elderly volunteers (53)	4·5	3 years	Double-blind, controlled vs placebo	Prophylactic effect on the cardiovascular symptom. HAWIE* total score and performance part, Raven coloured matrices, Maudsley Personality Inventory (MPI) extroversion	Spiegel et al., 1983
Cerebrovascular disorders (550)	3 (sublingual) 6 per oral)	12 weeks	Double-blind, 3 mg vs 6 mg	Global rating for psychiatric subjective symptoms (headache, dizziness, irritability); 3 mg significantly less effective than 6 mg	Yoshikawa et al., 1983
Mild to moderate forms of senile mental deterioration (30)	4·5, 12 weeks 7·5 for following 12 weeks	24 weeks	Double-blind, controlled vs placebo	All SCAG items except self-care	Van Loveren-Huyben et al., 1984

*HAWIE: Hamburg–Wechster Intelligence Test for Adults; English version WAIS.

Nicergoline (Table 44.5) was initially used to treat patients with cerebrovascular disorders. Most trials were open. A pattern of effects was observed that was usually summarised as improvement of subjective physical symptoms, namely headache, vertigo, dizziness, tiredness, etc. More recently, using the SCAG scale to evaluate the effects of nicergoline in patients with mild to moderate symptoms of senile dementia, a statistically significant improvement of all items covered was found in various double-blind trials.

Though all these results may be interesting from various points of view, they do not validate the dopamine hypothesis in ageing. They are too 'soft', even if statistically significant. Clinically, the results are not clear, and sometimes not even relevant. But before putting aside the dopamine hypothesis, some possible reasons for the meagre outcome should be discussed.

44.3. Problems with the diagnosis of senile dementia

One problem with which people doing research in this field are always confronted is that of diagnosis. The populations treated with dopamine agonists in the trials reported above ranged from patient groups with presenile dementia (the original Alzheimer's disease) to old volunteers—excluding patients with hepatic encephalopathy or alcohol withdrawal. Some improvement after treatment has been found in all except those with presenile dementia. Presenile dementia is the only disease that may be defined clearly in a dementing process: it begins before the age of 65, develops gradually, shows memory impairment as a leading symptom and may be combined with agnosia, apraxia or aphasia and other impaired cognitive functions. Morphologically, it is obligatory that cortical areas besides the hippocampus show a certain amount of neurofibrillary tangles and senile plaques.

Theoretically, senile dementia is separated from presenile dementia by an arbitrary cut at an age of onset of 65 years. Practically, presenile dementia is characterised by a severe course which brings the patients to a fully demented state within a few years and by biochemical post-mortem findings which, compared with age-matched controls, show a pattern that differs from that seen in milder forms with late onset (see below). Though there are severe cases with a late onset, senile dementia of late onset usually follows a mild course and the patient may die for various reasons before he is fully demented. Similarities between the two forms outweigh differences, and we believe nowadays that presenile and senile dementia are the same diagnostic entity.

According to the American DSM III (American Psychiatric Association, 1980), dementia is diagnosed only when the respective symptoms are 'sufficiently severe to interfere with social or occupational functioning'. This social category in the diagnosis may vary enormously from one patient to

Table 44.5. Trials with nicergoline.

Diagnosis and no. of patients	Nicergoline dose (mg/day^{-1})	Duration of therapy	Study design	Relevant positive effects on:	References
Cerebrovascular disease (39)	2 i.m. 15 p.o.	5–48 days	Open, uncontrolled	Headache, vertigo, memory, concentration, emotional lability	Guardamagna and Negri, 1972
Cerebrovascular disease (48)	15	3–4 weeks	Open, uncontrolled	Best results on headache, vertigo; concentration in hypertensive patients	Boudouresques et al., 1974
Cerebrovascular disease (39)	2 i.m. 15 p.o.	8–44 days	Open, uncontrolled	Headache, vertigo, etc.; memory, confusion; motor function	Alloro and Terenziani, 1973
Cerebrovascular disease (29)	15	5–80 days	Open, uncontrolled	Headache, vertigo, concentration	Mellini, 1973
Cerebral and peripheral arteriosclerosis (19)	15	20 days	Double-blind vs dihydroergotoxine	Headache, vertigo, memory; claudication; no difference between the drugs	Pazzaglia, 1972

Table 44.5 continued

Chronic cerebrovascular disease (38)	20	3 months	Double-blind, controlled vs placebo	Memory, vigilance, psychomotoric performance; no effect on EEG (age differences between groups)	Michelangeli et al., 1975
Old volunteers (10)	15 30 60	Acute, single dose	Double-blind, crossover; dihydroergotoxine 5 mg, placebo	Typical EEG effects 6–8 h after drug intake; learning capacity; increase of vigilance level	Grünberger and Saletu, 1980
Hypertension and chronic cerebrovascular insufficiency (359)	15	35 days	Open, uncontrolled	Headache, vertigo, ear noises, memory, orientation, social activities, mood	Dauverchain, 1979
Senile dementia (20)	8 i.m. 60 p.o.	14 weeks	Double-blind controlled vs placebo	All SCAG items except appetite and recent memory improved significantly	Arrigo et al., 1982
Senile cerebral insufficiency (44)	60 p.o.	24 weeks	Double-blind controlled vs placebo	Significant improvement in all factors of the SCAG scale	Granata et al., 1984

the next. However, in every patient the questions will arise as to when changes in cognition and behaviour are disturbing enough to be called pathological, and as to when the pattern of changes is so grave as to be diagnosed as dementia. It will not be a problem in severe cases, but in the mild cases there is no separation line but a continuous overlap between the patients with senile dementia and those persons who feel impaired and possibly show open deficits but are far from being considered as 'demented'. What I want to point out is that it may be the right solution, and for the time being the only possible method, to use an operant procedure for including patients in a clinical trial, i.e. patients to be included should present with a specified amount of pathological symptoms as rated by a certain rating scale instead of being diagnosed as suffering from a mild form of senile dementia.

Furthermore, if other pathogeneses of dementia such as certain cerebrovascular disorders, or possible risk factors of cerebrovascular disorders, (e.g. arterial hypertension or diabetes) are excluded, patient groups may result

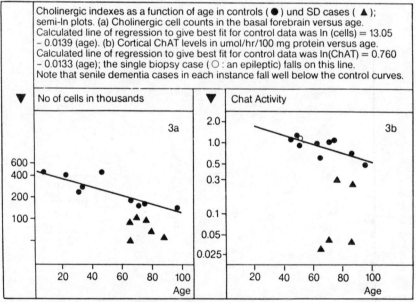

Cholinergic indexes as a function of age in controls (●) und SD cases (▲); semi-ln plots. (a) Cholinergic cell counts in the basal forebrain versus age. Calculated line of regression to give best fit for control data was ln (cells) = 13.05 − 0.0139 (age). (b) Cortical ChAT levels in umol/hr/100 mg protein versus age. Calculated line of regression to give best fit for control data was ln(ChAT) = 0.760 − 0.0133 (age); the single biopsy case (○ : an epileptic) falls on this line. Note that senile dementia cases in each instance fall well below the control curves.

From: P. L. McGeer et al., Neurology 1984, 34: 741 – 745

Fig. 44.2.   Cholinergic indices as a function of age in controls (●) and SD cases (▲); semi-log plots. (a) Cholinergic cell counts in the basal forebrain vs age. Calculated line of regression to give best fit for control data was ln (cells) $= 13\cdot05 - 0\cdot0139$ (age). (b) Cortical choline acetyl transferase (ChAT) levels in μmol h^{-1} per 100 mg protein vs age. Calculated line of regression to give best fit for control data was ln(ChAT) $= 0\cdot760 - 0\cdot0133$ (age); the single biopsy case (○ : an epileptic) falls on this line. Note that senile dementia cases in each instance fall well below the control curves. (From McGeer *et al.*, 1984.)

that are more similar to each other (though no diagnosis has been made) than groups with cerebrovascular insufficiency to those with symptoms of senile dementia (examples taken from the literature). Results from patient groups that have been recruited in such an operant procedure should be comparable and allow us to draw conclusions that lead to a re-evaluation of hypotheses. Some of the more recent trials with dihydroergotoxine, lisuride and nicergoline have been done in this way. The diagnostic labels used were symptoms of senile dementia, SDAT, pathological ageing, cerebral insufficiency, cognitive impairment, pathological symptoms in elderly patients, or possibly others. As long as a clear insight in basic pathogenetic mechanisms is missing, the clinical terminology cannot be more exact.

It is known that senile dementia of Alzheimer type (SDAT)—in its severe

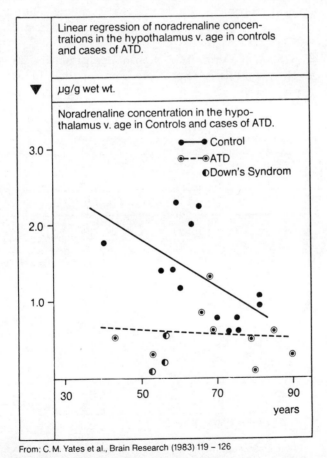

From: C. M. Yates et al., Brain Research (1983) 119 – 126

Fig. 44.3.   Linear regression of noradrenaline concentrations in the hypothalamus vs age in controls and cases of Alzheimer-type dementia (ATD). (From Yates *et al.*, 1983.)

NDS–P*

forms as well as in its mild or rudimentary forms—is a degenerative process in the central nervous system that causes impaired neural transmission and neuronal cell loss in various regions, in various systems and to a varying degree (for details and further literature see the reviews by Samorajski, 1977; Pradhan, 1980; Rossor, 1982; Gottfries *et al.*, 1983; Kopp and Tommasi, 1983). The decrease of cholinergic transmission (see Fig. 44.2) is thought to be specific for Alzheimer's disease. Figure 44.3 shows the correlation of the hypothalamic noradrenaline content with age in controls as well as in patients with SDAT. Figure 44.4 deals with the 5-hydroxytryptamine content in hypothalamus in controls and in SDAT patients. Figs 44.1–4 are examples from the literature which show, first, an age-related decline in all four transmitter systems, and, secondly, a decrease of the characteristics of transmitter function studied in SDAT patients (with the exception of dopamine where no corresponding illustration was available, but where, according to the literature, relative decreases were also observed).

In SDAT patients the decrease in transmitter parameters as compared with age-matched controls is much more pronounced in the severe, early onset cases, whereas in patients over 80 years old the data often show no

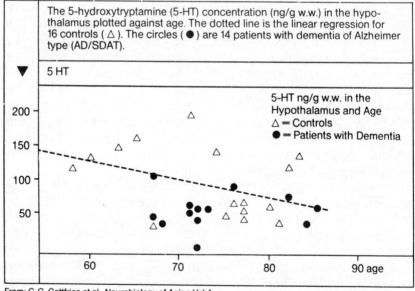

The 5-hydroxytryptamine (5-HT) concentration (ng/g w.w.) in the hypothalamus plotted against age. The dotted line is the linear regression for 16 controls (△). The circles (●) are 14 patients with dementia of Alzheimer type (AD/SDAT).

▼ 5 HT

5-HT ng/g w.w. in the Hypothalamus and Age
△ = Controls
● = Patients with Dementia

Fig. 44.4.   5-Hydroxytryptamine (5-HT) concentration (ng g^{-1} wet weight) in the hypothalamus plotted against age. The dotted line is the linear regression for 16 controls (△). The circles (●) are 14 patients with dementia of Alzheimer type (AD/SDAT). (From Gottries *et al.*, 1983.)

clear difference between SDAT and age. This seems to be true even for the cholinergic system, for, as Rossor *et al.* (1981) reported, in the frontal cortex there is no difference between patients with SDAT dying after the age of 80 and controls.

44.4. Memory impairment in senile dementia

Thus, there is enough evidence for us to accept the fact that there is a widespread defect of transmitter function in SDAT and in ageing. Morphological and biochemical studies support this hypothesis, but the question remains unsolved as to the exact clinical impact of these defects. The discovery of cholinergic deficits has prompted numerous trials using both acetylcholine precursors, cholinomimetic substances and cholinesterase inhibitors. One some occasions a temporary improvement of memory has been achieved. Disappointing as the result may be on the whole from a therapeutic point of view, the causal link between loss of cholinergic neurones and memory impairment in Alzheimer dementia is widely accepted. In the experimental situation, substances inhibiting cholinergic transmission produce memory deficits which can be abolished by substances that restore intact cholinergic transmission. So, undoubtedly, cholinergic function plays an important role in the mechanisms of memory, and its defects bring about an impairment of memory capacities. As, by definition, memory loss is the leading symptom in SDAT, undoubtedly biochemical and morphological findings may be biased towards deficits of structures and functions that are involved in memory. It is intriguing to speculate what would be the result if other symptoms, e.g. changes of alertness and arousal, had become the major criteria for SDAT. Would we know more about the importance of 5-hydroxytryptamine if research had focused on illuminating changes in functioning of serotonergic systems? Would we know more, apart from Parkinson's disease, about the role of dopamine in the regulation of emotional equilibrium? Or would noradrenaline have been the first transmitter to be implicated?

Loss of locus coeruleus neurones and their projections have been pointed out above all by Yates *et al.* (1983; see Fig. 44.3), Mann *et al.* (1984), and Berger and Alvarez (1983). The degree of the deficit seems to be related to the degree of dementia. Quantitatively the deficits, especially in SDAT of early onset, seem to be enormous and greater than those of the dopaminergic or the serotonergic system. No clinical trial has been done until now to elucidate the significance of these deficits. According to Berger and Alvarez (1983), noradrenaline is much more a neuromodulator of cholinergic and GABAergic transmission than a neurotransmitter in its own right. If this is true, its presence should be necessary for memory functions. Interestingly, Solomon *et al.* (1983) have found impairment of memory function in hypertensive subjects being treated with methyldopa or propranolol, substances that diminish adrenergic transmission. Further evidence in favour of the hypothesis that noradrenaline is involved in memory function stems from observations in Korsakoff's psychosis. Korsakoff patients have profound amnesia with relatively intact intellectual capabilities. It has been shown by CSF studies (McEntee *et al.*, 1984) that these patients have decreased levels of the metabolites of monoamines. The same authors have shown that the severity of anterograde amnesia in this disease can be lessened by treatment with clonidine, a direct α_2-adrenergic agonist (McEntee and Mair, 1980).

Dopamine has also been shown to interfere with memory. Newman *et al.* (1984) in a double-blind study succeeded in improving episodic effortful memory in old volunteers by treatment with L-dopa (800 mg day^{-1}). They point out that it is episodic effortful memory which is likely to be impaired in depression. The same group found cognitive dysfunction in untreated Parkinson's disease to resemble those of depression rather than memory–learning impairments seen in early stage SDAT patients (Weingartner *et al.*, 1983). It is known and has been mentioned earlier that cognitive impairment including memory dysfunction in parkinsonian patients can be reversed for some time by the initiation of L-dopa therapy. The potent dopamine agonist bromocriptine has been shown to possess therapeutic efficacy in the retarded phase of biphasic depression.

44.5. Impairment of arousal in senile dementia

Most of the trials conducted with dopamine agonists in patients with SDAT or symptoms of SDAT have shown as the most consistent result an improvement of mood and emotional lability. All these different findings lead to the assumption that intact dopaminergic transmission might be necessary for the basic function of arousal. It is impossible to assess directly and objectively such a basic function. No satisfactory model of indirect assessment of arousal or alertness has been established. Electrophysiologists pertain that ageing *per se* or in combination with pathological processes is related to a decrease of α-activity in the occipital electro-encephalogram (EEG). The presence of α-activity under certain conditions is considered to be an indicator for normal arousal or vigilance. The effects on the EEG of all the dopamine agonists discussed here with the exception of L-dopa have been studied in patient populations that are comparable to some extent. All these substances have been shown to increase α-activity (Itil *et al.*, 1975; Kugler *et al.*, 1978; Bente *et al.*, 1979; Borenstein *et al.*, 1981; Otomo *et al.*, 1981). This finding is interpreted as a sign of normalisation of pathologically impaired vigilance or arousal. Unfortunately, until now no satisfying correlation between EEG results and the results of exact psychometric data of improved cognition in the respective patient groups has been possible.

In a recent publication, Brown *et al.* (1984) report on their investigations which have been performed in parkinsonian patients during their on-phase and their off-phase. They found, besides the fluctuation of motor function, a fluctuation of cognitive function (general reasoning, ability for verbal, numerical and spatial material). The fluctuation of cognitive functioning, though mild in comparison to that seen in motor functioning, correlated much better with the level of affect and arousal as measured by a visual analogue scale than with age, the degree of motor disability, the duration of the disease or a combination of these. As it is now known that parkinsonian on–off phases are dependent on the intranigrostriatal dopamine level, these results demonstrate how deficits of dopamine or treatment with dopamine agonists might indirectly influence various cognitive functions.

Recent research on Parkinson's disease (Cools *et al.*, 1984) sheds more light on the possible mechanisms underlying cognitive impairment if dopamine systems are failing. The ability to shift from one task to another on cognitive and motor levels was found to be lower in parkinsonian patients than in controls matched for age and education. This result was interpreted as a central programming deficit. Similar results have been observed in laboratory animals with lesions of the dopaminergic system. Possibly this is one of the fundamental causes of what is observed as 'general slowing' in elderly people. And of course, this shifting capacity may also have a close relation to the basic state of arousal.

The therapeutic results of dopamine agonists in SDAT patients with parkinsonian symptoms, which have been reported here and were not limited to improvement of motor functions, may have been the consequence of deficits in the central dopaminergic systems that were great enough to cause obvious extrapyramidal symptoms as an indicator for the degree of impairment in dopaminergic transmission. In other cases, if the dopaminergic neuronal loss is less severe, so as not to reach the critical threshold (in the nigrostriatal system 50–80 % of cell loss is necessary to cause clinical symptoms!), the deficiency may either become manifest in periods of high demand or the whole pattern of neurotransmission is disturbed. It has been shown that memory functions can be influenced by manipulation of the cholinergic, noradrenergic and dopaminergic systems (too little is known about the serotinergic system, but it also may be involved).

44.6. Roles for dopamine agonists in therapies for senile dementia

In order to achieve positive results in trials following the ideas outlined in this chapter, the therapeutic intervention should consist of an enhancement of dopaminergic neurotransmission. Dopamine agonistic therapy often uses potent postsynaptic stimulants in doses that cause a reduction of intrinsic dopaminergic transmission. In animal experiments, levels of homovanillic acid, the metabolite of dopamine, decrease. In a putative population of dopaminergic neurones of an aged patient with symptoms of SDAT, the state of activity of presynaptic endings will extend from total degeneration to complete normal function. Possibly those neurones that are fully intact will not react to an overdose of dopaminergic substance (there are no hyperkinesias in normal subjects) and no harm can be done in synapses whose afferent part is totally degenerated. But in all places of transmission with impaired function, postsynaptic receptors will become subsensitive under a constant influence of potent stimulatory substances, and little is known about what counter-regulatory mechanisms will take their origin from the impaired presynaptic neurone which is suddenly cut off from its usual feedback interaction.

These worrying contemplations apply for the potent dopamine agonists bromocriptine, L-dopa and lisuride. The situation is different for dihydroergotoxine with its very limited dopaminergic action, and for nicergoline, which in the usual animal experiments and clinical trials has no dopaminergic effect but seems to produce an enhancement of mesolimbic dopaminergic transmission as it is signalled by an HVA increase in this region. Joseph et al. (1983) were able to increase the sensitivity of postsynaptic dopamine receptors in aged rats by chronic treatment with a neuroleptic substance. What is an unwanted side-effect in the treatment of psychosis may become a desirable approach in the treatment of symptoms in the aged.

In the evaluation of the dopamine agonistic effect of the ergoline derivatives lisuride, bromocriptine, dihydroergotoxine and nicergoline, we have neglected to mention that they are very 'dirty' substances. Their mixed pharmacological profiles do not allow us to isolate their dopamine-agonistic effects. But if they produce results which are similar to some extent in similar patient groups, these results may, with due caution, be linked to their common property: the interference with dopaminergic function.

Hopefully, future research will deal with dopamine-agonistic substances that are not muddled by other pharmacological properties; that stimulate dopaminergic transmission where necessary; and that do not interfere with those neurones whose transmitter processes are still functioning normally.

References

Aizawa, T. et al. (1980). Clinical utility of eunal in treatment of cerebrovascular disorders: multicenter double blind study in comparison with dihydroergotoxine mesylate. Clin. Eval. **8**, 577–628.
Alloro, L. and Terenziani, S. (1973). Esperienza clinica con un nuovo farmaco vasoattivo, la nicergolina, nelle vasculopatie cerebrali. Gaz. Med. Ital. **132**, 81–90.
American Psychiatric Association (1980). Diagnostic and Statistical Manual of Mental Disorders (DSM III). American Psychiatric Association, Washington, DC.

Arrigo, A., Braun, P., Kauchtschischwili, G. M., Moglia, A., Tartara, A. (1973). Influence of treatment on symptomatology and correlated electroencephalographic (EEG) changes in the aged. *Curr. Ther. Res.* **15**, 417–26.

Arrigo, A., Moglia, A., Borsotti, L., Massarini, M., Alfonsi, E., Battaglia, A. and Sacchetti, G. (1982). A double-blind, placebo-controlled crossover trial with nicergoline in patients with senile dementia. *Int. J. Clin. Pharm. Res.* Suppl. 1, **II** (4), 33–41.

Barbeau, A., Murphy, C. F. and Sourkes, T. L. (1961). Excretion of dopamine in disease of basal ganglia. *Science* **133**, 1706.

Bente, D., Glatthaar, G., Ulrich, G. and Lewinsky, M. (1979). Quantitative EEG-Untersuchungen zur vigilanzfördernden Wirkung von Nicergolin. *Arzneimittelforsch.* **29**, 1804–8.

Berger, B. and Alvarez, C. (1983). Pathologie de l'innervation catecholaminergique du cortex cérébral dans la démence de type Alzheimer. *La Presse Médicale* **12**, 3109–14.

Birkmayer, W. and Hornykiewicz, O. (1961). Der L-Dioxyphenylalanin (L-DOPA)—Effekt bei der Parkinson-Akinesie. *Wien. Klin. Wschr.* **73**, 787.

Borenstein, P., Soret, C., Graille, D. and Villejuif, Fr. (1981). Contribution à l'étude des syndromes dementiels sous l'action de la bromocriptine. *Sem. Hop.* **57**, 801–9.

Borg, V. and Weinhold, T. (1980). A preliminary double-blind study of two dopaminergic drugs, apomorphine and bromocriptine (Parlodel), in the treatment of the alcohol-withdrawal syndrome. *Curr. Ther. Res.* **27**, 170–7.

Boudouresques, J., Vigouroux, R. A., Boudouresques, G. and Monnier, M. C. (1974). Etude de la nicergoline dans l'insuffisance vasculaire cerebrale. *Lyon Mediter. Med.* **10**, 989–90.

Brown, R. G., Marsden, C. D., Quinn, N. and Wyke, M. A. (1984). Alterations in cognitive performance and affect-arousal state during fluctuations in motor function in Parkinson's disease. *J. Neurol. Neurosurg. Psychiat.* **47**, 454–65.

Carfagna, N. and Moretti, A. (1984). Neurochemical studies with chronic oral ergoline derivatives. Collegium Internationale Neuro-Psychopharmacologicum, *14th CINP Congress*, Florence. Abstract 661, p. 346.

Cools, A. R., van den Bercken, J. H. L., Horstink, M. W. I., van Spaendonck, K. P. M. and Berger, H. J. C. (1984). Cognitive and motor shifting aptitude disorder in Parkinson's disease. *J. Neurol., Neurosurg. Psychiat.* **47**, 443–53.

Dauverchain, J. (1979). Bedeutung von Nicergolin bei der symptomatischen Behandlung des arteriellen Hochdrucks und der chronischen zerebro-vaskulären Insuffizienz/Studie an 359 Beobachtungen. *Arzneimittelforsch.* **29**, 1308–10.

Drachman, D. A. and Stahl, St. (1975). Extrapyramidal dementia and levodopa. *Lancet*, **7910**, 809.

DSM III. (1980). *Diagnostic and Statistical Manual of Mental Disorders*. The American Psychiatric Association, Washington DC.

Fanchamps, A. (1983). Dihydroergotoxine in senile cerebral insufficiency. In *Aging Brain and Ergot Alkaloids*. Aging, vol. 23. Agnoli A. *et al.* (eds), p. 311. Raven Press, New York.

Gottfries, C. G. *et al.* (1983). Biochemical changes in dementia disorders of Alzheimer type. *Neurobiology of Aging* **4**, 261–71.

Granata, M., Dondero, D. and Tomasinelli, R. (1984). Studio clinico controllato della nicergolina nell'insufficienza cerebrale senile. *Atti della Accademia Medica Lombarda, Milano* **39**, 1–8.

Grünberger, J. and Saletu, B. (1980). Nicergolin beim älteren Menschen: Pharmako-EEG und psychometrische Studien zum Nachweis akuter und chronischer psychotroper und pharmakodynamischer Eigenschaften. *Therapiewoche* **30**, 3832–55.

Guardamagna, C. and Negri, S. (1972). Efficacia clinica della nicergolina nelle vasculopatie cerebrali. *Minerva Cardioangiol.* **62**, 636–41.

Halgin, R., Riklan, M. and Misiak, H. (1977). Levodopa, parkinsonism, and recent memory. *J. Nerv. Ment. Dis.* **164**, 268–72.

Itil, T. M., Herrmann, W. M. and Akpinar, S. (1975). Prediction of psychotropic properties of lisuride hydrogen maleate by quantitative pharmaco-electroencephalogram. *Int. J. Clin. Pharmacol.* **12**, 221–33.

Jacobi, P., Schneider, E. and Fischer, P. A. (1976). Störungen der Merkfähigkeit bei Parkinson-Kranken. *Fortschr. Neurol. Psychiatr.* **44**, 63–72.

Johnson, K., Presley, A. S. and Ballinger, B. R. (1978). Levodopa in senile dementia. *Brit. Med. J.* **6127**, 1625.

Joseph, J. A., Bartus, R. T., Clody, D., Morgan, D., Finch, C., Beer, B. and Sesack, S. (1983). Psychomotor performance in the senescent rodent: reduction of deficits via striatal dopamine receptor up-regulation. *Neurobiology of Aging* **4**, 313–19.

Kopp, N. and Tommasi, M. (1983). Les neuromédiateurs dans l'encéphale du vieillard. *La Presse Medicale* **48**, 3103–7.

Kristensen, V., Olsen, M. and Theilgaard, A. (1977). Levodopa treatment of presenile dementia. *Acta Psychiat. Scand.* **55**, 41–51.

Kugler, J., Oswald, W. D., Herzfeld, U., Seus, R., Pingel, J. and Welzel, D. (1978). Langzeittherapie altersbedingter Insuffizienzerscheinungen des Gehirns. *Dtsch. Med. Wochenschr.* **103**, 456–62.

Lewis, C., Ballinger, B. R. and Presley, A. S. (1978). Trial of levodopa in senile dementia. *Brit. Med. J.* **6112**, 550.

Loranger, A. W., Goodell, H. and Lee, J. E. (1972). Levodopa treatment of Parkinson's syndrome. *Arch. Gen. Psychiat.* **26**, 163–8.

Mann, D. M. A., Yates, P. O. and Marcyniuk, B. (1984). A comparison of changes in the nucleus basalis and locus coeruleus in Alzheimer's disease. *J. Neurol., Neurosurg. Psychiat.* **47**, 201–4.

McDonald, R. J. (1982). Drug treatment of senile dementia. In *The Psychopharmacology of Old Age*. D. Wheatley (ed.), pp. 113–38. Oxford University Press, Oxford.

McEntee, W. J. and Mair, R. G. (1980). Memory enhancement in Korsakoff's psychosis by clonidine: further evidence for a noradrenergic deficit. *Ann. Neurol.* **7**, 466–70.

McEntee, W. J. *et al.* (1984). Neurochemical pathology in Korsakoff's psychosis: implications for other cognitive disorders. *Neurology* **34**, 648–52.

McGeer, E. G., McGeer, P. L. and Wada, J. A. (1971). Distribution of tyrosine hydroxylase in human and animal brain. *J. Neurochemistry* **18**, 1647.

McGeer, P. L., McGeer, E. G. and Suzuki, J. S. (1977). Aging and extrapyramidal function. *Arch. Neurol.* **34**, 33–5.

McGeer, P. L. *et al.* (1984). Aging, Alzheimer's disease, and the cholinergic system of the basal forebrain. *Neurology* **34**, 741–5.

Marshall, J. S. and Berrios, N. (1979). Movement disorders in aged rats: Reversal by dopamine receptor stimulation. *Science* **206**, 447–9.

Meier, M. J. and Martin, W. E. (1970). Intellectual changes associated with levodopa therapy. *JAMA* **213**, 465–6.

Mellini, M. (1973). Considerazioni sull'impiego clinico di un nuovo farmaco vasoattivo: la nicergolina. *Minerva Med.* **64**, 1844–53.

Michelangeli, J., Sevilla, M., Lavagna, J. and Darcourt, G. (1975). Etude de l'action de la nicergoline (Sermion) dans la pathologie vasculaire chronique du 3^e age. *Ann. Med. Psychol. (Paris)* **4**, 449–510.

Misurec, J., Morávek, Z. and Nahunek, K. (1978). Lisuride (lysenyl spofa) in the treatment of organic psychosyndrome in involution. *Acta Nerv. Super.* (*Praha*) **20**, 87–8.

Morgan, M. Y., James, I. M. and Sherlock, S. (1980). The effect of bromocriptine on cerebral blood flow and metabolism in chronic hepatic encephalopathy. *Acta Neuroscan.* (*Suppl.*) **62**, 30–1.

Newman, R. P., Weingartner, H., Smallberg, S. A. and Calne, D. B. (1984). Effortful and automatic memory: effects of dopamine. *Neurology* **34**, 805–7.

Otomo, E. *et al.* (1981). Effects of lisuride hydrogen maleate on EEG in patients with cerebral vascular impairments and mild senile dementia: multi-center double blind study. *Jn J. Clin. Pharmacol. Ther.* **12**, 377–96.

Pazzaglia, P. (1972). Effetti clinici e poligrafici di un nuovo derivato ergolinico, la nicergolina, nei disturbi del circolo cerebrale. *Riv. Speriment. Freniatr. Med.* **96**, 348–63.

Pradhan, S. N. (1980). Minireview—central neurotransmitters and aging. *Life Sci.* **26**, 1643–56.

Rossor, M. N. (1982). Dementia. *Lancet*, 1200–4.

Rossor, M. N., Iversen, L. L., Johnson, A. L., Mountjoy, C. Q. and Roth, M. (1981). Cholinergic deficit in frontal cerebral cortex in Alzheimer's disease is age dependent. *Lancet*, 1422.

Samorajski, T. (1977). Central neurotransmitter substances and aging: a review. *J. Amer. Geriat. Soc.* **25**, 337–48.

Shader, R. I., Harmatz, J. S. and Salzman, C. (1974). A new scale for clinical assessment in geriatric populations: Sandoz Clinical Assessment Geriatric (SCAG). *J. Amer. Geriat. Soc.* **22**, 107–13.

Smith, A. H. W., Kay, D. S. G., Johnson, K. and Ballinger, B. R. (1979). Bromocriptine in senile dementia—a placebo controlled double blind trial. *IRCS Med. Sci.* **7**, 463.

Solomon, S., Hotchkiss, E., Saravay, S. M., Bayer, C., Ramsey, P. and Blum, R. S. (1983). Impairment of memory function by antihypertensive medication. *Arch. Gen Psychiat.* **40**, 1109–12.

Spiegel, R., Huber, F. and Köberle, S. (1983). A controlled long-term study with ergoloid mesylates (Hydergine®) in healthy, elderly volunteers: results after three years. *J. Amer. Geriat. Soc.* **31**, 549–55.

Suchy, I. (1983). Clinical effects of dopamine agonists in elderly patients. In *Lisuride and Other Dopamine Agonists*. D. B. Calne, R. Horowski, R. J. McDonald and W. Wuttke (eds). Raven Press, New York.

Suchy, I., Schneider, H. H., Riederer, P. and Horowski, R. (1983). Considerations on the clinical relevance of differences in receptor affinity of various dopaminergic ergot alkaloids. *Psychopharmacol. Bull.* **19** (4), 743–6.

Van Loveren-Huyben, C. M. S., Engelaar, H. F. W. J., Hermans, M. B. M., Van der Bom, J. A., Leering, C. and Munnichs, J. M. A. (1984). Double-blind clinical and psychologic study of ergoloid mesylates (Hydergine®) in subjects with senile mental deterioration. *J. Amer. Geriat. Soc.* **32**, 584–8.

Van Woert, M. H., Heninger, G., Rathey, U. and Bowers, M. B. (1970). L-DOPA in senile dementia. *Lancet*, 573–4.

Waehrens, J. and Gerlach, J. (1981). Bromocriptine and imipramine in endogenous depression. A double-blind controlled trial in out-patients. *J. Affect. Disord.* **3**, 193–202.

Weingartner, H., Burns, R. S. and Diebel, R. (1983). Defining the 'pseudodementia' of Parkinson's disease. *Neurology* **33** (Suppl. 2), 103.

Yates, C. M. *et al.* (1983). Catecholamines and cholinergic enzymes in pre-senile and senile Alzheimer-type dementia and Down's syndrome. *Brain Res.* **280**, 119–26.

Yesavage, J. A. (1981). Senile dementia: combined pharmacologic and psychologic treatment. *J. Amer. Geriat. Soc.* **24**, 164.
Yoshikawa, M., Hirai, S., Aizawa, T., Kuroiwa, Y., Goto, F., Sofue, I., Toyokura, Y., Yamamura, H. and Iwasaki, Y. (1983). A dose–response study with dihydroergotoxine mesylate in cerebrovascular disturbances. *J. Amer. Geriat. Soc.* **31**, 1–7.

Dopamine D-1 and D-2 antagonism: changes in fluid intake and saline preference in the water-deprived rat

D. B. Gilbert and S. J. Cooper

Department of Psychology, University of Birmingham, Birmingham B15 2TT, UK

45.1. Introduction

There is some experimental evidence that central dopaminergic pathways are involved in the control of drinking responses (Leibowitz, 1980; Dourish, 1983). An important recent development is the possibility that central dopamine projections to the amygdala may interact with gustatory projections to the same region to modulate responses to salt taste (Mogenson and Wu, 1982). In a two-bottle preference test, water-deprived rats were given a choice between water and a non-preferred sodium chloride (NaCl) solution. Injection of $0.2 \mu g$ spiperone (a dopamine receptor antagonist) into the central nucleus of the amygdala produced a significant increase in saline intake without affecting water intake. Dopamine receptors may, therefore, mediate drug-induced changes in saline preference.

We investigated this possibility in a series of experiments in which a number of independent variables were taken into account. First, two-bottle preference tests were conducted using a range of saline concentrations, since preference varies markedly as a function of NaCl concentration. Secondly, a variety of drugs have been studied. In addition to familiar neuroleptics (pimozide, haloperidol, clozapine), drugs that may act more selectively to block specific dopamine receptor subtypes were employed. Thus, the substituted benzamides (sulpiride, sultopride) are specific D-2 receptor antagonists; in contrast, the novel compound SCH 23390 acts as a D-1 receptor antagonist (Iorio *et al.*, 1983; Kebabian *et al.*, 1984). Thirdly, extensive dose–response data were collected for each drug.

45.2. Methods

Adult male rats (hooded General strain), bred in our laboratory, were adapted to a 22 h water-deprivation schedule, and were familiarised to a wooden test box in which they had access to two drinking-spouts. For a particular NaCl concentration there was a separate group of animals trained in the choice between the salt solution and water. Within each group, each animal was tested over a number of occasions with a series of doses of the compound under investigation. The salt preference test lasted 15 min, and saline and water intakes were measured separately. The results were analysed in terms of the total fluid intake (ml) and in terms of saline preference, i.e. (saline intake/total intake) × 100 %. Statistical analyses were carried out using the analysis of variance, followed by Dunnett's *t*-test. In Figs 45.1 and 2, data are shown as group means ($N = 8$ in each case), with standard errors represented by vertical lines. Significant differences between individual drug dose and corresponding control scores are denoted as follows: $*P < 0.05$; $**P < 0.01$.

45.3. Results

For the purpose of this report, data from the experiments carried out with sultopride and SCH 23390 are compared. Over all the saline concentrations

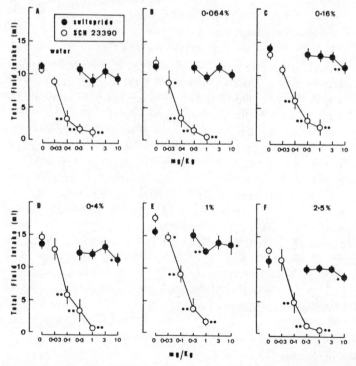

Fig. 45.1   Effects of sultropride ($0.3–10.0$ mg kg^{-1}, i.p.) and SCH 23390 ($0.03–1.0$ mg kg, s.c.) on the total fluid intake (ml) in salt preference tests in male rats. Five salt concentrations were tested in 15 min, two-bottle tests against water (0.064%, 0.16%, 0.4%, 1.0% and 2.5%; panels B–F, respectively). A water vs water condition was also tested (panel A).

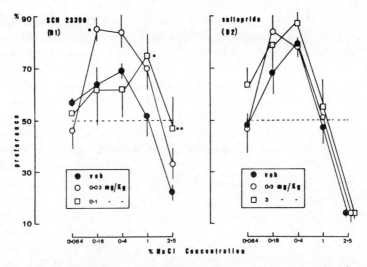

Fig. 45.2. Effects of sultopride and SCH 23390 on preference–aversion functions with saline concentrations from 0·064% to 2·5%, in 15-min two-bottle tests. Data for 0·03 and 0·1 mg kg^{-1} SCH 23390 are shown (left panel), and for 0·3 and 3·0 mg kg^{-1} sultopride (right panel).

tested in the two-bottle preference test, SCH 23390 (0·03–1·0 mg kg^{-1} s.c.) produced dose-related decrements in the total fluid consumption (Fig. 45.1). On the other hand, sultopride (0·3–10·0 mg kg^{-1} i.p.) produced only very modest effects on total fluid consumption (Fig. 45.1).

The pattern of changes in the saline preference measure was also different for the two compounds. Note that in the vehicle-control condition, preference rose to a peak at 0·4% NaCl solution, dropped to a point of indifference at 1% (i.e. 50% preference level, at which equal volumes of saline and water are consumed), and fell to a rejection of salt solution at a concentration of 2·5% (Fig. 45.2). SCH 23390 (0·03 mg kg^{-1}) clearly acted to raise the preference curve; at this dose, it had little effect on total fluid intake (Fig. 45.1). At 0·1 mg kg^{-1}, SCH 23390 significantly enhanced the preference for the 1% and 2·5% NaCl solutions (Fig. 45.2). On the other hand, sultopride (0·3 and 3·0 mg kg^{-1}) had no effect on the preference for the more concentrated salt solutions. Instead, preferences for the more dilute solutions were raised (Fig. 45.2).

45.4. Conclusions

Our data extend and qualify earlier reports that dopamine receptor antagonists reduce fluid consumption. Sultopride and sulpiride (data not shown) showed little effect on fluid intake in rehydrating rats. In contrast, the specific dopamine receptor antagonist, SCH 23390, produced complete suppression of drinking. Although D-1 receptor blockade appears sufficient to reduce intake, D-2 receptor blockade alone is probably not sufficient.

At some receptor sites, dopamine may act to modulate drinking responses to dilute preferred NaCl solutions. Sultopride and sulpiride (data not shown), caused increased preferences for such solutions without changing the response to less preferred, hypertonic, NaCl solutions. These receptors, we suggest, include the D-2 type. However, at other sites, dopamine may act to modulate drinking responses to more concentrated, preferred and non-preferred salt solutions. These receptors, we propose, include the D-1 type.

References

Dourish, C. T. (1983). *Progr. Neuro-Psychopharmacol. Biol. Psychiat.* **7**, 487–93.
Iorio, L. C., Barnett, A., Leitz, F. H., Houser, V. P. and Korduba, C. A. (1983). *J. Pharmacol. Exp. Ther.* **226**, 462–8.
Kebabian, J. W., Beaulieu, M. and Itoh, Y. (1984). *Can. J. Neurol. Sci.* **11**, 114–17.
Leibowitz, S. F. (1980). In: *Handbook of Hypothalamus, vol. 3, Part A.* P. J. Morgane and J. Panksepp (eds), pp. 299–437, Marcel Dekker, New York.
Mogenson, G. J. and Wu, M. (1982). *Brain Res. Bull.* **8**, 685–9.

46

Involvement of the striatum in dopamine agonist-induced yawning

C. T. Dourish[1], S. J. Cooper[2] and P. H. Hutson[1]

[1]*Department of Neurochemistry, Institute of Neurology, The National Hospital, Queen Square, London WC1N 2NS, UK*
[2]*Department of Psychology, University of Birmingham, PO Box 363, Birmingham B15 2TT, UK*

46.1. Introduction

Systemic administration of low doses of dopamine agonists produces a syndrome of yawning, stretching, chewing, reduced locomotion and sexual excitement in male rats (Mogilnicka and Klimek, 1977; Yamada and Furukawa, 1980). This yawning syndrome contrasts with the behavioural effects elicited by high doses of dopamine agonists which typically consist of stereotyped sniffing, gnawing, headbobbing and increased locomotion (Kelly, 1977). These latter effects are thought to be due to stimulation of postsynaptic dopamine receptors located in the striatum and nucleus accumbens (Kelly, 1977).

The yawning syndrome appears likely to be a consequence of the activation of inhibitory dopamine autoreceptors, and therefore it is claimed that yawning behaviour may be a useful screening test for drugs with dopamine autoreceptor agonist properties (Stahle and Ungerstedt, 1984).

The present study attempted to identify a central site of action of dopamine agonists in the production of the yawning syndrome. In the first experiment the behavioural effects of microinjections of piribedil and apomorphine into the striatum and nucleus accumbens were examined. In the second experiment, the effects of bilateral 6-hydroxydopamine (6-OHDA) lesions of the striatum on yawning elicited by a low dose of apomorphine (given s.c.) were examined.

46.2. Methods

46.2.1. Experiment 1

Adult, male Sprague-Dawley rats were implanted with bilateral guide cannulae aimed at either the caudate nucleus or the nucleus accumbens.

Following a one-week recovery period, animals were tested after bilateral intracerebral injections of saline and piribedil (50 and 100 μg) or apomorphine (5, 10 and 20 μg). During a 60-min test an observer recorded the incidence of yawning, stretching, chewing and penile grooming. In some cases scopolamine (1 mg kg^{-1} given 25 min after piribedil) or haloperidol (0·025 mg kg^{-1} given 30 min before piribedil) was administered in an attempt to block the yawning syndrome.

46.2.2. Experiment 2

Adult, male, Sprague-Dawley rats were given bilateral intrastriatal injections of 6-hydroxydopamine (16 μg free base per side, in 4 μlitre of saline containing ascorbic acid 1 mg ml^{-1}, infused at a rate of 1 μlitre min^{-1}) or vehicle (sham lesion). Two days later the infusion was repeated using the same dose of 6-OHDA. Two weeks later the yawning response of the rats to 0·1 mg kg^{-1} apomorphine (given s.c.) was recorded as described above. When testing was completed, the animals were sacrificed and the striatal concentration of dopamine determined using HPLC with electrochemical detection (see Dourish and Hutson, 1985, for details).

46.3. Results

Piribedil and apomorphine injected into the caudate nucleus of male rats produced a syndrome of yawning, stretching, chewing, penile grooming and reduced locomotion (see Tables 46.1 and 46.2). Similarly, intra-accumbens application of piribedil elicited the yawning syndrome although the effects were less pronounced than after equivalent intrastriatal injections (see Table 46.1). Systemic injection of haloperidol or scopolamine abolished the yawning syndrome induced by intrastriatal piribedil application.

Table 46.1. Differential effects of piribedil in the induction of yawning following microinjection into the caudate nucleus and nucleus accumbens of male rats.

| | Number of yawns during 1 h | |
Treatment	Caudate nucleus	Nucleus accumbens
Saline	2·3 ± 1·4 (13)	0·1 ± 0·1 (8)
Piribedil 50 μg	16·8 ± 5·7 (5)**	10·7 ± 4·3 (5)*
Piribedil 100 μg	27·6 ± 3·2 (13)**$\mathcal{N}$	11·1 ± 2·0 (8)*

Values are expressed as mean ± SE, with number of subjects tested in brackets. Piribedil or saline was injected bilaterally immediately prior to the 1-h test. Significant differences between drug and control treatments within each brain region were determined by a two-tailed *t*-test for correlated means following a significant ANOVA result: *$P < 0.05$, **$P < 0.005$. Significant differences in response between the two brain regions at the same drug dose were determined by a two-tailed *t*-test for independent means: $\mathcal{N} P < 0.005$.

Table 46.2. Effects of apomorphine on yawning following microinjection into the caudate nucleus of male rats.

Treatment	*Number of yawns during 1 h*
Saline	$2·2 \pm 1.3$
Apomorphine 5 μg	$16·5 \pm 9·4$
Apomorphine 10 μg	$12·2 \pm 4·0$**
Apomorphine 20 μg	$16·0 \pm 7·8$*

Values are expressed as mean $\pm$ SE of six subjects tested at each dose. Statistical comparisons between drug and control treatment were made by Wilcoxon test: *$P < 0·05$, **$P < 0·025$.

Unoperated controls and sham lesioned animals exhibited a pronounced yawning response after s.c. injection of $0·1$ mg kg^{-1} apomorphine. In contrast, the same drug treatment failed to elicit yawning in rats with extensive bilateral 6-OHDA lesions of the striatum. Thus, there was a clear correlation between striatal dopamine concentration and apomorphine-induced yawning (see Fig. 46.1).

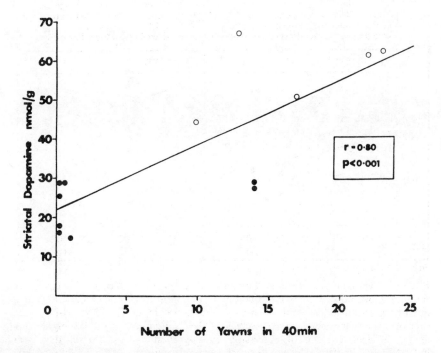

Fig. 46.1. Correlation (Pearson r) between the concentration of dopamine in the striatum and yawning elicited by $0·1$ mg kg^{-1} apomorphine injected subcutaneously.

46.4. Discussion

The present study demonstrated that a syndrome of yawning, stretching, chewing and sexual arousal is produced by central application of piribedil and apomorphine to dopamine terminal regions of the brain of male rats. This is consistent with evidence that systemic administration of low doses of dopamine agonists produces yawning in rats (Mogilnicka and Klimek, 1977; Yamada and Furukawa, 1980). It has been proposed that yawning elicited by low doses of dopamine agonists involves activation of presynaptic dopamine autoreceptors (Stahle and Ungerstedt, 1984) and the present data are compatible with this proposal. Thus, an extremely low dose of the dopamine antagonist haloperidol, which is insufficient to block postsynaptic dopamine receptors, abolished yawning induced by intra-striatal piribedil application. Furthermore, extensive damage to dopamin-ergic neuronal innervation of the striatum, by treatment with the neurotoxin 6-OHDA, prevented yawning elicited by s.c. apomorphine. These data strongly suggest that the yawning syndrome is mediated by dopamine autoreceptors.

Previous work has indicated that yawning may be produced by an inhibition of dopaminergic neurones together with an excitation of cholinergic neurones 'in series' (Holmgren and Urba-Holmgren, 1980; Yamada and Furukawa, 1980). Our results suggest that the yawning syndrome may involve a striatal dopamine–acetylcholine link since yawning induced by intrastriatal piribedil application was abolished by treatment with the muscarinic antagonist scopolamine.

References

Dourish, C. T. and Hutson, P. H. (1985). *Neuropharmacology*, in press.
Holmgren, B. and Urba-Holmgren, R. (1980). *Acta Neurobiol. Exp.* **40**, 633–42.
Kelly, P. H. (1977). In *Handbook of Psychopharmacology*, vol. 8. L. L. Iversen, S. D. Iversen and S. H. Snyder (eds), pp. 295–330. Plenum, New York.
Mogilnicka, E. and Klimek, V. (1977). *Pharmacol. Biochem. Behav.* **7**, 303–5.
Stahle, L. and Ungerstedt, U. (1984). *Eur. J. Pharmacol.* **98**, 307–10.
Yamada, K. and Furukawa, T. (1980). *Psychopharmacology* **67**, 39–43.

Part VI
Epilogue

47

The end of the beginning

W. Winlow[1] **and R. Markstein**[2]

[1]*Department of Physiology, School of Medicine, University of Leeds, Leeds L52 9NQ, UK;*
and [2]*Preclinical Research, Sandoz Ltd, CH-4002 Basle, Switzerland*

More than twenty-five years have passed since the occurrence of dopamine in the mammalian brain was first demonstrated by A. Carlsson. Its distribution in several neuronal pathways is now well established. Dopamine has a remarkable distribution in the brain. Approximately 70 % of the total amount occurs in the neostriatum, and the remainder is distributed in other brain areas such as the thalamus, the hypothalamus and limbic areas, including the hippocampus. Dopamine has attracted growing interest since its role in the pathophysiology of several brain diseases became obvious. In 1960, Ehringer and Hornykiewicz discovered that dopamine was almost completely depleted from the corpus striatum of patients with Parkinson's disease. This finding led to the successful use of L-dopa and, later on, of dopamine receptor agonists in the treatment of Parkinson's disease. Another important finding was that drugs with antipsychotic actions had the common pharmacological property to impair or block central dopaminergic neurotransmission. This observation led to the dopamine hypothesis of schizophrenia in which it is claimed that this disease is the result of an over-activity of central dopaminergic systems. Although this hypothesis is still a matter of debate, the many attempts undertaken to prove or to reject it have greatly contributed to our present knowledge of the central dopaminergic systems.

Meanwhile, the role of dopamine in the regulation of endocrine, cardiovascular and gastrointestinal systems has been established. Many drugs have been developed which directly affect central and peripheral dopaminergic systems. Some drugs have mainly therapeutic actions, and others are responsible for unwanted side-effects.

Thus much effort is therefore presently devoted to the development of new drugs with better therapeutic actions and less side-effects. A prerequisite for such new drugs is a better understanding of the physiology and pharmacology of the various dopaminergic systems. Over the past decade, the concept of multiple types of dopamine receptors emerged. Dopamine receptors have been subdivided, based on behavioural, biochemical and physicochemical criteria. Opinions about the validity of the various suggestions are still divided. Whereas some authors believe that there exist up to four dopamine receptor subtypes, others take the view that there is in fact only one receptor. The concept that there are two types designated D-1 and D-2 as proposed by Spano and elaborated by Kebabian and Calne in 1979 has gained great support from the recent development of D-1 selective drugs such as SK&F 38393 (D-1 agonist) and SCH 23390 (D-1 antagonist).

Another fascinating aspect of dopamine receptor research is the concept of dopamine autoreceptors. The existence of drugs with selectivity for these autoreceptors has been postulated, but much further research needs to be carried out in this area as in all others. It is, of course, vital that basic and clinical research on dopamine systems (and other neural systems) be carried out in a co-ordinated manner.

At our conference on dopamine systems we attempted to combine clinical and basic neuroscience and to bring neuroscientists from many sub-disciplines together in order to promote free discussion of their common problems and interests. We believe we have been successful and this book is the outcome. It seems to us important that the neurosciences should develop in a cohesive manner and that we should try to break down the artificial barriers that exist between those of us with interests in the neurosciences: between those who are medically qualified and those who are trained in the biological sciences; between physiologists, zoologists, pharmacologists, anatomists, biochemists, ethologists, psychologists, psychiatrists, neurologists. We owe much to the late Stephen Kuffler who first gave meaning to the concept of 'neurobiology' and created the first Department of Neurobiology at Harvard University. In Europe we have been slow to follow that example, but as we approach the year 2000 the intellectual drive towards creating Departments of Neuroscience (or Neurobiology and Behaviour) seems inevitable. This will mean some years of transition in our universities, as older disciplines become modified or cease to exist and new disciplines, including neuroscience, emerge. We hope that we are at the end of the beginning for the neurosciences.

Index

A23187 180
Acetylcholine release 46
Acromegaly 141–3, 146–53
  Bromocriptine 146–7
  GH secretion 146–53
  Pergolide 152
Adrenocorticotrophic hormone 129–30
Anti-psychotic drugs 242–50
Apomorphine headache 406–8

Bromocriptine 45–7, 186, 432–3, 440
  acromegaly 146–8
  senile dementia 440, 441
Bulbocapnine 43, 45, 46, 47

CF 25–397 186
CH 29–717 188, 190
CJ 201–678 344
Cortisol
  prolactin 199–201
  thyrotropin 199–201
CQ 32–084 186, 188–90
CQP 401–403 188
CU 32–085 188, 207–10

Deoxyglucose (DOG)-utilisation
  374–6
Dihydroergotoxine (see Hydergine)
D–1 receptors 41–2, 186–90, 344, 459–62,
  470
  ageing 104–6
D–2 receptors 42, 46, 178, 186–90, 440,
  459–62, 470
D–1, D–2, D–3, D–4 recognition sites
  34–5
DOPA
  therapy 323
  therapeutic complications 341–8
Dopamine
  abstinence syndrome 402–17
  adenylate cyclase stimulation 42–4,
    179

adenylate cyclase inhibition 49, 170–2,
  179
agonists 185–91, 287–9, 290–2
autoreceptors (soma-dendritic)
  18–20
biosynthesis 3–7
degradation 8–10
fluid intake 459–62
kidney 385–97
natriuretic effect 389–90
memory 427–36, 451–2
pituitary 123–34
postsynaptic potentials 53–65, 112–15
prolactin secretion 127–8, 156–63
receptors 25–37, 65, 108–11, 178–81,
  240–50
release 14–18, 20–1, 46
renin release 390–6
retina 68–86
sexual behaviour 211–14
storage 7–8
yawning 463–6
Dopamine neurons
  in invertebrates 53–65, 108–11,
    112–15, 116–19
  in vertebrate retina 68–72
  in vertebrate brain 27–8, 87–99
Dopamine receptors 25–37, 40–50
  chronic blockade 294–8
  chronic stimulation 254–65
  tolerance 294–8
Drug delivery 287–9
Dyskinesia 266–9, 344–5
Dihomo-gamma-linolenic acid (DHLA)
  303–6

Ergot derivatives 185–90

Forskolin
  Prolactin 163–4

GH-secretion 137–43, 132–3
  acromegaly 141–3, 147–53
  Bromocriptine 147–8
  dopamine 138–40
  Huntington's Chorea 140–1

Haloperidol
  chronic administration 294–8
Helix aspersa 108–11
Huntington's Chorea 140–1, 309–16
  animal model 315–16
  neurochemistry 310–15
  neuropathology 310
Hydergine
  senile dementia 440–5

Kidney 385–97

Levodopa 341–8
Lisuride 190
  migraine 420–1
  senile dementia 433–5, 439–40
Luteinising Hormone 129–30
LY141865 42, 65, 361, 362
Lymnaea stagnalis 62, 65, 112–15,
  116–19

Migraine 410–11, 415–24
MPTP (1-methyl-4-phenyl-1,2,3,6-
  tetrahydropyridine) 98, 346, 350–67
  biochemistry 363–5
  effect in man 351–3
  effect in monkeys 356–63
  effect in rodents 353–6, 378–81
Muscular rigidity 370–3

Neostriatum
  acetylcholine release 43–4
  dopamine release 43–4
  muscular rigidity 370–3
Neuroleptics 240–50
Nicergoline
  senile dementia 441–5
N-propyl-norapomorphine (NPA)
  252–65

Oestrogen
  prolactin 203–6

Parkinson's Disease 319–30
  adaptive changes 323
  correlation symptoms and dopamine
  loss 321–3

dementia 347
limbic dopamine 321
MPTP (1-methyl-4-phenyl-1,2,3,6-
  tetrahydropyridine) 350–67
  noradrenaline 325–7
  transplants 331–8
  striatal dopamine 319–21
  substitution therapy 323–8
Patterned discharge 117–19
Pituitary adenomas 130–3
Planorbis corneus 53–65
Prolactin 156, 178–81, 199–201
  Calcium 157–63, 180–1
  Calmodulin 157, 166
  chlorpromazine 299–302
  CH 29–717 188, 190
  Cortisol 199–201
  CU 32–085 207–10
  cyclic AMP 163–6
  forskolin 164–5
  pituitary 156–74, 127–8, 130–2
  tardive dyskinesia 299–302

Quasi-abstinence syndrome 402–12

Senile dementia 427–36, 439–54
  aetiology 445–51
  dopamine agonists 427–36, 439–45
Schizophrenia 230–8
  clinical aspects 219–28
  dopamine hypothesis 232
  limbic system 252–65
  management 224–7
  neuroleptics 232–6, 240–1, 245–50
  symptoms 219–23
SCH 23390 28, 36, 37, 42, 65, 460–2, 470
SK&F38393 28, 36, 37, 42, 65, 186, 362,
  470
Somatostatin (SRIF) 194–7
Sulpiride 45, 47, 417
  binding 28–37

Tardive dyskinesia 267–9
  neuroleptics 273–6
  prolactin 299–302
Thyroid Stimulating Hormone 128–9
TL 99 186
Transdihydrolisurid (TDHL) 190, 191
Transplants 331–8
TRH 158, 180, 194–7, 199–201

Water deprivation 459–62